A selection of some optical glasses with polished optical surfaces for various applications

The Properties of Optical Glass

Schott Series on Glass and Glass Ceramics
Science, Technology, and Applications

Low Thermal Expansion Glass Ceramics
ISBN 3-540-58598-2

Fibre Optics and Glass Integrated Optics
ISBN 3-540-58595-8

Thin Films on Glass
ISBN 3-540-58597-4

Electrochemistry of Glasses and Glass Melts
Including Glass Electrodes
ISBN 3-540-58608-3

Surface Analysis of Glasses
and Glass Ceramics, and Coatings
ISBN 3-540-58609-1

Analysis of the Composition and Structure
of Glass and Glass Ceramics
ISBN 3-540-58610-5

Hans Bach
Norbert Neuroth
Editors

The Properties of Optical Glass

With 156 Figures, 57 Tables, and 2 Fold-out Diagrams

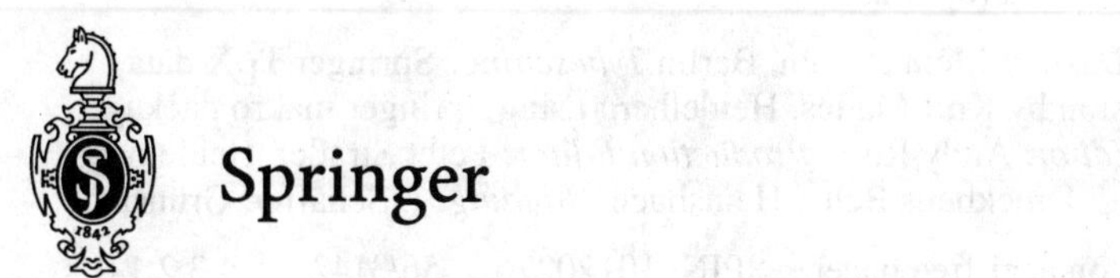

Dr. Hans Bach
Dr. Norbert Neuroth
Schott Glaswerke
Hattenbergstraße 10
D-55122 Mainz, Germany

ISBN 3-540-58357-2 Springer-Verlag Berlin Heidelberg New York
ISBN 0-387-58357-2 Springer-Verlag Berlin Heidelberg New York

CIP data applied for

Printed in Germany

Cover Design: Meta Design, Berlin *Typesetting:* Springer TEX data conversion by Kurt Mattes, Heidelberg using Springer makro package
Copy-Editor: Andy Ross *Production Editor:* Peter Straßer, Heidelberg
Printing: Druckhaus Beltz, Hemsbach *Binding:* J. Schäffer, Grünstadt

Printed on acid-free paper SPIN 10120276 56/3142 – 5 4 3 2 1 0

Foreword

This book, entitled *The Properties of Optical Glass*, is one of a series reporting on research and development activities on products and processes conducted by the Schott Group.

The scientifically founded development of new products and technical processes has traditionally been of vital importance at Schott and has always been performed on a scale determined by the prospects for application of our special glasses. The scale has increased enormously since the reconstruction of the Schott Glaswerke in Mainz. The range of expert knowledge required for that could never have been supplied by Schott alone. It is also a tradition in our company to cultivate collaboration with customers, universities, and research institutes. Publications in numerous technical journals, which since 1969 we have edited to a regular timeplan as *Forschungsberichte* – 'research reports' – formed the basis of this cooperation. They contain up-to-date information on the most various topics for the expert but are not suited as survey material for those whose standpoint is more remote.

This is the point where we would like to place our series, to stimulate the exchange of thoughts, so that we can consider from different points of view the possibilities offered by those incredibly versatile materials, glass and glass ceramics. We would like to show scientists and engineers, interested customers, and friends and employees of our firm the knowledge that has been won through our research and development at Schott in cooperation with the users of our materials.

The results documented in the volumes of the Schott Series are of course oriented to the tasks and targets of a company. We believe it will become quite clear that here readers can nevertheless – or rather for that reason – find demanding challenges for applied research, the development of process engineering, and characterization of measurement practice. Besides realizability, the profitability of solutions to customers' problems always plays a decisive role.

The first comprehensive presentation of research findings after the reconstruction of the factory in Mainz was edited by Prof. Dr. Dr. h.c. Erich Schott in 1959. It was entitled *Beiträge zur angewandten Glasforschung* – 'contributions to applied glass research' (Wissenschaftliche Verlagsgesellschaft m.b.H., Stuttgart 1959). Since then, there has been an extraordinary worldwide increase in the application of glass and glass ceramic materials. Glass fibres and components manufactured from them for use in lighting and traffic engineering or in telecommunications, high-purity and highly homogeneous glasses for masks and projection lenses in electronics, or glass ceramics with zero expansion in astronomy and in household appliance technology are only some examples. In many of these fields Schott has made essential contributions.

Due to the breadth and complexity of the field in which Schott is active, many volumes are needed to describe the company's research and development results. Otherwise it would be impossible to do full justice to the results of fundamental research work and technological development needed for product development. Furthermore, it is necessary to give an appropriate description of the methods of measurement and analysis needed for the development and manufacture of new products.

The next two volumes, which will be published within about a year, will be entitled *Low Thermal Expansion Glass Ceramics* and *Thin Films on Glass.* Another four volumes treating fibre optics and integrated optics, surface analysis, analysis of the composition and structure of glass and glass ceramics, and the electrochemistry of glasses and glass melts are in preparation. Descriptions of melting and processing technology and of glasses for various applications in industry and science and their properties are being considered.

With the presentation – in part detailed – of the work required for the development of successful products, Schott employees are giving all their interested colleagues who work in the field of science and technology an insight into the special experiences and successes in material science, material development, and the application of materials at Schott. Contributions from scientists and engineers who work in university and other research institutes and who played an essential role in Schott developments complete the survey of what has been achieved. At the same time such results show the need for the collaboration mentioned above.

In all the volumes of the series the fundamental issues from chemistry, physics, and engineering are dealt with, or at least works are cited that enable or assist the reader to work his or her way into the topics treated. We see this as indispensable because, with the series, Schott has a further goal in view. We aim to provide all future business partners from branches of industry where glasses and glass ceramics have not been applied so far with knowledge they can use in cooperation with Schott. Furthermore, the series may serve to fill gaps between the basic knowledge imparted by material science and the product descriptions published by Schott. Those who have already done business with our company may find the survey of fundamentals useful in extending collaboration to further business areas.

To make each volume sufficiently intelligible, the necessary fundamentals from chemistry, physics, and engineering are described or referred to via citations. We see this as the best way to enable all our potential business partners who are not already familiar with glass and glass ceramics to compare these materials with alternatives on a thoroughly scientific basis. We hope that this will lead to intensive technical discussions and collaborations on new fields of application of our materials and products, to our mutual advantage.

Every volume of the Schott Series will begin with a chapter providing a general idea of the current problems, results, and trends relating to the subjects treated. These introductory chapters and the reviews of the basic principles are intended to be useful for all those who are dealing for the first time with the special properties of glass and glass ceramic materials and their surface treatment in engineering, science, and education.

Many of our German clients are accustomed to reading scientific and technical publications in English, and most of our foreign customers have a better knowledge of English than of the German language. It was therefore mandatory to publish the Schott Series in English.

The publication of the Schott Series has been substantially supported by Springer-Verlag. We would like to express our special thanks to Dr. H.K.V. Lotsch and Dr. H.J. Kölsch for advice and assistance in this project.

The investment of resources by Schott and its employees to produce the Schott Series is, as already stated, necessary for the interdisciplinary dialogue and collaboration that are traditional at Schott. A model we still find exemplary today of a fruitful dialogue between fundamental research, glass research, and glass manufacture was achieved in the collaboration of Ernst Abbe, Otto Schott, and Carl Zeiss. It resulted in the manufacture of optical microscopes that realized in practice the maximum theoretically achievable resolution. It was especially such experiences that shaped the formulation of the founding statute of the Carl Zeiss Foundation, and the initiative for the Schott Series is in accord with the commitment expressed in the founding statute "to promote methodical scientific studies."

Mainz, January 1995

Dieter Krause
Vice President R & D

Preface

The main aim of the Schott Series volume *The Properties of Optical Glass* is to describe the properties of the optical glasses developed by Schott. The book is conceived as a monograph. However, the individual chapters have been written by different – sometimes by several – authors who are themselves active in the corresponding fields of research and development. Thus the reader is given direct access to the experience of these authors, some of whom are employed in our subsidiaries Schott Glass Technologies, Duryea, USA; Deutsche Spezialglas AG, Grünenplan; and also at Carl Zeiss, Oberkochen.

To give the reader a view of the extraordinarily broad range of applications of glasses and glass components, the volume opens with a general survey of the development of optical glasses and their important fields of application, and the aims, limits, and current state of new developments and trends in application.

The subsequent chapters treat the properties of optical glasses and other special glasses. Schott has significantly contributed to the development and production technology of these glasses since the establishment of the company more than 100 years ago. Information is provided, above all, on properties that are important for the application of optical glasses as classical optical elements such as lenses and prisms. The reader should develop a deeper understanding of optical glass as a material and at the same time find answers to particular questions such as – Within what limits can the properties of glasses for optical service be manufactured reproducibly? How does the material glass behave in processing? Which chemical and physical influences have to be considered in the application of optical glasses? In addition, an insight into special applications in optoelectronics is given. In these chapters, latest results from the Schott laboratories on the properties of optical glasses are reported.

Due to their vital importance for the construction and manufacture of optical instruments, the mechanical and thermal properties of optical glasses and the influence of thermal treatment on the optical constants are also considered. The influence of the tempering of glasses on their structure and the resulting change in their properties, however, will be treated together with the analysis of glasses in a subsequent volume. This division seems appropriate because for an adequate characterization of the properties of a glass, for many kinds of application, both an elemental analysis and a characterization of the structure through spectroscopy and other analytical methods are required.

For the production and utilization of optical surfaces the interaction between the surfaces of optical glasses with aqueous solutions and gases in the atmosphere, especially the water vapour contained in the air, is of great importance. Methods for the practical determination of chemical properties and the differences in the resistance of optical glasses resulting from them are therefore treated in a separate chapter. The following chapter describes results and experiences gathered during the production of optical surfaces. The chapter explains which mechanical and chemical interaction parameters must be taken into account in the manufacture of optical surfaces in order to obtain the best possible yields.

New developments and examples of the application of special optical glasses for opthalmics, the ultraviolet and infrared spectral regions, photochromic glasses, glasses for lasers, glasses for nuclear engineering, and coloured and protection glasses are the subjects of individual sections in the final chapter.

In all the contributions, descriptions in the text and tables and examples are chosen so as to assist the user of the material glass, especially the constructor of optical equipment, in his or her daily work with the material and components made from it. The form of the presentation and the selection of citations can also be useful for teaching and research, especially in those cases where data for measuring and analysis techniques are required. The cited literature should support the reader in gaining access to more detailed presentations to deepen his or her knowledge. The contributions in this volume fill the gap between the basic knowledge of glass and its properties provided by material science, on the one hand, and catalogue data supplied by glass manufacturers on the other.

In view of the amount of material to be covered, the description of the production of raw glass and of semimanufactured and finished products, including sol-gel techniques, requires its own volume in the Schott Series. For the same reason, separate volumes must also be reserved for the coating of glass surfaces and the production of optical fibres for light guides and image guides and for integrated optics.

We would like to thank all the authors of this book for their steady and pleasing cooperation.

We have received further valuable help from many colleagues. For critical reading of the manuscripts we thank Mrs. Marga Faulstich, Dr. habil. Walter Geffcken, Dipl.-Phys. Alfred Jacobsen, Dipl.-Phys. Hans Morian, Dr. Burkhard Speit, Schott Glaswerke, Mainz, Dr. Alexander J. Marker III, Schott Glass Technologies, Duryea, and Prof. Dr. Hans-Jürgen Hoffmann, Fachhochschule Hildesheim-Holzminden.

Additionally, we are indebted to several employees of Springer-Verlag, especially Barbara S. Hellbarth-Busch, Petra Treiber, and Peter Straßer, for helping us to overcome the difficulties involved in producing manuscripts ready for printing. For their help in solving text processing problems we are indebted to Urda Beiglböck and Frank Holzwarth, also of Springer-Verlag.

We are grateful to Dipl.-Math. Sieglinde Quast-Stein, Schott Glaswerke, who with her knowledge and experience provided substantial support in the implementation of the software guidelines supplied by Springer-Verlag.

We also thank Dipl.-Designer Andreas Jacobsen, Niedernhausen, for the creation of the numerous computer graphics needed to illustrate the texts.

We would especially like to thank Mrs. Angela Gamp-Paritschke, M.A., Schott Glaswerke, for the translation from German into English, for the correction of manuscripts submitted in English, and for her enthusiasm in performing all the hard work necessary to prepare manuscripts ready for printing.

January 1995

Hans Bach
Norbert Neuroth

Table of Contents

1. Overview – Optical Glass: An Engineered Material

Alexander J. Marker III, Norbert Neuroth

1.1 Glass: An Important Material for Optics

In the field of optics, the components utilized in imaging systems are mostly manufactured from glass. A faithful geometric and colour image of an object is generated by using the property of refraction. These imaging systems require combinations of various optical glass types in order to achieve this goal. Lens systems are designed to produce images that are either diminished in size, e.g. in photography, television and cinema technology, or enlarged in size, as in astronomy, telescopy and microscopy; or the images are projected. Ernst Abbe developed the modern basis of computing lens systems whereby the requirements for refractive index and dispersion, which is the dependence of the refractive index on wavelength, are specified. Optical glass types are then chosen which meet these required values.

Historically, optical glasses have been developed to provide the optical engineers and scientists with a variety of materials for use in the visible portion of the electromagnetic spectrum from approximately 380 to 780 nm. With the development of new, bright, narrow bandwidth sources in the form of lasers and better detectors, the classic concept of optical glass has been expanded to include new glasses for application in the ultraviolet and infrared spectral regions.

The prominent optical properties are refractive index, dispersion and transmission. These properties are mainly determined by the chemical composition and, to a lesser degree, by the melting process and the subsequent thermal treatment. These procedures fix the molecular structure of glass which is responsible for the interaction with the electromagnetic waves. Optics, more than any other field of application in technology and science, has taken advantage of the fact that glass is an engineered material [1.1–14]. Through a systematic variation of the glass composition, the selection of the proper manufacturing technique and the utilization of specialized post treatments, glass can be engineered to simultaneously optimize critical secondary properties as well as the primary properties of refractive index and dispersion. For many optical applications certain secondary properties such as optical homogeneity, absence of striae, bubbles and particles, low volume light scattering, chemical resistance, workability, coefficient of thermal expansion, birefringence and size limitations become critical design parameters.

Some examples of the quality of products from the field of optics illustrate the unique requirements which are attainable in many of today's optical glasses. Very homogeneous optical glass with good chemical durability and workability is required to produce extremely smooth surfaces. Surface roughness of the order of a few angstroms RMS has been achieved [1.15, 16]. Also glasses with high optical homogeneity should show very low volume light scattering [1.17–19]. Applications such as inertial confinement laser systems require very large optical elements of the order of one meter in diameter with high optical homogeneity so that their refractive index remains constant in the range of $\pm 1 \times 10^{-6}$ [1.20].

Today, great quantities of optical fibres are produced for the telecommunications industry with losses of the order of 0.2 dB/km for a wavelength of about 1 μm [1.21–24]. More than 10 years of process development were necessary to achieve this extraordinarily low level of attenuation. The telecommunications field including facsimile technology also requires very small optical components in the form of pick up lenses which are produced in lens arrays, whose size is a few millimetres [1.25], optical fibres whose diameters range from a few μm to several hundred μm and integrated optical components often with layers of only a few μm in thickness [1.26, 27].

The past hundred years have seen the development of hundreds of optical glasses, most of which are oxide based. However, new glasses for the short wavelength region of the spectrum from 0.3 μm down to the vacuum UV edge and the infrared spectral region, in particular above 4 μm, require an understanding of glass chemistry in such glass systems as fluorophosphates [1.28], heavy metal fluorides [1.29–31], heavy metal oxides [1.32, 33], and chalcogenide glasses [1.35].

1.2 Essential Properties

An optical component is designed to transmit, absorb, emit or direct light. Glass is utilized in many different optical elements; the most common and oldest examples being windows, lenses and prisms. In imaging systems the refractive index and dispersion are the most important parameters. Optical glass is classified according to its refractive index for the helium-d line (587.6 nm), n_d, and the Abbe number or value, ν_d, which characterizes the dispersion. The Abbe number is given by

$$\nu_d = \frac{n_d - 1}{n_F - n_C}, \tag{1.1}$$

where $n_F - n_C$ is the principal dispersion. The refractive indices n_F and n_C are measured at the hydrogen F line (486.1 nm) and the hydrogen C line (656.3 nm), respectively. The refractive indices n_d vary from approximately 1.4 to 2.4 while the Abbe number varies from 15 to 100.

It is essential to know the refractive index as accurately as possible in order to properly design imaging optics. Utilizing specially designed and constructed prism spectrometers, the method of the angle of minimum deviation is employed to measure the refractive index to an accuracy of a few parts per million. The measurements are performed at several wavelengths utilizing spectral lamps as light sources. The refractive index values at other wavelengths is computed from a dispersion formula [1.34, 36]. Another important piece of design information is the change in the refractive index with temperature. The temperature coefficient of refractive index is measured at a variety of spectral emission wavelengths for various temperature ranges [1.37].

The chemical resistance and the behaviour of a glass during processing such as grinding and polishing are very important data for the optical engineer. Because of the large range of chemical compositions required to achieve the desired properties of today's optical glasses, their behaviour during processing varies widely. A knowledge of the chemical resistance of the particular glass to water, acids and bases is vital for the successful processing and storing of the finished optical components. During the last 60 years, remarkable developments in the understanding of generating optical surfaces have been made due to the wide variety of methods of surface analysis now available [1.38–42]. New grinding and polishing materials and techniques are now available.

1.3 History of Development

Prior to 1880, optical components were made from simple crown and flint glasses. The crown glasses were soda-lime-silicates with low refractive indices and moderate Abbe numbers. The flint glasses were lead silicates with relatively low Abbe numbers. By 1884, approximately two dozen flint and crown glasses were available for optical system designs. Although different refractive indices and dispersions were available for designing systems, the secondary properties such as striae and bubble quality and the optical homogeneity were poor [1.43].

In the mid-1880s, Otto Schott began to develop new optical glasses specifically for the attainment of optical properties predetermined by Ernst Abbe, for the design of lens systems with the high imaging quality needed for microscopes and other optical instruments [1.44, 45]. This new scientific approach to glass development began the first phase of modern optical glass technology and the engineering of glass to specific applications. The glass types known at that time had a so-called normal behaviour of the dispersion: if one plots the relative partial dispersion versus the Abbe number (for details, see fold-out diagram of Sect. 2.1, Fig. 2.9), the old glasses are on one line (Abbe's "normal line" rule). For the design of a perfect lens system Abbe found that glass types are required which are located at a certain distance from this line. Otto Schott worked with boron oxide and phosphorous pentoxide as glass formers and, in 1884, introduced PK-50 and PSK-50. The new glasses based on boron and phosphorous oxides had low refractive indices and very low dispersions. Small amounts of boron oxide added to the glass composition also improved secondary properties such as chemical durability and manufacturability. These and other new glasses allowed for the design of apochromatic lens systems. In order to correct the secondary spectrum which is a deviation in focus over a small spectral region, at least one of the glasses must have a partial dispersion in the blue region of the visible spectrum that does not conform to Abbe's normal line rule. In 1911, Otto Schott introduced fluorine into borosilicate glasses. These FK glass types have low refractive indices and low dispersions. During this era barium oxide was also introduced into the compositions, especially as a replacement for lead oxide. The barium oxide exchange for lead oxide yielded glasses with lower mass densities, relatively high refractive indices and lower dispersions than the lead oxide glasses. In the 1940s and 50s, new optical glasses based on lanthanum oxide were developed. The lanthanum borate glasses and lanthanum silico-borate glasses have higher refractive indices and lower dispersions than the lead silicate glasses. In the 1960s and 1970s, fluoro- phosphate glasses went into production. These glasses have a very high fluoride content; many of the O^{2-} ions are replaced by two F^- ions in the glass structure. The only oxygen present in the chemical composition enters through the phosphorous pentoxide. These glasses have low refractive indices, very low dispersions and exhibit abnormally large partial dispersions in the blue region of the spectrum [1.43]. Beside this they have increased UV transmission. They are used in high precision lens systems for the UV and visible region.

1.4 Selected Applications

There are many applications of glass in the fields of optics and electronics which have resulted in a large development effort during the last forty years and much of this work will continue into the future.

The *ophthalmic* area is a specialized application of optical glass. Currently, glass with high refractive indices and low density and photochromic glasses are of importance. Lightweight glasses with mass densities of 3.0 to 3.5 g/cm^3 possessing high refractive indices of 1.7 or 1.8 have found acceptance in the ophthalmic market (Sect. 8.1). Recently an ophthalmic glass type with a refractive index of 1.9 and a mass density of 4.0 g/cm^3 has become available. Considerable development work in the area of photochromic glasses has resulted in glass types with faster recovery times. The recombination time of the silver and chlorine ions in glass has been reduced to minutes, typically 3 to 6 minutes. Photochromic glass is now also available with an increased refractive index of 1.6 and a recovery time of 3 minutes [1.46, 47].

The demand for better *ultraviolet transmitting glasses* has been driven by the electronics industry. The lithographic design rules for electronic circuits have decreased to submicron dimensions; thus, the wavelength of the radiation sources used in the lithography process has decreased correspondingly in order to retain diffraction limited imaging. Fluorophosphate glasses produced via a modified manufacturing procedure have resulted in shifting the UV-transmission edge near to that of fused silica. Lens systems can be constructed from these glasses for application in the 300 nm and longer wavelength spectral region. Because of the large abnormal dispersion in the blue end of the visible spectral region, these glasses are also used for apochromatic colour correction [1.28, 48].

Thermography, pyrometry, infrared spectroscopy, sensing and military applications require *infrared transmitting glasses.* The three most important glass families for infrared applications are the heavy metal oxide glasses, the heavy metal fluoride glasses and the chalcogenide glasses. The last two glass families are quite unique in that they do not contain oxygen as the anion for forming the glass structure. The heavy metal oxide glasses have infrared transmission which extends into the 6 to 10 μm spectral region and refractive indices as high as 2.4 [1.31, 32]. The refractive indices of these glasses are larger than those of most oxide glasses which have refractive indices of less than 2. The cut-on wavelength of these glasses is shifted well into the visible spectral region; typically transmission starts at 400 to 500 nm. The heavy metal fluoride glasses have infrared transmission extending to 8 μm. These glasses also show good ultraviolet transmission, extending down to 0.25 μm [1.29–31]. This extension of the transmission into both the ultraviolet and the infrared spectral regions is quite unusual when compared with other glass families. The chalcogenide glasses have infrared transmission extending into the 10 to 14 μm region with refractive indices as high as 3.1 [1.35]. During the last twenty years, the compositional space for stable glasses has been extended and there are more commercially available chalcogenide glasses.

Laser glass: Since the invention of the laser, many glass and crystal systems have been investigated to develop an understanding of the influence of the host material on the laser properties of the active ions. Today the most commonly utilized solid state laser host material is an yttrium aluminum garnet crystal doped with neodymium as the active ion [1.49]; however, glass has several advantages over crystals as a host material. Glass can be produced in large volumes with high optical homogeneity and free of absorbing particles; the nonlinear refractive indices can be low [1.50] and the concentration of the active ions in the glass can be much larger than in a crystal. These advantages made Nd:glass the material of choice for large laser systems such as those for inertial confinement fusion applications. Special phosphate laser glasses with very low temperature coefficients of optical pathlength are utilized in systems that require the beam quality to be independent of temperature gradients within the laser rod. Such an application is lasers for material working [1.51]. Recently optical fibres with erbium-doped core glasses have been used as in-line amplifiers for telecommunications applications [1.52].

In *atomic and nuclear technology* special optical glasses have found application. Terbium-activated silicate glasses are used as screens for X-ray imaging [1.53]. Cerium-activated barium silicate glasses or lead silicate glasses are utilized as elementary particle detectors. Ce-activated barium silicate glasses act as scintillating calorimeters [1.54]. Lead silicate glasses are used for Cerenkov counters. Fibre optic bundles with core glasses activated with cerium and terbium can be used to detect the tracks of high energy particles [1.56]. Also, drawn bundles of glass capillaries filled with an organic scintillator solution serve as particle detectors with high spatial resolution [1.57, 58].

Optical glasses irradiated with ultraviolet, X-ray, or gamma radiation show browning caused by induced colour centres. In many cases this browning can be greatly reduced or prevented by doping the base glass compositions with cerium oxide [1.59]. Glasses which contain heavy elements are utilized as window materials providing shielding against nuclear and X-ray radiation. High lead-containing silicate glasses doped with cerium oxide have been developed for such applications [1.60]. Here the cerium oxide also prevents the browning of the glass due to colour centres.

1.5 Progress in Manufacturing Technology

Before 1950, optical glass was commonly melted in ceramic crucibles heated in a gas-fired furnace. The volume of the crucibles could be as large as 800 litres. The chemical composition of the ceramic crucible was chosen so that the chemical attack on the crucible due to the activity of the melted glass was minimized. Though stirring was often used to improve the optical homogeneity, the yield of good quality glass was low.

As material development led to new glass families, improvements in melting technology were required. Some new glass types such as the lanthanum borates reacted so strongly with the existing ceramic crucible material that the noble metal platinum began to be used as a crucible material [1.61]. This resulted in specially designed tanks lined with platinum being developed for the melting of these very aggressive glasses. The concept of tank melting, i.e. the continuous production of optical glass, was extended to many of the older glass types [1.62]. The continuous production units are generally constructed of refractory materials such as high density fused silica, aluminum zirconium silicates or other special refractory materials. The choice of the ceramic refractory depends on the glass type to be melted. Critical parts of this melting unit such as the refiners and homogenization sections are made of platinum to achieve improved quality of the glass. Tank melting allows for the production of much larger volumes per piece of glass with better homogeneity and the application of automatic processing of glass into a variety of hot pressed forms such as lens blanks, prisms and discs or drawn forms such as rods, bars or sheets. Today, the entire production process from the handling of the raw materials to the forming and annealing of the glass can be automated.

A special challenge exists for the manufacture of optical glass and laser glass without absorbing particles [1.63, 64]. Optical elements for high fluence laser systems must not contain absorbing particles since these lead to the development of laser damage sites. Examples of such laser systems are those used for nuclear fusion experimentation. Particles as small as 2 to 5 μm can lead to damage sites in excess of 50 μm after repeated exposure to high fluence laser beams. Since optical and laser glasses are exposed to platinum during their production, special manufacturing techniques have been developed to reduce or eliminate platinum particles.

The sol-gel process has been an important production method for coatings for several decades [1.65–67]. In the last few years powders, fibres and solid pieces have also been produced by this method. These have mostly been organic materials (oxide glasses) but new types of non-crystalline solids intermediate between inorganic and organic materials have been produced too [1.68–72].

1.6 Today's Interests

Glass has been and will continue to be the *material of choice for imaging optics.* Optical engineers require glass types with extremely large or small values of the refractive index and the Abbe number, and with abnormal partial dispersions. Since high quality imaging systems often contain many elements, there is a strong desire to develop glasses with lower mass densities so that the total weight of the optical system can be reduced. This is particularly true for aircraft and space applications. Development work continues in search of glasses with improved blue and ultraviolet transmission, especially for glasses with high refractive indices. In the field of lithography, glasses with improved ultraviolet transmission are important for the design of lens systems utilized in the production of high capacity electronic integrated circuits. Schott has successfully demonstrated a manufacturing procedure for a fluorophosphate glass which has an internal transmission of 50 % percent at the wavelength 200 nm for a thickness of 5 mm. Glasses with improved chemical resistance are desirable because they show improved workability and long-term stability against surface attack. In special applications, glasses which are better stabilized against browning due to high intensity ultraviolet, gamma or nuclear particle irradiation are required.

In the area of process development, work continues to improve the optical quality of the glass and to reduce the production costs. During the last two years, Schott has developed a production process for manufacturing mirror blanks made of glass ceramics that are greater than eight metres in diameter.

There is a general desire of optical engineers to obtain new glasses having extreme optical properties with respect to n and ν. On the other hand, hundreds of glass types already exist in the conventional part of the Abbe diagram. For logistic reasons glass manufacturers would like to reduce the number of glasses produced. This has resulted in the producers of optical glass classifying glass types according to their availability.

There are several *new applications of glass in optics and optoelectronics.*

One example is *substrate glass* for masks used in the production of integrated electronic circuits by lithographic processes. These glasses must have good ultraviolet transmission and low coefficients of thermal expansion. Fused silica is a preferred glass type. Thin flat substrate glass sheets with thicknesses from 0.05 to 1.1 mm are required for liquid crystals, electroluminescent and ferroelectric liquid crystal displays, for covers of charge-coupled devices, for solar cells and hybrid circuits. The cover sheets for solar cells are normal window glasses; however, depending on the application, the substrate glasses must have either alkali barrier coatings or low or no alkali content and, in many cases, a low coefficient of thermal expansion [1.73, 74].

Glass or glass-ceramic substrates are used for computer magnetic discs with high density information storage capacity. The advantage of glass or glass-ceramics over aluminum is surface smoothness. Additionally, glass does not exhibit the warping effect during polishing that occurs for aluminum. Chemically strengthened glass discs can withstand rotation speeds of 30 000 revolutions per minute without breaking [1.75].

In the area of imaging optics, *microlenses* for reading of optical discs, laser printers, interfaces for laser diodes, coupling of optical fibres, integrated optics and other optical interconnections are produced in one- or two-dimensional arrays with element sizes varying from 10 μm to 1 mm. Several production methods are utilized to produce the arrays; these include molding of glass or plastics, deposition of thin films of silicon monoxide on a quartz glass substrate via lithographic techniques [1.76], laser chemical vapour deposition [1.77], planar lens with gradient refractive index structure via ion exchange, photothermal processing in photosensitive glasses [1.78], application of photoresist techniques [1.79], micro-controlled dip coating [1.80], monomer exchange diffusion in plastics, and composite lenses consisting of a semispherical glass lens with an aspheric plastic layer.

Gradient index optics: Traditionally the glasses for imaging optics have been homogeneous; however, during the last twenty years, glasses in which the refractive index varies as a prescribed function of location [1.81–90] have become useful for micro-optical systems, lens arrays and fibre optics. These gradient index glasses provide additional degrees of freedom in optical design. There are two basic types of gradient profiles: axial and radial. In axial gradients the refractive index varies along the optical axis. Optical components manufactured from these types of material are useful for the correction of spherical aberration [1.81, 82, 87, 89]. In a radial gradient material the refractive index varies continuously in every direction perpendicular to the optical axis. With this type of profile a rod with flat surfaces can focus light. These elements are often referred to as planar lenses; the effect is a function of the gradient and the rod length. Gradient index material can be produced by several methods including ion exchange processing, ion stuffing, sol gel, photocopolymerization of organic material, or monomer diffusion [1.88] and chemical vapour deposition processing. Gradient index fibres are generally produced via a chemical vapour deposition process. The ion exchange process is used commercially to produce rods from which optical elements are fabricated. An example of the commercial process is a thallium-containing glass which is placed in a melted potassium salt bath. The thallium ions in the glass rod are exchanged by potassium ions in the salt bath. This exchange of potassium for thallium results in a refractive index decrease; thus a zone with a gradient in the refractive index is generated. With small diameter rods, typically 0.5 to 4 mm, small lenses are produced for such optical systems as pick-up lenses in consumer electronic equipment, couplers for the optical telecommunication technique and interconnections for computer technology [1.91]. Two dimensional lens arrays have been produced by combining of photolithographic and ion exchange processing [1.92] or by photolytic techniques applied to photosensitive glasses [1.93]. Gradient index lens arrays are used in copiers, facsimile machines, optical printers [1.90, 94], liquid crystal devices, for projection television systems with increased brightness [1.95], for light coupling in devices for parallel optical fibre communication systems [1.90] and in multichannel matched filtering [1.90, 96].

Precision molding of optical glass: The conventional production of lenses is rather complicated. Glass which has been pressed to a rough form and annealed must have both sides processed. The lens blank is cleaned and mounted on a grinding block; the first side is then ground and polished. The procedure is repeated for the second side. The lens is then centred and cleaned. It would be much easier to hot press the glass into a precise finished form. For more than 20 years, work on developing a precision molding of optical glass has been in progress [1.97–108]. With a precision molding technique, aspherical lenses have been produced cost effectively. With the traditional lens manufacturing procedure only spherical surfaces have been produced. These surfaces result in errors of imaging such as spherical aberration, off axis coma and astigmatism. In order to reduce these errors, several spherical lenses with different properties are required. One aspherical lens is capable of performance equivalent to three or more spherical elements; thus, the number of lenses required for the optical system can be reduced leading to simplification in mounting, lower overall weight and light losses and a more compact design [1.84, 104, 109].

Although the precision molding process saves production steps and reduces the number of optical elements in comparison to the traditional production, there are difficulties in processing which must be overcome. The mold must have a very precise form and smooth surface which resists the attack from the hot glass for repeated pressing cycles. Secondly, the temperature control of the glass mold system is critical. The temperature of the glass must be high enough that the viscosity of the glass is in the range of 10^3 to 10^5 dPa·s. The temperature of the mold should be slightly above the transformation temperature of the glass. The glass must be kept under pressure for approximately 20 s; i.e., until the transformation temperature of the glass is reached, so that no shrinkage spots appear. The molded glass surface should not deviate from the desired shape by more than 0.1 μm. Today, pick-up lenses are produced with wavefront deviations of the order of 0.05 wavelength RMS. The molding process produces the lens complete with a mounting rim around the circumference of the lens for easy and precise mounting and alignment of the component.

Molded glass optics consist mostly of aspherical lenses which are used for cameras [1.103, 104], pick up lenses for discs [1.101, 102, 105, 106], and, in the telecommunications technology, for coupling of laser light into fibres, coupling the light from fibre to fibre, or coupling of light from fibre to receiver and in computer technology. The size of the precision optics is usually of the order of several millimetres up to 15 mm except for camera applications where the required sizes can be up to 30 mm.

Lenses with diameters up to 70 mm can be produced by molding techniques, but with less precision. The starting material for this molding process is a fire-polished glass rod which is heated until it becomes soft. Part of the rod is then pressed into a lens. The one side of the lens may have an ellipsoidal, parabolic, or hyperbolical form; the other side is flat or of a spherical shape. The aspheric surface is produced by the "finished lens pressing technology"; the second surface is produced by the conventional grinding and polishing technology [1.110]. These optical elements are used for automotive lighting and condensor lenses of projector systems.

Integrated optics: The use of fibre optics in telecommunications, sensor and computer technology requires new optical elements for light coupling processes. Consider the example where light from one fibre needs to be distributed to two other fibres. In the traditional approach the divergent beam coming from the first fibre would be collimated by a lens. The collimated beam is then divided into two beams by a semitransparent mirror, then the two beams are focused via lenses into the two fibres. Today this function is performed by an integrated optical chip of typically $30 \times 5 \times 2\,\mathrm{mm}^3$. In the glass plate a lithographic mask technique is used to produce very small channels which guide the light. The channel is produced by an ion exchange process resulting in the refractive index of the channel being higher than that of the base glass. The light is then guided within the channel via total internal reflection. The simplest integrated optical device is the "Y" splitter which takes light from a single input fibre and distributes it to two output fibres [1.111–115]. The superior advantage of this technique is the possibility to integrate and combine a number of these basic optical elements such as a splitter or a coupler on a single monolithic chip to form a complex optical circuit. The glasses developed for integrated optics chips, produced via ion exchange processes, should have the following characteristics ([1.111] see also the volume titled "Optical Fibres and Glass Integrated Optics" of this series): the refractive index should be close to that of the glass core of the fibre in order to suppress reflections at the fibre/device interface; the optical loss should be low in the visible and near infrared spectral regions and ion exchange properties should be good. Glasses used in these devices generally contain about 10 mole percent sodium or potassium ions which can be exchanged for caesium, silver or thallium, thus increasing the refractive index of the ion exchanged portion of the glass. These glasses must also be resistant to attack from aqueous solutions and melted salt baths. Silver ions have the highest mobility in ion exchange. The ion exchange process can be assisted by an electric field.

Glasses for light modulation: Glasses can be used to modulate or deflect the input light. This can be accomplished via the Faraday effect or the acousto-optical effect. The Faraday effect results from the refractive index of the glass being influenced by an applied magnetic field, whereas in the acousto-optical effect a sound field is utilized to modulate the refractive index of the material [1.116]. Glasses have been specifically developed to enhance these effects.

Recently, the nonlinear optical properties of glasses have provided the possibility to modulate light extremely fast (important for use in telecommunications and optical computers). Homogeneous glasses with high refractive indices and low Abbe numbers have demonstrated high nonlinear effects [1.117]. However, glasses containing small semiconductive crystals or small metallic colloids show much larger nonlinear effects in special spectral regions [1.118]. Interference patterns can be generated with a strong laser pump beam. This pattern results in regions of high and low refractive index via the nonlinear properties of the glass. Another signal beam incident on this material is diffracted by the laser grating pattern. The response time associated with the nonlinear effects in glasses is very fast, of the order of picoseconds [1.119]. Currently, the nonlinear effects in glass are only the objects of research.

These examples show that glass is still a very important material for modern optics and opto-electronics. There are many tasks to be performed and many possibilities for further development [1.9, 120–123].

In the following chapters scientists and engineers from the Schott Glaswerke and their subsidiaries report on the properties of glasses produced for optics and opto-electronics. The contributions mainly concentrate on those fields of research and development in which Schott has achieved special success. It is our hope that engineers and scientists involved in the applications of optical glass can advantageously use this book for their applications of this unique material.

For reading the manuscript and valuable discussions we thank Dr. H. Bach, Mrs. M. Faulstich, Mr. V. Fischer, Prof. Dr. H. J. Hoffmann, Mr. A. Jacobsen, W. W. Jochs, Dr. F.-T. Lentes, Mr. H. F. Morian, Dr. L. Roß, Dr. U. Pohl, Dr. B. Speit and Mr. K. Schilling.

References

1.1 D.R. Uhlmann, N.J. Kreidl (eds.): *Glass Science and Technology*, Vol. 1: "Glass forming systems", (Academie Press, New York 1983)

1.2 W. Vogel: *Glaschemie*, 3rd edition (Springer-Verlag, Berlin, Heidelberg 1992)

1.3 H. Rawson: "Properties and Application of Glass" in *Glass Science and Technology 3*, (Elsevier Scientific Publishing Co., Amsterdam 1980)

1.4 I. Fanderlik: *Optical Properties of Glass* (Elsevier, Amsterdam 1983)

1.5 C. Hoffmann, W. Norwig: "Zu den Anforderungen der Optik an die Entwicklung optischer Medien", Wiss. Zeitschrift Pädagogische Hochschule Erfurt **18**, 5–18 (1982)

1.6 S. Musikant: "Optical Materials", in *Optical Engineering,* ed. by B.J. Thomson, Vol. 6 (chapter 2: Glass) (M. Decker Inc., New York 1985) pp. 23–58

1.7 T.S. Izumitani: *Optical Glass* (Kyoritsu Shuppan Company, Ltd., Tokyo 1984); Lawrence Livermore National Laboratory, Livermore (USA); American Institute of Physics. Translation Series New York, 1986

1.8 J.W. Fleming: "Glasses" in *CRC Handbook of Laser Science and Technology,* Vol. IV, Part 2: Optical Materials, (CRC Press Inc. Boca Rato, Florida, USA 1986)

1.9 Gan Fuxi: "New Glass-Forming Systems and Their Practical Application", J. Non-Cryst. Sol. **123**, 385–399 (1990)

1.10 H.F. Morian: "New Glasses for Optics and Optoelectronics", SPIE Proc. Vol. **1400** (1991) pp. 146–157

1.11 A.J. Marker III: "Optical Properties: a trip through the glass map", SPIE Proc. Vol. **1535**, 60–77 (1991)

1.12 J. Petzoldt, F. Th. Lentes: "Glasses and glass ceramic materials for modern optical application", in *From Galileo's "ocialino" to Optoelectronics* by P. Mazzoldi (ed.) Internat. Conference Padova, June 9–12, 1992 (World Scientific Publishing, Singapore 1993) pp. 229–246

1.13 M.J. Weber: "Optical Properties of Glasses", in *Materials Sciences and Technology*, Vol. 9: Glasses and amorphous materials, ed. by R.W. Cahn et. al.; volume-editor J. Zarzycki (VCH-Verlagsges. Weinheim, Germany 1991) pp. 619–664

1.14 I. Fanderlik: "Structure et caractéristiques optiques de verres", Rivista della Stazione Sperimentale del Vetro, 65–71 (1989)

1.15 J.M. Benett, J.J. Schaffer, Y. Shibano, Y. Namba: "Float polishing of optical materials", Appl. Opt. **26**, 696–703 (1987)

1.16 J. van Vingerden, H.J. Frankena, B.A. van der Zwan: "Production and measurement of superpolished surfaces", Opt. Eng. **31**, 1086–1092 (1992)

1.17 H.N. Daglish: "Light scattering in selected optical glasses", Glass Technol. **11**, 30–35 (1970)

1.18 J. Schroeder: "Light scattering of glass" in *Treatise on materials science and technology,* Vol. 17, ed. by M. Tomozawa, R.H. Doremus (Academic Press, New York 1979) pp. 115–172

1.19 J. Steinert, G. Tittelbach: "Streulichtuntersuchungen an optischen farblosen Gläsern", Silikattechnik **35**, 186–190 (1984)

1.20 F. Reitmayer, E. Schuster: "Homogeneity of optical glasses", Appl. Optics **11**, 1107–1111 (1972)

1.21 C.K. Kao: *Optical Fiber Systems* (Mc Graw-Hill Book Co., New York 1982)

1.22 R.T. Kersten: *Optische Nachrichtentechnik* (Springer Verlag, Berlin, Heidelberg 1983)

1.23 T. Li (ed.): *Optical Fiber Communications*, Vol. 1, Fiber Fabrication" (Academic Press, Orlando, Fl., USA 1985)

1.24 M.F. Yan: "Optical fiber processing: Science and Technology", Amer. Ceram. Soc. Bull. **72**, No. 5, 107–119 (1993)

1.25 K. Iga, S. Misawa, M. Oikawa: "Stacked planar optics by the use of planar microlens array", Proc. 10th European Conference on optical communication, Stuttgart, Sept. 1984, Redaktion R. Werner (VDE-Verlag, Berlin, Offenbach) pp. 30–31

1.26 L. Roß: "Integrated optical components in substrate glasses", Glastech. Ber. **62**, 285–297 (1989)

1.27 S.I. Najafi (ed.): *Introduction to glass integrated optics* (Artech House, Boston 1992)

1.28 M.J. Liepmann, A.J. Marker III, J.M. Melvin: "Optical and physical properties of UV-transmitting fluorcrown glasses" SPIE Proc. Vol. **1128**, 213–224 (1989)

1.29 M. Poulain, J. Lucas: "Une nouvelle classe de materiaux – Les verres fluorés au tetrafluore de zirconium", Verres et Réfract. **32**, 505–513 (1978)

1.30 J. Lucas, J.L. Adam: "Halide glasses and their optical properties", Glastechn. Ber. **62**, 422–440 (1989)

1.31 A.E. Comys (ed.): "Fluoride Glasses" in *Critical Reports on Applied Chemistry*, Vol. 27 (J. Wiley & Sons, Chichester, New York 1989)

1.32 J.C. Lapp, W.H. Dumbaugh, M.L. Powley, "Recent advances in heavy metal oxide glass research", SPIE-Proc. Vol. **1327**, 162–170 (1990)

1.33 H. Zheng, P. Lin, J.D. Mackenzie "Properties of the infrared transmitting Bi-Ca-Sr (Cu, Zn)-O glasses doped with lead oxide "SPIE-Proc. Vol. **1327**, 171–179 (1990)

1.34 Schott Catalogue of Optical Glasses No. 10.000e 1992 and Technical Information Optical Glass No. 23, available on request from Schott Glaswerke, Mainz

1.35 J.A. Savage: "Infrared optical materials and their antireflection coatings" (Hilger, Bristol 1985)

1.36 H.J. Hoffmann, W.W. Jochs, G. Westenberger: "Use of the Sellmeier dispersion formula for optical glasses and practical implications", SPIE Proc. Vol. **1780**, 303–314 (1992)

1.37 A dispersion formula for the temperature coefficient of the refractive index of optical glasses is given in H.J. Hoffmann, W.W. Jochs, and G. Westenberger: "A dispersion formula for the thermo-optic coefficient of optical glasses", SPIE Proc. Vol. **1327**, 219–231 (1990) and in the Technical Information Optical Glass No. 19 (1988), available on request from Schott Glaswerke Mainz

1.38 E. Berger: "Über die Fleckenbildungsgeschwindigkeit säurelöslicher Gläser und ihre Abhängigkeit von der thermischen Vorgeschichte", Glast. Ber. **12**, 189–198 (1934)

1.39 L. Holland: "The properties of glass surfaces" (Champman and Hall, London 1966)

1.40 C.M. Jantzen, M.J. Plodinec: "Thermodynamic mode of natural, medieval and nuclear waste glass durability", J. Non Cryst. Sol. **67**, 207–223 (1984)

1.41 K.H. Fiedler: "Problems with polishing and its quality assessment" in *Optical fabrication and testing*, Technical digest Series, Vol. 13, pp.: WC6-1-WC6-6, Optical Society of America 1988

1.42 H. Bach: "The use of surface analysis to optimize the production parameter of the surface of glasses and glass ceramics", Glass Technology **30**, 75–82 (1989)

1.43 A.J.: Marker, R.J. Scheller: "A technological history of optical glass" in *Advances in the Fusion of Glass,* ed. by Amer. Ceram. Soc. (1988) pp. 4.1–4.24

1.44 E. Abbe: "Über neue Mikroskope", Sitzungsber. der Jenaischen Ges. f. Medizin u. Naturwiss., Jahrgang 1886, pp. 107–128

1.45 C. Hoffmann: "Die Auswirkung der Zusammenarbeit von E. Abbe und O. Schott für die Entwicklung optischer Gläser für den Jenaer optischen Präzisionsgerätebau am Ende des 19. Jahrhunderts", Augenoptik **101**, 163–168 (1986)

1.46 R.J. Araujo "Ophthalmic glass, particularly photochromic glass", J. Non-Cryst. Sol. **47**, 69–86 (1982)

1.47 H-J. Hoffmann: "The use of silver salts in photochromic glasses", in *Photochromism – Molecules and Systems*, ed. by H. Dürr, H. Bonas-Laurent, Studies in organic chemistry, Vol. 40 (Elsevier Science Publisher B.V., Amsterdam, The Netherlands 1990) pp. 822–854

1.48 T.I.J. Al-Baho, R.C.M. Learner, J. Maxwell: "The design of high performance aplanatic achromatics for the near ultra violet waveband", SPIE Proc. Vol. **1354**, 417–428 (1990)

1.49 M.J. Weber (ed.): *Handbook of Laser Science and Technology,* Vol. 1: Lasers and Masers (CRC Press, Boca Raton, Fl, USA 1982)

1.50 M.J. Weber: "Science and technology of laser glass", J. Non-Cryst. Solids **123**, 208–222 (1990)

1.51 R. Iffländer: *Festkörperlaser zur Materialbearbeitung* (Springer-Verlag, Berlin, Heidelberg 1990) (Serie: Laser in Technik u. Forschung)

1.52 M.J.F. Digonnet (ed.): "Rare-earth-doped fiber laser sources and amplifier" SPIE Proc. Volume **MS 37** (1991)

1.53 W.P. Siegmund: "Fiber optic tapers in electronic imaging", (prepared for Electronic Imaging West, Pasadena 1989) available at Schott Fiber Optics, Southbridge, MA. U.S.A.

1.54 P. Hartmann: "Untersuchungen an Glas-Szintillationen"; Ph. D. Thesis, Universität Mainz (1983)

1.55 H.F. Morian: "Teilchenforschung mit OPAL und DELPHI", Schott-Information No. **2**, 20–24 (1986) available on request from Schott Glaswerke, Mainz

1.56 M. Heming, P.A. Naß, R.R. Strack, H.M. Voyagis: "Scintillating Glass Fibers", in *Proceedings of the Workshop on Application of Scintillating Fibers in Particle Physics*, ed. By R. Nahnhauer (Blossin 1990) pp. 121–125

1.57 W.P. Siegmund, R.R. Strack, H.M. Voyagis, M. Heming and P.A. Naß: "Glass capillary production possibilities", in *Proceedings of the Workshop on Application of Scintillating Fibers in Particle Physics*, ed. By R. Nahnhauer (Blossin 1990) pp. 127–139

1.58 W.P. Siegmund, P. Naß: "Fiber detectors for particle research", SCHOTT-Information No. **62**, 6–9 (1992) available on request from SCHOTT Glaswerke, Mainz

1.59 B. Speit, E. Rädlein, G. H. Frischat, A. Marker III, J. Hayden: "Radiation resistant optical glasses", Nucl. Instr. Methods **B65**, 384–386 (1992)

1.60 B. Speit, S. Grün: "Irradiation energy dependence of discolouration in radiation shielding glasses", SPIE Proc. Vol **1327**, 92–99 (1990)

1.61 C. Eden, W. Geffcken: "Verfahren zur Herstellung tyndalleffektfreier optischer Gläser", German Patent No. 1019804 (1952)

1.62 C. Eden: "Verfahrenstechnische Fortschritte beim Schmelzen von optischem Glas", Schott-Information No. **4**, 1–5 (1968) available on request from Schott Glaswerke, Mainz

1.63 J. Heinz, W. Pannhorst, G.H. Frischat: "Formation of Platinum particles during the production of optical glasses", Proc. of the Internat. Glass Congress Madrid, Vol. **6**, 185–190 (S. E. de Ceramica y Vidrio 1992)

1.64 J.S. Hayden, H.J. Hoffmann "Laser-Gläser ohne Platin-Teilchen", Werkstoff und Innovation **4**, 47–49 (1991)

1.65 H. Schröder: "Oxide Layers Deposited from Organic Solutions", in *Physics of Thin Films*, Vol. 5, ed. by G. Hass, H.E. Thun (Academic Press Inc., New York 1969) pp. 87–141

1.66 H. Dislich: "Neue Wege zu Mehrkomponentenoxidgläsern", Angewandte Chemie **83**, pp. 427 ff (1971), also in Angewandte Chemie International Edition **10**, 363–370 (1971)

1.67 H. Dislich, P. Hinz, N.-J. Arfsten, E.K. Hußmann: "Sol-Gel Yesterday, Today, and Tomorrow", Glastechn. Ber. **62**, 46–51 (1989)

1.68 H. Schmidt: "New Type of Non-Crystalline Solids between Inorganic and Organic Materials", J. Non-Cryst. Solids **73**, 681–691 (1985)

1.69 L.C. Klein (ed.): *Sol-Gel Technology for Thin Films, Fibers, Preforms, Electronics, and Special Shapes* (Noyes Publications, Park Ridge, New Jersey, USA 1988), especially pp. 199–246

1.70 L.L. Hench, M.J.R. Wilson: "Processing of Gel-Silica Monoliths for Optics", J. Non-Cryst. Solids **121**, 234–243 (1990)

1.71 S. Sakka (ed.): "Glasses and Glass Ceramics from Gels", J. Non-Cryst. Solids **100**, 1–545 (1988), especially pp. 241–478

1.72 L. Esquivias: "Advanced Materials from Gels", Proc. 6th Internat. Workshop on Glasses and Glass Ceramics from Gels, J. Non-Cryst. Solids **147** and **148**, 1–823 (1992), especially pp. 508–652

1.73 J.M. Gallego, L. Greasley, I. Molyneux: "Requirements for low cost high quality thin glass substrates for flat panel displays" in *Science and Technology of Glasses for Optoelectronics* (ed. by Ceram. Soc. Japan and Internat. Comm. Glass 1990) pp. 35–38

1.74 Y. Takahashi: "Ultra-thin-glass production of low or non-alkali glass by float process", in *Science and Technology of Glasses for Optoelectronics* (ed. by Ceram. Soc. Japan and Internat. Comm. Glass 1990) pp. 9–15

1.75 T. Matsudaira, K.M. Ishizaki, C. Krishnan: "Glass disks for magnetic recording", in *Science and Technology of Glasses for Optoelectronics* (ed. by Ceram. Soc. Japan and Internat. Comm. Glass 1990) pp. 39–42

1.76 J. Jahns, S.J. Walker: "Two dimensional array of diffractive microlenses fabricated by thin film deposition", Appl. Opt. **29**, 931–936 (1990)

1.77 M. Kubo, M. Hanabusa: "Fabrication of microlenses by laser chemical vapor deposition", Appl. Opt. **29**, 2755–2759 (1990)

1.78 D. Baranowski, L.G. Mann, R.H. Bellman, N.F. Borelli: "Photothermal technique generates lens arrays", Laser Focus World **25,** 139–143 (1989)

1.79 N.P. Eisenberg, A. Karsenty, J. Broder: "A new process for manufacturing arrays of microlenses", SPIE Proc. Vol. **1038**, 388–8 (1989)

1.80 D.W. Hewak, J.W.Y. Lit: "Fabrication of tapers and lenslike waveguides by a microcontrolled dip coating procedure", Appl. Opt. **27**, 4562–4565 (1988)

1.81 P.J. Sands: "Third order aberrations of inhomogeneous lenses" J. Opt. Soc. Am. **60**, 1436–1443 (1970)

1.82 E. Marchand: *Gradient index optics* (Academic Press, New York 1978)

1.83 D.T. Moore: "Gradient index optics: a review", Appl. Opt. **29**, 1035–1038 (1980)

1.84 L.G. Atkinson III, S.N. Houde-Walter, D.T. Moore, D.P. Ryan, J.M. Stagaman: "Design of a gradient-index photographic objective", Appl. Opt. **21**, 993–998 (1982)

1.85 I. Kitano: "Current status of gradient index rod lenses" in *Japan Annual Reviews in Electronics, Computer and Telecommunications,* ed. by Suematsu, Vol. 5 (Ohnsha Ltd. and North-Holland Publ. Co. 1983) pp. 151–166

1.86 J.B. Caldwell, D.T. Moore: "Design of gradient-index lens systems for disk format cameras" Appl. Opt. **25**, 3351–3355 (1986)

1.87 L. Pugliese, D.T. Moore: "Gradient index optics", Photonics Spectra, Vol. **21**, 71–76 (1987)

1.88 S. Houde-Walter: "Lens designers: gradient-index optics are in your future", Laser Focus World **25**, 151–160 (1989)

1.89 D.Y.H. Wang, D.T. Moore: "Third order aberration theory for weak gradient-index lenses", Appl. Opt. **29**, 4016–4025 (1990)

1.90 K. Koizumi: "Gradient index optical systems", in *From Galileo's "ocialino" to Optoelectronics* by P. Mazzoldi (ed.) Internat. Conference Padova, June 9–12, 1992 (World Scientific Publishing, Singapore 1993) pp. 125–149

1.91 W.L. Tomlinson: "Application of GRIN-rod lenses in optical fiber communication systems", Appl. Optics **19**, 1127–1138 (1980)

1.92 M. Oikawa, K. Iga, T. Sanada: "Distributed-index planar microlens array from deep electromigration", Electr. Lett. **17**, 452–454 (1981)

1.93 N.F. Borelli, D.L. Morse, R.H. Bellmann, W.L. Morgan: "Photolytic technique for producing microlenses in photosensitive glass", Appl. Opt. **24**, 2520–2525 (1985)

1.94 K. Matsushita, N. Akazawa, M. Toyama: "Development of gradient index fiber array", Topical Meeting on Gradient-Index Optical Imaging Systems (Optical Society of America, Rochester 1979) Techn. Dig. WC2-1 to WC2-4

1.95 T. Kishimoto, H. Nemoto, K. Hamanaka, H. Imai, S. Tanigushi, M. Oikawa, E. Okuda, K. Nishizawa: "Application of planar microlens arrays to liquid crystal devices", Proc. of the 16. Internat. Congr. on Glass Oct. 1992, Madrid Vol. 3, 103–108 (S. E. de ceramica y Vidrio 1992)

1.96 K. Hamanaka: Japan J. Appl. Phys. **29**, pp. L 1277 (1990)

1.97 M.A. Angle, G.E. Blair, C.C. Maier: "Method for molding glass lenses", US Patent 3833 347 (1974)

1.98 C.K. Wu: "Optical articles prepared from hydrated glasses", US Patent 4073654 (1980)

1.99 H.J. Joormann, J. Verweii, J. Haisma: "Precision pressed glass object, method of preparing glass", European Patent 19342 (1980)

1.100 G.A.A. Menden-Piesslinger, J.H.P. Van de Heuvel: "Precision pressed optical components made of glass and glass suitable therefore" US Patent 4391915 (1983)

1.101 R.O. Maschmeyer, C. A. Andrysick, T. W. Geyer, H. E. Meissner, C. J. Parker, L. M. Sanford: "Precision molded glass optics", Appl. Opt. **22**, 2410–2412 (1983)

1.102 R.O. Maschmeyer, R. M. Hujar, L. L. Carpenter, B. W. Nicholson, E. F. Volzenilek: "Optical performance of a diffraction-limited molded-glass biaspheric lens", Appl. Opt. **22**, 2413–2415 (1983)

1.103 T. Izumitani: "Precision molded lenses", 13th Conference of the Internat. Com. Opt., Conference Dig, 516–517 (1984)

1.104 T. Agnivilina, D. Richards, H. Pollicove: "Finished lens molding saves time and money", Photonics Spectra **20**, 73–80 (1986)

1.105 P.I. Kingsburg: "Molded glass collimator/objective lens pair for pick up applications", SPIE Proc. Vol. **740**, 59–61 (1987)

1.106 M.A. Fitch: "Molded Optics: mating precision and mass production", Photonics Spectra **25**, 83–87 (1991)

1.107 V. Fischer, T. Matsuura, S. Hirota: "Asphärische optische Oberflächen blankpressen", Feinwerktechnik & Meßtechnik **99**, 243–245 (1991)

1.108 O. Buschmann, V. Fischer: "Moulded Optics, Fertigpreßlinge mit asphärischen Oberflächen" in *Jahrbuch für Optik und Feinmechanik 1993,* ed. by H. Zarm (Fachverlag Schiele & Schön GmbH, Berlin) pp. 108–121

1.109 J. Kross, R. Schuhmann: "Zur Korrektur optischer Systeme mit asphärischen Flächen", Optik **70**, 76–85 (1985)

1.110 DOCTER Optic-Handbuch, 1992, pp. 4.0.0/B1–B3 (Verlag Docter Optic Wetzlar)

1.111 Roß "Integrated optical components in substrate glasses", Glast. Ber. **62**, 285–297 (1989)

1.112 M. Börner, R. Müller, R. Schiek: "Elemente der integrierten Optik" (B.G. Teubner, Stuttgart 1990)

1.113 R.G. Hunsperger: *Integrated Optics: Theory and Technology* (Springer-Verlag, New York 1991)

1.114 K.J. Ebeling: *Integrierte Optik* (Springer-Verlag, Berlin, Heidelberg 1991)

1.115 R.Th. Kersten (ed.): *Integrierte Optik,* Fachausschußbericht Nr. 74, Verlag der Deutschen Glastechn. Ges. Frankfurt 1992

1.116 J. Xu, R. Stroud: *Acousto-Optic Devices: Principles, Design and Applications*, Wiley Series on Pure and Applied Optics, ed. by J.W. Goodman (J. Wiley, New York 1992)

1.117 H. Nasu, J.D. Mackenzie: "Nonlinear optical properties of glasses and sol gel-based composites", Opt. Eng. **26**, 102–106 (1987)

1.118 E. M. Vogel, M.J. Weber, D.M. Krol: "Nonlinear optical phenomena in glass", Physics and Chemistry of Glasses **32**, 231–254 (1991)

1.119 B. Van Wonterghem, S.M. Saltiel, P.M. Rentzepis: "Relationship between phase-conjugation efficiency and grating response time in semiconductor-doped glasses", J. Opt. Sos. Am. **B 6**, 1823–1827 (1989)

1.120 N.J. Kreidl: "Recent highlights of glass science, a selective topical review", part 1, Glast. Ber. **60**, 249–260 (1987), part 2, Glast. Ber. **63**, 277–287 (1990)

1.121 N.J. Kreidl: "Recent application of glass science", J. Non-Cryst. Sol. **123**, 377–384 (1990)

1.122 J. Zarzycki (editor) "Glasses and Amorphous Materials" in *Materials Science and Technology*, Vol. 9 (VCH-Verlagsgesellschaft, Weinheim Germany, New York 1991)

1.123 W. Vogel: "Möglichkeiten und Grenzen der Entwicklung optischer Gläser von Otto Schott bis zur Gegenwart", Wiss. Ztschr. Fr. Schiller-Univ. Jena, Math. Naturwiss. Reihe **32**, 495–508 (1983)

2. Optical Properties

2.1 Refractive Index and Dispersion

Frank-Thomas Lentes

2.1.1 Introduction

Optical glass is traditionally used to transmit visible light, including its neighbouring spectral regions, for a variety of purposes (e.g., optical windows, lenses). Optical imaging systems, whether individual lenses, photographic lenses, microscopes, telescopes, or other instruments, utilize glass in order to deflect rays of light emitted from a single point in the object plane in such a way that most of them are focused to a single point in the image plane. This deflection in transmission optics is caused by *refraction.* The refraction is different for different glasses and the refractive index depends on the wavelength of the light. This phenomenon is called *dispersion.* Both refractive index and dispersion are considered in the following.

2.1.2 Law of Refraction

If light enters a non-absorbing homogeneous material, *reflection* (discussed in Sect. 2.3) and *refraction* occur at the boundary surface. The *refractive index* n is given by the ratio of the velocity of light in the vacuum c to that in the medium v

$$n = \frac{c}{v} \,. \tag{2.1}$$

The ratio of the index in the medium 1 to that of the medium 2 is called the *relative index* of refraction. An important special case is the refractive index n_{rel} defined by the equation

$$n_{\mathrm{rel}} = \frac{n_{\mathrm{glass}}}{n_{\mathrm{air}}} \,. \tag{2.2}$$

The refractive index data quoted in this book and in the catalogues of glass manufacturers are relative to air rather than to vacuum.

At the surface an incident ray is refracted; this phenomenon is described by *Snell's* law [2.1] (sometimes called law of Descartes [2.2, 3]) which is given here in its vectorial form [2.4]

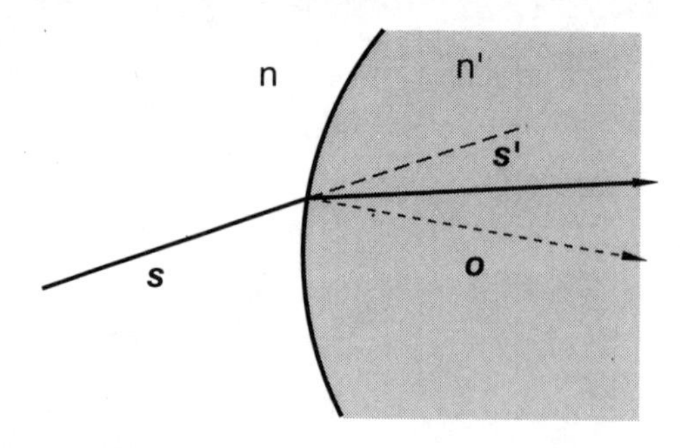

Fig. 2.1. Refraction at a surface between two optical media

$$\boldsymbol{s} \times \boldsymbol{o} = \boldsymbol{s}' \times \boldsymbol{o}. \tag{2.3}$$

$\boldsymbol{s}$ and $\boldsymbol{s}'$ are vectors with the direction cosines of the incident and of the refracted rays and with lengths corresponding to the refractive index in the media 1 and 2, respectively (cf. Fig. 2.1). The vector $\boldsymbol{o}$ designates a unit vector in the direction normal to the surface. The formulation using the vector cross product automatically takes into account that the incident ray, the refracted ray and the normal vector of the surface are coplanar.

In the following we will consider optical glasses as *homogeneous* and *isotropic* optical materials (i.e., the refractive index varies neither spatially nor directionally) with a high light transmission in the wavelength range used. Deviations from this assumption will be discussed in the chapters "Differential Changes of the Refractive Index", and Chap. 3, "Optical Quality" (Chaps. 2, Sect. 2.4).

2.1.3 Dispersion and the Various Dispersion Formulae

Basic Dispersion Theory

In general the refractive index and therefore the phase velocity in the medium depend on the wavelength used. This phenomenon is called *dispersion.*

The dispersion is called *normal* if the refractive index *increases* with shorter wavelengths. Otherwise *anomalous* dispersion occurs. Normal dispersion is observed far away from the absorption bands of the glass, whereas anomalous dispersion occurs at or near the absorption bands (see Fig. 2.2).

In principle the quantum-mechanical dispersion theory should be considered, but the classic electromagnetic theory is sufficient for our purpose. The variation of the refractive index n with the wavelength λ or the corresponding frequency ν can be explained by the electromagnetic theory applied to the molecular structure of matter. If an electromagnetic wave impinges on an atom or a molecule, the bound charges vibrate at the frequency of the incident wave. The bound charges have eigenfrequencies or *resonances* at the wavelengths λ_j (cf. *Born* and *Wolf* [2.5]). Then one can obtain the following dispersion relation which applies to media of *low density*:

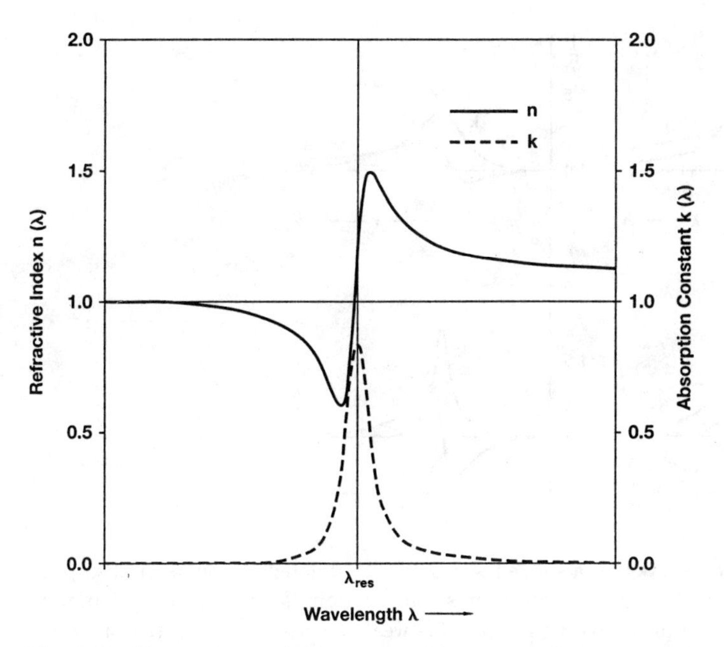

Fig. 2.2. Absorption and dispersion of light in a dielectric medium at the resonance wavelength $\lambda_{\rm res}$

$$\hat{u}^2 - 1 = (n - {\rm i}k)^2 = \frac{Ne^2}{2\pi c\varepsilon_0 m}\sum_j \frac{f_j\lambda^2\lambda_j^2}{2\pi c(\lambda^2-\lambda_j^2) + {\rm i}\gamma_j\lambda_j^2\lambda}\,. \tag{2.4}$$

In (2.4) $\hat{u}$ represents the complex refractive index in vacuum, where n is the real and k the imaginary part, N the number of molecules per unit volume, f_j the oscillator strength of the resonance with wavelength λ_j and damping constant γ_j, and m and e the mass and the charge of the electron. The constants f_j determine the strength of the absorption, and the constants γ_j may be considered as a measure of the frictional force at the resonance wavelength. The pair n and k are called the optical constants.

In the regions of negligible absorption, far from resonances where $k = 0$, (2.4) reduces to

$$n^2 - 1 = \frac{Ne^2}{4\pi^2c^2\varepsilon_0 m}\sum_j \frac{f_j\lambda_j^2\lambda^2}{\lambda^2-\lambda_j^2}\,. \tag{2.5}$$

With the abbreviation $A_j = \frac{Ne^2 f_j\lambda_j^2}{4\pi^2c^2\varepsilon_0 m}$ we can write (2.5) in a simplified form:

$$n^2 - 1 = \sum_j \frac{A_j\lambda^2}{\lambda^2-\lambda_j^2}\,. \tag{2.6}$$

In *dense* media where an interaction of atoms with the effective electric field cannot be neglected, the term n^2-1 in (2.5) and (2.6) should be replaced by $\frac{n^2-1}{n^2+2}$ according to *Lorentz* [2.7] and *Lorenz* [2.8]:

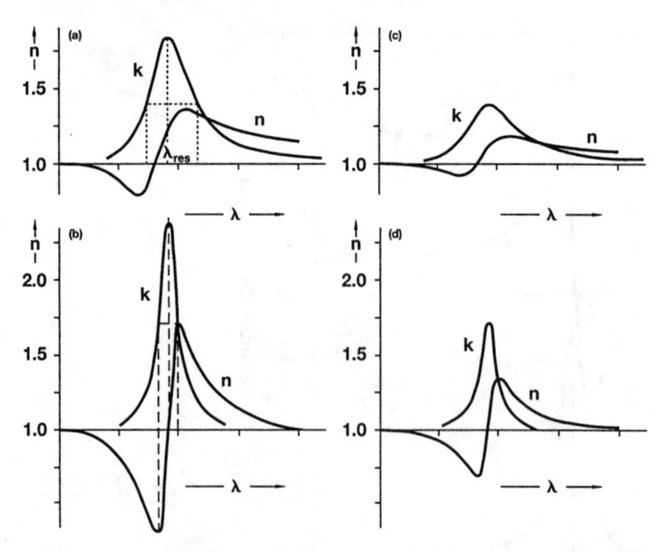

Fig. 2.3. Dispersion curves for a single resonance with different amounts of absorption and friction ($k = n\kappa$): (*a*) strong absorption, strong friction; (*b*) strong absorption, weak friction; (*c*) weak absorption, strong friction; (*d*) weak absorption, weak friction [2.6]

$$\frac{n^2 - 1}{n^2 + 2} = \sum_j \frac{A_j^2 \lambda^2}{\lambda^2 - \lambda_j^2} . \tag{2.7}$$

Kramers and *Kronig* [2.9] showed that under very general conditions a relationship between the real part, Re, and the imaginary part, Im, of a complex function

$$F(\tilde{\nu}) = \mathrm{Re}(\tilde{\nu}) - \mathrm{i}\,\mathrm{Im}(\tilde{\nu})$$

can be derived, where $\tilde{\nu} = \omega/c = 2\pi/\lambda$ designates the free-space wave number.

$$\mathrm{Re}(\tilde{\nu}) = \frac{2}{\pi} \int_0^\infty \frac{\tilde{\nu}' \cdot \mathrm{Im}(\tilde{\nu}')}{\tilde{\nu}'^2 - \tilde{\nu}^2} d\tilde{\nu}' \tag{2.8}$$

Functions that fulfil this relation are

$$F(\tilde{\nu}) = N(\tilde{\nu}) - 1 = \underbrace{n(\tilde{\nu}) - 1}_{\mathrm{Re}(\tilde{\nu})} - \mathrm{i} \underbrace{k\,(\tilde{\nu})}_{\mathrm{Im}(\tilde{\nu})} \tag{2.9}$$

and

$$F(\tilde{\nu}) = \varepsilon - 1 = N^2(\tilde{\nu}) - 1 = \underbrace{n^2(\tilde{\nu}) - k^2(\tilde{\nu}) - 1}_{\mathrm{Re}(\tilde{\nu})} - \mathrm{i}\, \underbrace{2n(\tilde{\nu})k(\tilde{\nu})}_{\mathrm{Im}(\tilde{\nu})} \tag{2.10}$$

where ε denotes the complex dielectric function and N the complex refractive index. Further possible expressions for $F(\tilde{\nu})$ exist but they will not be considered here.

In the literature the form (2.10) is preferred. However, the form (2.9) is just as valid as (2.10) but more convenient for various analytic considerations. Therefore we can write (2.6) in a simplified form

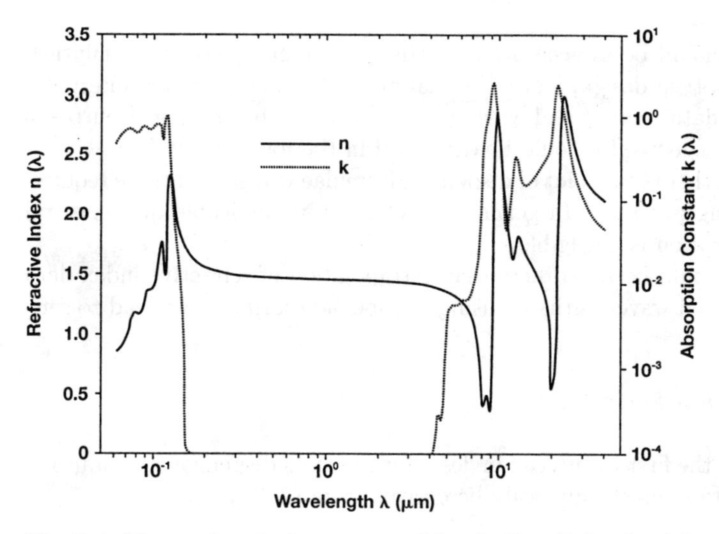

Fig. 2.4. Measured optical constants of fused silica (SiO_2 glass) [2.10]

$$n - 1 = \sum_j \frac{\tilde{A}_j \lambda^2}{\lambda^2 - \tilde{\lambda}_j^2} \; . \tag{2.11}$$

Real Dispersion Curves of Glasses

The dispersion curves of Figs. 2.2 and 2.3 indicate the variation of n and k for the case of a single resonator model. In reality the resonance phenomena are much more complex. This is demonstrated in Fig. 2.4 where the measured values of the optical constants for fused silica are shown (data from *Philipp* [2.10]). Even this simple compound has three main absorption features (one in the UV at 0.1 μm and two in the IR at 9 μm and 21 μm) with a complicated fine-structure. If one wants to express the variation of n with the wavelength in terms of a dispersion formula, it is useful to define an *effective* resonance wavelength corresponding to the centre of gravity of the respective absorption band.

The dispersion curves of optical glasses are fixed by the number, position and strengths of the absorption bands of the glass components and their structure. Within certain limits the designer of an optical glass can vary these properties in order to obtain a desired dispersion curve (Chap. 3).

Requirements for Dispersion Formulae

Often one would like to know the refractive index for unmeasured wavelengths. An analytical representation of the variation of the refractive index n with the wavelength λ is also necessary for computational purposes during the design process of optical systems. Many scientists in the fields of optics and glass science have therefore been trying to derive a *dispersion formula* in order to calculate the refractive index using a few constants derived from measurements. The formula should cover the whole useful spectral range of a glass (from the UV absorption edge to IR) and should be as accurate as needed (4 decimal places for analytical purposes during optical design, 5 or 6 decimal places for high precision interpolation of measured data points) and as simple as possible. For analytical purposes it would be advantageous if n could be expressed in the form of $n = \sum a_j \lambda^j$.

Consequently, there exist a lot of dispersion formulae which fulfil these requirements with varying precision. In general, they are only applicable in a spectral range where absorption is negligible.

Warning: It should be noted here that extrapolation of refractive index data outside the measured wavelengths by using a dispersion formula can lead to serious errors!

Cauchy Dispersion Formula

Cauchy [2.11] was the first to succeed in describing normal dispersion algebraically in 1830. His equation which can easily be derived from (2.6), reads:

$$n = a + \frac{b}{\lambda^2} + \frac{c}{\lambda^4} . \tag{2.12}$$

The accuracy within the visible spectrum is of the order of 10^{-4}.

Hartmann Dispersion Formula

In the visible range the dispersion formula of *Hartmann*

$$n = n_0 + \frac{A}{(\lambda - \lambda_0)^B} \tag{2.13}$$

is valid. The constants A, B, n_0 and λ_0 must be derived empirically. The exponent B varies between 0.5 and 2, but can be replaced by unity in most cases. The accuracy is 3 to 4 decimal places.

Sellmeier Dispersion Formula

This dispersion formula was derived by *Sellmeier* in 1871 [2.12] and can be written in the following form:

$$n^2 = 1 + \sum_{j=1}^{N} \frac{a_j \lambda^2}{\lambda^2 - \lambda_j^2} . \tag{2.14}$$

The λ_j designate the centre of gravity of the jth absorption band in the relevant spectral range (called the *effective resonance wavelength*). It can be shown that virtually all dispersion curves of optical glasses can be expressed by using two effective resonance wavelengths in the UV (typically in the region 50 nm to 150 nm) and one in the IR (roughly at 7 to 10 μm).

The main advantage of the Sellmeier dispersion formula is that it is based on physical reasoning and therefore has excellent fitting properties in the data reduction process of measured refractive indices. Furthermore, interpolation errors are minor and even the extrapolation errors are significantly smaller than the Laurent series expansion formula (2.16).

In 1992 a Sellmeier formula containing three effective resonance wavelengths was adopted as the standard dispersion formula for the *Schott* glass catalogue [2.13]. The accuracy in the refractive index is given below:

$404\,\text{nm} \leq \lambda \leq 706\,\text{nm}$ $\quad \pm 3 \times 10^{-6}$

$365\,\text{nm} \leq \lambda \leq 404\,\text{nm}$ and
$706\,\text{nm} \leq \lambda \leq 1014\,\text{nm}$ $\quad \pm 5 \times 10^{-6}$

$1014\,\text{nm} \leq \lambda \leq 2.3\,\mu\text{m}$ $\quad \pm 1 \times 10^{-5}$

Helmholtz-Ketteler-Drude Dispersion Formula

The *Helmholtz-Ketteler-Drude* [2.14, 15] dispersion formula is a modification of the Sellmeier formula (for a derivation see *Born* and *Wolf* [2.5])

$$n^2 = a_0 + \sum_{j=1}^{N} \frac{a_j}{\lambda^2 - \lambda_j^2} \; . \tag{2.15}$$

The λ_j are the effective resonance wavelengths of the oscillators. The accuracy is expected to be comparable to the Sellmeier formula.

Schott Dispersion Formula

This dispersion formula which gives an accuracy of 5 or 6 decimal places was proposed by *Schott* in 1966 [2.16] as well suited for optical glasses and was accepted by most glass manufacturers. It can be derived as a Laurent series expansion of *finite* order of the Sellmeier dispersion formula with an *arbitrary* number of effective resonance wavelengths (see e.g. *Born* and *Wolf* [2.5]). The λ^2 term comes from the IR absorption band and the λ^{-j} terms model the UV absorption bands.

$$n^2 = A_0 + A_1\lambda^2 + A_2\lambda^{-2} + A_3\lambda^{-4} + A_4\lambda^{-6} + A_5\lambda^{-8} \; . \tag{2.16}$$

The constants A_i are determined by a least squares fit to the measured refractive index data. The accuracy is given below:

$400\,\text{nm} \leq \lambda \leq 750\,\text{nm}$ $\quad \pm 3 \times 10^{-6}$

$355\,\text{nm} \leq \lambda \leq 400\,\text{nm}$ and
$750\,\text{nm} \leq \lambda \leq 1014\,\text{nm}$ $\quad \pm 6 \times 10^{-6}$.

Since this formula takes into account an arbitrary number of effective resonance wavelengths, the interpolation accuracy between the measured refractive indices is somewhat poorer than that of the Sellmeier formula. The validity of this dispersion formula can be extended further into the UV by adding a λ^{-10} term and into the IR by adding a λ^4 term and higher terms.

Herzberger Dispersion Formula

Based on an analytical study, *Herzberger* [2.17] came to the conclusion that in the visible range the manifold of optical glasses can be explained by 4 parameters

$$n = A_0 + A_1\lambda^2 + \frac{A_2}{\lambda^2 - \lambda_0^2} + \frac{A_3}{(\lambda^2 - \lambda_0^2)^2}, \tag{2.17}$$

with $\lambda_0 = 168\,\text{nm}$. This dispersion formula is accurate to 4 decimal places in the visible spectral range.

Geffcken Dispersion Formula

Geffcken [2.18–20] sought to express the dependence of the refractive index on the wavelength λ as a combination of one function describing the spectral behaviour of the so-called "normal glasses" (cf. below in 2.1.5) and a δ-function which accounts for the abnormal dispersion of a specific glass. By definition the δ-function becomes zero for all normal glasses.

$$n(\lambda) = 1 + [1 - D(\lambda)]\,[n(\lambda_1) - 1] + B(\lambda)\,[n(\lambda_2) - n(\lambda_1)] + D(\lambda)\delta(\lambda) \tag{2.18}$$

In this equation λ_1 and λ_2 are standard wavelengths (e.g. $\lambda_1 = 644$ nm, $\lambda_2 = 480$ nm), n_1 and n_2 the corresponding refractive indices, $B(\lambda)$ and $D(\lambda)$ tabulated functions for a fixed pair of λ_1 and λ_2 (independent of the glass type) and finally δ a tabulated function specific for a given glass. The δ-function describes the deviation of the relative partial dispersion from the behaviour of "normal glasses".

For the wavelength range from 365 nm to 1014 nm an accuracy of about 10^{-6} can be achieved.

Buchdahl Dispersion Formula

In terms of analytical algebraic expressions it is very desirable to have a dispersion model in the form $n = \sum a_j\omega^j$ for the design of highly colour-corrected lens systems. The a_j should be independent of the wavelength and the variable ω should be a function of the wavelength, $\omega = \omega(\lambda)$. None of the dispersion formulae mentioned so far fulfils this requirement. *Buchdahl* [2.21] introduced a change in variables from the wavelength λ to a chromatic coordinate ω defined as

$$\omega(\lambda) = \frac{\lambda - \lambda_0}{1 + \alpha(\lambda - \lambda_0)}\,. \tag{2.19}$$

λ_0 is a reference (or base) wavelength and α is a constant for all glasses having a value of 2.5. For applications in the visible spectrum λ_0 is chosen as 587.6 nm.

The refractive index n can now be written as an implicit function of the wavelength in terms of the chromatic coordinate ω

$$n(\omega) = n_0 + \nu_1\omega + \nu_2\omega^2 + \nu_3\omega^3 + \ldots + \nu_n\omega^n\,. \tag{2.20}$$

n_0 is the index of refraction at λ_0, and the ν_j correspond to the dispersion of the glass. This series expansion converges very quickly so that only a few terms are necessary. The accuracy for an expansion with powers in ω of $j = 2$ and $j = 3$ is given below (cf. *Robb* and *Mercado* [2.22])

$$\begin{array}{llll} j = & 2: 400\ \text{nm} & \leq \lambda \leq & 700\ \text{nm} & \pm 1 \times 10^{-4} \\ j = & 3: 355\ \text{nm} & \leq \lambda \leq & 1014\ \text{nm} & \pm 2 \times 10^{-4}. \end{array}$$

This model was used by *Robb* [2.23] for the selection of glass doublets and triplets having excellent colour correction.

2.1.4 Measurement and Limitations of the Accuracy of Refractive Index Data

General Remarks

There are a lot of measuring techniques deriving the optical constants of glasses (cf. *Bell* [2.24] and *Ward* [2.25] and references therein). Usually the refractive index is measured by determining the angle of the minimum deviation of light in a prism with a spectrometer (accurate to $\pm 1 \times 10^{-6}$) or by determination of the critical angle with the help of a *Pulfrich* refractometer (accurate to $\pm 1 \times 10^{-5}$) for routine work. Despite great efforts the accuracy of the measured index data can barely exceed $\pm 1 \times 10^{-6}$. There are several reasons for these limitations which we shall discuss further. Interferometric techniques can be applied to obtain relative measurements such as for checking homogeneity (spatial index variation). In this case the relative accuracy may reach $\pm 1 \times 10^{-7}$.

The refractive indices given in catalogues and handbooks are obtained by measuring pieces of glass samples in air under controlled environmental conditions (i.e., temperature, air pressure, humidity, etc.). The differences in the refractive index are up to $\pm 1 \times 10^{-3}$ from melt to melt. Single selected pieces of glass (for a few glasses up to 1 m in diameter) can be manufactured with a relative index variation of about $\pm 1 \times 10^{-6}$.

Preparation of Samples

In order to measure the refractive index with high accuracy, the preparation of the samples to be measured is crucial. One needs highly homogeneous and striae-free pieces of glass. Much greater efforts are necessary to achieve a surface flatness compatible with the desired index accuracy. If an accuracy of $\pm 1 \times 10^{-6}$ is required, the variation of the angle of incidence must not exceed a few tenths of an arcsecond due to surface imperfections. The influence of irregular deviations is reduced by integration over the prism aperture. On the other hand, if the angular deviation of a prism is measured in a spectral range where the transmission is low, the influence of the surface curvature is increased. Because of the different lengths of light paths across the aperture of the prism, the highest transmission is near the thin edge which may easily have surface imperfections (e.g. turned down edge). Therefore a systematic error may occur.

Refractive Index of Air

Since the refractive indices are relative to air, it is necessary to know the refractive index of air very accurately [2.26–28]. More details will be given in Sect. 2.4 below.

As an example, at the wavelength of sodium ($\lambda_0 = 589.46$ nm, temperature $T = 20°$ C, air pressure $p = 760$ Torr, partial pressure of water vapour $w = 10$ Torr) one obtains the following expressions for the differential variation of the refractive index of air [2.26]:

$$\frac{dn_{\text{air}}}{dT} = -\alpha\frac{n_{\text{air}}-1}{1+\alpha T} = -1.071\times 10^{-6}\ \text{K}^{-1}$$

$$\frac{dn_{\text{air}}}{dp} = \frac{n_{\text{air}}-1}{p} = +0.268\times 10^{-6}\ \text{hPa}^{-1}$$

$$\frac{dn_{\text{air}}}{dw} = -\frac{41\times 10^{-9}\ \text{hPa}^{-1}}{1+\alpha T} = -0.039\times 10^{-6}\ \text{hPa}^{-1} \qquad (2.21)$$

where α designates the thermal expansion of air ($\alpha = 0.00367\ \text{K}^{-1}$). Consequently variations of the temperature of $dT = \pm 5\,\text{K}$ and the pressure of $dp = \pm 20\,\text{hPa}$ have a noticeable effect on the refractive index of air.

Wavelengths of Spectral Lines for the Determination of the Refractive Index

For the determination of refractive indices one needs highly intensive monochromatic light sources. It is convenient to use *low pressure* spectral discharge lamps; the implicit assumption is that the wavelengths of the spectral lines are fixed and known as accurately as necessary. Neither of these assumptions is correct.

In general, a spectral line consists of several multiplet components (due to fine and hyperfine structure) with various intensities which depend on pressure, isotope composition and other conditions in a discharge lamp. Therefore, if it is measured by a spectrometer with limited resolution, the centre of gravity of such a multiplet may noticeably shift in wavelength.

Additionally, the *absolute* accuracy of the wavelength of a spectral line may be questioned. It is true that high resolution spectroscopy (e.g. laser spectroscopy) allows a highly accurate measurement of the multiplet structure of a spectral line. However, the data obtained are only *relative* to a given reference wavelength. In general, most existing data on absolute wavelengths are measured *relative to air* and afterwards reduced to those *in vacuum*. These data are often quite old and not as accurate as actually needed today; this status is not expected to change quickly since the task of measuring absolute wavelengths is extremely laborious.

Table 2.1 shows the spectral lines which are generally used to determine the refractive indices of glasses.

Table 2.1. Wavelength of spectral lines for the determination of refractive indices

Wavelength nm	Designation	Spectral Line	Element
2325.4		IR mercury line	Hg
1970.1		IR mercury line	Hg
1529.6		IR mercury line	Hg
1060.0		neodymium glass laser	Nd
1013.98	t	IR hydrogen line	Hg
852.1101	s	IR caesium line	Cs
706.5188	r	red helium line	He
656.2725	C	red hydrogen line	H
643.8469	C'	red cadmium line	Cd
632.8		helium-neon gas laser	He-Ne
589.2938	D	yellow sodium line center of doublet	Na
587.5618	d	yellow helium line	He
546.0740	e	green mercury line	Hg
486.1327	F	blue hydrogen line	H
479.9914	F'	blue cadmium line	Cd
435.8343	g	blue mercury line	Hg
404.6561	h	violet mercury line	Hg
365.0146	i	UV mercury line	Hg

2.1.5 Definition of the Characteristic Properties of Optical Glasses

Characteristic Values of Glass Types

For the design of optical systems a range of optical glasses is indispensable. Consequently, a classification of a specific glass according to its optical properties *refraction* and *dispersion* is necessary. The following numbers and values characterize these properties for optical applications in the visible spectral range: the main refractive index n, the main dispersion $n_{\lambda_1} - n_{\lambda_2}$, the Abbe number ν [2.29] and the relative partial dispersion P.

a) The *main refractive index* n_d or n_e is the index for the helium line at $\lambda = 587.56$ nm or mercury line at $\lambda = 546.07$ nm. This value varies between 1.4 and 2.3 for oxide glasses.

b) The main dispersion $n_F - n_C$ or $n_{F'} - n_{C'}$ describes the variation of the refractive index between the hydrogen lines F ($\lambda = 486.13$ nm) and C ($\lambda = 656.27$ nm) or the cadmium lines F' (479.99 nm) and C' (643.85 nm). In the case of optical glasses, n_F is always larger than n_C (normal dispersion).

c) The *reciprocal dispersive power* or *Abbe* number ν is basically the ratio between the refractive power and the main dispersion of a glass. It is defined by

$$\nu_d = \frac{n_d - 1}{n_F - n_C}$$

$$\nu_e = \frac{n_e - 1}{n_{F'} - n_{C'}} \quad . \tag{2.22}$$

In most cases the refractive index at the d-line n_d or e-line n_e is used as a reference wavelength; the corresponding Abbe numbers are designated as ν_d or ν_e. A small number corresponds to large dispersion and vice versa. A generalized differential version of equation (2.22), covering a continuous wavelength range, can be written

$$\nu^*(\lambda) = \frac{1 - n(\lambda)}{dn/d\lambda}$$

$$[\nu^*] = \text{length} \ . \tag{2.23}$$

d) The *relative partial dispersion* P or P' is needed to control the colour correction in more sophisticated optical systems (correction of the secondary spectrum). It can be considered as a number describing the similarity of the dispersion of two glasses.

$$P_{\lambda 1,\lambda 2} = \frac{n_{\lambda 1} - n_{\lambda 2}}{n_F - n_C}$$

$$P'_{\lambda 1,\lambda 2} = \frac{n_{\lambda 1} - n_{\lambda 2}}{n_{F'} - n_{C'}} \ . \tag{2.24}$$

Basically it is a measure of the rate of refractive index change dependent on the wavelength (i.e. the second derivative).

Traditionally optical glasses in the range of

$$\begin{array}{llll} n_d < 1.60, & \nu_d > 55 & n_e < 1.6028, & \nu_e > 54.7 \\ n_d > 1.60, & \nu_d > 50 & n_e > 1.6028, & \nu_e > 49.7 \end{array}$$

are designated as *crown* glasses, the other ones as *flint* glasses. Although these names have no other significance today they roughly divide the Abbe diagram (see fold-out diagram Fig. 2.7) into glasses with relatively low and high dispersive powers.

Figure 2.5 shows helium spectra produced by a crown and flint prism in comparison to a grating (normal) spectrum.

It can be seen that the dispersions of the two glasses and especially the relative spacings of the lines are different even if the crown glass spectrum is so enlarged that the two selected lines coincide with those in the flint glass spectrum. Obviously the relative partial dispersions of both glasses are not equal. We will discuss this topic later in greater detail.

In Fig. 2.6 the variation of the refractive index n with the wavelength λ is shown for glasses with similar values of n_d but different ν_d. Additionally, the first derivative $dn/d\lambda$ is plotted. Clearly, at the wavelength of the d-line (587 nm) the refractive indices of the glasses coincide, whereas the glass with the *lower* Abbe number has a *higher* refraction at shorter wavelengths and vice versa. The corresponding plot of the dispersion $dn/d\lambda$ stresses the different behaviour of both glasses.

A specific glass type has a designation based on its glass composition and a running number (e.g., boro-crown 7 as BK 7), see Table 2.2. There exist a lot of different types as can be seen in the *Schott* nomenclature system:

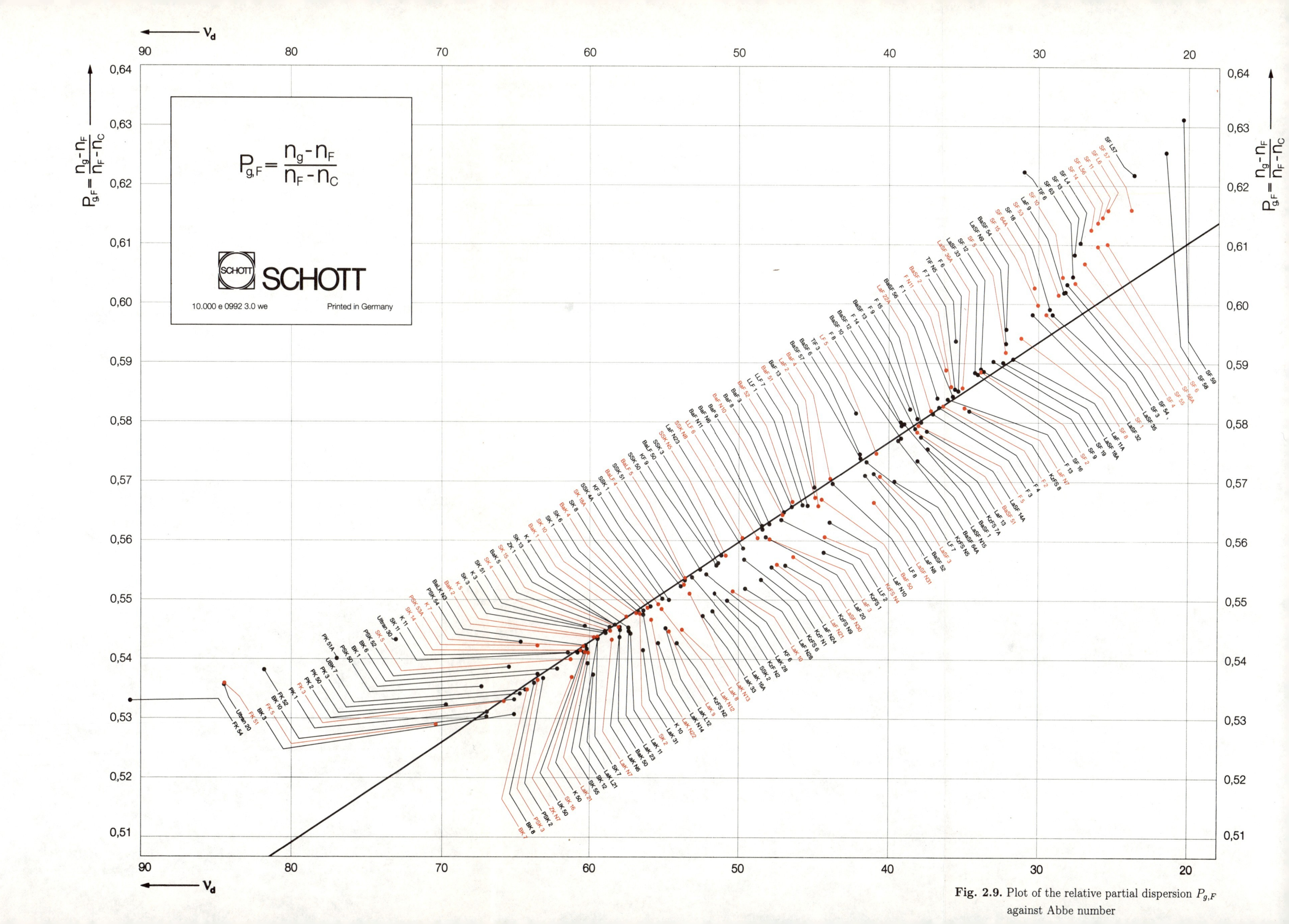

Fig. 2.9. Plot of the relative partial dispersion $P_{g,F}$ against Abbe number

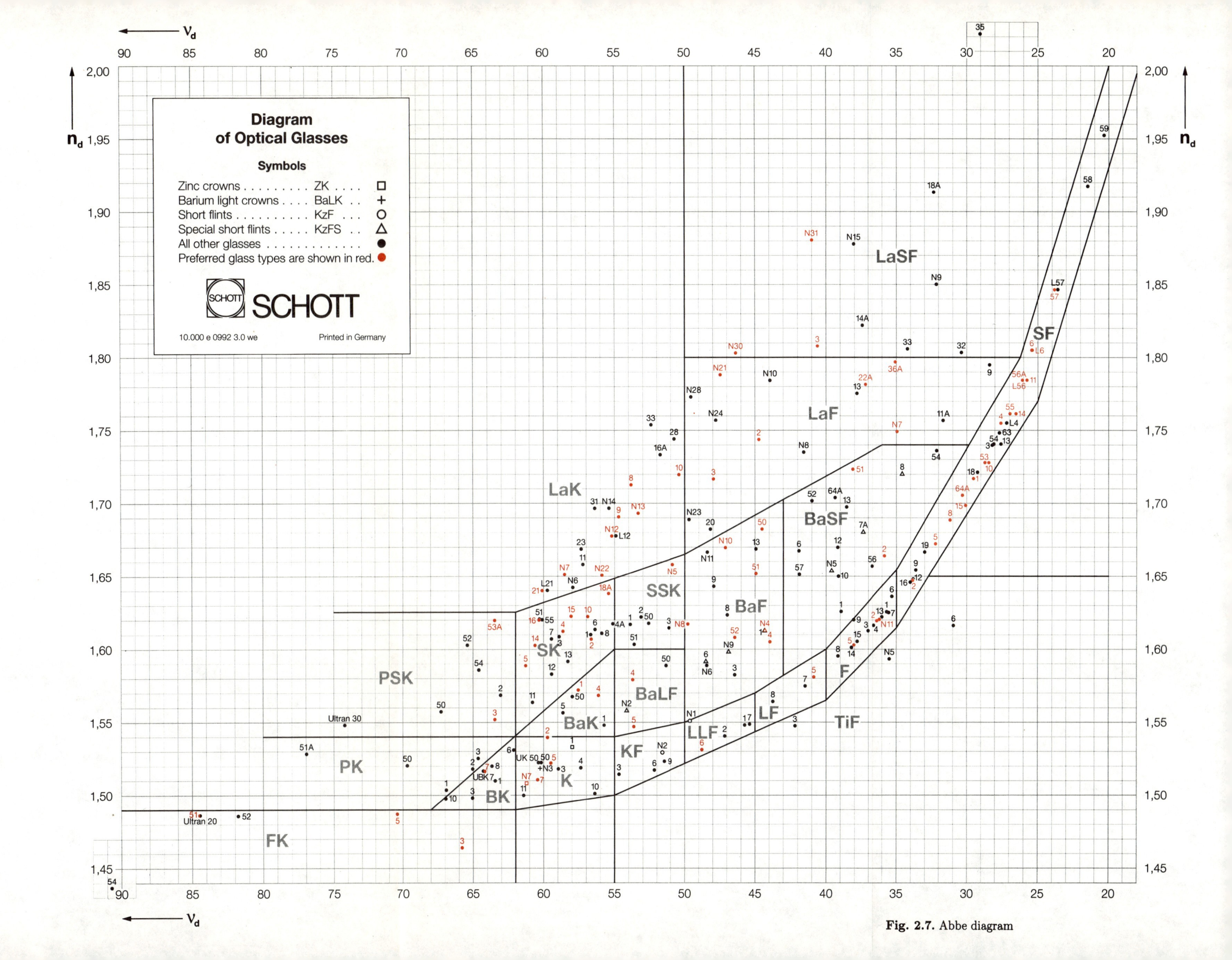

Fig. 2.7. Abbe diagram

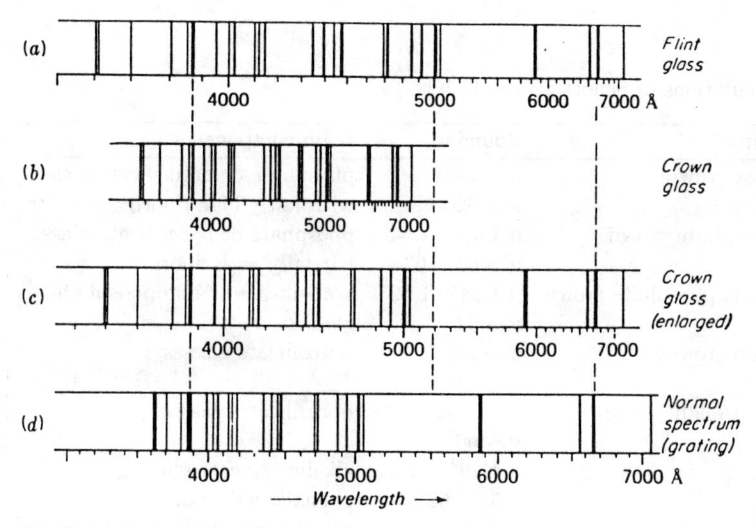

Fig. 2.5. Comparison of the helium spectrum produced by flint glass and crown glass prisms with a normal spectrum [2.6]

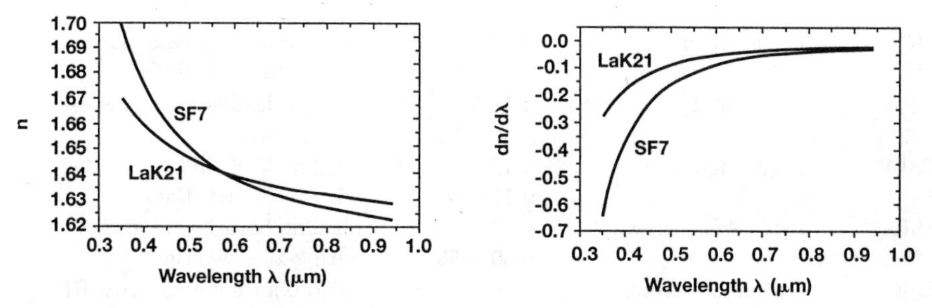

Fig. 2.6. Comparison of dispersion curves $n(\lambda)$ for glasses with similar n_d but different ν_d and the corresponding first derivative $dn/d\lambda$

Some newer glasses with similar optical properties differ significantly in composition (e.g. light weight glasses).

In order to give a glass type a unique designation a 6-digit code number was introduced:

$$1000\ (n-1) + 10\nu$$

where n was rounded to 3 decimal places and ν to 1 decimal place. For example the glass type BK 7 with $n_d = 1.5168$ and $\nu_d = 64.17$ was designated BK 7-517642. In Table 2.3, optical glasses of the four major manufacturers (i.e., Corning, Hoya, Ohara, Schott) are listed. They are sorted according to increasing code numbers.

Table 2.2. Designations of Schott glasses [2.30]

Glass type	Name	Boundary	Main components
FK	fluor crown	$n < 1.49$ $\nu > 62$	phosphate or borosilicate glass with fairly high fluorine content
PK	phosphate crown	n 1.49 – 1.54 $\nu > 68 - 62$	phosphate or borosilicate glass, partially with fluorine
PSK	dense phosphate crown	n 1.54 – 1.625 $\nu > 62$	borosilicate or borophosphate glasses
BK	boro-crown	$n \sim 1.51$* $\nu \sim 64$	borosilicate glasses
BaLK	barite light crown	$n \sim 1.52$ $\nu \sim 60$	borosilicate glasses
K	crown	n 1.495 – 1.54 ν 55 – 62	alkaline-silicate glasses, partially with boron
ZK	zinc crown	n 1.51 – 1.53 ν 57 – 61	silicate or borosilicate glasses with > 10% wt. ZnO
BaK	barite crown	n 1.55 – 1.60* ν 55 – 62	borosilicate glasses with > 10% wt. BaO
SK	dense crown	n 1.55 – 1.60* ν 55 – 62	alumino-borosilicate glasses with >15% wt. BaO
KF	crown flint	n 1.55 – 1.545* ν 50 – 55	alkaline lead silicate glasses
BaLF	barite light flint	n 1.55 – 1.59* ν 51 – 54	alkaline lead silicate glasses with >10% wt. BaO
SSK	extra-dense crown	n 1.60 – 1.66 ν 50 – 55	alkaline borosilicate glasses with >29% wt. BaO
LaK	lanthanum crown	$n > 1.64$* $\nu > 50$	silico-borate glasses with RE oxides
LLF	extra-light flint	n 1.53 – 1.57* ν 45 – 50	alkaline silicate glasses with 20–27% wt. PbO
BaF	barite flint	n 1.58 – 1.69* ν 43 – 50	alkaline silicate glasses with $\sum$ PbO and BaO <40% wt.
LF	light flint	n 1.56 – 1.59* ν 40 – 45	alkaline silicate glasses with 28–35% wt. PbO
F	flint	n 1.595 – 1.64* ν 35 – 40	alkaline silicate glasses with 39–47% wt. PbO
BaSF	dense barite flint	n 1.60 – 1.74* ν 32 – 43	alkaline silicate with $\sum$ PbO and BaO >40% wt.
LaF	lanthanum flint	n 1.63 – 1.80* ν 28 – 50	borosilicate or silico-borate glasses with RE oxides
LaSF	dense lanthanum flint	$n > 1.80$* ν 27 – 50	RE-oxide-borate or RE-oxide alkaline-earth-silicate glasses
SF	dense flint	$n > 1.63$* $\nu < 35$	alkaline silicate glasses with >47% wt. PbO

Table 2.2. (Cont.)

Glass type	Name	Boundary	Main components
TiK	low crown	$n < 1.495$ ν 55 – 62	alkaline-alumino-borosilicate glass
TiF	low flint	n 1.50 – 1.65* $\nu < 33$ @ 1.65 $\nu < 55$ @ 1.50	titanium-alkaline-alumino-borosilicate or phosphate glass
TiSF	dense low flint	$n > 1.65$* $\nu < 30$	titanium-alkaline-silicate glass
KzF	short flint	$n \sim 1.54$ $\nu \sim 50$	antimony-borosilicate glass
KzFS	short flint special glass	n 1.59 – 1.72 ν 34 – 54	alumino-lead-borate glass

* only approximate due to non-rectangular boundaries in the $n_d - \nu_d$ diagram

Glass Diagrams

In order to provide an overview of the range of optical properties for the large (>200) number of optical glasses their positions are plotted in so-called *glass diagrams*. They are useful for designing optical systems.

The n-ν-diagram (often called Abbe diagram [2.29]) is a plot of refractive index against the Abbe number for the various glass types (see fold-out diagram Fig. 2.7).

Occcasionally a plot of n against $1/\nu$ is preferred in order to stress the importance of $1/\nu$ for the colour correction of optical systems (see *Zimmer* [2.31]: The reciprocal relative dispersion ν has been introduced only for practical reasons – handier numbers – but already Abbe preferred the representation $1/\nu$. Therefore one can introduce a *dispersion number* m according to

$$m_d = 1831 \frac{n_F - n_C}{n_d - 1} = \frac{1831}{\nu_d} \tag{2.25}$$

where $1831 = 1/0.000546$ corresponds to the reciprocal wavelength of the e-line expressed in millimetres.

Figure 2.8 shows the resulting diagram where the deviations of the relative partial dispersions from the line of "normal glasses" in the blue $\Delta P_{g,F}$ and red $\Delta P_{C,s}$ are designated as vectors.

The majority of glasses (the so-called "normal glasses") show a more or less linear relationship between the Abbe number and the relative partial dispersion as already found by Abbe:

$$P_{x,y} \approx a_{xy} + b_{xy}\nu_d = \tilde{P}_{x,y} \,. \tag{2.26}$$

The glasses K7 and F2 are often chosen as reference glasses, defining the line (cf. fold-out diagram Fig. 2.9). A somewhat better representation which takes the small curvature of the line of "normal glasses" into account was given by *Izumitani* and *Nakagawa* [2.32]

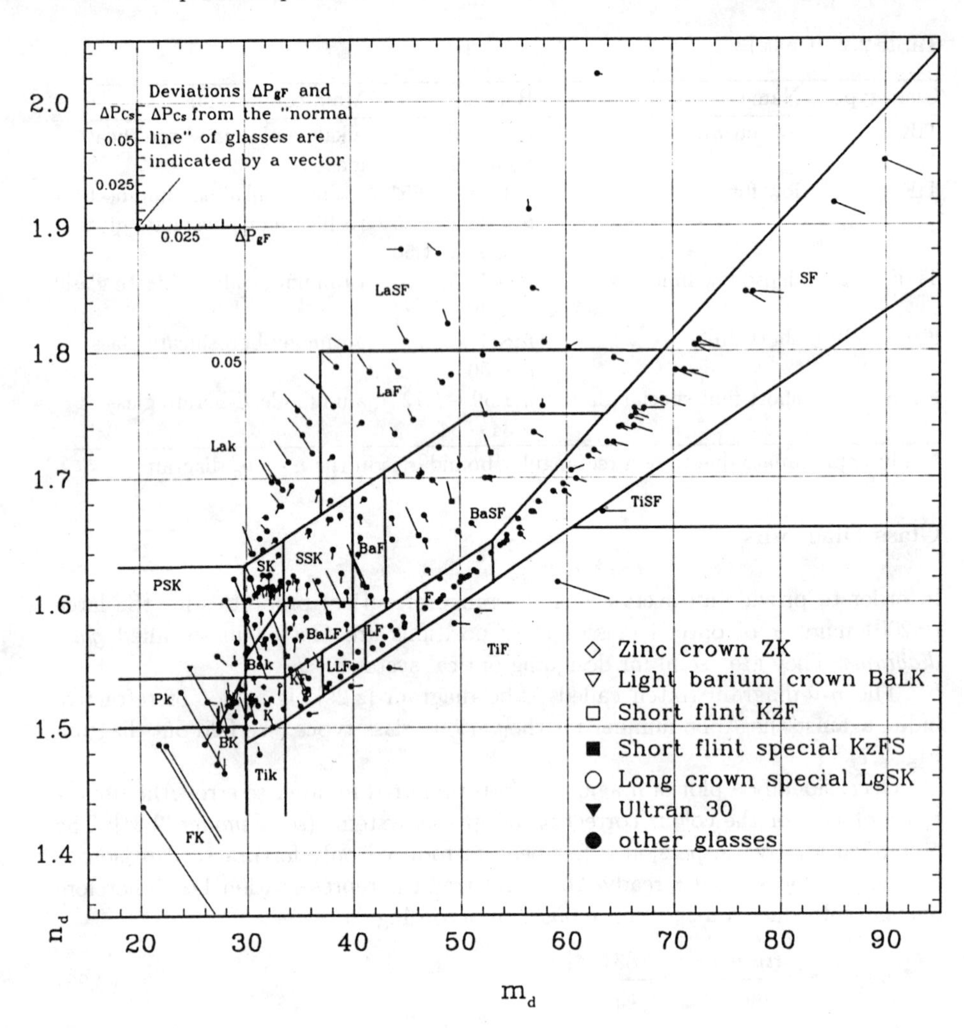

Fig. 2.8. Refractive index plotted against dispersion number [2.31] and indication of the partial dispersion

$$P_x = a_x + b_x \nu_d + c_x \nu_d^2 \ . \tag{2.27}$$

In order to correct the secondary spectrum (i.e. colour correction for more than two wavelengths) glasses are required which do *not* fulfil relation (2.26). A plot of the relative partial dispersion in a specific wavelength range (e.g. $P_{g,F}$ in the blue, $P_{C,s}$ in the red) against the Abbe number ν facilitates the selection of interesting glasses (see fold-out diagram Fig. 2.9). For a doublet in the thin-lens approximation both glasses should have the same relative partial dispersions P (possibly in a wide spectral range), but the difference of the $1/\nu$-values should be as large as possible to keep the individual refractive powers of both lenses as low as possible.

If one requires that a triplet lens system should bring four colours to a common focus (e.g. red, green, blue and violet), the three glasses must additionally lie on a straight line in a plot of the red P_r against the blue P_b as shown by *Herzberger* [2.17].

Colour Correction and Relative Partial Dispersions

Generally, a nearly perfect correction of polychromatic aberrations is a laborious task and it essentially depends on the available glass types with abnormal relative partial dispersions. From the thin-lens approximation one can derive the following condition for perfect colour correction for a lens doublet with vanishing separation d

$$n_2(\lambda) = \alpha + \beta n_1(\lambda) . \tag{2.28}$$

In other words, the dispersion curves of the two glasses must be related by a *linear* transformation. In general, however, this cannot be attained for an arbitrary wide wavelength range as will now be shown. If we express both dispersion curves by Sellmeier formulae, we obtain, assuming a fixed number N of resonance wavelengths,

$$n_1(\lambda) = 1 + \sum_{j=1}^{N} a_j \frac{\lambda^2}{\lambda^2 - \lambda_{1j}^2}$$

$$n_2(\lambda) = 1 + \sum_{j=1}^{N} b_j \frac{\lambda^2}{\lambda^2 - \lambda_{2j}^2} . \tag{2.29}$$

If we expand $n_2(\lambda)$ using (2.28), we can write

$$n_2(\lambda) = \alpha + \beta + \beta \sum_{j=1}^{N} a_j \frac{\lambda^2}{\lambda^2 - \lambda_{1j}^2} . \tag{2.30}$$

Then the following conclusions can be drawn:

a) The resonance wavelengths of both glasses must be the **same**. With some effort this can be achieved in practice.

b) All oscillator strengths of glass 2 must be decreased/increased by the **same** factor β. This is virtually impossible.

c) The constants α and β must fulfil the condition $\alpha + \beta = 1$ or $\alpha = 1 - \beta$.

The generalized Abbe numbers for both glasses are given by

$$\nu_1(\lambda) = \frac{1 - n_1(\lambda)}{dn_1/d\lambda}$$

$$\nu_2(\lambda) = \frac{1 - n_2(\lambda)}{dn_2/d\lambda} . \tag{2.31}$$

If ν_2 is expanded by using (2.28), we obtain

$$\nu_2(\lambda) = \frac{1-\alpha-\beta n_1(\lambda)}{d(\alpha+\beta n_1(\lambda))/d\lambda} = \frac{1-\alpha-\beta n_1(\lambda)}{\beta dn_1/d\lambda} . \tag{2.32}$$

By using c) this results in $\nu_1 = \nu_2$! As is well known, this condition does not allow achromatization.

A viable way to solve this problem could be the introduction of additional resonance wavelengths, a method which makes an approximate solution for a limited spectral range possible. If we add a single resonance λ_{2r}, only the following condition must be fulfilled:

$$1 + b_r \frac{\lambda^2}{\lambda^2 - \lambda_{2r}^2} = \alpha + \beta . \tag{2.33}$$

If $\lambda \gg \lambda_{2r}$, we can approximate (2.33) by

$$1 + b_r \approx \alpha + \beta , \tag{2.34}$$

which is independent of λ. For visible applications the new resonance wavelength must be as far in the UV as possible.

On the other hand, if $\lambda_{2r} \gg \lambda$, i.e., the new resonance is in the IR, one obtains

$$1 - b_r \frac{\lambda^2}{\lambda_{2r}^2} = \alpha + \beta , \tag{2.35}$$

which depends quadratically on λ. This explains why it is much easier to develop glasses with abnormal partial dispersions such as the fluor crown glasses having resonances in the far UV.

A key idea for the development of new glasses could be the designing of glass pairs (doublets) with relative partial dispersions adapted to each other. If we form the first and second derivative of (2.28) with respect to the wavelength, we can create a differential partial dispersion $P^*(\lambda)$:

$$P^*(\lambda) = \frac{n'(\lambda)}{n''(\lambda)} . \tag{2.36}$$

In the case of adapted glasses the following condition must be fulfilled

$$\frac{n_2'(\lambda)}{n_2''(\lambda)} = \frac{n_1'(\lambda)}{n_1''(\lambda)} . \tag{2.37}$$

A plot of $P^*(\lambda)$ has the advantage of being independent of a chosen reference wavelength system (e.g. F- and C-lines) so that the spectral variation of the relative partial dispersions and their similarity between glasses can easily be checked.

In Fig. 2.10 a plot of several optical glasses is shown. Obviously BK 7 and FK51 have similar relative partial dispersions in the visible and near UV, whereas in the near IR both glasses do not go well together. On the other hand, the relative partial dispersions of F2 and SF59 differ significantly from the two other glasses.

If one were to succeed in developing optical glasses which fulfil (2.37), at least approximately over a limited spectral range, a lot would be gained.

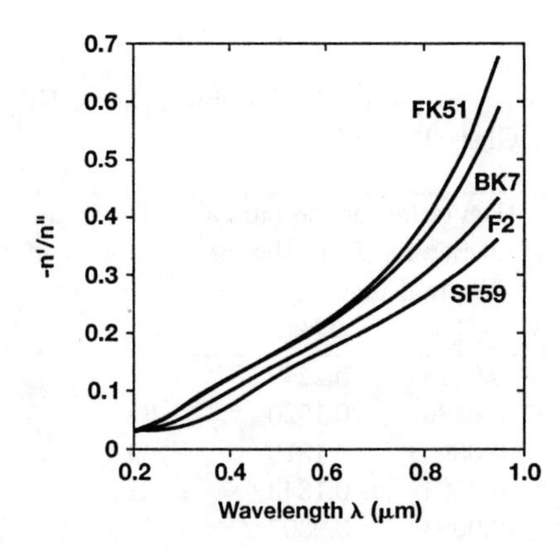

Fig. 2.10. Spectral variation of the differential partial dispersions of several glasses

Acknowledgement

I would like to thank Dr. habil. W. Geffcken (formerly at Schott Glaswerke), Dr. W. Vollrath (LEICA Mikroskopie und Systeme GmbH, Wetzlar) and Dr. H. Zügge (Carl Zeiss, Oberkochen) for reading the manuscript and making valuable suggestions.

Table 2.3. Optical glass cross reference by index. (Producers: C: Corning France; H: Hoya; O: Ohara; S: Schott Glaswerke and Schott Glass Technologies)

Code	Glass type	Refractive index	Abbe number	Deviation of the partial dispersion from the "normal line"		Producer
		n_d	v_d	$\Delta P_{g,F}$	$\Delta P_{C,t}$	
437-907	FK 54	1.43700	90.70	0.0418	–0.1249	S
438-950	FPL 53	1.43875	94.97	0.0480	–0.1520	O
456-903	FPL 52	1.45600	90.31	0.0390	–0.1310	O
457-903	FCD10	1.45650	90.77	0.0504	–0.1343	H
464-658	FK 3	1.46450	65.77	–0.0003	0.0207	S
465-657	FC A63-65	1.46450	65.70			C
465-658	FC3	1.46450	65.77	0.0055	0.0271	H
465-659	FSL 3	1.46450	65.94	–0.0020	0.0260	O
471-673	FC1	1.47069	67.31	0.0052	0.0207	H
471-674	FSL 1	1.47069	67.39	0.0010	0.0170	O
486-817	PFC A86-82	1.48605	81.70			C
486-818	FK 52	1.48605	81.80	0.0319	–0.1114	S
487-702	FSL 5	1.48749	70.21	0.0020	0.0140	O
487-704	FC A87-70	1.48725	70.40			C
487-704	FC5	1.48749	70.44	0.0091	0.0238	H
487-704	FK 5	1.48749	70.41	0.0036	0.0202	S
487-845	FK 51	1.48656	84.47	0.0342	–0.1112	S
487-845	ULTRAN 20	1.48656	84.47	0.0340	–0.1097	S
488-658	FC A88-66	1.48750	65.80			C
497-816	FCD1	1.49700	81.61	0.0374	–0.0967	H
497-816	FPL 51	1.49700	81.61	0.0280	–0.0990	O
498-650	BSL 3	1.49831	65.03	–0.0040	0.0336	O
498-651	BSC3	1.49831	65.13	–0.0021	0.0427	H
498-651	BK 3	1.49831	65.06	–0.0036	0.0394	S
498-668	BSL 10	1.49782	66.83	–0.0040	0.0300	O
498-670	BK 10	1.49782	66.95	–0.0008	0.0314	S
500-614	K 11	1.50013	61.44	0.0006	0.0130	S
500-660	BSC4	1.50048	66.00	0.0016	0.0384	H
500-660	BSL 4	1.50048	65.99	–0.0031	0.0319	O
501-563	C10	1.50137	56.30	0.0038	0.0109	H
501-564	FTL 10	1.50137	56.40	0.0035	–0.0019	O
501-564	K 10	1.50137	56.41	–0.0015	0.0094	S
504-668	BSL 21	1.50378	66.81	–0.0030	0.0220	O
504-669	PC1	1.50378	66.89	0.0041	0.0244	H
504-669	PK 1	1.50378	66.92	–0.0001	0.0258	S
508-608	ZSL 7	1.50847	60.83	–0.0063	0.0263	O
508-612	ZKN 7	1.50847	61.19	–0.0039	0.0266	S
508-613	ZnC7	1.50847	61.27	–0.0017	0.0283	H
510-621	NSL 1	1.50977	62.10	–0.0016	–0.0021	O

Table 2.3. (cont.)

Code	Glass type	Refractive index	Abbe number	Deviation of the partial dispersion from the "normal line"		Producer
		n_d	v_d	$\Delta P_{g,F}$	$\Delta P_{C,t}$	
510-634	BSC1	1.51009	63.43	0.0028	0.0136	H
510-635	BSC B10-63	1.51000	63.50			C
510-635	BK 1	1.51009	63.46	0.0004	0.0101	S
510-636	BSL 1	1.51009	63.63	–0.0001	0.0112	O
511-510	FF1	1.51118	51.03	0.0072	0.0092	H
511-510	FTL 8	1.51118	51.02	0.0058	0.0035	O
511-604	C B11-60	1.51100	60.40			C
511-604	K 7	1.51112	60.41	0.0000	0.0000	S
511-605	C7	1.51112	60.49	0.0000	0.0000	H
511-605	NSL 7	1.51112	60.49	0.0000	0.0000	O
512-581	ZSL 4	1.51190	58.14	–0.0020	–0.0079	O
515-546	CF3	1.51454	54.63	0.0014	0.0057	H
515-547	NSL 33	1.51454	54.71	–0.0021	–0.0014	O
515-547	KF 3	1.51454	54.70	–0.0018	0.0039	S
516-568	C2	1.51602	56.77	0.0016	0.0058	H
516-568	NSL 2	1.51602	56.80	–0.0033	0.0053	O
516-642	BSL 7	1.51633	64.15	–0.0026	0.0199	O
517-522	CF6	1.51742	52.15	0.0005	0.0084	H
517-522	KF 6	1.51742	52.20	–0.0017	0.0073	S
517-524	NSL 36	1.51742	52.41	–0.0020	0.0076	O
517-642	BSC B16-64	1.51680	64.20			C
517-642	BSC7	1.51680	64.20	0.0015	0.0223	H
517-642	BK 7	1.51680	64.17	–0.0009	0.0216	S
517-643	UBK 7	1.51680	64.29	–0.0008	0.0189	S
517-696	APL 1	1.51728	69.56	0.0034	–0.0017	O
517-697	PCS1	1.51728	69.68	0.0084	–0.0040	H
518-590	C3	1.51823	58.96	0.0035	0.0010	H
518-590	NSL 3	1.51823	58.96	–0.0004	–0.0045	O
518-590	K 3	1.51823	58.98	0.0001	–0.0017	S
518-603	C B18-60	1.51820	60.30			C
518-603	NSL 23	1.51835	60.27	–0.0019	0.0025	O
518-603	BaLK N3	1.51849	60.26	–0.0001	0.0037	S
518-604	BaCL3	1.51835	60.40	0.0021	–0.0068	H
518-650	BSL 22	1.51821	65.04	–0.0036	0.0284	O
518-651	BSC B18-65	1.51820	65.10			C
518-651	PK 2	1.51821	65.05	–0.0012	0.0255	S
518-652	PC2	1.51821	65.16	0.0025	0.0249	H
519-574	K 4	1.51895	57.40	–0.0020	0.0086	S
519-645	BSL 12	1.51874	64.48	–0.0032	0.0188	O
520-637	BK 8	1.52015	63.69	–0.0007	0.0152	S

Table 2.3. (cont.)

Code	Glass type	Refractive index	Abbe number	Deviation of the partial dispersion from the "normal line"		Producer
		n_d	v_d	$\Delta P_{g,F}$	$\Delta P_{C,t}$	
521-526	SSL 5	1.52130	52.55	–0.0072	0.0302	O
521-528	SbF5	1.52130	52.78	–0.0053	0.0337	H
521-697	PK 50	1.52054	69.71	0.0058	0.0009	S
522-595	C5	1.52249	59.45	0.0033	–0.0035	H
522-595	K 5	1.52249	59.48	0.0000	–0.0025	S
522-598	NSL 5	1.52249	59.79	0.0003	–0.0020	O
523-508	NSL 35	1.52310	50.84	–0.0024	0.0054	O
523-510	CF5	1.52310	50.95	0.0003	0.0088	H
523-515	KF 9	1.52341	51.49	–0.0014	0.0064	S
523-585	NSL 51	1.52307	58.49	–0.0005	0.0014	O
523-586	C12	1.52307	58.64	0.0030	0.0005	H
523-594	C B23-59	1.52300	59.40			C
523-600	NSL 50	1.52257	60.03	–0.0001	0.0054	O
523-602	K 50	1.52257	60.18	–0.0009	0.0068	S
523-604	C40	1.52300	60.40	0.0014	0.0073	H
523-604	UK 50	1.52257	60.38	–0.0010	0.0064	S
525-646	PC3	1.52542	64.62	0.0027	0.0230	H
525-646	BSL 23	1.52542	64.55	–0.0014	0.0242	O
525-647	PK 3	1.52542	64.66	–0.0009	0.0218	S
526-511	CF2	1.52630	51.05	–0.0019	0.0023	H
526-512	NSL 32	1.52630	51.17	–0.0018	0.0021	O
526-600	NSL 21	1.52642	59.99	–0.0024	–0.0021	O
526-601	BaCL1	1.52642	60.11	0.0048	–0.0043	H
527-511	SSL 6	1.52682	51.13	–0.0075	0.0283	O
529-516	SbF2	1.52944	51.64	–0.0036	0.0321	H
529-516	KzF N2	1.52944	51.63	–0.0058	0.0304	S
529-517	SSL 2	1.52944	51.72	–0.0067	0.0297	O
529-518	CHD B29-52	1.52900	51.80			C
529-770	PK 51A	1.52855	76.98	0.0258	–0.0991	S
531-621	BSC6	1.53113	62.07	0.0017	0.0057	H
531-622	BK 6	1.53113	62.15	–0.0009	0.0101	S
531-625	BSL 6	1.53113	62.45	–0.0025	0.0085	O
532-488	FEL6	1.53172	48.84	–0.0030	0.0020	H
532-488	LLF 6	1.53172	48.76	–0.0013	0.0054	S
532-489	PBL 6	1.53172	48.91	–0.0027	–0.0013	O
533-459	FF2	1.53256	45.94	0.0095	0.0116	H
533-459	FTM 8	1.53256	45.91	0.0077	0.0044	O
533-580	ZSL 1	1.53315	57.98	–0.0022	–0.0048	O
533-580	ZK 1	1.53315	57.98	–0.0008	–0.0009	S
533-581	ZnC1	1.53315	58.06	–0.0006	0.0019	H

Table 2.3. (cont.)

Code	Glass type	Refractive index	Abbe number	Deviation of the partial dispersion from the "normal line"		Producer
		n_d	v_d	$\Delta P_{g,F}$	$\Delta P_{C,t}$	
534-515	CF4	1.53358	51.54	0.0012	0.0027	H
534-516	NSL 34	1.53358	51.56	–0.0025	–0.0068	O
534-555	ZSL 5	1.53375	55.53	–0.0026	–0.0174	O
540-510	NSL 31	1.54041	51.00	–0.0010	–0.0050	O
540-595	BAL 12	1.53996	59.45	–0.0010	–0.0098	O
540-597	BCL B39-59	1.53950	59.70			C
540-597	BaC2	1.53996	59.72	0.0039	–0.0051	H
540-597	BaK 2	1.53996	59.71	0.0004	–0.0089	S
541-472	FEL2	1.54072	47.20	–0.0004	0.0043	H
541-472	PBL 2	1.54072	47.23	–0.0019	0.0019	O
541-472	LLF 2	1.54072	47.17	–0.0010	0.0032	S
541-476	FeL B41-48	1.54099	47.62			C
542-734	PFC B42-73	1.54220	73.37			C
547-536	BAL 5	1.54739	53.55	–0.0006	–0.0022	O
547-536	BaLF 5	1.54739	53.63	0.0001	–0.0122	S
548-422	TiF 3	1.54765	42.20	0.0087	0.0052	S
548-457	FeL B48-46	1.54810	45.70			C
548-458	FEL1	1.54814	45.82	–0.0016	0.0014	H
548-458	PBL 1	1.54814	45.78	–0.0020	0.0006	O
548-458	LLF 1	1.54814	45.75	–0.0009	0.0025	S
548-535	FBL B48-53	1.54775	53.50			C
548-628	BAL 21	1.54771	62.84	–0.0045	0.0121	O
548-742	ULTRAN 30	1.54830	74.24	0.0224	–0.0901	S
549-454	FEL7	1.54883	45.44	–0.0015	0.0025	H
549-454	LLF 7	1.54883	45.41	–0.0016	0.0054	S
549-456	PBL 7	1.54869	45.56	–0.0027	0.0022	O
551-495	SbF1	1.55115	49.52	–0.0012	0.0186	H
551-496	KzF N1	1.55115	49.64	–0.0035	0.0153	S
552-634	PCD3	1.55232	63.42	0.0042	0.0100	H
552-635	PSK 3	1.55232	63.46	–0.0005	0.0118	S
552-638	BAL 23	1.55232	63.75	–0.0036	0.0184	O
557-486	BAM 1	1.55690	48.55	0.0001	–0.0102	O
557-586	BaC5	1.55671	58.56	–0.0006	–0.0093	H
557-587	BAL 15	1.55671	58.69	–0.0023	0.0016	O
557-587	BaK 5	1.55671	58.65	0.0003	–0.0127	S
558-525	FBLS B58-5	1.55750	52.50			C
558-542	KzFS N2	1.55836	54.16	–0.0100	0.0601	S
558-673	PSK 50	1.55753	67.28	0.0048	–0.0052	S
560-470	PBL 3	1.56013	46.99	–0.0011	–0.0030	O
560-471	FEL3	1.56013	47.09	0.0009	0.0018	H

Table 2.3. (cont.)

Code	Glass type	Refractive index	Abbe number	Deviation of the partial dispersion from the "normal line"		Producer
		n_d	v_d	$\Delta P_{g,F}$	$\Delta P_{C,t}$	
560-612	BAL 50	1.55963	61.17	–0.0033	0.0058	O
561-452	FEL4	1.56138	45.23	0.0001	0.0024	H
561-452	PBL 4	1.56138	45.19	–0.0007	0.0011	O
564-438	PBL 28	1.56444	43.79	–0.0012	–0.0008	O
564-438	LF 8	1.56444	43.75	–0.0006	0.0000	S
564-607	BAL 41	1.56384	60.70	–0.0029	0.0045	O
564-608	BaCD11	1.56384	60.83	0.0020	–0.0024	H
564-608	SK 11	1.56384	60.80	–0.0004	–0.0024	S
564-609	BCD B64-61	1.56400	60.90			C
565-530	ADF1	1.56500	52.96	–0.0084	0.0458	H
567-428	FL6	1.56732	42.84	–0.0004	0.0067	H
567-428	PBL 26	1.56732	42.83	–0.0007	–0.0012	O
568-580	BaK 50	1.56774	57.99	–0.0026	0.0025	S
569-560	BCL B69-56	1.56880	56.00			C
569-560	BaC4	1.56883	56.04	0.0011	0.0003	H
569-561	BaK 4	1.56883	56.13	–0.0007	–0.0068	S
569-563	BAL 14	1.56883	56.34	–0.0025	–0.0047	O
569-631	PCD2	1.56873	63.10	0.0014	0.0145	H
569-631	PSK 2	1.56873	63.08	–0.0009	0.0119	S
569-632	BAL 22	1.56873	63.16	–0.0029	0.0120	O
570-481	SSL 4	1.57041	48.13	–0.0030	0.0108	O
570-493	BAM 2	1.56965	49.33	–0.0013	–0.0141	O
570-494	BaF2	1.56965	49.39	0.0030	–0.0165	H
571-508	BAL 2	1.57099	50.80	–0.0007	–0.0129	O
571-509	BaFL2	1.57099	50.86	0.0015	–0.0139	H
571-530	BaFL3	1.57135	52.98	0.0002	–0.0150	H
571-530	BAL 3	1.57135	52.97	–0.0007	–0.0207	O
573-426	FL1	1.57309	42.59	0.0010	0.0019	H
573-426	PBL 21	1.57309	42.58	–0.0007	–0.0021	O
573-575	BCL B72-57	1.57250	57.50			C
573-575	BaC1	1.57250	57.49	0.0043	–0.0110	H
573-576	BaK 1	1.57250	57.55	0.0002	–0.0167	S
573-578	BAL 11	1.57250	57.76	–0.0018	–0.0021	O
574-564	BaC6	1.57444	56.36	0.0026	–0.0029	H
574-565	BAL 16	1.57444	56.48	–0.0024	–0.0083	O
575-415	FL7	1.57501	41.51	–0.0009	0.0001	H
575-415	PBL 27	1.57501	41.49	–0.0015	0.0011	O
575-415	LF 7	1.57501	41.49	–0.0007	0.0024	S
576-414	FL57	1.57616	41.39	–0.0002	0.0034	H
578-415	PBL 24	1.57845	41.52	0.0001	0.0006	O

Table 2.3. (cont.)

Code	Glass type	Refractive index	Abbe number	Deviation of the partial dispersion from the "normal line"		Producer
		n_d	v_d	$\Delta P_{g,F}$	$\Delta P_{C,t}$	
578-417	FL4	1.57845	41.71	–0.0024	0.0064	H
580-537	BaFL4	1.57957	53.71	0.0000	–0.0113	H
580-537	BAL 4	1.57957	53.71	–0.0014	–0.0100	O
580-537	BaLF 4	1.57957	53.71	–0.0009	–0.0075	S
581-408	FL B81-41	1.58140	40.80			C
581-408	PBL 25	1.58144	40.75	–0.0005	0.0029	O
581-409	FL5	1.58144	40.89	–0.0002	0.0014	H
581-409	LF 5	1.58144	40.85	–0.0003	–0.0006	S
582-420	FL3	1.58215	42.03	0.0007	0.0027	H
582-421	PBL 23	1.58215	42.09	–0.0012	0.0024	O
583-464	BAM 3	1.58267	46.39	–0.0004	–0.0066	O
583-465	BaF3	1.58267	46.46	–0.0032	–0.0160	H
583-465	BaF 3	1.58267	46.47	0.0001	–0.0095	S
583-594	BAL 42	1.58313	59.38	–0.0022	–0.0063	O
583-595	BaCD12	1.58313	59.46	–0.0002	–0.0018	H
583-595	SK 12	1.58313	59.45	–0.0003	–0.0078	S
586-646	PSK 54	1.58599	64.60	0.0077	–0.0440	S
589-410	FL2	1.58921	40.96	–0.0008	0.0032	H
589-411	PBL 22	1.58921	41.08	–0.0005	–0.0026	O
589-485	BaF6	1.58900	48.54	0.0024	–0.0063	H
589-485	BaF N6	1.58900	48.45	0.0002	–0.0015	S
589-486	BAM 6	1.58900	48.59	–0.0010	–0.0115	O
589-512	BAL 7	1.58875	51.18	–0.0012	–0.0180	O
589-514	BaLF 50	1.58893	51.37	–0.0011	–0.0051	S
589-529	BaFL6	1.58904	52.93	–0.0034	–0.0084	H
589-532	BAL 6	1.58904	53.20	–0.0016	–0.0135	O
589-612	BCD B89-61	1.58900	61.20			C
589-612	BAL 35	1.58913	61.18	–0.0021	0.0017	O
589-613	BaCD5	1.58913	61.25	0.0023	0.0046	H
589-613	SK 5	1.58913	61.27	–0.0007	0.0008	S
592-485	KzFS 6	1.59196	48.51	–0.0088	0.0460	S
592-583	BaCD13	1.59181	58.31	0.0026	–0.0104	H
592-583	SK 13	1.59181	58.30	0.0001	–0.0140	S
593-353	FTM 16	1.59270	35.30	0.0090	0.0061	O
593-355	FF5	1.59270	35.45	0.0081	0.0078	H
594-355	TiF N5	1.59355	35.51	0.0096	0.0069	S
596-392	F8	1.59551	39.22	–0.0002	0.0024	H
596-392	PBM 8	1.59551	39.21	–0.0009	–0.0007	O
596-392	F 8	1.59551	39.18	–0.0006	0.0022	S
597-553	ADC2	1.59700	55.29	0.0077	–0.0020	H

Table 2.3. (cont.)

Code	Glass type	Refractive index	Abbe number	Deviation of the partial dispersion from the "normal line"		Producer
		n_d	v_d	$\Delta P_{g,F}$	$\Delta P_{C,t}$	
599-469	KzFS N9	1.59856	46.90	–0.0091	0.0452	S
600-425	BaFD16	1.60000	42.46	0.0026	0.0174	H
601-382	F 14	1.60140	38.23	–0.0006	0.0022	S
603-380	F5	1.60342	38.01	–0.0010	–0.0010	H
603-380	PBM 5	1.60342	38.01	–0.0004	–0.0007	O
603-380	F 5	1.60342	38.03	–0.0003	0.0017	S
603-423	BAM 25	1.60323	42.32	0.0000	–0.0081	O
603-424	BaFD5	1.60323	42.39	0.0014	–0.0026	H
603-606	SK 14	1.60311	60.60	–0.0003	–0.0033	S
603-607	BaCD14	1.60311	60.69	0.0020	0.0024	H
603-607	BSM 14	1.60311	60.70	–0.0024	–0.0024	O
603-654	PSK 52	1.60310	65.41	0.0049	–0.0183	S
603-655	PHM 53	1.60300	65.48	0.0046	–0.0273	O
604-381	FD C04-38	1.60350	38.10			C
604-536	SSK 51	1.60361	53.63	–0.0003	–0.0132	S
604-640	PCDS C04-6	1.60350	64.00			C
606-378	F 15	1.60565	37.83	–0.0002	0.0013	S
606-379	F15	1.60565	37.90	–0.0007	0.0022	H
606-437	BAM 4	1.60562	43.72	0.0004	–0.0128	O
606-439	FB C06-44	1.60560	43.90			C
606-439	BaF4	1.60562	43.88	0.0014	–0.0083	H
606-439	BaF 4	1.60562	43.93	0.0006	–0.0118	S
607-403	BAM 23	1.60717	40.26	0.0008	–0.0040	O
607-404	BaFD3	1.60717	40.36	0.0010	–0.0059	H
607-492	BAM 5	1.60729	49.19	–0.0010	–0.0112	O
607-493	BaF5	1.60729	49.34	0.0032	–0.0105	H
607-550	BaCD50	1.60700	54.98	0.0000	0.0013	H
607-567	BaCD2	1.60738	56.71	0.0019	–0.0132	H
607-567	SK 2	1.60738	56.65	–0.0008	–0.0162	S
607-568	BSM 2	1.60738	56.81	–0.0010	–0.0179	O
607-594	BSM 7	1.60729	59.38	–0.0017	–0.0101	O
607-595	BaCD7	1.60729	59.47	0.0018	–0.0083	H
607-595	SK 7	1.60729	59.46	–0.0004	–0.0087	S
608-462	BaF7	1.60801	46.21	0.0015	–0.0085	H
608-566	BCD C07-57	1.60750	56.60			C
609-464	BaF 52	1.60859	46.44	0.0009	–0.0122	S
609-589	BaCD3	1.60881	58.86	0.0025	–0.0098	H
609-589	SK 3	1.60881	58.92	–0.0003	–0.0085	S
609-590	BCD C09-59	1.60875	59.00			C
609-590	BSM 3	1.60881	58.95	–0.0021	–0.0098	O

Table 2.3. (cont.)

Code	Glass type	Refractive index n_d	Abbe number v_d	Deviation of the partial dispersion from the "normal line" $\Delta P_{g,F}$	$\Delta P_{C,t}$	Producer
610-565	BSM 1	1.61025	56.51	–0.0021	–0.0174	O
610-567	BaCD1	1.61025	56.65	0.0009	–0.0149	H
610-567	SK 1	1.61025	56.71	–0.0005	–0.0145	S
611-558	BaCD8	1.61117	55.77	0.0005	–0.0104	H
611-559	BSM 8	1.61117	55.92	–0.0021	–0.0135	O
611-559	SK 8	1.61117	55.92	–0.0008	–0.0134	S
613-370	F3	1.61293	36.96	–0.0001	0.0021	H
613-370	PBM 3	1.61293	37.01	0.0000	0.0005	O
613-370	F 3	1.61293	37.04	0.0000	0.0012	S
613-438	BPM 4	1.61340	43.84	–0.0080	0.0352	O
613-442	BPM 51	1.61340	44.24	–0.0084	0.0376	O
613-443	FSB C13-44	1.61340	44.30			C
613-443	ADF40	1.61340	44.29	–0.0079	0.0421	H
613-443	KzFS 1	1.61310	44.34	–0.0112	0.0546	S
613-443	KzFS N4	1.61340	44.29	–0.0087	0.0406	S
613-444	ADF10	1.61310	44.36	–0.0100	0.0564	H
613-449	ADF4	1.61250	44.87	–0.0056	0.0446	H
613-574	BSM 19	1.61342	57.41	–0.0012	–0.0094	O
613-585	BCD C13-58	1.61270	58.50			C
613-586	BaCD4	1.61272	58.58	0.0019	–0.0078	H
613-586	SK 4	1.61272	58.63	–0.0004	–0.0082	S
613-588	BSM 4	1.61272	58.75	–0.0014	–0.0130	O
614-550	BSM 9	1.61405	54.95	–0.0020	–0.0145	O
614-551	BaCD9	1.61405	55.12	0.0014	–0.0101	H
614-564	BCD C13-56	1.61350	56.40			C
614-564	BaCD6	1.61375	56.38	0.0002	–0.0141	H
614-564	BSM 6	1.61375	56.36	–0.0021	–0.0154	O
614-564	SK 6	1.61375	56.40	–0.0007	–0.0169	S
615-512	BaCED3	1.61484	51.15	0.0004	–0.0152	H
615-512	BSM 23	1.61484	51.17	–0.0012	–0.0120	O
615-512	SSK 3	1.61484	51.16	–0.0002	–0.0142	S
616-444	FSB C16-44	1.61600	44.40			C
617-310	TiF 6	1.61650	30.97	0.0306	–0.0081	S
617-366	F4	1.61659	36.61	0.0001	0.0011	H
617-366	PBM 4	1.61659	36.63	0.0005	–0.0005	O
617-366	F 4	1.61659	36.63	0.0003	0.0004	S
617-539	BaCED1	1.61720	53.94	0.0011	–0.0131	H
617-539	SSK 1	1.61720	53.91	–0.0008	–0.0128	S
617-540	BSM 21	1.61720	54.04	–0.0026	–0.0163	O
617-628	PCD5	1.61700	62.83	0.0059	–0.0104	H

Table 2.3. (cont.)

Code	Glass type	Refractive index n_d	Abbe number v_d	Deviation of the partial dispersion from the "normal line" $\Delta P_{g,F}$	$\Delta P_{C,t}$	Producer
617-628	PHM 51	1.61700	62.80	0.0031	–0.0127	O
618-498	BSM 28	1.61772	49.83	0.0005	–0.0028	O
618-498	SSK N8	1.61772	49.77	0.0004	–0.0040	S
618-526	SSK 50	1.61795	52.61	–0.0002	–0.0139	S
618-551	BCD C17-55	1.61760	55.10			C
618-551	BSM 24	1.61765	55.05	–0.0021	–0.0135	O
618-551	SSK 4A	1.61765	55.14	–0.0008	–0.0149	S
618-552	BaCED4	1.61765	55.16	0.0005	–0.0115	H
618-634	PCD4	1.61800	63.39	0.0059	–0.0154	H
618-634	PHM 52	1.61800	63.39	0.0052	–0.0346	O
620-363	FD C20-36	1.62000	36.30			C
620-363	F2	1.62004	36.30	0.0000	0.0000	H
620-363	PBM 2	1.62004	36.26	0.0000	0.0000	O
620-364	F 2	1.62004	36.37	0.0002	0.0007	S
620-381	F9	1.62045	38.09	0.0001	0.0081	H
620-381	PBM 9	1.62045	38.12	–0.0010	0.0044	O
620-381	F 9	1.62045	38.08	0.0009	0.0035	S
620-496	BSM 29	1.62012	49.55	–0.0005	–0.0080	O
620-498	BaCED9	1.62012	49.82	0.0015	–0.0040	H
620-601	SK 55	1.62041	60.12	–0.0033	0.0162	S
620-603	BCD C20-60	1.62040	60.30			C
620-603	BaCD16	1.62041	60.34	–0.0003	0.0105	H
620-603	BSM 16	1.62041	60.28	–0.0012	–0.0022	O
620-603	BSM 16C	1.62041	60.26	–0.0014	–0.0017	O
620-603	SK 16	1.62041	60.32	–0.0011	0.0016	S
620-622	ADC1	1.62000	62.19	0.0088	–0.0394	H
620-622	ATC1	1.62000	62.19	0.0088	–0.0394	H
620-635	PCD53	1.62014	63.52	0.0059	–0.0265	H
620-635	PSK 53A	1.62014	63.48	0.0053	–0.0274	S
621-359	PBM 11	1.62096	35.88	0.0058	0.0136	O
621-360	F11	1.62096	35.95	0.0054	0.0141	H
621-362	F N11	1.62096	36.18	0.0060	0.0139	S
621-603	SK 51	1.62090	60.31	0.0032	–0.0205	S
622-360	F 13	1.62237	36.04	0.0008	–0.0006	S
622-531	BaCED2	1.62230	53.11	0.0005	–0.0105	H
622-532	BSM 22	1.62230	53.20	–0.0018	–0.0157	O
622-532	SSK 2	1.62230	53.15	–0.0006	–0.0135	S
623-531	BCDD C23-5	1.62250	53.10			C
623-569	BCD C23-57	1.62280	56.90			C
623-569	BaCD10	1.62280	56.91	0.0008	–0.0144	H

Table 2.3. (cont.)

Code	Glass type	Refractive index n_d	Abbe number v_d	Deviation of the partial dispersion from the "normal line" $\Delta P_{g,F}$	$\Delta P_{C,t}$	Producer
623-569	SK 10	1.62280	56.90	–0.0003	–0.0153	S
623-571	BSM 10	1.62280	57.06	–0.0025	–0.0105	O
623-581	BCD C23-58	1.62300	58.10			C
623-581	BaCD15	1.62299	58.12	0.0008	–0.0068	H
623-581	SK 15	1.62299	58.06	–0.0008	–0.0090	S
623-582	BSM 15	1.62299	58.15	–0.0017	–0.0094	O
624-365	PBM 10	1.62364	36.54	0.0002	0.0011	O
624-469	FB C24-47	1.62400	46.90			C
624-470	BaF 8	1.62374	47.00	0.0001	–0.0130	S
624-471	BaF8	1.62374	47.05	0.0005	–0.0109	H
624-471	BAM 8	1.62374	47.10	–0.0019	–0.0041	O
625-356	F7	1.62536	35.58	0.0007	–0.0022	H
625-356	F 7	1.62536	35.56	0.0016	–0.0031	S
626-356	FD C26-36	1.62590	35.60			C
626-357	F1	1.62588	35.74	–0.0016	0.0017	H
626-357	PBM 1	1.62588	35.70	0.0005	0.0001	O
626-357	F 1	1.62588	35.70	0.0007	–0.0001	S
626-390	BaSF 1	1.62606	38.96	0.0014	–0.0077	S
626-391	BaFD1	1.62606	39.09	0.0000	–0.0135	H
626-392	BAM 21	1.62606	39.21	0.0001	–0.0008	O
636-353	F6	1.63636	35.34	–0.0008	–0.0013	H
636-353	F 6	1.63636	35.34	0.0009	–0.0013	S
636-354	PBM 6	1.63636	35.38	0.0009	–0.0012	O
637-353	FD C37-35	1.63650	35.30			C
639-449	BAM 12	1.63930	44.88	–0.0001	–0.0125	O
639-451	BaF12	1.63930	45.07	0.0016	–0.0108	H
639-554	BSM 18	1.63854	55.38	–0.0019	–0.0048	O
639-554	SK 18A	1.63854	55.42	–0.0013	–0.0047	S
639-555	BCD C39-56	1.63850	55.50			C
639-555	BaCD18	1.63854	55.45	–0.0011	–0.0057	H
640-345	PBM 27	1.63980	34.48	0.0018	–0.0012	O
640-346	FD7	1.63980	34.57	0.0009	0.0020	H
640-598	LaK L21	1.64048	59.75	–0.0059	0.0365	S
640-601	BSM 81	1.64000	60.09	–0.0077	0.0348	O
640-601	LaK 21	1.64049	60.10	–0.0017	0.0052	S
640-602	LaCL6	1.64000	60.15	–0.0023	0.0142	H
640-602	LaCL60	1.64000	60.20	–0.0039	0.0334	H
641-568	LaCL1	1.64085	56.83	–0.0028	0.0141	H
641-569	BSM 93	1.64100	56.93	–0.0024	–0.0015	O
641-601	BCS C41-60	1.64050	60.10			C

Table 2.3. (cont.)

Code	Glass type	Refractive index	Abbe number	Deviation of the partial dispersion from the "normal line"		Producer
		n_d	v_d	$\Delta P_{g,F}$	$\Delta P_{C,t}$	
642-580	LaK N6	1.64250	57.96	–0.0014	–0.0078	S
643-479	BaF9	1.64328	47.94	0.0007	–0.0074	H
643-479	BAM 9	1.64328	47.85	0.0000	–0.0122	O
643-480	BaF 9	1.64328	47.96	–0.0003	–0.0129	S
643-580	BCS C42-58	1.64250	58.04			C
643-580	LaC6	1.64250	57.96	–0.0030	0.0144	H
643-584	BSM 36	1.64250	58.37	–0.0055	0.0148	O
645-408	ADF355	1.64450	40.76	–0.0053	0.0363	H
645-408	BPH 35	1.64450	40.82	–0.0067	0.0292	O
646-341	SF 16	1.64611	34.05	0.0016	–0.0008	S
648-338	FD2	1.64769	33.84	0.0014	0.0006	H
648-338	FD12	1.64831	33.79	0.0016	0.0008	H
648-338	PBM 32	1.64831	33.83	0.0017	–0.0013	O
648-338	PBM 22	1.64769	33.80	0.0015	–0.0022	O
648-338	SF 12	1.64831	33.84	0.0021	0.0011	S
648-339	FDD C48-34	1.64800	33.90			C
648-339	SF 2	1.64769	33.85	0.0017	–0.0010	S
648-462	FB C48-46	1.64775	46.20			C
649-530	BaCED20	1.64850	53.03	0.0012	–0.0123	H
649-530	BSM 71	1.64850	53.03	–0.0018	–0.0112	O
650-392	FBD C51-39	1.65020	39.20			C
650-392	BaSF 10	1.65016	39.15	0.0015	–0.0114	S
650-393	BaFD10	1.65016	39.34	0.0005	–0.0094	H
650-394	BAH 30	1.65016	39.39	0.0015	–0.0141	O
650-557	LaCL2	1.65020	55.74	–0.0009	–0.0036	H
651-380	ATF2	1.65052	38.04	0.0193	–0.0214	H
651-383	BaFD4	1.65128	38.31	0.0013	–0.0055	H
651-383	BAH 24	1.65128	38.25	0.0024	–0.0100	O
651-419	BaSF 57	1.65147	41.90	0.0006	–0.0027	S
651-559	BCS C51-56	1.65113	55.90			C
651-559	LaK N22	1.65113	55.89	–0.0031	–0.0058	S
651-562	LAL 54	1.65100	56.15	–0.0023	–0.0058	O
652-449	BaF23	1.65224	44.92	0.0009	–0.0069	H
652-449	BaF 51	1.65224	44.93	–0.0009	–0.0080	S
652-584	LaC7	1.65160	58.40	–0.0035	0.0075	H
652-585	BCS C52-58	1.65160	58.50			C
652-585	LAL 7	1.65160	58.52	–0.0041	0.0052	O
652-585	LaK N7	1.65160	58.52	–0.0021	0.0010	S
654-336	PBM 29	1.65446	33.62	0.0014	–0.0006	O
654-337	FD9	1.65446	33.72	0.0004	0.0002	H

Table 2.3. (cont.)

Code	Glass type	Refractive index n_d	Abbe number v_d	Deviation of the partial dispersion from the "normal line" $\Delta P_{g,F}$	$\Delta P_{C,t}$	Producer
654-337	SF 9	1.65446	33.65	0.0014	0.0003	S
654-396	ADF50	1.65412	39.62	–0.0071	0.0384	H
654-396	KzFS N5	1.65412	39.63	–0.0071	0.0371	S
654-397	BPH 5	1.65412	39.70	–0.0069	0.0345	O
654-447	ATF4	1.65376	44.72	0.0041	–0.0201	H
656-401	FBDS C55-4	1.65550	40.10			C
657-367	BaSF 56	1.65715	36.74	0.0021	–0.0072	S
658-509	BaCED5	1.65844	50.85	0.0007	–0.0093	H
658-509	BSM 25	1.65844	50.86	–0.0038	–0.0063	O
658-509	SSK N5	1.65844	50.88	–0.0007	–0.0090	S
658-534	LAL 55	1.65830	53.44	–0.0022	–0.0094	O
658-572	BCS C57-57	1.65750	57.20			C
658-573	LaC11	1.65830	57.26	–0.0036	0.0113	H
658-573	LAL 11	1.65830	57.33	–0.0037	–0.0049	O
658-573	LaK 11	1.65830	57.26	–0.0032	0.0041	S
659-510	BCDD C58-5	1.65850	51.00			C
664-358	BAH 22	1.66446	35.81	0.0013	–0.0054	O
664-358	BaSF 2	1.66446	35.83	0.0026	–0.0071	S
664-359	BaFD2	1.66446	35.89	0.0015	–0.0102	H
664-489	BaF21	1.66422	48.94	0.0027	–0.0193	H
665-534	LaCL3	1.66480	53.40	–0.0011	–0.0008	H
667-330	SF 19	1.66680	33.01	0.0020	0.0004	S
667-331	FDD C67-33	1.66700	33.10			C
667-331	FD19	1.66680	33.06	0.0013	–0.0038	H
667-331	PBM 39	1.66680	33.05	0.0014	0.0011	O
667-483	BaF11	1.66672	48.30	0.0021	–0.0119	H
667-483	BAH 11	1.66672	48.32	–0.0013	–0.0128	O
667-484	FB C67-48	1.66670	48.40			C
667-484	BaF N11	1.66672	48.42	–0.0005	–0.0080	S
668-419	BaFD6	1.66755	41.93	0.0021	–0.0076	H
668-419	BAH 26	1.66755	41.93	0.0011	–0.0105	O
668-419	BaSF 6	1.66755	41.93	0.0013	–0.0095	S
669-449	BaF13	1.66892	44.91	0.0022	–0.0050	H
669-450	BAH 13	1.66892	44.99	0.0005	–0.0093	O
669-450	BaF 13	1.66892	44.96	0.0008	–0.0093	S
669-574	LaK 23	1.66882	57.38	–0.0026	–0.0002	S
670-392	BaSF 12	1.66998	39.20	0.0022	–0.0083	S
670-393	BAH 32	1.66998	39.28	–0.0004	–0.0071	O
670-471	FB C70-47	1.67000	47.10			C
670-471	BaF N10	1.67003	47.11	–0.0001	–0.0071	S

Table 2.3. (cont.)

Code	Glass type	Refractive index	Abbe number	Deviation of the partial dispersion from the "normal line"		Producer
		n_d	v_d	$\Delta P_{g,F}$	$\Delta P_{C,t}$	
670-472	BaF10	1.67003	47.20	0.0007	0.0035	H
670-473	BAH 10	1.67003	47.25	–0.0006	–0.0095	O
670-516	LAL 53	1.67000	51.63	–0.0017	–0.0127	O
670-517	LaCL4	1.66960	51.66	–0.0047	–0.0002	H
670-573	LaCL7	1.67000	57.31	–0.0049	0.0206	H
670-573	LAL 52	1.67000	57.33	–0.0049	0.0044	O
673-321	PBM 25	1.67270	32.10	0.0027	–0.0025	O
673-322	FD5	1.67270	32.17	0.0014	–0.0019	H
673-322	SF 5	1.67270	32.21	0.0023	–0.0010	S
673-323	FeD C72-32	1.67270	32.30			C
677-375	ADF405	1.67650	37.48	–0.0050	0.0251	H
677-375	BPH 40	1.67650	37.54	–0.0045	0.0207	O
678-506	LaCL9	1.67790	50.55	–0.0018	–0.0115	H
678-507	LAL 56	1.67790	50.72	–0.0028	–0.0136	O
678-534	LaCL8	1.67790	53.42	–0.0009	–0.0053	H
678-534	LAL 57	1.67790	53.36	–0.0049	–0.0100	O
678-549	LaK L12	1.67790	54.92	–0.0062	0.0143	S
678-552	LaK N12	1.67790	55.20	–0.0024	–0.0126	S
678-553	LAL 12	1.67790	55.33	–0.0055	–0.0063	O
678-555	BCS C78-56	1.67800	55.50			C
678-555	LaC12	1.67790	55.52	–0.0032	–0.0021	H
681-374	KzFS 7A	1.68064	37.39	–0.0054	0.0300	S
682-482	LaFL20	1.68248	48.20	–0.0012	–0.0121	H
682-482	LaF 20	1.68248	48.20	–0.0021	–0.0096	S
683-445	BaF 50	1.68273	44.50	–0.0020	–0.0075	S
683-447	BaF22	1.68250	44.67	0.0040	–0.0011	H
683-447	BAH 51	1.68250	44.65	–0.0005	–0.0048	O
686-439	BaF20	1.68578	43.90	–0.0002	–0.0065	H
686-440	BAH 53	1.68578	44.00	0.0010	–0.0034	O
686-492	LAM 56	1.68600	49.16	–0.0023	–0.0116	O
686-494	LaFL1	1.68600	49.35	0.0008	–0.0215	H
687-429	ADF8	1.68650	42.86	–0.0055	0.0300	H
689-311	PBM 28	1.68893	31.08	0.0031	–0.0014	O
689-312	FeD C89-31	1.68900	31.20			C
689-312	FD8	1.68893	31.16	0.0022	–0.0005	H
689-312	SF 8	1.68893	31.18	0.0029	–0.0013	S
689-495	LaFL23	1.68900	49.48	–0.0015	–0.0103	H
689-496	FBS C90-50	1.68900	49.60			C
689-497	LaF N23	1.68900	49.71	–0.0015	–0.0152	S
691-547	LaC9	1.69100	54.70	–0.0079	0.0225	H

Table 2.3. (cont.)

Code	Glass type	Refractive index	Abbe number	Deviation of the partial dispersion from the "normal line"		Producer
		n_d	v_d	$\Delta P_{g,F}$	$\Delta P_{C,t}$	
691-547	LaK 9	1.69100	54.71	–0.0071	0.0223	S
691-548	BCS C90-55	1.69100	54.80			C
691-548	LAL 9	1.69100	54.84	–0.0077	0.0204	O
694-508	LaCL5	1.69350	50.79	–0.0049	–0.0152	H
694-508	LAL 58	1.69350	50.81	–0.0042	–0.0119	O
694-532	LAL 13	1.69350	53.23	–0.0082	0.0157	O
694-533	LaC13	1.69350	53.34	–0.0067	0.0206	H
694-533	LaK N13	1.69350	53.33	–0.0030	–0.0139	S
695-422	BAH 54	1.69500	42.16	–0.0039	0.0068	O
696-365	FBDS C97-3	1.69600	36.50			C
697-485	LaFL2	1.69700	48.51	–0.0023	–0.0121	H
697-485	LAM 59	1.69700	48.51	–0.0025	–0.0110	O
697-554	BCS C97-55	1.69700	55.40			C
697-554	LaK N14	1.69680	55.41	–0.0079	0.0273	S
697-555	LaC14	1.69680	55.46	–0.0059	0.0282	H
697-555	LAL 14	1.69680	55.53	–0.0085	0.0263	O
697-561	LaC15	1.69680	56.11	–0.0057	0.0271	H
697-562	BCS C97-56	1.69700	56.20			C
697-564	LaK 31	1.69673	56.42	–0.0074	0.0287	S
697-565	LAL 64	1.69680	56.47	–0.0085	0.0279	O
698-386	BaSF 13	1.69761	38.57	0.0033	–0.0045	S
699-301	FeD C99-30	1.69900	30.10			C
699-301	FD15	1.69895	30.05	0.0052	0.0007	H
699-301	PBM 35	1.69895	30.12	0.0063	–0.0023	O
699-301	SF 15	1.69895	30.07	0.0066	–0.0014	S
700-349	BaFD14	1.69968	34.92	0.0035	–0.0019	H
700-478	LaFL3	1.70030	47.84	–0.0028	–0.0119	H
700-481	LAM 51	1.70000	48.08	–0.0016	–0.0062	O
702-401	BAH 71	1.70200	40.10	0.0003	–0.0101	O
702-402	BaFD15	1.70200	40.20	–0.0015	0.0097	H
702-410	BaSF 52	1.70181	41.01	–0.0035	0.0164	S
702-411	FBD D01-41	1.70180	41.10			C
702-412	BaFD7	1.70154	41.15	0.0027	–0.0092	H
702-412	BAH 27	1.70154	41.24	0.0024	–0.0047	O
702-412	BAH 77	1.70154	41.21	0.0003	–0.0032	O
704-394	BaSF 64A	1.70400	39.38	–0.0006	0.0069	S
706-303	SF 64A	1.70585	30.30	0.0099	0.0066	S
713-432	LAM 57	1.71285	43.19	–0.0005	–0.0048	O
713-433	LaFL4	1.71270	43.29	–0.0046	0.0212	H
713-538	BCS D13-54	1.71300	53.80			C

Table 2.3. (cont.)

Code	Glass type	Refractive index n_d	Abbe number v_d	Deviation of the partial dispersion from the "normal line" $\Delta P_{g,F}$	$\Delta P_{C,t}$	Producer
713-538	LAL 8	1.71300	53.84	–0.0089	0.0190	O
713-538	LaK 8	1.71300	53.83	–0.0083	0.0266	S
713-539	LaC8	1.71300	53.94	–0.0070	0.0228	H
717-295	FeD D17-29	1.71740	29.50			C
717-295	FD1	1.71736	29.50	0.0034	0.0003	H
717-295	PBH 1	1.71736	29.51	0.0038	–0.0027	O
717-295	SF 1	1.71736	29.51	0.0042	–0.0018	S
717-479	LAM 3	1.71700	47.94	–0.0035	–0.0111	O
717-480	FBS D17-48	1.71700	48.00			C
717-480	LaF30	1.71700	47.98	–0.0041	0.0029	H
717-480	LaF3	1.71700	47.98	–0.0062	–0.0085	H
717-480	LaF 3	1.71700	47.96	–0.0028	–0.0054	S
719-335	ADF455	1.71852	33.46	–0.0030	0.0144	H
719-335	BPH 45	1.71850	33.52	–0.0009	0.0145	O
720-293	FD20	1.72022	29.30	0.0053	0.0005	H
720-293	PBH 51	1.72022	29.31	0.0045	–0.0038	O
720-346	KzFS 8	1.72047	34.61	–0.0037	0.0217	S
720-347	BPH 8	1.72047	34.72	–0.0032	0.0200	O
720-420	LaFL6	1.72000	42.02	0.0005	–0.0027	H
720-420	LAM 58	1.72000	41.98	–0.0003	–0.0051	O
720-437	LAM 52	1.72000	43.71	–0.0008	–0.0086	O
720-439	LaFL5	1.72000	43.90	–0.0052	–0.0046	H
720-460	LAM 61	1.72000	46.03	–0.0024	–0.0179	O
720-503	BCS D20-50	1.72000	50.30			C
720-503	LaC10	1.72000	50.34	–0.0075	0.0061	H
720-503	LAL 10	1.72000	50.25	–0.0069	–0.0009	O
720-504	LaK 10	1.72000	50.41	–0.0075	0.0240	S
722-292	FD18	1.72151	29.24	0.0031	–0.0004	H
722-292	PBH 18	1.72151	29.24	0.0046	–0.0033	O
722-293	SF 18	1.72151	29.25	0.0045	–0.0021	S
723-380	BaFD8	1.72342	37.99	0.0020	–0.0033	H
723-380	BAH 28	1.72342	37.95	0.0035	–0.0025	O
723-380	BAH 78	1.72342	38.03	–0.0005	–0.0044	O
724-380	FBD D23-38	1.72350	38.00			C
724-381	BaSF 51	1.72373	38.11	–0.0013	0.0013	S
726-534	TaC1	1.72600	53.42	–0.0094	0.0182	H
726-536	LAL 60	1.72600	53.57	–0.0091	0.0245	O
728-283	FD10	1.72825	28.32	0.0073	–0.0001	H
728-284	FeD D28-28	1.72830	28.40			C
728-284	SF 10	1.72825	28.41	0.0085	–0.0012	S

Table 2.3. (cont.)

Code	Glass type	Refractive index n_d	Abbe number v_d	Deviation of the partial dispersion from the "normal line" $\Delta P_{g,F}$	$\Delta P_{C,t}$	Producer
728-285	PBH 10	1.72825	28.46	0.0088	–0.0020	O
728-287	FDS2	1.72830	28.67	0.0048	–0.0012	H
728-287	SF 53	1.72830	28.69	0.0061	–0.0026	S
729-547	TaC8	1.72916	54.67	–0.0046	0.0259	H
729-547	LAL 18	1.72916	54.68	–0.0082	0.0208	O
732-284	FeD D31-28	1.73151	28.41			C
734-265	FDS D34-26	1.73400	26.50			C
734-511	TaC4	1.73400	51.05	–0.0064	0.0209	H
734-514	BCS D34-51	1.73350	51.40			C
734-515	LAL 59	1.73400	51.49	–0.0098	0.0173	O
734-518	LaK 16A	1.73350	51.78	–0.0087	0.0208	S
735-411	LAM 8	1.73520	41.08	–0.0020	–0.0042	O
735-416	LaF N8	1.73520	41.59	–0.0029	0.0072	S
735-498	NbF3	1.73500	49.79	–0.0089	0.0189	H
735-498	LAM 53	1.73500	49.76	–0.0093	0.0241	O
736-322	BaSF 54	1.73627	32.15	0.0059	–0.0040	S
740-281	FeD D40-28	1.74000	28.10			C
740-282	FD3	1.74000	28.24	0.0041	–0.0027	H
740-282	SF 3	1.74000	28.20	0.0056	–0.0032	S
740-283	PBH 3	1.74000	28.29	0.0060	–0.0052	O
740-317	ADF505	1.74000	31.66	–0.0009	0.0096	H
740-317	BPH 50	1.74000	31.71	0.0000	0.0080	O
741-276	SF 13	1.74077	27.60	0.0109	–0.0021	S
741-278	FD13	1.74077	27.76	0.0099	–0.0019	H
741-278	PBH 13	1.74077	27.79	0.0097	–0.0040	O
741-281	SF 54	1.74080	28.09	0.0068	–0.0025	S
741-526	TaC2	1.74100	52.60	–0.0092	0.0171	H
741-527	LAL 61	1.74100	52.65	–0.0092	0.0223	O
743-492	NbF1	1.74330	49.22	–0.0103	0.0142	H
743-493	LAM 60	1.74320	49.31	–0.0094	0.0184	O
744-447	LaF 2	1.74400	44.72	–0.0027	–0.0075	S
744-448	FBS D44-45	1.74400	44.80			C
744-448	LAM 2	1.74400	44.79	–0.0036	–0.0066	O
744-449	LaF20	1.74400	44.87	–0.0052	0.0092	H
744-449	LaF2	1.74400	44.90	–0.0045	–0.0098	H
744-508	LaK 28	1.74429	50.77	–0.0085	0.0189	S
748-277	SF 63	1.74840	27.71	0.0074	–0.0031	S
750-350	LaF7	1.74950	35.04	–0.0043	–0.0065	H
750-350	LAM 65	1.74950	34.96	0.0031	–0.0106	O
750-350	LaF N7	1.74950	34.95	–0.0025	0.0174	S

Table 2.3. (cont.)

Code	Glass type	Refractive index n_d	Abbe number ν_d	Deviation of the partial dispersion from the "normal line" $\Delta P_{g,F}$	$\Delta P_{C,t}$	Producer
750-353	LAM 7	1.74950	35.27	0.0032	–0.0101	O
751-277	FD41	1.75084	27.69	0.0034	–0.0038	H
751-277	PBH 54	1.75084	27.69	0.0067	–0.0040	O
754-524	LaK 33	1.75398	52.43	–0.0083	0.0182	S
755-272	SF L4	1.75520	27.21	0.0123	0.0043	S
755-275	FD4	1.75520	27.53	0.0042	–0.0017	H
755-275	PBH 4	1.75520	27.51	0.0074	–0.0044	O
755-276	FeD D56-27	1.75520	27.60			C
755-276	SF 4	1.75520	27.58	0.0062	–0.0032	S
755-523	TaC6	1.75500	52.32	–0.0068	0.0195	H
755-523	YGH 51	1.75500	52.33	–0.0088	0.0188	O
756-251	TPH 55	1.75550	25.07	0.0197	0.0044	O
757-317	LaF 11A	1.75693	31.70	0.0002	0.0078	S
757-318	NbFD9	1.75690	31.80	0.0001	0.0040	H
757-363	NbFD6	1.75670	36.34	0.0015	–0.0119	H
757-477	NbF2	1.75700	47.71	–0.0076	0.0120	H
757-478	LAM 54	1.75700	47.82	–0.0076	0.0188	O
757-478	LaF N24	1.75719	47.81	–0.0079	0.0217	S
762-265	TIH 14	1.76182	26.52	0.0151	0.0032	O
762-265	SF 14	1.76182	26.53	0.0134	–0.0037	S
762-266	FD140	1.76182	26.61	0.0118	0.0060	H
762-266	FD14	1.76182	26.55	0.0111	–0.0047	H
762-266	PBH 14	1.76182	26.55	0.0125	–0.0050	O
762-269	FeD D62-27	1.76180	26.90			C
762-269	FDS5	1.76180	26.91	0.0064	–0.0037	H
762-270	SF 55	1.76180	26.95	0.0084	–0.0039	S
762-271	PBH 25	1.76180	27.11	0.0085	–0.0059	O
762-401	LAM 55	1.76200	40.10	–0.0006	–0.0014	O
762-403	NbFD5	1.76200	40.26	–0.0005	–0.0082	H
773-496	TaF1	1.77250	49.62	–0.0086	0.0212	H
773-496	LAH 66	1.77250	49.60	–0.0097	0.0164	O
773-496	LaF N28	1.77314	49.57	–0.0085	0.0126	S
776-378	LaF 13	1.77551	37.84	–0.0026	0.0056	S
782-372	LaF 22A	1.78179	37.20	0.0008	0.0005	S
783-361	NbFD7	1.78300	36.07	0.0015	0.0138	H
783-362	LAM 62	1.78300	36.15	0.0042	–0.0015	O
784-440	LaF N10	1.78443	43.95	–0.0068	0.0194	S
785-257	FD11	1.78472	25.70	0.0098	–0.0053	H
785-257	FD110	1.78472	25.72	0.0138	0.0036	H
785-257	TIH 11	1.78472	25.68	0.0160	0.0018	O

Table 2.3. (cont.)

Code	Glass type	Refractive index	Abbe number	Deviation of the partial dispersion from the "normal line"		Producer
		n_d	v_d	$\Delta P_{g,F}$	$\Delta P_{C,t}$	
785-257	PBH 11	1.78472	25.71	0.0129	–0.0056	O
785-258	FeD D85-25	1.78475	25.80			C
785-258	SF 11	1.78472	25.76	0.0142	–0.0043	S
785-259	FeD D85-26	1.78500	25.90			C
785-261	FDS30	1.78470	26.10	0.0142	–0.0044	H
785-261	FDS3	1.78470	26.06	0.0090	–0.0033	H
785-261	SF 56A	1.78470	26.08	0.0098	–0.0042	S
785-261	SF L56	1.78470	26.08	0.0138	0.0048	S
785-262	PBH 23	1.78470	26.22	0.0092	–0.0066	O
785-263	TIH 23	1.78470	26.30	0.0142	0.0014	O
786-439	NbFD11	1.78590	43.93	–0.0081	0.0120	H
786-442	LAH 51	1.78590	44.19	–0.0069	0.0087	O
787-500	YGH 52	1.78650	50.00	–0.0093	0.0000	O
788-474	FBS D88-47	1.78800	47.40			C
788-474	LAH 64	1.78800	47.38	–0.0086	0.0157	O
788-475	TaF4	1.78800	47.49	–0.0090	0.0157	H
788-475	LaF N21	1.78831	47.47	–0.0079	0.0158	S
789-428	NbFD1	1.78900	42.76	–0.0073	0.0107	H
795-284	LaF 9	1.79504	28.39	0.0059	–0.0015	S
795-453	LAH 67	1.79500	45.29	–0.0083	0.0163	O
795-454	TaF2	1.79450	45.39	–0.0094	0.0114	H
796-432	FBS D96-43	1.79619	43.20			C
797-351	LaSF 36A	1.79712	35.08	0.0011	0.0008	S
797-411	NbFD2	1.79720	41.14	–0.0055	0.0083	H
800-422	LAH 52	1.79952	42.24	–0.0069	0.0192	O
800-423	NbFD12	1.79950	42.34	–0.0072	0.0077	H
800-456	FBS E00-46	1.80000	45.60			C
801-350	LAM 66	1.80100	34.97	0.0011	0.0129	O
802-443	NbFD14	1.80170	44.32	–0.0090	0.0138	H
803-304	LaSF 32	1.80349	30.40	0.0055	0.0094	S
803-464	FBS E03-47	1.80300	46.40			C
803-464	LaSF N30	1.80318	46.38	–0.0084	0.0099	S
803-467	LAH 62	1.80300	46.66	–0.0096	0.0153	O
804-396	LAH 63	1.80440	39.58	–0.0046	0.0116	O
804-465	TaF3	1.80420	46.50	–0.0067	0.0172	H
804-466	LAH 65	1.80400	46.58	–0.0090	0.0139	O
805-254	PBH 6	1.80518	25.43	0.0097	–0.0073	O
805-254	TIH 6	1.80518	25.43	0.0150	0.0017	O
805-254	SF L6	1.80518	25.39	0.0148	0.0032	S
805-254	SF 6	1.80518	25.43	0.0092	–0.0048	S

Table 2.3. (cont.)

Code	Glass type	Refractive index	Abbe number	Deviation of the partial dispersion from the "normal line"		Producer
		n_d	v_d	$\Delta P_{g,F}$	$\Delta P_{C,t}$	
805-255	FeD E05-25	1.80510	25.50			C
805-255	FD6	1.80518	25.46	0.0075	–0.0049	H
805-255	FD60	1.80518	25.46	0.0132	0.0041	H
805-396	NbFD3	1.80450	39.64	–0.0056	0.0083	H
806-333	NbFD15	1.80610	33.27	0.0000	0.0046	H
806-342	LaSF 33	1.80596	34.24	0.0022	–0.0034	S
806-407	NbFD13	1.80610	40.73	–0.0079	0.0101	H
806-410	LAH 53	1.80610	40.95	–0.0055	0.0114	O
807-355	NbFD8	1.80740	35.54	0.0005	0.0038	H
808-406	LaSF 3	1.80801	40.61	–0.0046	0.0184	S
816-444	LAH 54	1.81554	44.36	–0.0084	0.0136	O
816-445	TaFD10	1.81550	44.54	–0.0074	0.0134	H
816-466	TaF5	1.81600	46.57	–0.0083	0.0084	H
816-466	LAH 59	1.81600	46.62	–0.0090	0.0031	O
822-375	LaSF 14A	1.82223	37.45	–0.0023	0.0174	S
829-365	NbFD40	1.82900	36.50	0.0002	–0.0061	H
834-372	LAH 60	1.83400	37.17	–0.0039	0.0106	O
834-373	NbFD10	1.83400	37.34	–0.0021	0.0077	H
835-427	LAH 55	1.83481	42.72	–0.0087	0.0115	O
835-430	TaFD5	1.83500	42.98	–0.0082	0.0122	H
836-302	FBS E37-30	1.83640	30.20			C
847-236	SF L57	1.84666	23.62	0.0177	0.0034	S
847-238	FDS9	1.84666	23.83	0.0096	–0.0067	H
847-238	FDS90	1.84666	23.78	0.0136	0.0029	H
847-238	TIH 53	1.84666	23.78	0.0181	0.0014	O
847-238	SF 57	1.84666	23.83	0.0123	–0.0065	S
847-239	PBH 53	1.84666	23.89	0.0134	–0.0086	O
848-430	TaFD6	1.84750	43.03	–0.0123	0.0029	H
850-322	TaFD9	1.85030	32.18	0.0030	0.0069	H
850-322	LaSF N9	1.85025	32.17	0.0036	–0.0031	S
850-323	LAH 71	1.85026	32.29	0.0027	0.0004	O
855-366	TaFD13	1.85540	36.56	–0.0044	0.0090	H
863-415	LAH 57	1.86300	41.53	–0.0074	0.0049	O
865-401	FBS E65-40	1.86500	40.10			C
874-353	LAH 75	1.87400	35.26	–0.0017	0.0075	O
878-381	LaSF N15	1.87800	38.07	–0.0063	0.0115	S
878-382	TaFD17	1.87800	38.20	–0.0048	0.0053	H
878-385	FBS E78-38	1.87800	38.48			C
881-410	LaSF N31	1.88067	41.01	–0.0084	–0.0046	S
883-408	TaFD30	1.88300	40.80	–0.0093	0.0016	H

Table 2.3. (cont.)

Code	Glass type	Refractive index	Abbe number	Deviation of the partial dispersion from the "normal line"		Producer
		n_d	v_d	$\Delta P_{g,F}$	$\Delta P_{C,t}$	
883-408	LAH 58	1.88300	40.78	–0.0087	0.0012	O
903-298	TaFD20	1.90315	29.84	0.0063	0.0024	H
913-324	LaSF 18A	1.91348	32.36	0.0008	0.0116	S
915-212	PBH 72	1.91536	21.17	0.0220	–0.0123	O
918-215	SF 58	1.91761	21.51	0.0180	–0.0095	S
923-209	FDS1	1.92286	20.88	0.0174	–0.0095	H
923-209	PBH 21	1.92286	20.88	0.0230	–0.0124	O
923-213	PBH 71	1.92286	21.29	0.0199	–0.0111	O
923-360	TaFD31	1.92250	35.95	–0.0056	0.0001	H
933-279	TaFD21	1.93260	27.92	0.0081	0.0077	H
952-204	SF 59	1.95250	20.36	0.0217	–0.0102	S
1003-235	TaFD43	2.00294	23.51	0.0060	–0.0042	H
1022-291	LaSF 35	2.02204	29.06	0.0033	–0.0009	S

Additional References

Abelès, F. (ed.): *Optical properties of solids* (Elsevier, New York 1972)

Conrady, A.E.: *Applied optics and optical design* (Dover, New York 1960)

Corning-France: Optical glass catalog (1991)

Driscoll, W.G. and Vaughan, W. (eds.): *Handbook of optics* (McGraw-Hill, New York 1978)

Fanderlik, I.: Optical properties of glass (Elsevier, Amsterdam 1983)

Fraunhofer, J.: "Bestimmung des Brechungs- und Farbzerstreuungsvermögens verschiedener Glasarten in bezug auf die Vervollkommnung achromatischer Fernrohre" in *Ostwalds Klassiker der exakten Wissenschaften Nr. 150*, herausgegeben von A. v. Oettingen (Verlag von Wilhelm Engelmann, Leipzig 1905) pp. 3-36

Haferkorn, H.: *Optik* (Deutscher Verlag der Wissenschaften, Berlin 1981)

Hofmann, Ch.: *Die optische Abbildung* (Akadem. Verlagsgesellschaft Geest & Portig, Leipzig 1980)

Hoya: *Optical glass catalog*, Tokyo (1990)

Izumitani, T.S.: *Optical glass* (American Institute of Physics, Translation series, New York 1986)

Kingslake, R.: *Lens design fundamentals* (Academic Press, New York 1978)

Kittel, C.: Introduction to solid state physics, 4th ed. (Wiley, New York 1971)

Kreidl, N.J. and Rood, J.L.: "Optical materials" in *Applied Optics and Optical Engineering,* ed. by R. Kingslake, Vol. I (Academic Press, New York 1965) pp. 153-200

Ohara: Optical glass catalog (1990)

Reitmayer, F.: Die Dispersion einiger optischer Gläser im Spektralgebiet von 0,33 µm bis 2,0 µm und ihre Abhängigkeit von den Eigenfrequenzen der Gläser, Glastechn. Ber. 34, 122-130 (1961)

Schott: Optical glass. Technical information No. 1 - No. 25, Mainz (1972-1988)

Smith, W.J.: *Modern optical engineering* (2nd ed., McGraw-Hill, New York 1990)

Uhlmann, D.R., Kreidl, N.J.: *Optical properties of Glass* (The American Ceramic Society, Westerville 1992)

Fleming, J. W.: "Optical glasses", in *CRC Handbook of Laser Science*, ed. by M. J. Weber, Vol. 4, Part 2 (CRC Press, Boca Raton 1986) pp. 69-83

2.2 The Chemical Composition of Optical Glasses and Its Influence on the Optical Properties

Marc K. Th. Clement

2.2.1 General Remarks

General Characteristics

The term "glass" is commonly applied to a fusion product of an inorganic material which is cooled to a solid state without crystallizing. It is often associated with fusion products made from silicates which are used for windows, bottles, lenses, etc.. Beside silicates a lot of other substances can also be obtained in the glassy state. Nevertheless, a lot of non-silicate substances have been obtained glassy. It is widely understood that the possibility of a material to form glass is not an atomic or molecular property and cannot be measured by chemical or physical properties, but rather is a state of the solid state which can be reached by a sufficiently high cooling rate.

Glasses are normally specified by a certain set of defined properties which are typical of them, but distinguish them from other forms of the solid state. In contrast to crystals, glasses do not have a defined melting point; the viscosity of glasses decreases continuously provided that the temperature increases without discontinuity. Like liquids, glasses are isotropic, which is a very important characteristic for their optical applications. Unlike other non-crystalline materials, glasses are usually made by a fusion process. Other production methods are the chemical vapour deposition and sol-gel processes. By the latter methods it is possible to produce several materials in the glassy state that cannot be made by a melting process.

Optical glass is defined as a group of glasses having the following characteristics:

- well defined optical properties,
- optical homogeneous performance,
- absence of bubbles, striae, strain and inclusions.

These characteristics must be taken into consideration during the production of optical glasses.

Oxides, some of which have been known for more than 5000 years, are the classic glass-forming materials.

The oxides B_2O_3, SiO_2, GeO_2, P_2O_5, As_2O_3 and Sb_2O_3 form glasses even without the addition of further oxides. They are usually known as glass-formers. Also oxides like TeO_2, SeO_2, MoO_3, WO_3, Bi_2O_3, Al_2O_3, Ga_2O_3, TiO_2, and V_2O_3 are known to form glasses if they are melted with a sufficient quantity of at least one different oxide ("conditional glass-formers" [2.33, 34]). Figure 2.11 shows a part of the periodic table whose oxides are glass-formers (hatched area) and conditional glass-formers (crossed area). Reviews on the glass formation, the structures and

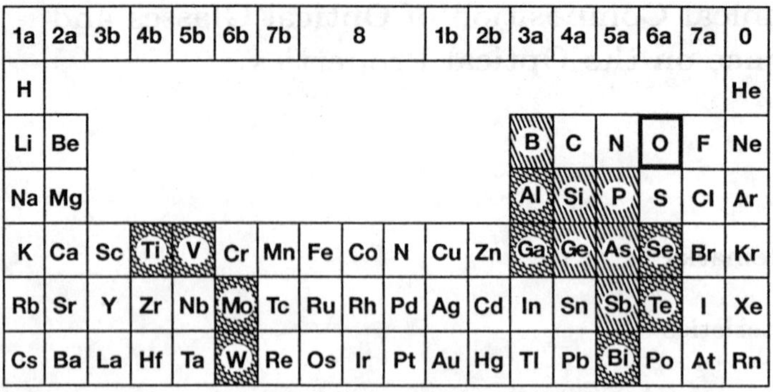

1a	2a	3b	4b	5b	6b	7b	8			1b	2b	3a	4a	5a	6a	7a	0
H																	He
Li	Be											B	C	N	O	F	Ne
Na	Mg											Al	Si	P	S	Cl	Ar
K	Ca	Sc	Ti	V	Cr	Mn	Fe	Co	N	Cu	Zn	Ga	Ge	As	Se	Br	Kr
Rb	Sr	Y	Zr	Nb	Mo	Tc	Ru	Rh	Pd	Ag	Cd	In	Sn	Sb	Te	I	Xe
Cs	Ba	La	Hf	Ta	W	Re	Os	Ir	Pt	Au	Hg	Tl	Pb	Bi	Po	At	Rn

Fig. 2.11. Elements whose oxides act as glass-formers (hatched) and conditional glass-formers (crossed) for oxide glass systems [2.34]

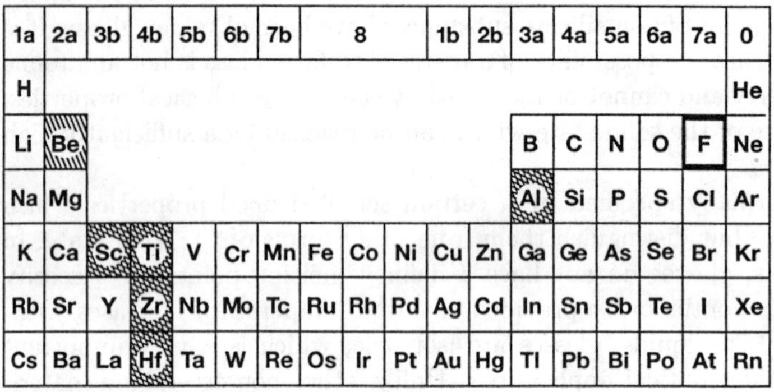

1a	2a	3b	4b	5b	6b	7b	8			1b	2b	3a	4a	5a	6a	7a	0
H																	He
Li	Be											B	C	N	O	F	Ne
Na	Mg											Al	Si	P	S	Cl	Ar
K	Ca	Sc	Ti	V	Cr	Mn	Fe	Co	Ni	Cu	Zn	Ga	Ge	As	Se	Br	Kr
Rb	Sr	Y	Zr	Nb	Mo	Tc	Ru	Rh	Pd	Ag	Cd	In	Sn	Sb	Te	I	Xe
Cs	Ba	La	Hf	Ta	W	Re	Os	Ir	Pt	Au	Hg	Tl	Pb	Bi	Po	At	Rn

Fig. 2.12. Elements whose fluorides act as glass-formers (hatched) and conditional glass-formers (crossed) for fluoride glass systems [2.48]

properties of the oxide systems of glass-formers and conditional glass-formers have been given by *Kreidl* [2.35–38].

Also chalcogenides [2.39–41] and halides [2.42–47] can form glasses to be used in optics. Among the fluorides only BeF_2 is a glass-former. There are several conditional glass formers which produce glasses with interesting optical properties, see Fig. 2.12 [2.47, 48]. Some mixtures of oxides and fluorides also form glasses.

The understanding of the structure of glasses has been strongly influenced by the "network-theory" of *Zachariasen* [2.49, 50] and *Warren* [2.51] who explained that the short-term order of the constituents of a glass, i.e. the order of the atoms relative to their neighbours, is nearly the same as the order found in crystals. The long-term order, however, which is typical of crystalline materials, is missing in glasses. Unlike crystals, glasses show a "random network". The constituents of a glass can be divided into cations which have the function of "network formers", i.e. they contribute to the basic structure of the glass (e.g. SiO_2, B_2O_3, P_2O_5, ...) and cations which have the function of "network modifiers", i.e. cations which break up the network (e.g. Na_2O, CaO, PbO). The group of cations which change the network was later extended by the group of the "intermediate" cations which can act as network formers and network modifiers.

The introduction of a large number of network modifiers into a glass leads to an increase in the splitting of the network. An increased mobility of the structural units causes a decrease in the viscosity. Moreover, other physical properties, such as the electrical conductivity, vary greatly. As the network modifiers are irregularly distributed in the whole network, the properties of the glass normally change continuously according to the change of the composition.

Nevertheless, such a continuous change of the properties is not found in all glasses, since the type of network structure is not the same for different compositions. For example, if the content of Na_2O is increased in a SiO_2 glass, a continuous change in the properties of the glass can be found. If, however, the Na_2O content is increased in a B_2O_3 glass, abnormal changes in the properties of the glass can be found. This can be explained by the fact that an increasing alkaline content causes changes in the structural units of the network of the B_2O_3 glass. Normally a B_2O_3 glass contains units of $[BO_3]$ which form the network of the glass. If alkaline oxides are introduced into this network, the coordination of the boron changes from $[BO_3]$ units to $[BO_4]$ units; this leads to a network with different physical properties. If the alkaline content increases further, the coordination number of boron is reduced again and the network is changed once more [2.52, 53]. Similar structural changes which depend on the glass composition are also reported for other glass constituents [2.54].

Description of the Composition

Most of the optical glasses are based on oxide glass systems which comprise a certain number of cations (e.g. Na^+, Ca^{2+}, Pb^{2+}, La^{3+}). The composition of a mixture, e.g. of a quaternary system consisting of the elements silicon, sodium, boron and oxygen, is shown by a regular tetrahedron where any composition is represented by a point inside the tetrahedron. The pure elements are represented by the corners of the tetrahedron. On the sides which combine oxygen with the elements silicon, sodium and boron, the position of the binary compositions of the cations with oxygen (SiO_2, B_2O_3, Na_2O) can be fixed (Fig. 2.13).

In case of limitation to a subsystem of the quaternary system, where the following stoichiometric condition is fulfilled (n_i is the number of moles of the component i):

$$n_O = 2n_{Si} + 3n_B/2 + n_{Na}/2 \, , \tag{2.38}$$

a quasiternary system can be defined whose compositions may be shown using a triangular cross-section. The corners of this triangle denote the three binary compounds SiO_2, B_2O_3 and Na_2O. The sides of the triangle represent the binary systems SiO_2-B_2O_3, SiO_2-Na_2O and B_2O_3-Na_2O (Fig. 2.14). Because of condition (2.38) the compositions of this quaternary system can be characterized by the mole fractions x_{SiO_2}, $x_{B_2O_3}$ and x_{Na_2O}. For a sufficient description of the quasiternary system only two of the three mole fractions are needed as the third mole fraction is defined by the condition that the sum of all mole fractions has to be unity, i.e.

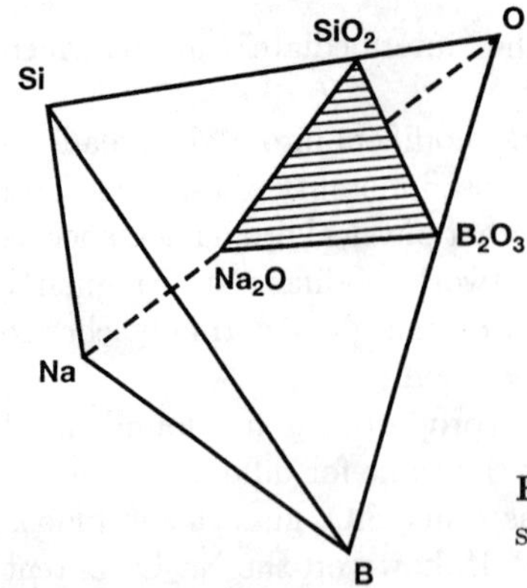

Fig. 2.13. Quaternary system of boron, silicon, sodium and oxygen

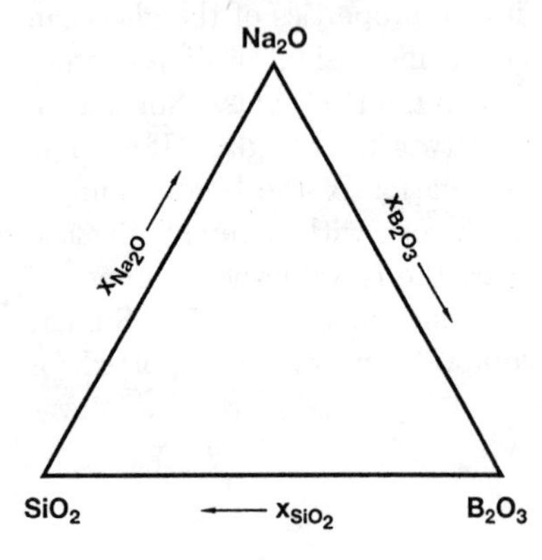

Fig. 2.14. Binary subsystems SiO_2-B_2O_3, B_2O_3-Na_2O and SiO_2-Na_2O and the quasiternary system SiO_2-B_2O_3-Na_2O

$$x_{SiO_2} + x_{B_2O_3} + x_{Na_2O} = 1 . \tag{2.39}$$

So this system can be described by:

$$Si_aB_{2b}Na_{2(1-a-b)}O_{1+a+2b} \text{ with } 0 \leq a, b \leq 1. \tag{2.40}$$

Analogous to the treatment of quasibinary sections as edges of a quasiternary plane, the quasiternary systems can be treated as surfaces of a quasiquaternary body which contains 5 elements.

A very common method to describe the composition of a glass is to quote the content of each oxide by weight (weight%, wt%) or mole (mole%). In a system with i oxides, only $i-1$ concentration values have to be known. The value of the oxide concentration i is fixed by a condition similar to (2.39).

The transformation of the concentration values of a set of oxides i from wt% (w_i) to mole% (m_i) can be realized by

$$m_i = \frac{100\, w_i}{M_i \sum_{j=1}^{n} \frac{w_j}{M_j}}, \quad 1 \leq i \leq n \tag{2.41}$$

where M_i represents the molar mass of the oxide i (2.42). For the transformation of concentration values from mole% (m_i) to wt% the following equation can be used:

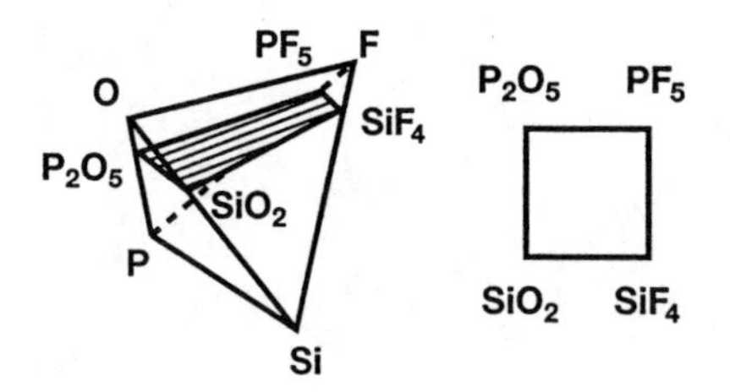

Fig. 2.15. Quaternary system with fluorine

$$w_i = \frac{100\, m_i\, M_i}{\sum_{j=1}^{n} m_j M_j}, \quad 1 \le i \le n\,. \tag{2.42}$$

Sometimes it is convenient to express the composition of a glass by cations with only the oxygen content being constant. Thus the system $SiO_2{-}B_2O_3{-}Na_2O$ can be described by:

$$Si_{3x}B_{4y}Na_{12(1-x-y)}O_6 \text{ with } 0 \le x, y \le 1\,. \tag{2.43}$$

If the composition of a quasiternary system is described in this way, the quantity of cations is always related to a fixed amount of oxygen. By this method the molar fractions of a quaternary system can be used to describe the composition of a quasiternary system. The concentration unit of this description is known as equivalent-percent (eq%).

To convert the values a and b of (2.40) into the values x and y of (2.43) the content has to be related to the same amount of oxygen. Thus

$$x = \frac{2a}{1+a+2b} \tag{2.44}$$

and

$$y = \frac{3b}{1+a+2b} \tag{2.45}$$

and

$$z = (1-x-y) = \frac{1-a-b}{1+a+2b}\,. \tag{2.46}$$

These are three methods used to describe the composition of a glass; the first two methods, however, are more common.

Sometimes problems arise with the description of glass systems, especialli if other anions besides oxygen, e.g. fluorine, are introduced into a glass. In general, these glasses can be described as discussed above. The only difference, however, is that now a different quaternary system has to be described.

Starting again from a regular tetrahedron (Fig. 2.15), which represents the quaternary system, the quasiternary system is expressed by a square cross section.

For glasses containing anions other than oxygen, *Geffcken* [2.55] has proposed a special description. An oxide glass, consisting of sodium and silica is normally described by the contents of the corresponding oxides (Na_2O, SiO_2). This description can be derived from the following chemical reaction which starts with the batch materials, e.g.:

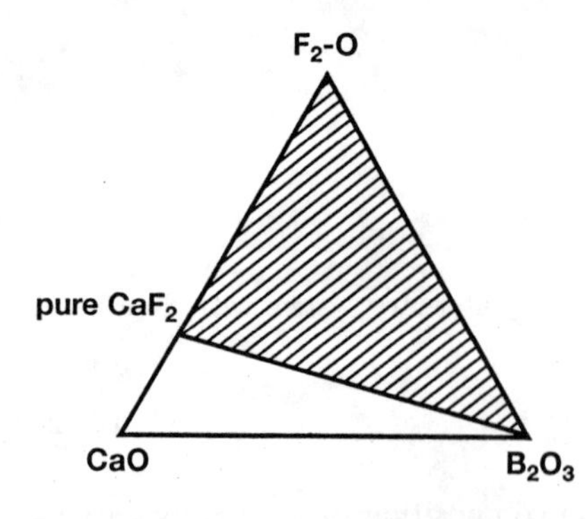

Fig. 2.16. Introduction of the component (F_2-O)

$$2Na_2CO_3 + SiO_2 \rightarrow Na_4SiO_4 + 2CO_2 \ . \tag{2.47}$$

The product of the reaction Na_4SiO_4 is transformed into $2Na_2O{\cdot}SiO_2$ and represents the composition of the glass expressed in oxides of the cations.

If fluorine is introduced into the glass and the amounts of sodium and silica are not changed, the batch reaction is as follows:

$$Na_2CO_3 + 2NaF + SiO_2 \rightarrow Na_4SiO_3F_2 + CO_2 \ . \tag{2.48}$$

If the product of the reaction $Na_4SiO_3F_2$ is also expressed in oxides of the cations as in (2.47), a description like "$2Na_2O{\cdot}SiO_2{\cdot}F_2$" is not correct, as it refers to the material "$Na_4SiO_4F_2$", which is not the same as in (2.48).

If, however, the concentration of a glass is to be described by the indication of the oxides of cations, the fictitious component "(F_2-O)" (= fluorine minus oxygen) has to be introduced. "(F_2-O)" represents the part of oxygen which is replaced by fluorine. Using this fictitious component the product of the reaction (2.48) would be expressed as $2Na_2O{\cdot}SiO_2{\cdot}(F_2\text{-}O)$, which leads to the correct composition of the reaction product $Na_4SiO_3F_2$.

Similarly, a glass consisting for example of CaF_2-B_2O_3, is described as a ternary system comprising the compounds CaO-B_2O_3 and the fictitious component (F_2-O). The fictitious component (F_2-O) can be deducted from (2.49):

$$CaF_2 \rightarrow CaO + (F_2\text{-}O) \ . \tag{2.49}$$

In the above mentioned system the content of fluorine is evidently fixed by the concentrations of CaO and B_2O_3 (Fig. 2.16). If this kind of description is applied, only a part of the concentration triangle can be used to describe the concentration.

The introduction of the component (F_2-O) is useful, since with the help of this component the condition that the sum of the concentrations of all components is unity (or 100%), can be fulfilled in a glass which contains more than one type of anion. Therefore even glasses that contain different concentrations of fluorine can be very easily compared [2.55, 56].

To evaluate the region of a system which forms glassy phases, usually a series of compositions is melted and casted. The casted melts are characterized. Figures 2.17–19 give some examples of glassy regions of various systems [2.55–58]. A review of the optical properties of different glass systems has been given by *Polukhin*

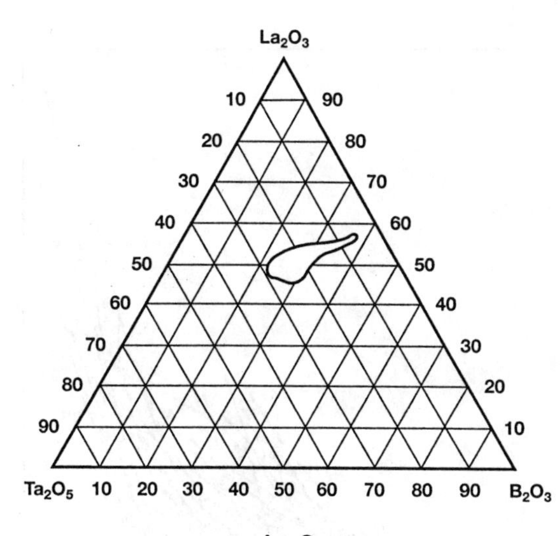

Fig. 2.17. Glass system B_2O_3-La_2O_3-Ta_2O_5 (wt%) [2.57]

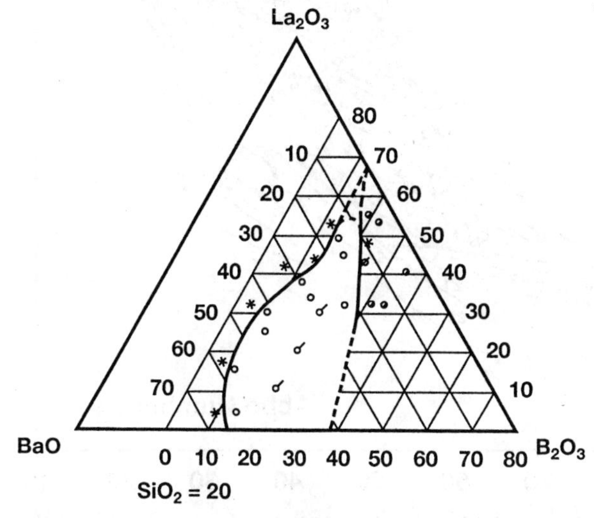

Fig. 2.18. Glass system B_2O_3-La_2O_3-BaO-20 SiO_2 (wt%) [2.55] (circle = clear glass, circle with outside dash = surface crystallization, circle with dash = few crystals in the glass, half-filled circle = phase separated glass, star = glass mixed with crystals)

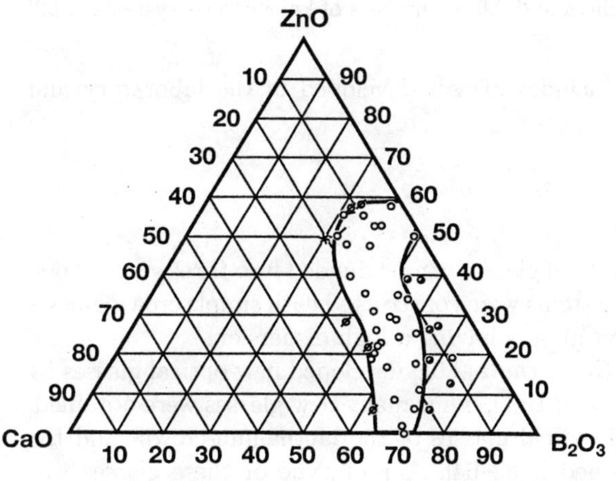

Fig. 2.19. Glass system B_2O_3-ZnO-CaO (wt%) [2.55] (Symbols as in Fig. 2.18)

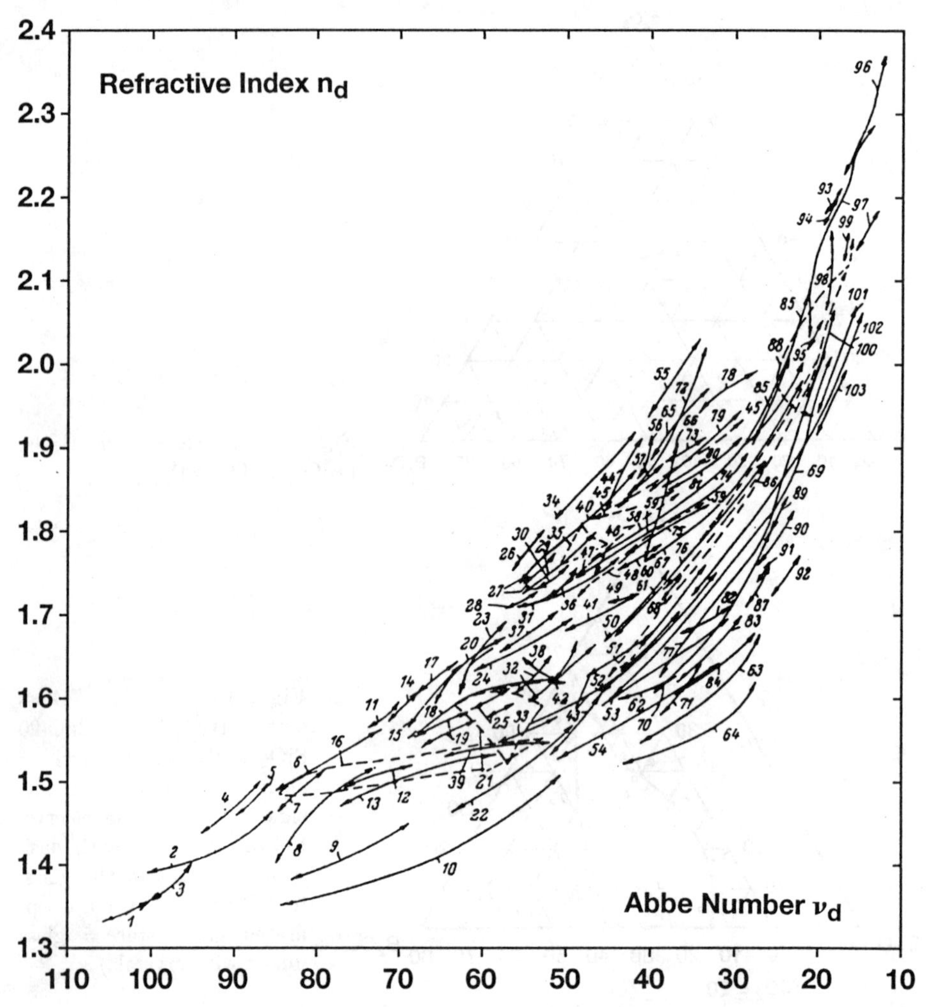

Fig. 2.20. Survey of refractive indices and Abbe numbers of known glass systems [2.59]

[2.59] (Fig. 2.20). This figure includes glasses developed in the laboratory and commercially available glasses.

2.2.2 Glass Types

The historical development of optical glasses can be divided into three steps [2.60]:

Step 1: Before 1880 optical systems were constructed with simple crown glasses having a low dispersion and flint glasses having a high dispersion.

Step 2: From 1880 to 1895 *Otto Schott* [2.61] developed new optical glasses by the introduction of B_2O_3, P_2O_5 and BaO. Thus many new glasses were obtained.

Step 3: From 1930 to 1940 optical glasses of the lanthanum-crown- and the lanthan-flint-group were developed [2.62–64]. A prototype of these glasses was the glass reported by *Brewster* and *Kreidl* [2.63]. Other glasses based on boron-lanthanum-systems [2.55, 59] or fluorine-containing phosphate glasses [2.58] were later developed.

The designation of optical glasses traditionally consists of a combination of the term "crown" (German: "Kron") and "flint" (German: "Flint") with the terms "dense" (German: "Schwer") denoting a high and "light" (German: "Leicht") a low refractive index. The term "crown" originally described glasses belonging to the glass system SiO_2-CaO-Na_2O, which were blown as crown-like glasses, whereas the term "flint" was used for glasses of the system SiO_2-PbO. In the past only "flint stones" from the British south coast could be used for these flint glasses in order to achieve a transmission that was acceptable for that time.

Subgroups of optical glasses are classified by the chemical components which are characteristic of the optical properties, e.g., "barium-dense-crown" (BaSF) or "fluorine-crown" (FK).

For the fabrication of high-quality optical systems achromatic and apochromatic systems are required. For an achromatic system the chromatic aberration is minimized for two colours, e.g., red and blue. The other part of the visible spectrum remains uncorrected ("secondary spectrum"). The chromatic aberration in an apochromatic system, however, is corrected through the whole visible region [2.65, 66].

Most of the optical glasses show normal dispersion. These glasses are not suitable for the construction of apochromatic systems. Therefore, for the construction of apochromatic systems, glasses with a short spectrum in the blue range of the visible spectrum have been developed and called KzF and KzFS glasses. The abbreviations denote "short-flint" and "short-flint-special" (German: "Kurz-Flint", "Kurz-Flint-Sonder", respectively). Special optical glasses with a long spectrum in the blue range of the visible spectrum are called LgSK, which stands for "long-dense-crown" (German: "Lang-Schwer-Kron"). KzF and KzFS glasses are glasses with Abbe numbers smaller than 55, whereas the glasses with the characteristics of LgSK glasses exceed the Abbe number 55. A combination of glasses with short-flint and long-crown characteristics is suitable for reducing the chromatic aberration.

Today we know more than 250 different types of optical glasses that can be classified by their chemical composition (Fig. 2.21).

The following is an outline of the various glass systems:

1. SiO_2-B_2O_3-M_2O:

This system is formed by the network formers SiO_2 and B_2O_3. Alkaline oxides (M_2O) have been introduced as network modifiers. The presence of B_2O_3 and alkaline oxides is essential as the glass-forming oxide SiO_2 has melting temperatures which are too high for conventional melting techniques. The addition of B_2O_3 and alkaline oxides facilitates the meltability of these glasses. The optical glass BK 7® is one of the main representative glasses of this system. It is the optical glass which is used most frequently and it can be produced with extreme homogeneity.

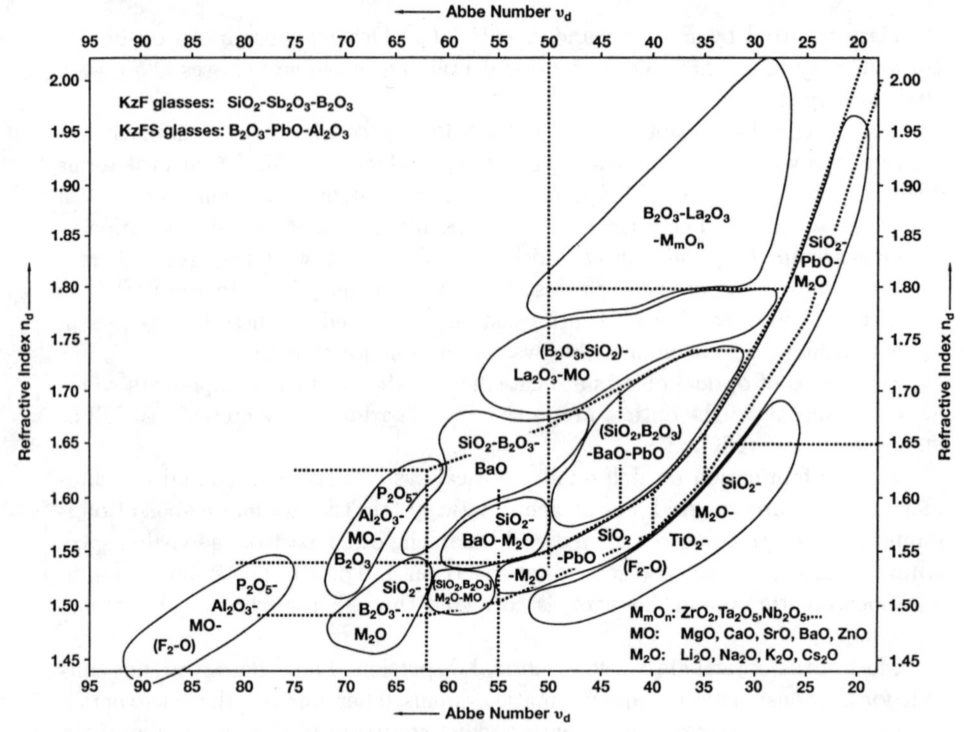

Fig. 2.21. Glass systems of optical glasses in the n_d/ν_d-plot (KzF and KzFS glasses are not considered)

2. SiO_2-B_2O_3-BaO and
3. SiO_2-BaO-M_2O:

If BaO is used as the main network modifier instead of the alkaline oxide or B_2O_3, this results in two large glass systems. In contrast to other earth alkaline network modifiers, the introduction of BaO has various advantages. Beside PbO there is no other divalent oxide which increases the refractive index as strongly as BaO. Additionally, BaO neither decreases the Abbe number nor makes it shift the UV-transmission edge significantly to greater wavelengths as PbO does. BaO-containing glasses normally show a good abrasion hardness. In some glasses BaO is partly replaced by ZnO.

4. (SiO_2, B_2O_3)-M_2O-MO:

To obtain glasses of this glass system divalent oxides (MO) are exchanged for B_2O_3. From among the divalent oxides only CaO, ZnO and PbO are used in this system. For the production of glasses with high chemical resistance especially ZnO is particularly important since it increases the durability with respect to water and acids much more than CaO, provided that the ZnO content exceeds approximately 10wt%. It is also possible to reduce the melting temperatures by a simultaneous exchange of ZnO for SiO_2 and to increase the alkaline content without harming the durability against weathering. Regarding these characteristics, ZnO is much more effective than B_2O_3.

The introduction of CaO leads to the production of glasses with good mechanical properties and increased abrasion hardness. The chemical durability of these glasses increases with growing CaO content.

5. (SiO_2, B_2O_3)-BaO-PbO:
This glass system is a mixture of the glass systems SiO_2-BaO-M_2O and SiO_2-PbO-M_2O. Thus these glasses represent a transition from the one glass system to the other. Coming from the system SiO_2-BaO-M_2O, the exchange of PbO for BaO leads to the production of glasses with a lower tendency to crystallize. Compared with similar BaO-containing glasses, PbO-containing glasses are characterized by lower softening points and higher expansion coefficients. In some glasses of this system BaO is partly replaced by ZnO.

6. SiO_2-PbO-M_2O:
Glasses of this system have been known for a very long time. They are widely used for optical glasses and crystal glasses. As a glass component PbO plays a vital role due to the fact that on the one hand it increases the refractive index and on the other hand it decreases the Abbe number very strongly. Furthermore, PbO significantly effects the partial dispersion of glasses since a high PbO-content leads to a positive deviation from the "normal-line". High amounts of PbO lead to a high glass density and a rather small chemical resistance.

PbO can be introduced into glasses to more than 80 wt%, which indicates that it does not only have network modifying properties.

Normally, if a network modifier is introduced into a glass, e.g. Na_2O into a SiO_2-glass, bridging oxygen atoms are converted into non-bridging oxygen atoms:

$$\text{–Si–O–Si–} \quad + \quad Na_2O \quad \rightarrow \quad \text{–Si–}O^-Na^+ \quad + \quad Na^+\text{-O–Si–} \tag{2.50}$$

With an increasing content of network modifiers the network progressively weakened and the viscosity of the glass melt is decreased. If the content of the network modifiers is increased further, more and more bridging oxygen atoms are converted into non-bridging atoms. For a ratio of network modifiers to network formers of about 1:1, a three-dimensional network cannot be formed any longer and glass formation becomes impossible. However, the glass-forming tendency of binary PbO-SiO_2 systems leads to a PbO content of more than 70 mole%. According to *Zachariasen* and *Warren* these glasses would normally not exist if PbO was only a network modifier.

Fajans [2.67] and *Stanworth* [2.68] explained how the introduction of such a high amount of PbO is possible. In small concentrations PbO has the same function as any other network modifier. High concentrations of PbO, however, cause an increase in the formation of [PbO_4]-tetrahedra. These [PbO_4]-tetrahedra can act as network formers and are incorporated in the three-dimensional network.

Glasses which contain large amounts of PbO usually show a more or less yellow colour. Impurities, such as Fe_2O_3 or Cr_2O_3, have very strong effects on the colour of PbO containing glasses, even stronger than in soda-potash-lime glasses [2.69].

Vogel and *Heindorf* [2.70] found that, firstly, beside the different structures of the PbO-containing glasses, an equilibrium between Pb^{2+} and Pb^{4+} exists, and, secondly, brown PbO_2 is formed if the melt is supersaturated with oxygen. Unlike the yellow colour that is due to impurities, the yellow tint which is caused by the equilibrium Pb^{2+}/Pb^{4+} can be reversed by a variation of the redox-condition of the melt.

For a long time attempts have been made to increase the chemical stability of optical glasses with high PbO contents. Replacing parts of the PbO content with TiO_2 (in the range of some wt%) has been a common method to increase the chemical durability. A sufficient TiO_2 content not only improves the chemical durability but, unfortunately, also leads to a transmission loss in the UV-range. A solution to this problem is reached by replacing parts of the PbO contents with both TiO_2 and ZrO_2 [2.71]. The chemical durability is improved and the optical properties of the original glass (SiO_2-PbO-M_2O) are obtained again.

7. P_2O_5-MO-B_2O_3-Al_2O_3:
Glasses in which P_2O_5 acts as a network former have been known for a long time [2.72]. P_2O_5 forms glasses with rather low dispersions (high Abbe numbers) and high refractive indices compared with the classic glasses based on B_2O_3 or SiO_2. Glasses with a high P_2O_5 content show long dispersion spectra in the blue range. These glasses are normally used to correct chromatic aberration. The second and very important property of glasses containing P_2O_5 is their excellent UV-transparency.

A disadvantage of phosphate glasses is that their chemical durability is poor compared with the durability of SiO_2-based glasses.

8. P_2O_5-Al_2O_3-MO-(F_2-O):
In order to obtain glasses with lower dispersions than those of phosphate glasses, one can introduce fluorine. Most of these glasses are based on a P_2O_5-Al_2O_3-MO glass system which contains high amounts of fluorine. This glass system shows a relatively high chemical durability and a relatively low devitrification tendency. Fluorine takes the places of oxygen atoms so that formally one oxygen is replaced by two fluorine atoms. The formation of the fluoro-phosphate glasses of this system is rather complex and can be described with amorphous fluoro-alumino-phosphates [2.73].

As fluorine has a smaller ionic radius and a higher electronegativity than oxygen, the polarization of the cations in these glasses is increased so that the refractive index and the dispersion are decreased. The replacement of oxygen by fluorine leads to glasses which have distinct LgK characteristics.

9. (B_2O_3, SiO_2)-La_2O_3-MO:
In connection with B_2O_3, glasses containing large amounts of La_2O_3 form a very extensive glass system. Glasses belonging to this system normally show a high refractive index with a relatively low dispersion. As their chemical resistance is rather poor, their durability can be increased by replacing SiO_2 with B_2O_3 and by

the introduction of divalent oxides. In addition the introduction of ZrO_2 and TiO_2 is another possibility to achieve better chemical properties. In the past, ThO_2 and CdO have also been used to achieve this aim, but due to environmental protection these oxides have been eliminated from these glasses.

10. B_2O_3-La_2O_3:
These glasses are based on the B_2O_3-La_2O_3 glass system. They all show high refractive indices and high dispersion. In order to obtain high refractive indices large numbers of oxides that are known to cause a high refractive index, such as Gd_2O_3, Y_2O_3, TiO_2, Nb_2O_5, Ta_2O_5, WO_3, and ZrO_2 are incorporated in the glasses of this system. The large number of network modifiers leads to glasses whose viscosities depend to a high degree on temperature, i.e., they are called short [2.74] and which show a strong tendency to crystallize. On the other hand, these glasses show a distinct hardness and chemical resistance, properties which are even increased with higher refractive indices. In this respect, the glasses differ from those of the system SiO_2-PbO-M_2O, which also show a high refractive index.

11. SiO_2-Sb_2O_3-B_2O_3:
The replacement of PbO by Sb_2O_3 leads to an extensive glass system. Sb_2O_3 reduces the dispersion of the blue part of the visible spectrum. Consequently, short-flint glasses are obtained. The introduction of Sb_2O_3 into a glass shows results similar to those we have in the case of PbO, although Sb_2O_3 glasses seem to form long chains of -Sb-O-Sb-O- units [2.75].

12. B_2O_3-PbO-Al_2O_3:
If SiO_2 is replaced by B_2O_3, glasses with primarily very poor chemical properties are obtained. Glasses of this system have an intensive short-flint characteristic, since the PbO content induces absorption in the blue range of the visible spectrum, and B_2O_3 causes strong absorption in the infrared range of the spectrum. Al_2O_3 is added to improve the chemical properties to an acceptable level.

13. SiO_2-M_2O-TiO_2-(F_2-O):
The PbO content in glasses of the system SiO_2-M_2O-PbO can be replaced by TiO_2. As TiO_2 causes an increase of the refractive index, glasses of the system SiO_2-M_2O-TiO_2-(F_2-O) are characterized by high refractive indices. The dispersion of these glasses is increased to a higher degree than that of glasses which contain the same amount of PbO. The introduction of TiO_2 into glasses is limited, because by increasing the TiO_2 content, the glasses strongly tend to crystallize. This should be avoided by the introduction of alkaline oxides into the glass. In spite of the large content of alkaline oxides, these glasses are highly resistant to influences of water and acids. Glasses with high TiO_2 contents show a strong absorption in the UV-range. This absorption is stronger than that of glasses containing PbO or Sb_2O_3. Both glass types are characterized by a positive deviation of the partial dispersion from the normal line.

2.2.3 Reduction of the Density

The PbO content of glasses of the SiO_2-PbO-M_2O system can also be replaced by the oxides WO_3, TiO_2, ZrO_2 and Nb_2O_5. The glasses obtained normally show excellent chemical durability, low density and good mechanical properties which allow good grinding and polishing. The production of these glasses, however, is much more difficult than that of the traditional glasses with large PbO content. Because of the low density, the glasses of the SF-group and the LaK-group are called "lightweight glasses" in the Schott programme, where they are indicated by the letter "L" preceding the glass number.

Additionally, in glasses which originally do not contain any PbO, a decreased density may be achieved by optimizing the glass compositions. The compounds responsible for a high density are replaced by compounds which contribute little to the glass density (e.g., by replacing BaO with CaO and MgO). Table 2.4 gives a summary of the lightweight glasses of the Schott programme. Other glass manufacturers have also developed lightweight glasses.

Table 2.4. Lightweight glasses of the Schott programme compared with the original "heavy" glasses (ρ: glass density, wr: weight reduction, SR: acid resistance (ISO 8424), AR: alkali resistance (ISO in draft), T_g: transformation temperature, HK: Knoop hardness)

Type	n_d	ν_d	ρ g/cm^3	Weight reduct.	SR	AR	T_g	H_K
F N11	1.62096	36.18	2.66	–26%	1	1.0	577	610
F2	1.62004	36.37	3.61		1	2.3	438	420
SF L4	1.75520	27.21	3.17	–34%	1.0	1.3	569	580
SF 4	1.75520	27.58	4.79		4.3	2.3	424	390
SF L6	1.80518	25.39	3.37	–35%	2	1	605	570
SF 6	1.80518	25.43	5.18		51.3	2.3	426	370
SF L56	1.78470	26.08	3.28	–33%	1	1.3	592	570
SF 56A	1.78470	26.08	4.92		3.2	2.2	433	380
SF L57	1.84666	23.62	3.55	–36%	1	1	598	580
SF 57	1.84666	23.83	5.51		52	2.3	402	350
SF 64A	1.70585	30.30	3.00	–26%	1	1.2	580	590
SF 15	1.69895	30.07	4.06		1	1.2	455	420
BaSF 64A	1.70400	39.38	3.20	–19%	3.2	1.2	582	650
BaSF 13	1.69761	38.57	3.97		52.3	1.2	584	530
LaSF 32	1.80349	30.40	3.52	–28%	1	1.0	552	660
LaSF 8	1.80741	31.61	4.87		52	1.2	476	420
LaSF 36A	1.79712	35.08	3.60	–20%	3.3	1.0	651	670
LaSF 33	1.80596	34.24	4.48		51	1.2	543	540
LaK L12	1.67790	54.92	3.32	–19%	58.3	2.2	651	700
LaK N12	1.67790	55.20	4.10		53.3	3.3	614	560
LaK L21	1.64048	59.75	2.97	–21%	53.3	3.3	641	710
LaK 21	1.64049	60.10	3.74		53.2	4.3	639	600

2.2.4 Development of Optical Glasses

Calculation of the Optical Properties

In modern glass development, the computation of properties from the chemical composition has been realized with the help of models which relate the composition to the properties. The experiments necessary for the evaluation of these models are based on statistical methods which allow the experiments to be kept to a minimum.

For the development of a glass it is necessary to know the concentration range of the components which will be introduced into the glass. On the one hand, the model is evaluated only for this concentration range, but on the other hand, it makes it possible to forecast the values of the properties within this concentration range with high accuracy.

Thus, as some data are known, we are able to forecast the optical properties. These data are based on the assumption that it is possible to calculate the optical properties with constant or nearly constant influences of the glass components within a certain concentration range.

Refractive Index and Main Dispersion

One method to calculate the refractive indices of an optical glass is based on the assumption that the refractive index can be calculated if the nature and the concentration of the various glass components are known. In this case the term "component" is used rather arbitrarily and may refer to oxides or to structural units.

A simple but nevertheless very efficient method to calculate the refractive index n_d has been introduced by *Appen* [2.76–85]:

$$n_d = \frac{\sum_{i=1}^{n} n_{d,i} c_i}{100} , \tag{2.51}$$

where $n_{d,i}$ is a factor which refers to a glass oxide and c_i is the oxide concentration in mole%. Some factors are shown in Table 2.5. In addition Appen also introduced factors for the main dispersion:

$$D = n_F - n_C \tag{2.52}$$

which is calculated similarly to (2.51):

$$D = \frac{\sum_{i=1}^{n} D_i c_i}{100} . \tag{2.53}$$

Some factors of this calculation are also given in Table 2.5.

The Abbe number ν_d can be calculated from equations (2.51) and (2.53) by:

$$\nu_d = \frac{n_d - 1}{D} . \tag{2.54}$$

Another method for the calculation of the refractive index n_d by means of the composition has been proposed by *Demkina* [2.86]:

Table 2.5. Factors necessary to calculate the refractive index n_d and the dispersion D of glasses by the composition [2.79–85]

Component	$n_{d,i}$ [a]	D_i [a]
SiO_2	1.4585...1.475	0.00675
B_2O_3	1.470...1.710	0.0066...0.0090
Al_2O_3	1.520	0.0085
Li_2O	1.695 (1.6555)	0.0138 (0.0130)
Na_2O	1.590 (1.575)	0.0142 (0.0140)
K_2O	1.575 (1.595)	0.0130 (0.0132)
MgO	1.610 (1.570)	0.0111
CaO	1.730	0.0148
SrO	1.770	0.0163
BaO	1.880	0.0189
ZnO	1.710	0.0165
TiO_2	2.08...2.23	0.050...0.062
PbO	2.5...2.35	0.0528...0.0744

[a] n_{R_2O}, D_{R_2O}: values in parentheses are valid for glasses in the M_2O-SiO_2 system

n_{K_2O}, D_{K_2O}: values in parentheses are valid for those glasses which contain more than 1 mole% Na_2O

n_{MgO}: the value in parentheses is valid for parts of the glasses of the systems Na_2O-MgO-SiO_2 and K_2O-MgO-SiO_2

n_{SiO_2}: $n_{SiO_2} = 1.5085 - 0.0005 \cdot c_{SiO_2}$ for $100 \geq c_{SiO_2} \geq 67$; $n_{SiO_2} = 1.475$ for $c_{SiO_2} \leq 67$

n_{TiO_2}, D_{TiO_2}: $n_{TiO_2} = 2.480 - 0.005 \cdot c_{SiO_2}$, $D_{TiO_2} = 0.082 - 0.0004 \cdot c_{SiO_2}$ } for $\sum c_{R_2O} < 15$ and $80 \geq c_{SiO_2} \geq 50$

n_{PbO}, D_{PbO}: with $\alpha = c_{SiO_2} + c_{B_2O_3} + c_{Al_2O_3}$ is:
- $n_{PbO} = 2.685 - 0.0067 \cdot \alpha$, $D_{PbO} = 0.1104 - 0.00072 \cdot \alpha$ } for $80 \geq \alpha \geq 50$
- $n_{PbO} = 2.350$, $D_{PbO} = 0.0528$ } for $\alpha \geq 80$

$n_{B_2O_3}$, $D_{B_2O_3}$: with $\beta = (\sum c_{R_2O} + \sum c_{RO} + \sum c_{Al_2O_3})/c_{B_2O_3}$ is:

– for $64 \geq c_{SiO_2} \geq 44$:
- $n_{B_2O_3} = 1.710$, $D_{B_2O_3} = 0.0090$ } $\beta > 4$
- $n_{B_2O_3} = 1.518 + 0.048 \cdot \beta$, $D_{B_2O_3} = 0.0064 + 0.00065 \cdot \beta$ } $4 > \beta > 1$
- $n_{B_2O_3} = 1.616 + 0.048 \cdot \beta$, $D_{B_2O_3} = 0.0064 + 0.00065 \cdot \beta$ } $1 > \beta > 1/3$
- $n_{B_2O_3} = 1.470$, $D_{B_2O_3} = 0.0066$ } $\beta < 1/3$

– for $80 \geq c_{SiO_2} \geq 71$:
- $n_{B_2O_3} = 1.710$, $D_{B_2O_3} = 0.0090$ } $\beta > 1.6$
- $n_{B_2O_3} = 1.518 + 0.12 \cdot \beta$, $D_{B_2O_3} = 0.0090$ } $1.6 > \beta > 1$
- $n_{B_2O_3} = 1.760 - 0.12/\beta$, $D_{B_2O_3} = 0.0064 + 0.00065 \cdot \beta$ } $1 > \beta > 1/2$
- $n_{B_2O_3} = 1.614 - 0.48/\beta$, $D_{B_2O_3} = 0.0064 + 0.00065 \cdot \beta$ } $1/2 > \beta > 1/3$
- $n_{B_2O_3} = 1.470$, $D_{B_2O_3} = 0.0066$ } $\beta < 1/3$

Table 2.6. Data for the calculation of optical properties. After [2.86]

Oxide	t_i	$n_{d,i}$	D_i	ν_i
SiO_2 (I)	60	1.475	0.00695	68
SiO_2 (II)	60	1.458	0.00678	68
B_2O_3 (I)	43	1.61	0.00750	81
B_2O_3 (II)	70	1.464	0.00670	69
Al_2O_3	59	1.49	0.00850	58
Na_2O	62	1.59	0.01400	42
K_2O	94	1.58	0.01200	48
MgO	140	1.64	0.01300	49
CaO	86	1.83	0.01750	47
BaO	213	2.03	0.02280	45
ZnO	223	1.96	0.02850	34
PbO (I)	343	2.46	0.07700	19
PbO (II)	223	2.46	0.07700	19
PbO (III)	223	2.50	0.11600	13

$$n_d = \frac{\sum_{i=1}^{n} \frac{c_i n_{d,i}}{t_i}}{\sum_{i=1}^{n} \frac{c_i}{t_i}} . \tag{2.55}$$

In (2.55) c_i is the concentration of the oxide i in wt% in the glass, t_i is a special conversion coefficient used to transform weight units into so-called "volume fractions" so that it will be possible to apply (2.55) to numerous glass compositions. $n_{d,i}$ is a coefficient determined by experimental measurements and characterizes the contribution of the oxide i to the refractive index. A similar equation can also be used for the main dispersion ($D = n_F - n_C$) and the Abbe number (ν_d). Some coefficients are shown in Table 2.6.

The presence of two values for one glass component indicates that the contribution of this component to the optical properties depends on special concentration ranges. In order to select the correct values from Table 2.6, *Demkina* drew up a number of rules which have to be obeyed.

Huggins and *Sun* [2.87, 88] have developed a method for calculating the optical properties of a glass. This method is related to the concept of the specific volume of glass. The refractive index n_d is calculated from

$$n_d = 1 + \rho \sum_{i=1}^{n} n_{d,i} c_i \tag{2.56}$$

where ρ is the density of the glass, $n_{d,i}$ the contribution of the oxide i to the refractive index and c_i the concentration of the oxide i in wt%. The main dispersion is calculated by (2.57):

$$D = \rho \sum_{i=1}^{n} D_i c_i \tag{2.57}$$

The density of the glass can be calculated according to

Table 2.7. Coefficients for the calculation of the optical properties. After [2.87, 88]

Oxide	$R_i \times 10^2$	ρ_i for various N_{Si} 0.270–0.345	0.345–0.400	0.400–0.435	0.435–0.500
SiO_2	3.3300	0.4043	0.4281	0.4409	0.4542
B_2O_3 (BO_4)	4.3079	0.5900	0.5260	0.4600	0.3450
B_2O_3 (BO_3)	4.3079	0.7910	0.7270	0.6610	0.5460
Al_2O_3	2.9429	0.4620	0.4180	0.3730	0.2940
Li_2O	3.3470	0.4520	0.4020	0.3500	0.2610
Na_2O	1.6131	0.3730	0.3490	0.3240	0.2810
K_2O	1.0617	0.3900	0.3740	0.3570	0.3290
MgO	2.4800	0.3970	0.3600	0.3220	0.2560
CaO	1.7832	0.2850	0.2590	0.2310	0.1840
SrO	0.9650	0.2000	0.1850	0.1710	0.1450
BaO	0.6521	0.1420	0.1320	0.1220	0.1040
TiO_2	2.5032	0.3190	0.2820	0.2430	0.1760
ZrO_2	1.6231	0.2220	0.1980	0.1730	0.1300
PbO	0.4480	0.1060	0.0955	0.0926	0.0807

$$\frac{1}{\rho} = \sum_{i=1}^{n} \rho_i c_i \, , \tag{2.58}$$

where ρ_i is the influence of the oxide i on the density and c_i the concentration of the oxide in wt%. The values of ρ_i depend on the number of silicon atoms N_{Si} per oxygen atom in the glass. For the determination of this number (2.59) is used:

$$N_{Si} = \frac{c_{SiO_2}}{60.06 \sum_{i=1}^{n} R_i c_i} \, , \tag{2.59}$$

where c_i is the weight fraction of the oxide i, R_i is the ratio of the number of oxygen atoms in one formula unit of the oxide to the molecular weight of the oxide, c_{SiO_2} is the concentration of SiO_2 in the glass in wt% and 60.06 is the molecular mass of SiO_2. Table 2.7 shows some values for R_i and ρ_i (dependent on the value of N_{Si}). Table 2.8 shows some values for $n_{d,i}$ and D_i.

Gilard and *Dubrul* [2.89] have proposed a non-linear model for the calculation of the refractive index:

$$n_d = \sum_{i=1}^{n} \left\{ a_{1,i} c_i + a_{2,i} (c_i - a_{3,i})^2 \right\} \tag{2.60}$$

where the concentration c_i is used in wt%. The coefficients for several oxides are shown in Table 2.9.

Table 2.8. Coefficients for $n_{d,i}$ and D_i using the method given in [2.87, 88]

Oxide	n_d	$D_i \times 10^2$
SiO_2	0.2083	0.3095
B_2O_3 (BO_4)	0.2150	0.2580
B_2O_3 (BO_3)	0.2530	0.4310
Al_2O_3	0.2038	0.3630
Li_2O	0.3080	0.7028
Na_2O	0.1941	0.5013
K_2O	0.2025	0.4320
MgO	0.2100	0.4540
CaO	[a]	[a]
SrO	0.1592	0.3140
BaO	[a]	[a]
TiO_2	0.3130	0.6800
ZrO_2	0.2090	0.6900
PbO	[a]	[a]

[a]: $n_{d,i}$ and D_i of CaO, BaO and PbO are calculated as follows:

– CaO:

$$n_{d,\mathrm{CaO}} = 0.2257 + 0.4770 \times 10^{-4} \cdot c_{\mathrm{CaO}} / \sum_{i=1}^{n} R_i c_i$$
$$D_{\mathrm{CaO}} = 0.4636 + 0.0013 \cdot c_{\mathrm{CaO}} / \sum_{i=1}^{n} R_i c_i$$

– BaO:

$$n_{d,\mathrm{BaO}} = 0.1290 + 2.1300 \times 10^{-4} \cdot c_{\mathrm{BaO}} / \sum_{i=1}^{n} R_i c_i$$
$$D_{\mathrm{BaO}} = 0.2200 + 0.0630 \cdot c_{\mathrm{BaO}} / \sum_{i=1}^{n} R_i c_i$$

– PbO:

$$n_{d,\mathrm{PbO}} = 0.1272 + 2.0440 \times 10^{-6} \cdot c^2{}_{\mathrm{PbO}} / (\sum_{i=1}^{n} R_i c_i)^2$$
$$D_{\mathrm{PbO}} = 0.582 + 0.5044 \times 10^{-4} \cdot c^2{}_{\mathrm{PbO}} / (\sum_{i=1}^{n} R_i c_i)^2$$

Partial Dispersion

As mentioned above, the refractive index is a function of the wavelength. The partial dispersion is used to express the variation of the the refractive index in different parts of the optical spectrum.

Normally a nearly linear relationship between the partial dispersion and the Abbe number is found for most optical glasses. Glasses which show large deviations from this linear relationship are called glasses with an abnormal partial dispersion.

The oscillation frequency in the IR-range of the main network formers influences the dispersion of the glass significantly. *Reitmayer* [2.90] and *Izumitani* [2.91] explained the negative values of the deviation from the "normal line" of the glasses KzFS 2, KzFS 4, and LaK 9 by a shift of the oscillation wavelength to lower values due to the very high B_2O_3 contents of these glasses. *Brewster* [2.92] found that a negative deviation occurs if SiO_2 is replaced by B_2O_3. The replacement of SiO_2 by P_2O_5 leads to a positive deviation from the "normal-line". Measurements on silicate glasses with different water concentrations show an increase in the deviation of the partial dispersion from the "normal-line" with increasing water-content [2.93].

Table 2.9. Data for the calculation of the refractive index achieved by the method given in [2.89] (in wt%)

Oxide	$a_{1,i}$	$a_{2,i}$	$a_{3,i}$	Conc. range of oxide [a]
SiO_2	0.01458	0	0	0–100
B_2O_3	0.01725	-50×10^{-6}	0	0–40
Al_2O_3	0.01510	0	0	0–20
Li_2O (I)	0.01820	0	0	0–25
Li_2O (II)	0.01980	-64×10^{-6}	0	25–45
Na_2O (I)	0.01618	0	0	0–20
Na_2O (II)	0.01646	-14×10^{-6}	0	20–55
K_2O (I)	0.01575	0	0	0–20
K_2O (II)	0.01575	$+130 \times 10^{-6}$	−20	20–60
MgO	0.01700	0	0	0–15
CaO	0.01785	0	0	0–40
BaO	0.01690	0	0	0–50
ZnO	0.01675	0	0	0–18
TiO_2	0.0200	0	0	0–10
PbO (I)	0.01760	0	0	0–40
PbO (II)	0.01760	$+140 \times 10^{-6}$	−40	40–80

[a]: in weight %

The reasons for the abnormal partial dispersion may be explained by Lorentz's theoretical equation. With this equation the refraction index n can be expressed as a function of the wavelength λ in a region where no absorption occurs:

$$n^2 - 1 = \frac{\sum_{i=1}^{n} K N_i f_i}{\frac{1}{\lambda_i{}^2} - \frac{1}{\lambda^2}} \,. \tag{2.61}$$

Here λ_i are the wavelengths of the inherent absorption, f_i is the oscillator strength of the inherent absorption, N_i is the number of atoms per volume and K is a constant. For optical glasses a model with 4 inherent absorptions [2.94] can be used, namely, for the absorption by bridging oxygen atoms (index 1), the absorption by non-bridging oxygen atoms (index 2), the absorption in the infrared region (index 3) and the absorption in the ultraviolet region (index 4), provided that the glass contains PbO:

$$n^2 - 1 = \frac{\sum_{i=1}^{4} K N_i f_i}{\frac{1}{\lambda_i{}^2} - \frac{1}{\lambda^2}} \,. \tag{2.62}$$

Using (2.62) *Izumitani* explained that the shift of the absorption peaks resulting from different structural groups in a glass causes a change in the run of the refractive index depending on the wavelength. For a calculation of the partial dispersion of a glass system, the constants λ_i, K and f_i must be determined.

If the constants are determined, the dispersion of the refractive index can be calculated by the number of atoms Ni which cause absorption bands.

Linear models represent another way to calculate the partial dispersion of glasses. In principle there are two ways to determine the partial dispersion, e.g. $P_{g,F}$:

$$P_{g,F} = \frac{n_g - n_F}{n_F - n_C} . \tag{2.63}$$

One possibility is to use linear models to determine first the refractive indices n_j similarly to *Appen's* model:

$$n_j = \sum_{i=1}^{n} n_{j,i} c_i, \quad j \in \{g, F, C\} . \tag{2.64}$$

From these refractive indices n_j the partial dispersion is determined with (2.63).

The other possibility is to calculate the partial dispersion by using a linear model for the partial dispersion itself:

$$P_{x,y} = \sum_{i=1}^{n} P_{x,y,i} c_i \tag{2.65}$$

where $P_{x,y,i}$ is the contribution of the oxide i to the partial dispersion $P_{x,y}$ and c_i is the concentration of the oxide. The index "$_{x,y}$" indicates the wavelength combination of the partial dispersion which is to be calculated.

Glass systems with high PbO contents might show a negative deviation value from the "normal-line". This is shown in Fig. 2.22, where the ratio of B_2O_3 to PbO, starting with KzFS N4, is changed continuously and the partial dispersion is calculated by a model similar to (2.65). This "simple" variation of the B_2O_3-PbO ratio leads to values of the partial dispersion which can also be obtained by the other KzFS glasses of the Schott programme [2.95].

Although the partial dispersion can be calculated by a linear model, in a plot of $P_{g,F}$ versus ν_d the curve does not follow a straight line, but occurs as the typical curve indicated in Fig. 2.22.

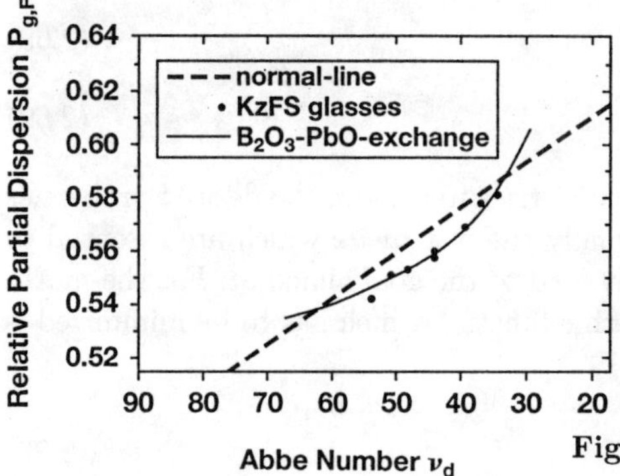

Fig. 2.22. Change in the values of $P_{g,F}$ and ν_d upon exchanging B_2O_3 and PbO

Optimization of the Glass Composition

During recent years a lot of data on the properties of optical glasses have been collected. They can be used for the regression analysis and the determination of regression coefficients. These coefficients, e.g. for linear models, represent a mathematical relationship between the glass properties and the glass composition, and they provide information about the influence of the different glass compositions on the glass properties. Moreover, it is possible to calculate the optimum glass composition even if more than one property of the glass is desired.

In this case one needs to know the regression coefficients, a procedure for optimization and an error function. For linear models the glass properties can be calculated with:

$$Q_{j,\mathrm{cal}} = \sum_{i=1}^{n} Q_{j,i} c_i \tag{2.66}$$

where $Q_{j,\mathrm{cal}}$ is the calculated glass property j, c_i the concentration of the component i and $Q_{j,i}$ the regression coefficient of the component i for the property j.

In practice, however, (2.66) is unsuitable, as normally the development of a glass starts from a given glass with defined properties. For this reason, it is more convenient to use the deviation of the desired properties from the properties of the known glass for the optimization [2.96]:

$$Q_{j,\mathrm{cal}} = Q_{j,d} + \sum_{i=1}^{n} Q_{j,i} \Delta c_i \ , \tag{2.67}$$

where $Q_{j,d}$ is the property j of the given glass and Δc_i is the deviation of the concentration of the component i of the desired glass from the concentration of the component i of the given glass. Due to this transformation, errors arising from the use of simple statistical models may be reduced.

Before starting the optimization, the order of magnitude of the properties $Q_{j,d}$ which are calculated must be considered. Normally, for the optimization procedure only properties with a large magnitude and not those with a small magnitude are considered. For this reason the properties have to be normalized to unity:

$$F_j = 1/Q_{j,d} \tag{2.68}$$

$$Q'_{j,\mathrm{cal}} = F_j Q_{j,\mathrm{cal}} \ . \tag{2.69}$$

In order to develop a glass, the scientist tries to achieve the desired properties by changing the glass composition. Usually the test melts which are essential to glass development, will not immediately lead to the goal aimed at. For the mathematical treatment of this process an error function which has to be minimized is defined [2.97]:

$$E = \sum_{j=1}^{n} (Q'_{j,\mathrm{cal}} - Q'_{j,\mathrm{a}})^2 w_j \tag{2.70}$$

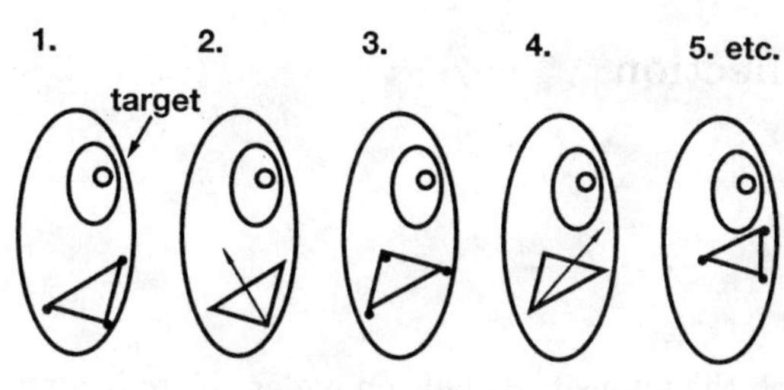

Fig. 2.23. Operation of the simplex method for a two-component system

where $Q'_{j,\mathrm{a}}$ is the aspired final value of the glass property j and w_j a weighting factor for the property j. The weighting factor is used to influence the error function with respect to the special glass properties. For a new glass, the demanded property profile is normally not expressed by an exact value, but by upper and lower limits of the properties. Consequently, (2.70) has to be changed to:

$$E_j = \begin{cases} (Q'_{j,\mathrm{cal}} - Q'_{j,\mathrm{a,u}})^2 w_j, & \text{for } Q'_{j,\mathrm{cal}} > Q'_{j,\mathrm{a,u}} \\ 0, & \text{for } Q'_{j,\mathrm{a,l}} < Q'_{j,\mathrm{cal}} < Q'_{j,\mathrm{a,u}} \\ (Q'_{j,\mathrm{cal}} - Q'_{j,\mathrm{a,l}})^2 w_j, & \text{for } Q'_{j,\mathrm{cal}} < Q'_{j,\mathrm{a,l}} \end{cases} \tag{2.71}$$

and

$$E = \sum_{j=1}^{n} E_j \tag{2.72}$$

where $Q'_{j,\mathrm{a,l}}$ is the normalized value of the lower limit of the property j and $Q'_{j,\mathrm{a,u}}$ the normalized value of the upper limit of the property j.

The error function of (2.72) has to be minimized to meet the required properties of the glass. For the minimization of the error function various mathematical tools can be used.

One method which can be applied to various tasks and with which a personal computer can easily be programmed, is the Simplex-method [2.98, 99]. For the optimization of a glass with n components, $n+1$ different glass compositions are set up. During the calculation an error function, see (2.72), is calculated for every starting composition. The starting composition with the highest error function is now replaced by a new starting composition by reflection (Fig. 2.23). Thus, a new set of $n+1$ different glass compositions is obtained for which the calculation of the error function and a new replacement can be performed.

After having carried out several replacements (iteration steps), one of the calculated starting compositions reaches an optimum and the error function is minimized.

The disadvantage of the simplex method is the large number of iteration steps necessary to reach the optimum. However, since a common iteration procedure is done within a few minutes it presents an acceptable method for the identification of melt experiments.

Acknowledgements

For their critical reading of the manuscript I thank Mrs. Marga Faulstich and Dr. habil. W. Geffcken.

2.3 Transmission and Reflection

Norbert Neuroth

2.3.1 General Relations

Let us assume that a light beam with the intensity I_0 falls on a glass plate having the thickness d. At the entrance surface a part of the beam is reflected (Fig. 2.24).

After that the intensity of the light beam is $I_e = I_0(1 - r)$ with r being the reflectivity. Inside the glass the beam is attenuated according to the exponential function. At the exit surface the beam intensity is

$$I_d = I_e \exp(-kd) \tag{2.73}$$

where k is the absorption constant. At the exit surface another reflection occurs. The transmitted beam has the intensity

$$I_t = I_d(1 - r) \; . \tag{2.74}$$

These formulae give the following relation:

$$I_t = I_e(1 - r) \exp(-kd) = I_0(1 - r)^2 \exp(-kd) \; . \tag{2.75}$$

The beam reflected at the exit surface returns to the entrance surface and is divided into a transmitted and a reflected part (Fig. 2.25). This process is repeated infinitely. Taking the multiple reflections into account the transmission of the glass plate is

$$\tau = \frac{I_t}{I_0} = P\tau_i \tag{2.76}$$

with

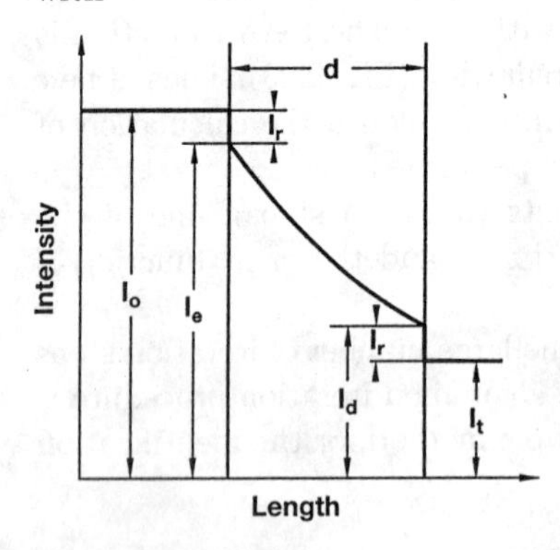

Fig. 2.24. Change in the intensity of a light beam passing through a glass plate

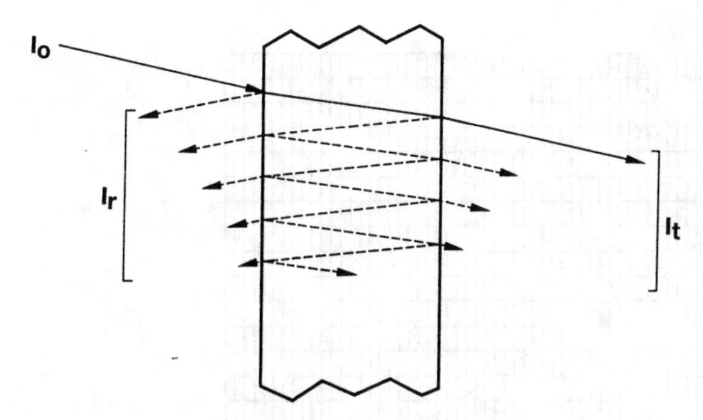

Fig. 2.25. Multiple reflection of a light beam at the surfaces of a glass plate

$$\tau_\mathrm{i} = \frac{I_\mathrm{d}}{I_\mathrm{e}} = \textit{internal} \text{ transmission} \tag{2.77}$$

and

$$P \approx \frac{(1-r)^2}{1-r^2} = \frac{2\,n}{n^2+1} \,. \tag{2.78}$$

The internal transmission is that part of the beam attenuation which is caused by the glass itself, whereas the "reflection factor" P is caused by the reflections. The last formula has been derived from Fresnel's formula which describes the relation between the reflectivity r and the refractive index n. The reflected intensity of the beam incident perpendicularly upon a plate consisting of dielectric material with the refractive index n is

$$r = \left[\frac{n-1}{n+1}\right]^2 \,. \tag{2.79}$$

This relation is used in formula (2.78). The approximation of the first part of the formula (2.78) causes deviations from the exact value in the order of 10^{-3} to 10^{-2} (depending on the r- and τ_i-value) [2.100]. In Fig. 2.26 the reflection factor P is given as a function of the refractive index n.

Usually, the transmission τ of the plate is measured by a spectrophotometer [2.101–105]. In order to calculate the internal transmission τ_i and the absorption constant k it is essential to know the refractive index, which is used to calculate the reflection factor P according to formula (2.78). Knowing τ_i the absorption constant

$$k = \frac{1}{d} \ln \frac{1}{\tau_\mathrm{i}} \tag{2.80}$$

can be computed.

When d is given in cm, k has the dimension cm^{-1}. If the refractive index is unknown, it is necessary to measure the transmission of another plate that is of the same material but has a smaller thickness d_2. The ratio of the transmission

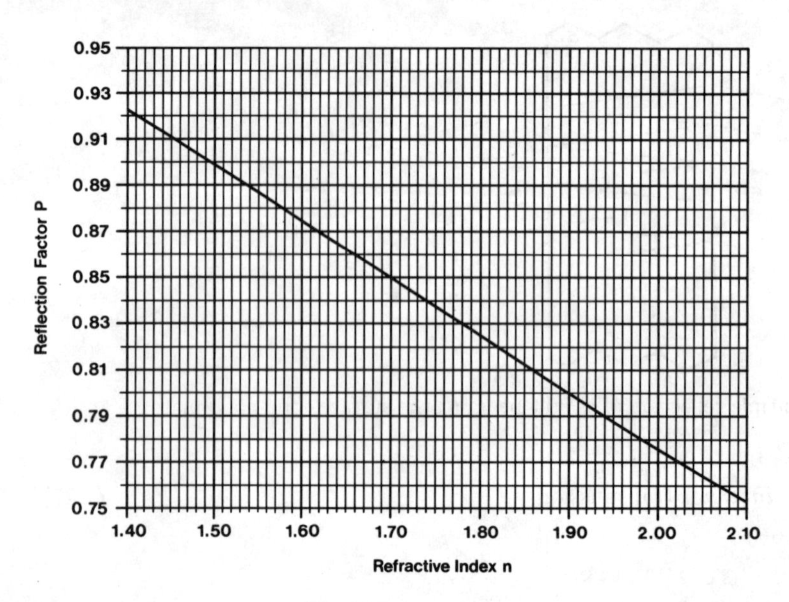

Fig. 2.26. Reflection factor P of a glass plate as a function of its refractive index

τ_1 of the plate with the thickness d_1 to the transmission τ_2 of the plate with the thickness d_2 is the internal transmission of a plate with the thickness $d_1 - d_2$:

$$\tau_{i1-2} = \frac{\tau_1}{\tau_2} . \tag{2.81}$$

Then the absorption constant is

$$k = \frac{1}{d_1 - d_2} \ln \frac{1}{\tau_{i1-2}} = \frac{1}{d_1 - d_2} \ln \frac{\tau_2}{\tau_1} , \quad d_2 < d_1 . \tag{2.82}$$

k is the natural absorption constant. It is also common to use 10 as a base of the exponential function:

$$I_\mathrm{d} = I_\mathrm{e} \cdot 10^{-Ad} . \tag{2.83}$$

The absorbance A is calculated from

$$A = \frac{1}{d} \log \frac{1}{\tau_\mathrm{i}} . \tag{2.84}$$

Note that k is 2.3 times larger than A.

Sometimes the attenuation of a glass plate is expressed as

$$^{10}\log \frac{I_0}{I_\mathrm{t}} = \text{extinction or optical density} . \tag{2.85}$$

This expression includes losses of absorption and reflection.

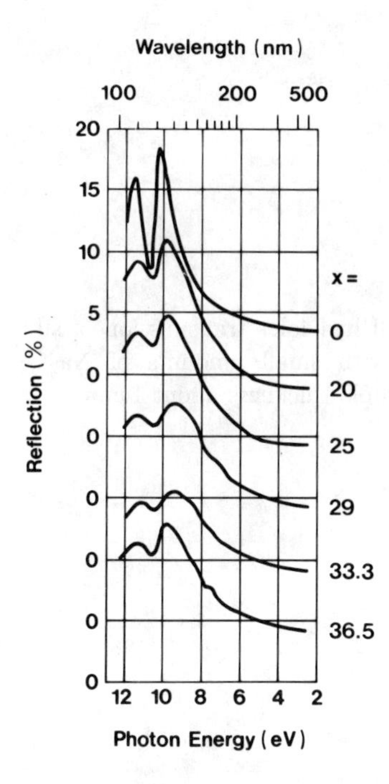

Fig. 2.27. Ultraviolet reflection spectra of glasses of the system $x\mathrm{Na_2O}$-(1-x)$\mathrm{SiO_2}$ [2.106]

If the internal transmission τ_{i1} of a plate with the thickness d_1 is known, it is possible to calculate the internal transmission τ_{i2} of a plate with a thickness d_2:

$$\log\ \tau_{i2} = \frac{d_2}{d_1} \log\ \tau_{i1}\ . \tag{2.86}$$

The transmission τ_2 is obtained by multiplying τ_{i2} with the reflection factor P.

2.3.2 Sources of Optical Loss in Glass

Intrinsic Absorption

Glasses are dielectric materials. Normally (in the case of oxide glasses) they are transparent in the near ultraviolet, the visible and the near infrared. Unlike metals, they are transparent because the electrons of the atomic shells are bound to the individual atom. With metals there are free electrons which can oscillate with the frequency of an incident electromagnetic wave. Therefore the light is strongly reflected. Although the electrons of the glass are bound they can be excited to higher energy levels by ultraviolet radiation. In Fig. 2.27 the ultraviolet reflection spectra of glasses of the system $x\mathrm{Na_2O}$-(1-x)$\mathrm{SiO_2}$ are given [2.106]. Pure $\mathrm{SiO_2}$ glass has relatively strong and narrow reflection peaks at 122 and 108 nm and shorter wavelengths. If sodium oxide is added, these bands become broader. According to *Sigel* [2.107, 108] this is due to a change of the oxygen surrounding. In a pure $\mathrm{SiO_2}$ glass $\mathrm{SiO_4}$-tetrahedra are the basis of the molecular structure. They are

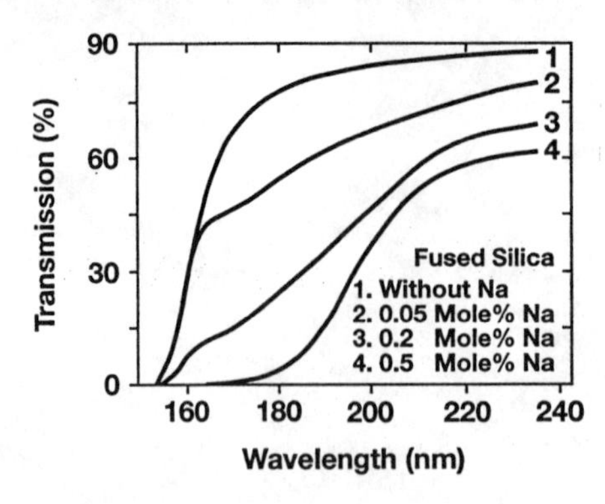

Fig. 2.28. Ultraviolet transmission of silica glass with small amounts of Na_2O [2.107]; sample thickness about 1 mm

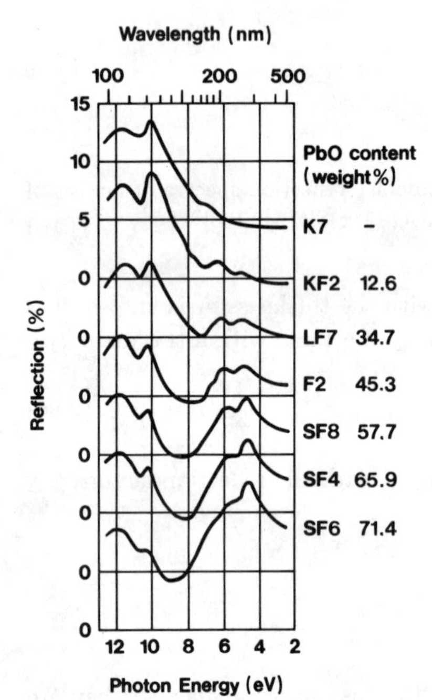

Fig. 2.29. Ultraviolet reflection of silicate glasses [2.106]

forming a special network. Each oxygen atom has two neighbouring Si atoms (bridging oxygens). The presence of sodium in glass produces oxygen atoms which have on their sides a Si atom and a Na atom, respectively. They are called non-bridging oxygen atoms. The excitation energy levels of their electrons are varied. Consequently, with an increasing sodium content, the UV reflection bands are broadened and an additional band appears at about 170 nm. Figure 2.28 shows the change of the UV-transmission of Na_2-SiO_2 glasses. Even a very small Na_2O content reduces the UV-transmission considerably. If the glass contains heavier elements, e.g. lead oxides, two other reflection bands appear between 180 and 280 nm; see Fig. 2.29 [2.106].

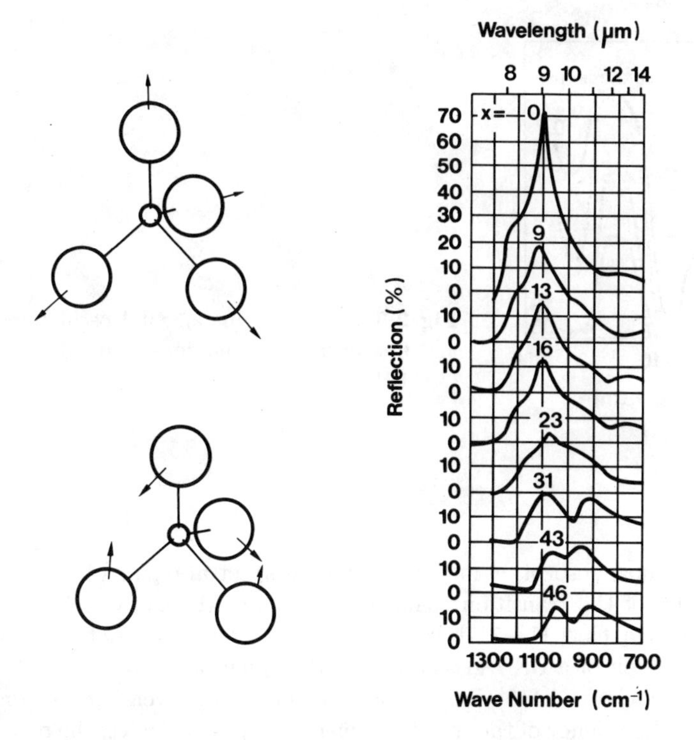

Fig. 2.30. (a) Two vibration modes of a SiO_4 tetrahedron [2.109] (b) infrared reflection spectra of glasses of the system $x Na_2O$-(1-x)SiO_2 [2.110]

The other source of intrinsic loss is molecular vibrations. Figure 2.30a presents two vibration modes of the SiO_4-tetrahedra: – the Si-O stretching and bending vibrations of the SiO_4-tetrahedra [2.109]. In Fig. 2.30b the infrared reflection spectra of the SiO_2 glass and of the $x Na_2O$-(1-x)SiO_2 glass system is shown [2.110]. The stretching vibration Si→O causes the 8.9 μm reflection peak. If Na_2O is added, the peak becomes weaker as well as broader and is shifted to longer wavelengths. Another peak occurs in the range of 10.4 to 10.8 μm; it is due to vibrations of non-bridging oxygens. In this range the absorption of glass is very strong: the absorption constant is between 10^4 and 10^5 cm^{-1}. Such a strong absorption of the basic oscillation also causes attenuations at the overtone frequencies. In Fig. 2.31 the transmission of fused silica with thickness values of 0.7 μm to 1 mm is shown [2.111, 112]. The fundamental oscillations induce absorption in the range of 8 to 13 μm (maxima at 8.2 [*shoulder*], 9.4 and 12.4 μm). Even the first overtones at 5.0, 5.3 and 6.2 μm lead to strong absorption (up to 200 cm^{-1}). Therefore a glass sample that is 1 mm thick becomes transparent at wavelengths smaller than 4.5 μm, i.e., at half of the wavelength of the fundamental oscillation.

The strong UV and IR absorption phenomena that have been described determine the limitations of the spectral transmission. They are caused by the material itself (intrinsic loss).

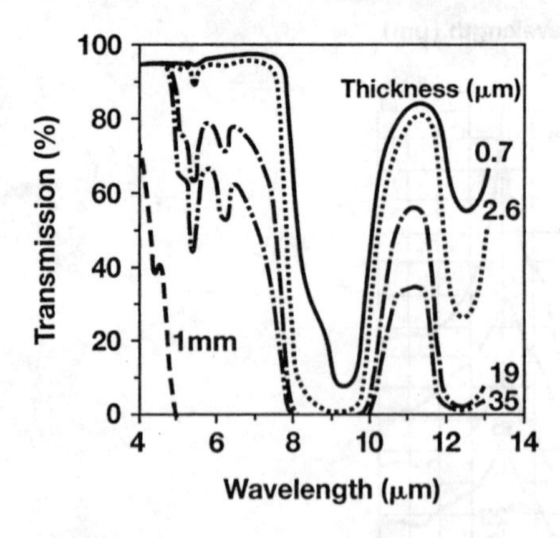

Fig. 2.31. Infrared fundamental oscillations and first overtone oscillations of fused silica [2.111, 112]

Impurities

Some special oxides show absorption in the transparent spectral region of a glass. These are the oxides of the transition elements Ti, V, Cr, Mn, Fe, Co, Ni, Cu and many of the rare earths [2.113–117]. Figure 2.32, for instance, shows the spectral internal transmission of a BK 7 glass doped with 1 ppm of the oxides of the different transition elements [2.118]. The absorption spectra of several transition elements depend on the valency of the specific element. In optical glasses, the presence of colouring elements has to be avoided although they may be useful for the production of a coloured glass. Usually, the content of iron oxide in optical glasses is in the range of 1 to 10 ppm whereas the concentrations of the other transition elements are less than 1 ppm. As mentioned above, these impurity levels are still too high to be acceptable for glass fibres used in telecommunication, since for this usage impurity levels within the range of ppb are required. The traditional production techniques are unsuitable. By means of the chemical-vapour-deposition technique, however, it is possible to produce optical fibres with a loss that is 10^{-4} times lower than that arising in the production of standard optical glasses.

Scattering

The phenomenon of light scattering in glass may be caused by undissolved particles (such as stones, metals, crystals), or by bubbles, phase separation, and/or density fluctuations. The first three reasons for scattering given here are due to the melting technique. The density fluctuations are provoked by the Brownian movement of the molecules in the melt [2.119]. Heavy components (e.g., PbO) and light components (e.g., SiO_2, Na_2O) may separate due to the thermal movements of the molecules in the melt. These small separations of lighter and heavier components cause fluctuations of the density and of the refractive index. If the melt is annealed, these fluctuations of the refractive index are frozen in.

The size of the fluctuations is much smaller than that of the wavelengths. According to Rayleigh, the scattering in this case depends on the wavelength λ

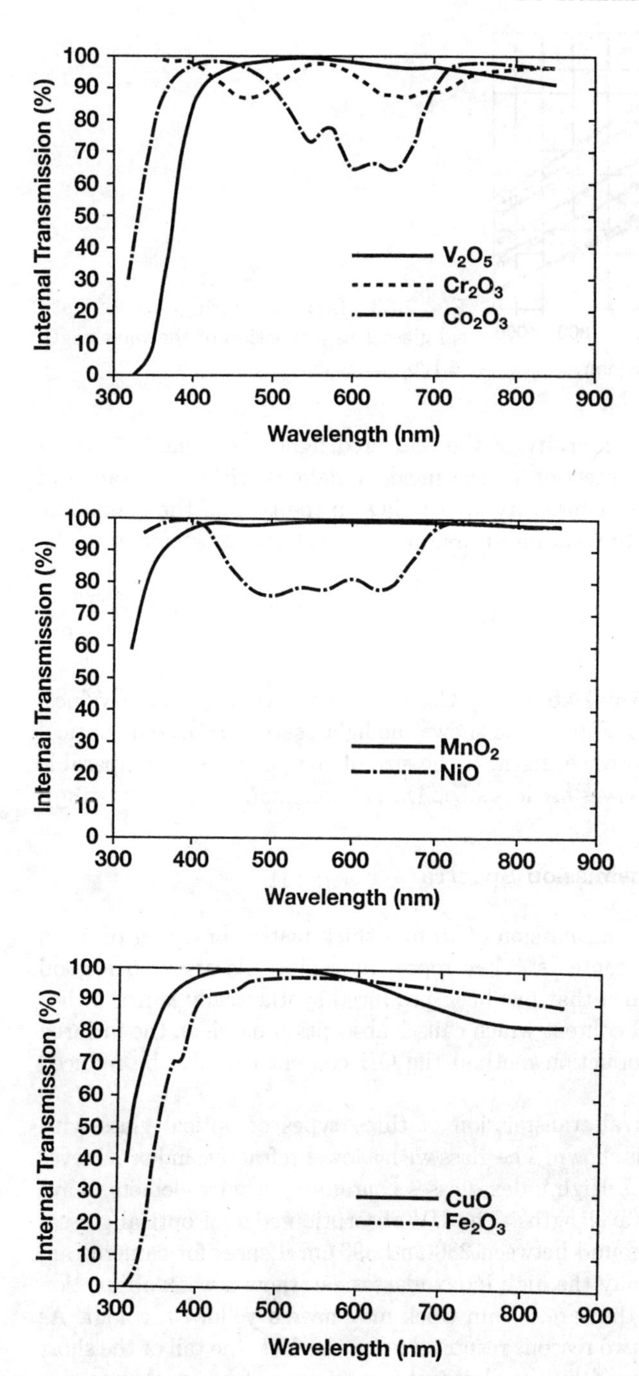

Fig. 2.32. (a–c) Spectral internal transmission of a coloured BK 7 glass by different oxides; sample thickness 100 nm [2.118]

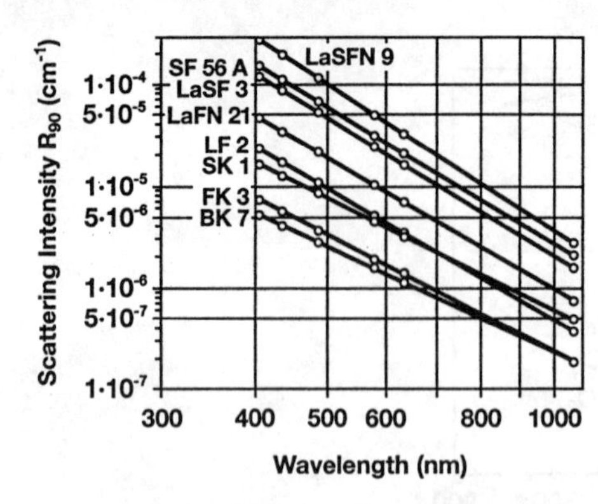

Fig. 2.33. Light scattering R_{90} of optical glasses as a function of the wavelength [2.120]

corresponding to λ^{-4}. The intensity of the scattered light (wavelength 546 nm) at an angle of 90° to the direction of the incident light is within the range of 2 to 70×10^{-6} of the incident intensity, depending on the type of the glass. The chemical composition and the melting technique influence the exponent m in the relationship

$$I_{\mathrm{scat}} \approx \frac{1}{\lambda^m}, \tag{2.87}$$

where m varies between 3.5 and 4.8 [2.120]. Crown glasses have the lower m-values, flint glasses the higher ones. Figure 2.33 shows the light scattering of some optical glasses as a function of the wavelength. If the size of the particles is comparable with the wavelength, the curves are less steep (m is lower than 4, Mie scattering).

2.3.3 Examples of Transmission Spectra

Figure 2.34 illustrates the transmission of 10 mm thick plates consisting of fused silica with different water contents. They represent typical glasses with a good UV-transmission. Fused silica that has been produced synthetically is free of impurities apart from the OH content which causes absorption bands in the infrared region. Using a special production method the OH content can also be reduced significantly.

In Fig. 2.35 the spectral transmission of three types of optical glass with different refractive indices is shown. The glass with a lower refractive index achieves the better UV-transmission. High index glasses containing heavier elements have UV-absorption at longer wavelengths. The UV-absorption edge of optical glasses being 5 mm thick is to be found between 280 and 390 nm. Lenses for cameras are up to 20 mm thick. Especially the high index glasses can show a weak absorption in the blue region. Pieces that are 20 mm thick may have a yellowish colour. As described above, there are two reasons responsible for this loss: the tail of the short wavelength absorption, caused by the electronic transitions of heavy elements in the glass and/or impurities like iron (Fe^{3+}), platinum, titanum or cerium.

If the glass is to have a good UV transparency, it must contain neither heavy elements nor any of the impurities mentioned. In Fig. 2.36 the transmission of

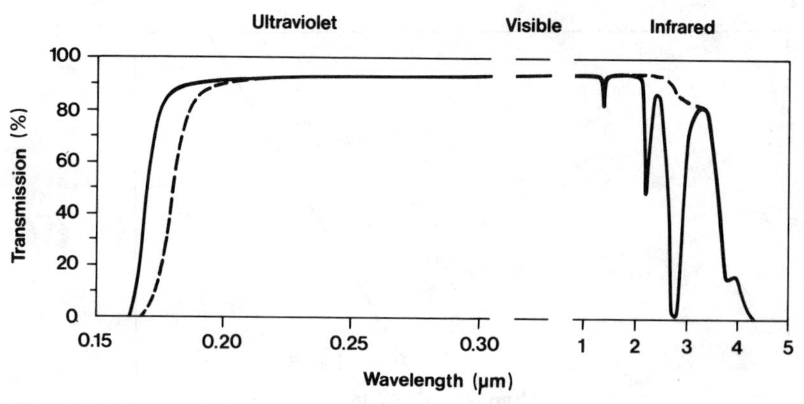

Fig. 2.34. Transmission of fused silica; thickness 10 mm (— OH containing, - - - with a low OH content) [2.112]

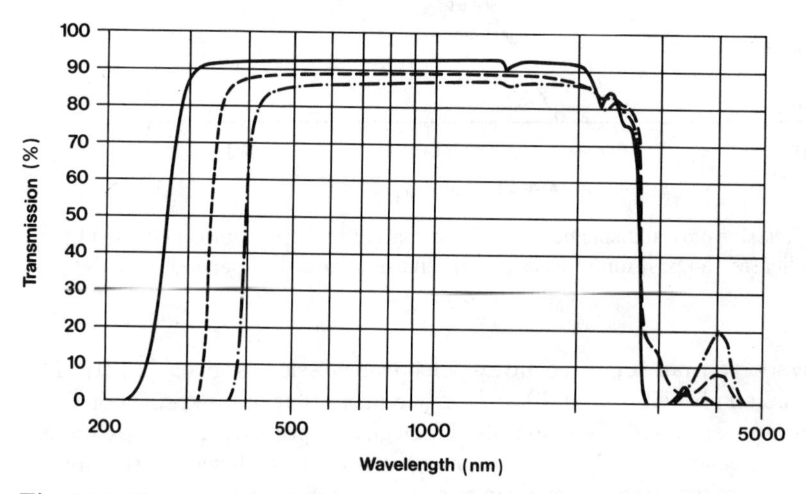

Fig. 2.35. Transmission of three types of optical glass; thickness 5 mm (— FK5, - - - SF2, - · - · SF11) [2.118]

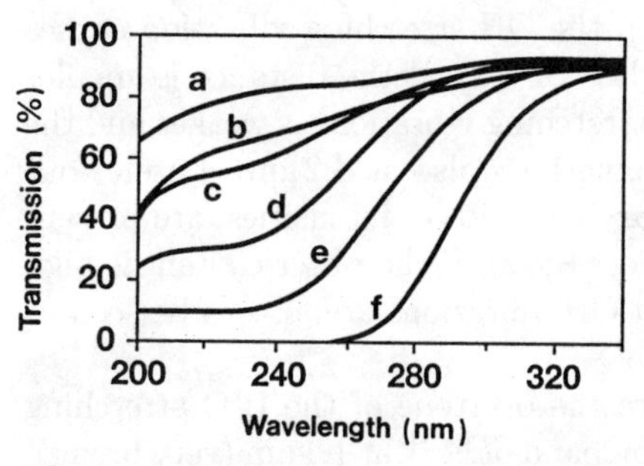

Fig. 2.36. Transmission of optical glasses with improved ultraviolet transmission; thickness 5 mm a) Ultran® 10, b) Ultran® 30, c) Ultran® 20 (preliminary values, see Sect. 8.5), d) Ultran® 15, e) Ultran® 25, f) UBK-7 [2.121]

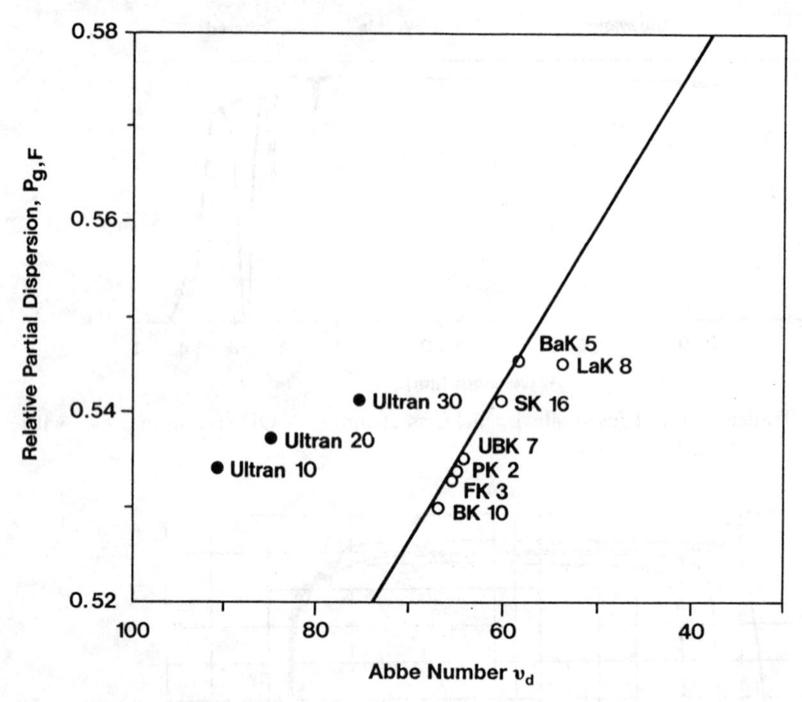

Fig. 2.37. Relative partial dispersion of UV transparent and other optical glasses [2.121]. The line indicates the position of glasses with "normal partial dispersion"

optical glasses with an improved ultra-violet transmission is given [2.121]. The UV-transmission of the types a), b), c) nearly reaches that of fused silica or CaF_2 crystal which are standard UV materials. These glass types have also an extremely anomalous dispersion (see Fig. 2.37), a low temperature coefficient of the optical pathlength and a low stress optical constant (see also Sect. 8.3).

The impurity most crucial for the infrared transmission of glasses is the OH content. In Fig. 2.35 typical OH-absorption bands are to be seen. The OH absorption is sensitive to the atomic surroundings [2.122]. If the distance between the proton and another oxygen is great (≥ 3 Å), the OH stretching vibration causes an absorption band at 2.9 µm (Fig. 2.35, glass SF 11). If the distance is smaller (about 2.5 Å), the force constant of the OH stretching vibrations is weaker and the absorption bands occur at about 3.6 µm (sometimes also at 4.2 µm). Often both types of OH bands (2.9 and 3.6 µm) with varying ratios of intensities, are present at the same time in a glass (see Fig. 2.35, glass SF 2). If the water content is high and/or the sample is thick, the overtones of OH vibrations are also to be seen at 2.2 and 1.4 µm.

Other sources of infrared absorption are the overtone of the B-O stretching vibration at 3.7 µm and the electron transition band of Fe^{2+} at 1.1 µm (very broad).

One possibility to improve the infrared transmission of glasses is the reduction of water content as well as the minimization of the Fe content. A further step is

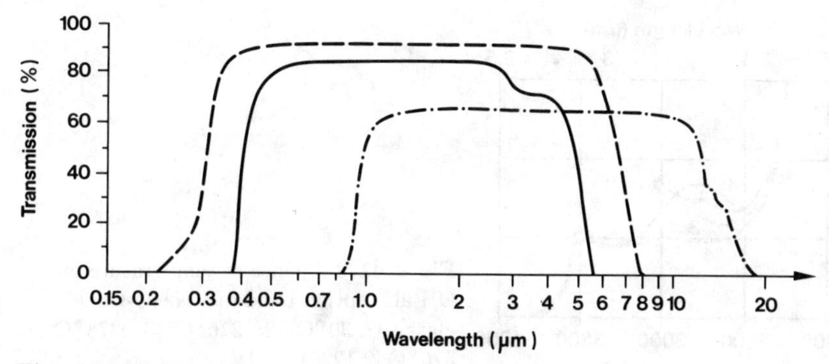

Fig. 2.38. Three types of infrared glasses; thickness 5 mm (— germanium oxide, - - - fluoride, – · – · chalcogenide)

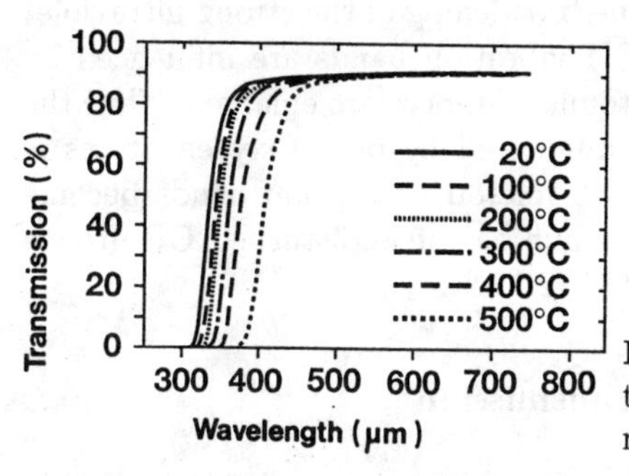

Fig. 2.39. Influence of temperature on the ultraviolet transmission of flint glass F2; thickness 10 mm [2.118]

the selection of glass formers with lower molecular vibration frequencies e.g. the oxides of Ge, As, Te, Bi, Pb, Ga, the fluorides of Zr, Y, Hf, the chalcogenides of As, Ge, Sb, Ga. In Fig. 2.38, a selection of typical infrared glasses is given (for further information see Chap. 8.4). It is possible to produce glasses with a transmission up to 35 μm (not yet commercially available).

2.3.4 Influence of Temperature

If the temperature of a glass is increased, the ultraviolet absorption edge is shifted to longer wavelengths (Fig. 2.39) and the infrared absorption edge to smaller

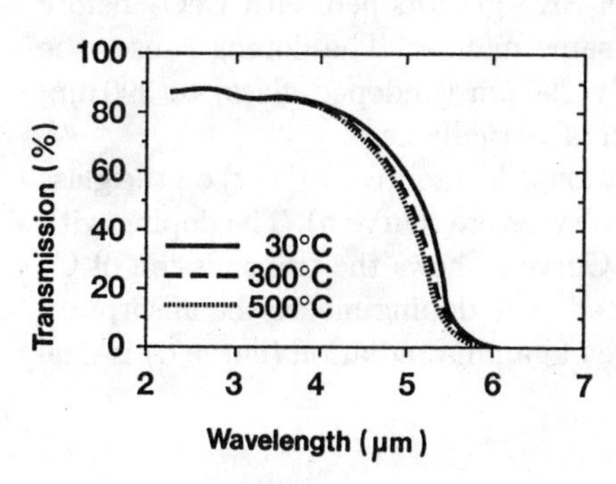

Fig. 2.40. Influence of temperature on the infrared transmission of the glass IRG 2; thickness 5 mm [2.118]

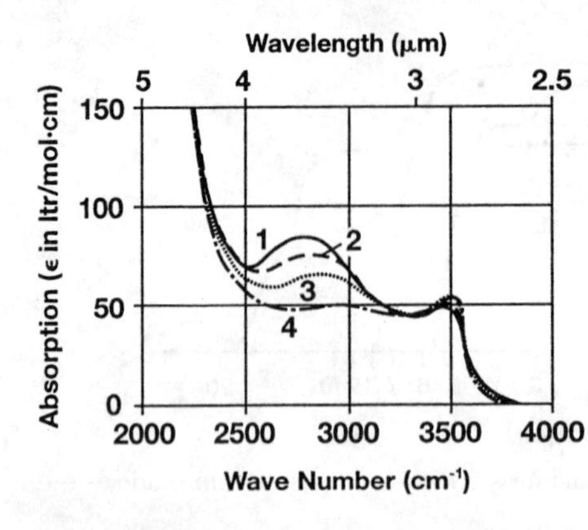

Fig. 2.41. Influence of temperature on the OH-absorption band in a Na_2O-CaO-SiO_2-glass. 1: 30 °C, 2: 276 °C, 3: 378 °C, 4: 576 °C [2.122] part IV

wavelengths (Fig. 2.40). This results from the broadening of the strong ultraviolet and infrared absorption bands. Also the OH vibration bands are influenced by temperature. At higher temperatures the atomic distances are enlarged. Thus the number of OH groups whose protons are influenced by other oxygen atoms is diminished. That means that the 3.6 and 4.2 micron absorption bands become less intense at higher temperatures and the number of undisturbed OH-groups with absorption at 2.9 µm increases (Fig. 2.41 [2.122]).

2.3.5 Influence of Radiation on the Transmission

Electromagnetic radiation may influence the transmission of optical glasses. Highly intensive visible light radiation can reduce the transmission especially in the blue region. The irradiation with electromagnetic radiation with shorter wavelengths generally has a greater influence: ultraviolet radiation, X-rays, gamma rays and particle beams (electrons, protons or other ions and neutrons). The irradiation can generate defect centres of different nature: ionization, trapped electrons, trapped holes, ruptured Si-O bonds, non-bridging oxygens [2.123]. Impurities in the glass can also be influenced by the radiation and therefore induce absorption (for instance by changing the valency). In Fig. 2.42 left, the transmission of SK 5 glass is shown before and after irradiation with ultraviolet light (200–400 nm) of a xenon lamp, whose power density is $1.2\,W/cm^2$ [2.124]. Polyvalent ions in glass, especially cerium, are able to diminish or to prevent the discolouration by irradiation. Figure 2.42 right shows the transmission of SK5 glass doped with CeO_2 before and after irradiation with UV light in the same manner. The doping causes the absorption edge to shift from approximately 290 nm (undoped glass) to 390 nm. After that the influence of the UV-radiation is virtually zero.

In Fig. 2.43 the influence of gamma radiation (^{60}Co source, 10^4 Gy) on the glass BK 7 is illustrated [2.125]. The darkening is very severe (curve b). The doping with CeO_2 protects the glass against darkening. Curve c shows the transmission of Ce doped BK 7-G 25 that has not been irradiated. The doping makes the absorption edge shift from 310 to 370 nm. The influence of gamma radiation (curve d) is now

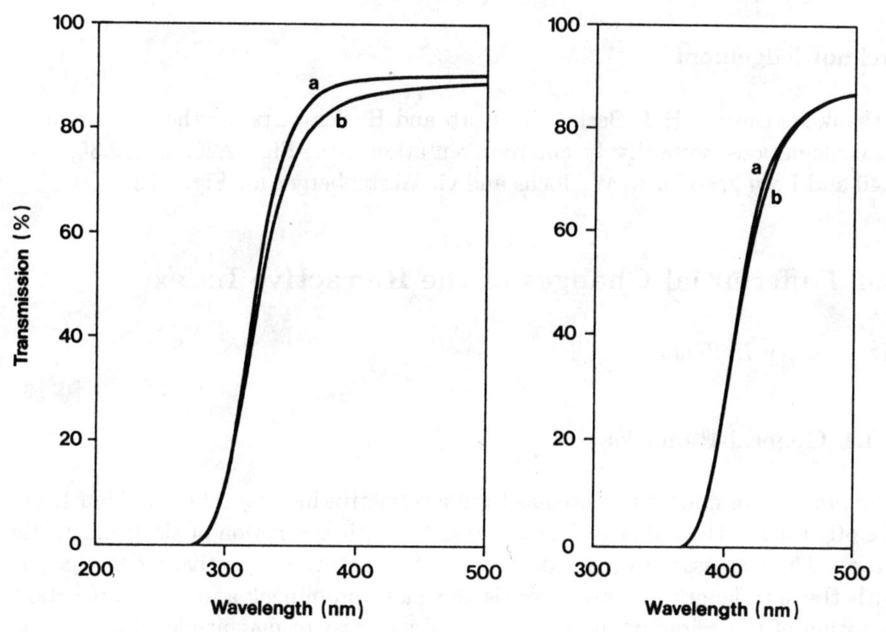

Fig. 2.42. Influence of UV radiation of a xenon lamp (200–400 nm; 1.2 W/cm^2) on the transmission of left) SK5 glass, right) SK5-G6 glass, protected by CeO_2: a) unsolarized, b) solarized [2.124]; sample thickness 10 mm

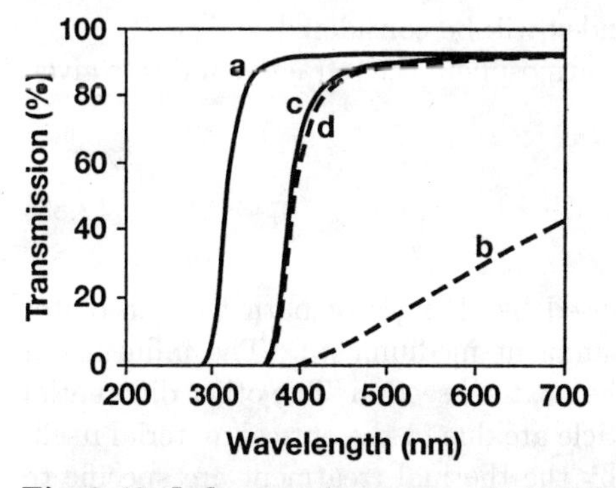

Fig. 2.43. Influence of gamma radiation (source ^{60}Co, 10^4 Gy) on the spectral transmission of BK 7 glass, sample thickness 10 mm, without Ce doping: a) not irradiated; b) irradiated; with Ce doping (BK 7-G25); c) not irradiated; d) irradiated [2.125]

very small. A CeO_2 doped BK 7 glass shows similarly small changes in transmission after having been irradiated with proton or electron beams. A stronger darkening is caused by neutron irradiation [2.125]. More details are given in the literature [2.124–133].

Acknowledgement

I thank R. Haspel, H.J. Becker, T. Korb and H. Schwartz for the measurement and calculations, respectively, and representation of the Figs. 2.26, 2.32, 2.35, 2.39, 2.40 and I am grateful to W. Jochs and G. Westenberger for Fig. 2.33.

2.4 Differential Changes of the Refractive Index

Hans-Jürgen Hoffmann

2.4.1 General Remarks

The presence of matter is prerequisite to a refractive index n different from 1. On the other hand, the use of matter is correlated with absorption of electromagnetic waves. This necessarily causes dispersion of n, i.e., the refractive index changes with the wavelength λ. Dispersion is the most prominent and most important variation of the refractive index because of its order of magnitude which can be as large as 0.1 over the visible spectral region. It is the main reason for the rich variety of optical glasses used to achieve a good quality of achromatic imaging which has been discussed in Sect. 2.1. In the following, other causes for minor — differential — changes of the refractive index will be considered.

Since for applications of glasses in the atmosphere the refractive index is given relative to air by

$$n_{\mathrm{rel}}(\lambda) = \frac{n_{\mathrm{abs}}(\lambda)}{n_{\mathrm{air}}(\lambda)} \tag{2.88}$$

differential changes of $n_{\mathrm{rel}}(\lambda)$ can be caused by changes of both the (absolute) index of the material, n_{abs}, and of the ambient medium, n_{air}. The influence of possible changes of n_{air} is considered in the next subsection. The other differential changes to be discussed in the present article are due to the optical material itself. Changes of the refractive index caused by the thermal treatment are specific to glasses (we do not consider changes due to small variations of the chemical compositions, to different raw materials or accidental impurities or to diverse melting parameters). Knowledge of the modifications induced by thermal treatment of glasses is important to achieve reproducible properties of optical elements. In order to obtain homogeneous materials, all parts of a glass sample have to undergo the same temperature–time programme.

Another homogeneous change of n can be caused by a shift of the ambient temperature. If the optical elements are to be used in an environment with changing temperature, the refractive index shifts because of both a shift of the intrinsic material properties and of the refractive index of the ambient medium. To predict this shift, the temperature coefficient of n or thermo-optical coefficient as a function of λ and of the temperature T has to be known.

Some external intensive parameters ("forces") cause anisotropies in a glass. Since optical glasses are (statistically) isotropic materials, the anisotropic behaviour can be characterized by a minimum number of independent tensor parameters. Stress-optical effects are well known in this respect. Stress changes n for the different planes of vibration of linearely polarized electromagnetic waves. Thus, in order to keep the good homogeneity and isotropy of optical glasses, it is mandatory to avoid mechanical stress in optical elements.

Electric fields, too, can cause anisotropy in glasses for linearly polarized electromagnetic waves. In order to effect appreciable changes, however, large electric field strengths are required. Under normal circumstances this hardly occurs. Therefore, this effect is of minor importance.

Anisotropies for circularly polarized electromagnetic waves are caused by magnetic fields. The difference between the changes for right- and left-handed circularly polarized electromagnetic waves can be detected by investigations of the Faraday rotation. This effect can be used for the modulation of light beams and for non-reciprocal optical elements.

The contributions of the phenomena mentioned so far will be considered in the following.

2.4.2 The Refractive Index of Air and Its Variations

The absolute refractive index of air has been the subject of many studies in the past. Its value and its dispersion are relevant for imaging optics in the atmospheric environment (see, [2.134–139] and references cited therein). Since the refractive index of air, n_{air}, can be determined very precisely by interferometric methods with an accuracy in the order of 10^{-7} or better, whereas the accuracy for the refractive index of glass is usually limited to the range of 10^{-5} to 10^{-6}, many dispersion formulae for n_{air} are useful. In the following we refer to the relationship

$$n_{air}(15\ °C, p_0) =$$

$$1 + 10^{-8}\left(6432.8 + \frac{2\,949\,810\ \mu m^{-2}\cdot\lambda^2}{(146\ \mu m^{-2}\cdot\lambda^2 - 1)} + \frac{25\,540\ \mu m^{-2}\cdot\lambda^2}{(41\ \mu m^{-2}\cdot\lambda^2 - 1)}\right) \tag{2.89}$$

which was accepted by the Joint Commission for Spectroscopy in Rome in September 1952 as cited in [2.134]. Relation (2.89) applies in the range of wavelengths between 0.2 μm and 1.35 μm for dry air containing 0.03 % CO_2 by volume at a temperature of 15 °C and a pressure $p_0 = 0.101325 \times 10^6$ Pa (corresponding to 760 Torr). Curve a in Fig. 2.44 shows n_{air} as a function of the wavelength for these parameters.

Its value is roughly 1.0003 in the visible part of the spectrum. Consequently, the contribution of n_{air} to n_{rel} of any glass cannot be neglected if high precision is required. The refractive index of air changes with the temperature, T_c (in degrees centigrade), the atmospheric pressure, p, and the partial pressure of additional gaseous components, of which for example the partial pressure w of water vapour is important. The changes due to T_c, p and w can be described according to [2.134]

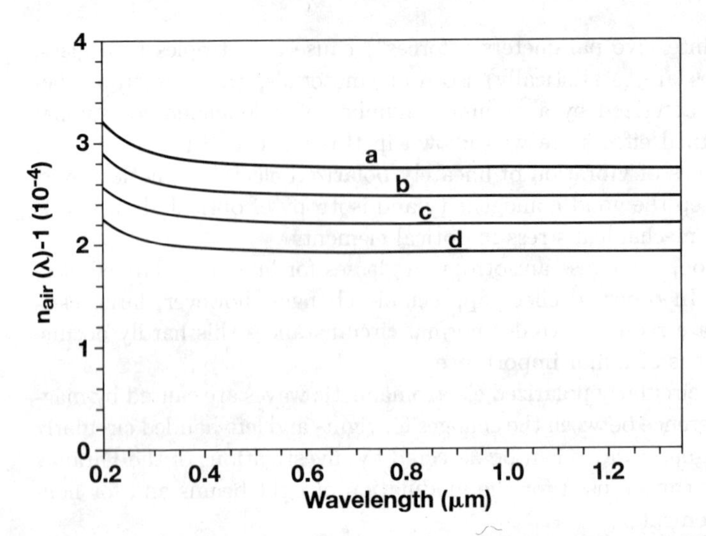

Fig. 2.44. $n(\lambda) - 1$ (n: refractive index) of dry air containing 0.03% CO_2 as a function of the wavelength according to (2.89) and (2.90) at 20 °C. a) $p_0 = 0.101325 \times 10^6$ Pa, b) 0.9 p_0, c) 0.8 p_0, d) 0.7 p_0

by

$$n_{\mathrm{air}} = 1 + \frac{n_{\mathrm{air}}(15\,°\mathrm{C}, p_0) - 1}{1 + \frac{\alpha}{1+15\,°\mathrm{C}\cdot\alpha}(T_c - 15\,°\mathrm{C})} \cdot \frac{p}{p_0} - \frac{413 \times 10^{-12}\,\mathrm{Pa}^{-1} \cdot w}{1 + \alpha T_c} \tag{2.90}$$

with $\alpha = 3.67 \times 10^{-3}/°\mathrm{C}$. The last term in (2.90) applies only to the visible spectral region [Ref. 2.134, p. 408].

If there are other constituents i in the ambient atmosphere, their contribution to the refractive index of the surrounding medium can be taken into account by the relation

$$\Delta n_i(T, p) = n_i(T, p) - 1 = \frac{n_i(T_0, p_0) - 1}{\rho_i(T_0, p_0)} \rho_i(T, p) \tag{2.91}$$

where $\rho_i(T_0, p_0)$ and $\rho_i(T, p)$ are the mass densities of additional constituents at the reference partial pressure p_0 and the temperature T_0, and at the partial pressure p and temperature T, respectively. Equation (2.91) is based on the dispersion formula of *Gladstone-Dale* [2.140]

$$n - 1 = \mathrm{const.} \times \rho \tag{2.92}$$

which was developed to describe the refractive index of liquids. It applies also to less dense materials, such as gases. However, there are no general a priori rules for the constant of proportionality or its dispersion. Therefore, the respective data have to be determined experimentally or taken from the literature.

Any variation of the composition of the ambient atmosphere, its temperature or the atmospheric pressure changes not only the relative refractive index of the optical glass but also its dispersion to a small extent. Fluctuations of the atmospheric pressure by about 2% correspond to changes of n_{rel} by (5–6) $\times 10^{-6} n_{\text{rel}}$.

Other variations may be due to changes of the height, h, above sea level. The atmospheric pressure of each component of the air decreases – in the isothermal case and considered as an ideal gas – according to the barometric pressure equation

$$p_i(h) = p_i(0) \exp\left(-\frac{M_i g}{k_B T} h\right) \tag{2.93}$$

(for a derivation of this formula see, e.g. [2.141]; M_i, mass of a molecule of kind i; g, acceleration of gravity; k_B Boltzmann's constant; T, absolute temperature). In addition, the temperature and the relative composition of the atmosphere change. Neglecting these changes, a moderate shift of 300 m in the position above sea level causes a variation of about 10^{-5} of n_{air} and — as a consequence — a shift of the relative refractive index of optical glasses by $10^{-5} \times n_{\text{rel}}$.

2.4.3 Thermal Treatment of Glasses and the Refractive Index

The stress relief as a function of time, t, has been described by a linear and a quadratic relaxation model. In a linear model one assumes that the rate of stress relief is proportional to the stress σ, which is left, namely

$$\frac{d\sigma}{dt} = -\frac{\sigma(t)}{\tau(T)} \tag{2.94}$$

with time variable t and the relaxation time constant τ, which depends on the temperature, T. If τ does not depend on the stress, the solution of (2.94) is

$$\sigma(t) = \sigma_0 \, \exp\left(-\frac{t}{\tau}\right) \tag{2.95}$$

with the initial value of the stress σ_0.

In a quadratic model one assumes for the rate of stress relief [2.142]

$$\frac{d\sigma}{dt} = -a\,\sigma^2 \tag{2.96}$$

with the constant of proportionality a. The solution of (2.96) is

$$\sigma(t) = \frac{\sigma_0}{1 + \sigma_0 \, a \, t} \tag{2.97}$$

if $\sigma = \sigma_0$ for $t = 0$.

Both equations (2.95) and (2.97) are approximations to the relaxations of the stresses as pointed out by *Berger* [2.143].

Stresses in glass relax within about 15 minutes if the viscosity is 10^{12} dPas [2.144]. The viscosity reaches this value in many cases at about 20 to 80 °C above the transformation temperature T_g. Technical glasses that are 1 cm or less thick are annealed from those temperatures with constant rates in the order of 1 °C/min.

This is sufficient for the stress relief in technical glasses. On the other hand, even small stresses due to thermal gradients in the annealing range cause permanent strains and inhomogeneities of the refractive index. This can be detrimental to optical glasses of high quality. Thus, optical glasses have to be annealed very carefully in order to provide the good optical homogeneity required for transmissive optical elements. The effects of precision annealing on the homogeneity of optical glasses have been investigated by *Reitmayer* and *Schuster* [2.145]. They succeeded in reducing the inhomogeneity of the refractive index to values as low as $\pm 3 \times 10^{-7}$ in a BK 7® disk (diameter: 400 mm, thickness: 47 mm).

Even if there are no inhomogeneities due to stresses, there still exist driving forces for the glass to relax. The reason for this is that the glass reaches a non-equilibrium state in the process of solidification from the melt, since the minimum of Gibbs' free energy is for solids obviously in the ordered (i.e. crystalline) state. Glasses are solids without long-range order. Therefore, there are driving forces on an atomic scale to reach a state of advanced order. As a consequence, glasses tend to relax to a state with less Gibbs' free energy. This relaxation is impeded by kinetic parameters which depend on the composition of the glass. Since the relaxation requires structural rearrangements of the constituents, such effects are predominantly induced within the temperature range near T_g (but they can also occur at much lower temperatures, as will be shown later). The fact that the specific volume of a glass depends on the annealing rate near T_g (see textbooks on glass or ceramics such as [2.146, 147]) is very well known. Corresponding to the variation of the specific volume, many other parameters of a glass depend more or less on the state of relaxation. Among these parameters are the mass density, the coefficient of linear thermal expansion, the refractive index, the hardness, the transformation temperature, thermodynamic data as the specific heats, etc. *Berger* [2.143] gives a survey of the early work on the dependence of physical parameters of glass as a function of the respective thermal history.

There are numerous experimental investigations of the changes of parameters by thermal treatment in the transformation range. The following is an arbitrary selection of publications:

Brandt has determined the refractive index of a borosilicate crown glass as a function of the annealing time at a given temperature in order to derive annealing schedules for predetermined index homogeneity [2.148].

Lillie and *Ritland* have studied the influence of different thermal treatments on a borosilicate glass [2.149]. In particular they applied the methods of "constant-temperature soaking" and "constant cooling rate". They recommended the method of constant cooling rate, because this procedure insures a uniform thermal history of the glass and it requires less total time. Each part of a glass sample experiences to a very good approximation the same temperature-time programme if one starts the annealing at a constant (slow) rate sufficiently above T_g. The temperature differences have to be as small as possible in order to avoid thermal stresses during and after the annealing process. For an infinite plate of thickness d which is annealed at a constant rate $v = dT/dt$, *Lillie* and *Ritland* published the relation [2.149]

$$\Delta T = \frac{v}{8\,C} d^2 \tag{2.98}$$

for the maximum difference in temperature between the centre and the surface of the plate with the thermal diffusivity $C = k/\rho\, c_p$, where k is the coefficient of thermal conductivity, ρ is the mass density and c_p is the specific heat capacity at constant pressure. Inserting typical values for optical glasses yields, with $d = 1$ cm and $v = 1\,°\text{C/h}$, temperature differences ΔT in the order of 0.01 °C and 1 °C if v is 100 times as fast.

Stresses σ caused by temperature differences ΔT can be estimated using [2.144]

$$\sigma = \frac{\alpha\, E}{1 - \mu} \Delta T \tag{2.99}$$

where the coefficient of linear thermal expansion is α, Young's modulus is E and Poisson's ratio is μ. Inserting $\alpha = 10^{-5}/°\text{C}$, $E = 7.5 \times 10^4\,\text{N/mm}^2$, $\mu = 0.25$ and $\Delta T = 0.01$ °C yields stresses of about $10^{-2}\,\text{N/mm}^2$. (In this respect it is worth mentioning that for large pieces, the weight of the glass itself can cause even larger stresses). Stresses of that order of magnitude, however, do not cause the large differences of the refractive index observed upon varying the annealing rates.

From a practical point of view it is important to predict the changes of the refractive index, e.g. with the parameters of the thermal treatment in the transformation range. For this purpose many correlations between experimental quantities have been investigated.

Tool, *Tilton* and *Saunders* [2.150] and *Spinner* and *Waxler* [2.151] have determined the change of the mass density, ρ, and of the refractive index, n, of various optical glasses induced by annealing and by hydrostatic pressure. According to their results, the increase of n was larger for a given increase of ρ upon annealing than for the same increase of ρ upon compression. Thus, the change of n upon annealing is not caused by a variation of the density alone. Instead one must consider that some restructuring also takes place during annealing.

Prod'homme has determined the temperature coefficient of the refractive index, n, the mass density, ρ, and the coefficient of linear thermal expansion, α, of various optical glasses and their modifications caused by thermal treatment near T_g [2.152]. In his comprehensive work he reported on measurements of n and ρ as a function of time at different temperatures and discussed several possibilities to interpret the relaxation. In particular, the relaxation was correlated with the change of the viscosity as a function of time. The increase of the viscosity with time had been reported earlier by *Lillie* [2.153, 154].

To describe the relaxation behaviour of glasses near T_g, *Tool* has introduced the concept of the *fictive temperature* [2.155]. Some limitations of this concept have been discussed by *Ritland* [2.156]. More recent models also take advantage of the concept of the fictive temperature (see [2.157–159]). As these models unfortunately require too many parameters they are in most cases not useful for predictions.

Since the refractive index n depends on the parameters of the thermal treatment, such as temperature, duration and annealing rate, one would like to profit from this effect to shift and correct n by small amounts, as can be necessary to

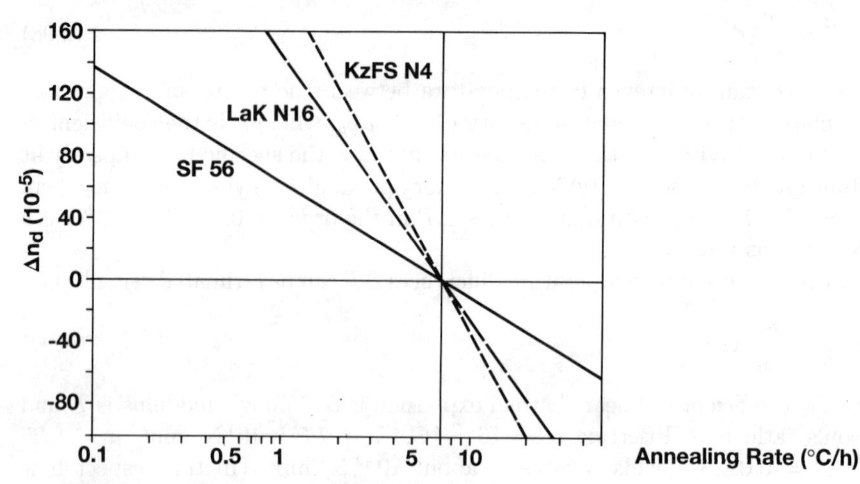

Fig. 2.45. Change of the refractive index n_d ($\lambda_d = 587.56$ nm) as a function of the annealing rate v. Reference annealing rate: $v_1 = 7$ °C/h; data from [2.164]

meet the requirements in production. One thus needs accurate prescriptive data on how much one has to change the annealing parameters in order to reproduce a given value of n. For the special case of a constant annealing rate v this can be achieved. *Lillie* and *Ritland* published for a special glass the relation [2.149]

$$\Delta n_d(v/v_1) = m_d \log\left(\frac{v}{v_1}\right) \tag{2.100}$$

with the reference annealing rate v_1 and the constant of proportionality m_d for the d-line. Relation (2.100) has been applied to other glasses by *Rötger* and *Besen* [2.160, 161] and *Danyushevskii* ([2.162] as cited in [2.160, 161, 163]).

Figure 2.45 shows for an arbitrary selection of optical glasses the variation of Δn_d with the (uniform) annealing rate, v.

For a standard annealing rate v_2 different from v_1 the corresponding curves in Fig. 2.45 are just shifted by $-m_d \log(v_2/v_1)$, since one obtains from (2.100)

$$\Delta n_d(\frac{v}{v_2}) = \Delta n(\frac{v}{v_1}\cdot\frac{v_1}{v_2}) = m_d \log(\frac{v}{v_1}\cdot\frac{v_1}{v_2})$$

$$= m_d \log(\frac{v}{v_1}) - m_d \log(\frac{v_2}{v_1}) \, . \tag{2.101}$$

In addition to n_d one can also adjust the Abbe number ν_d of the optical glass by an appropriate choice of the annealing rate v [2.161, 164]

$$\Delta\nu_d(\frac{v}{v_1}) = m_{\nu_d} \log(\frac{v}{v_1}) \tag{2.102}$$

which is similar to (2.100) (see Fig. 2.46). However, it is not possible to adjust n_d and ν_d independently by variation of the annealing rate v.

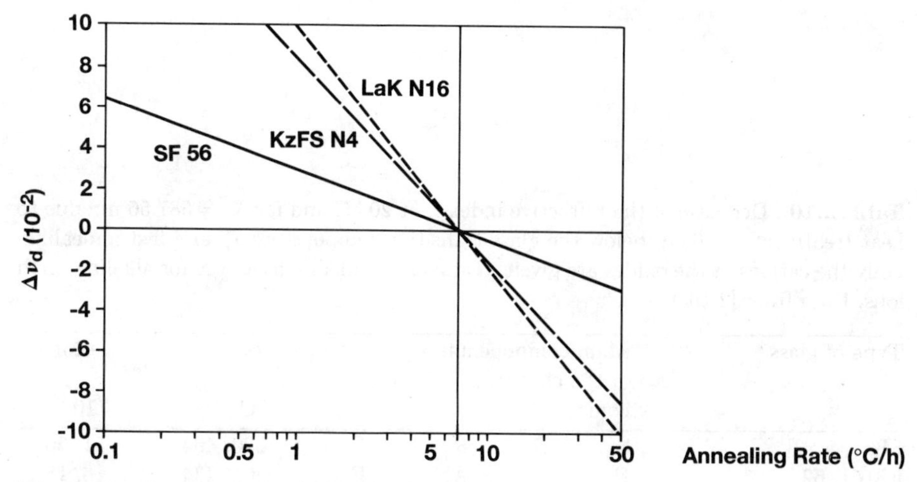

Fig. 2.46. Change of the Abbe number ν_d (λ_d = 587.56 nm) as a function of the annealing rate v. Reference annealing rate: $v_1 = 7$ °C/h;, data from [2.164]

Equations (2.100) and (2.102) cannot be applied to the limits $v \to 0$ and $v \to \infty$, since both relations diverge in both cases. In principle one can remove these divergences by suitable factors. These limits, however, are not relevant in practice; for $v \to 0$ one needs impractically slow annealing rates and for $v \to \infty$ the temperature gradients in the glass samples are too large.

It has to be emphasized that n_d and ν_d can be changed with the annealing rate according to (2.100) and (2.102) if the annealing starts above the transformation temperature T_g. However, glasses can relax also at temperatures far below T_g. Therefore, variations of the refractive indices must be taken into account, if the glasses are subjected to high temperatures with subsequent fast annealing during the manufacturing process of optical elements. Such components may be subjected, for example during coating processes, to temperatures well above 300 °C, which is within 100 °C of the T_g of some types of glasses. Very often the components are fast annealed from the maximum process temperature. In this case small changes of the refractive index may occur. Table 2.10 taken from Ref. [2.165] illustrates the changes Δn_d of the refractive index that occur if the optical component is fast annealed from 100 °C below T_g. In all cases, n_d decreases due to fast annealing. The main cationic components are also included in that table. It is worth noting that glasses containing large amounts of TiO_2 show a rather large decrease of n_d upon fast annealing. The fast annealing process may also change the dispersion as can be seen from Fig. 2.47 [2.165].

In order to avoid detrimental changes of the refractive index and its dispersion, it is mandatory to reduce the maximum process temperature and the annealing rates.

Table 2.10. Decrease of the refractive index n at 20 °C and for $\lambda_d = 587.56$ nm due to heat treatment at 100 K below the glass transition temperature T_g and fast annealing. Only the cations of the oxides are given. R stands for alkali ions and E for alkaline earth ions. Data from [2.165]

Type of glass	Main components of the chemical composition				T_g °C	$-\Delta n_d$ (10^{-6})
FK 1, 5	Si	B	R	F	390, 464	51, 56
FK 51, 52	P	E	Al	F	405, 434	16, 18
PK 2	Si	B	R		568	41
PK 50	P	Al	R		496	24
PSK 3	Si	B	Ba		602	51
BK 6, 7	Si	B	R		537, 559	41, 48
K 5, 7, 10, 50	Si	R			459–553	43–71
ZK 1, N7	Si	R	Zn		528–562	40, 18
BaK 4	Si	Ba	Zn		552	52
SK 16, 52	Si	B	E		638, 624	20, 4
KF 1, 9	Si	Pb	R		478, 445	58, 39
BaLF 4	Si	Ba	Zn		569	48
SSK 52	Si	Ba	B		639	18
LLF 1	Si	Pb	R		448	61
BaF N10, 53	Si	Ba			630	14, 22
LF 2	Si	Pb	R		440	43
F 2, 9	Si	Pb			432, 468	36
F N11	Si	Ti			572	64
BaSF 1, 50	Si	Pb	E		493, 500	1–9
LaSF N9	Si	Ba	La	Ti	698	54
SF 6, 13, 15, 58, 59	Si	Pb			362–472	6–39
SF L6, L56, N64	Si	Ti	Ba		578–585	120–134
TiF 2, 3, N5	Si	Ti	R(F)		419–472	135–201
TiF 6	P	Ti	R		410	96
TiSF 1	Si	Ti	K		495	230
TiK 1	Si	B	Al	R	340	55
KzFS 1, N4 N7	B	Pb	Al		472–512	13–16
LgSK 2	As	B	Al	F	512	10

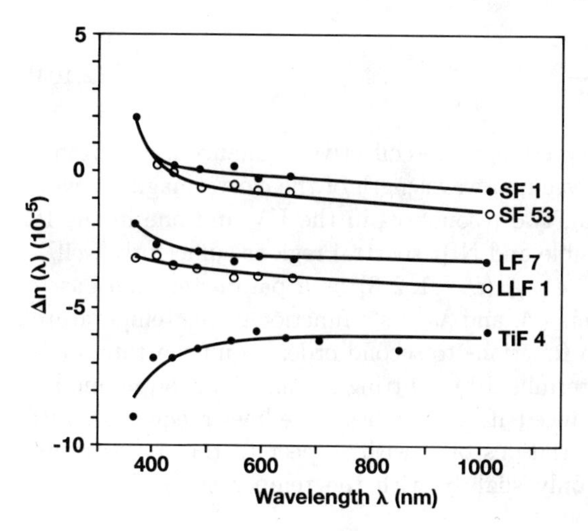

Fig. 2.47. Changes of the refractive indices Δn as a function of wavelength after thermal cycling of different optical glasses (heating to 250 °C for 30 minutes and subsequent fast annealing). Data from [2.165]

2.4.4 The Thermo-Optical Coefficient

The refractive indices of optical glasses and their dispersion, as well as the parameters of the respective dispersion equations in the catalogues, refer to a fixed temperature (usually 20 °C, see [2.164]). If the optical elements are used at various ambient temperatures, it is necessary to know how much the refractive index and its dispersion changes with the temperature. Such variations are described very often by the average thermo-optic coefficients

$$\frac{n(\lambda_k, T_{j+1}) - n(\lambda_k, T_j)}{T_{j+1} - T_j} = \frac{\Delta n(\lambda_k, T_{j,j+1})}{\Delta T_{j,j+1}} \tag{2.103}$$

which are determined from the refractive indices measured at different temperatures T_j for a given set of wavelengths λ_k or directly from the incremental or decremental differences $\Delta n(\lambda_k, T_{j,j+1})$ according to the method of *Pulfrich* [2.166]. Data of many optical glasses are compiled in the catalogues for the spectral lines $h = 404.66$ nm, $g = 435.83$ nm, $F' = 479.99$ nm, $e = 546.07$ nm, $d = 587.56$ nm, $C' = 643.85$ nm, $s = 852.11$ nm, $t = 1013.98$ nm and 1060 nm and for steps of typically 20 °C in temperature ranges from −40 °C to +80 °C and in some cases within even larger intervals [2.167–173].

Instead of data determined for a small number of discrete wavelengths and averaged over a given temperature interval, one prefers for practical applications a dispersion formula of the thermo-optical coefficient. In this respect it is reasonable to assume that the parameters of the dispersion formulae given in Sect. 2.1 depend on the temperature [2.174]. Among these dispersion formulae, the three-term Sellmeier equation provides a very good fit of n from the UV to NIR with an accuracy better than 10^{-5}:

$$n^2(\lambda) - 1 = a \sum_{i=1}^{3} \frac{N_i}{V} f_i \frac{\lambda^2 \lambda_{0i}^2}{\lambda^2 - \lambda_{0i}^2} \tag{2.104}$$

where N_i/V, f_i, λ_{0i} are the density, strength and effective resonance wavelength of the oscillators of kind i; λ is the vacuum wavelength of the electromagnetic wave and "a" is a constant. In general, two resonances in the UV and one in the IR describe the dispersion in the visible and NIR spectral regions sufficiently well.

With the abbreviation $A_i = a \frac{N_i}{V} f_i$ $(i = 1, 2, 3)$ as a parameter, one has to determine a total of six parameters (A_i and λ_{0i}) as a function of the temperature. The expansion of these unknown functions to second order requires a total of 18 parameters which have to be determined by a fitting routine. This large number of fitting parameters can be reduced if one applies a Sellmeier equation with only one term to describe the variations of n with respect to the temperature. This is justified since n varies only slightly with the temperature. Under this approximation one obtains [2.168]:

$$\frac{dn(\lambda, T)}{dT} = \frac{n^2(\lambda, T_0) - 1}{2n(\lambda, T_0)} \times \left(D_0 + 2D_1(T - T_0) + 3D_2(T - T_0)^2 + \frac{E_0 + 2E_1(T - T_0)}{\lambda^2 - \lambda_0^2} \right) . \tag{2.105}$$

This equation can be integrated in order to give the increments or decrements Δn of the refractive index as a function of both the temperature and the wavelength, with respect to a standard dispersion equation for the temperature T_0:

$$\Delta n(\lambda, T - T_0) = \frac{n^2(\lambda, T_0) - 1}{2n(\lambda, T_0)} \left(D_0(T - T_0) + D_1(T - T_0)^2 + D_2(T - T_0)^3 + \frac{E_0(T - T_0) + E_1(T - T_0)^2}{\lambda^2 - \lambda_0^2} \right) . \tag{2.106}$$

Both equations (2.105) and (2.106) can be simplified further, since the dependence of the prefactor on λ can be taken into account — at least to a good approximation — by slightly different fitting parameters yielding

$$\frac{dn(\lambda, T)}{dT} = D_{s0} + 2D_{s1}(T - T_0) + 3D_{s2}(T - T_0)^2 + \frac{E_{s0} + 2E_{s1}(T - T_0)}{\lambda^2 - \lambda_{s0}^2} \tag{2.107}$$

and

$$\Delta n(\lambda, T) = D_{s0}(T - T_0) + D_{s1}(T - T_0)^2 + D_{s2}(T - T_0)^3 + \frac{E_{s0}(T - T_0) + E_{s1}(T - T_0)^2}{\lambda^2 - \lambda_{s0}^2} \tag{2.108}$$

Equations (2.105) to (2.108) have been derived to describe the thermo-optic coefficient and the increments or decrements of the refractive index Δn_{abs} with respect

Table 2.11. Coefficients to be inserted into the dispersion formulae (2.105) and (2.106). Data from [2.167]

Type of glass	$D_0 \times 10^6$	$D_1 \times 10^8$	$D_2 \times 10^{11}$	$E_0 \times 10^7$	$E_1 \times 10^{10}$	λ_0 (μm)
FK 52	−20.9	−0.855	3.41	3.11	4.02	0.204
PK 3	1.08	1.29	−2.65	4.37	7.07	0.138
PK 51A	−19.8	−0.606	1.60	4.16	5.01	0.134
BK 7	1.81	1.20	−2.30	4.95	7.71	0.149
K 5	−0.413	1.03	−3.40	4.73	5.19	0.213
K 7	−1.67	0.880	−2.86	5.42	7.81	0.172
K 10	4.86	1.72	−3.02	3.82	4.53	0.260
BaK 4	2.90	1.33	−3.10	5.14	7.37	0.213
SK 15	0.669	1.24	−2.62	3.76	5.03	0.233
SK 16	−0.0951	1.13	−1.19	3.86	5.92	0.192
BaF 51	−0.130	0.891	−0.290	7.94	11.1	0.196
F 2	1.51	1.56	−2.78	9.34	10.4	0.250
BaSF 10	−0.433	1.24	−2.20	9.40	9.51	0.230
BaSF 64	1.60	1.02	−2.68	7.87	9.65	0.229
LaF 2	−4.09	0.904	−1.06	6.67	7.07	0.213
LaF 3	−2.35	1.07	−0.938	5.72	6.01	0.220
SF 3	3.72	1.74	−3.21	14.9	14.1	0.260
SF 6	7.04	1.93	−4.03	17.8	18.8	0.272
SF 14	8.85	1.81	−4.12	13.9	15.3	0.279
SF L56	−4.30	0.840	−2.28	10.0	15.2	0.287
SF 58	4.98	1.68	−4.44	23.7	23.9	0.280
SF N64	−2.05	1.03	−3.15	9.26	12.3	0.269

to vacuum. These quantities can without difficulty be recalculated for other ambient media of the glasses by use of the relationships:

$$n_{\text{abs}} = n_{\text{rel}} n_{\text{med} } \tag{2.109}$$

and

$$\frac{d}{dT} n_{\text{abs}} = n_{\text{med}} \frac{d}{dT} n_{\text{rel}} + n_{\text{rel}} \frac{d}{dT} n_{\text{med}} \tag{2.110}$$

where n_{med} and n_{rel} are the refractive indices of the ambient medium with respect to vacuum and of the glass with respect to the medium. For air as the ambient medium one can use (2.89) and (2.90) cited above.

Representative mean values of the coefficients to be inserted into the dispersion formulae (2.105) and (2.106) are given for a selection of optical glasses in Table 2.11. The fitting parameters of further types of optical glasses can be found in [2.167]. Figures 2.48–50 show the thermo-optic coefficients dn_{rel}/dT and dn_{abs}/dT and the increments Δn_{rel} for a special sample of the optical glass type LaSF N9. Comparing Figs. 2.48 and 2.49 one can see how much the thermo-optic coefficient can be changed when air is the ambient medium. In addition to quantitative modifications, even qualitative changes of the dispersion of the thermo-optic coefficient can be observed.

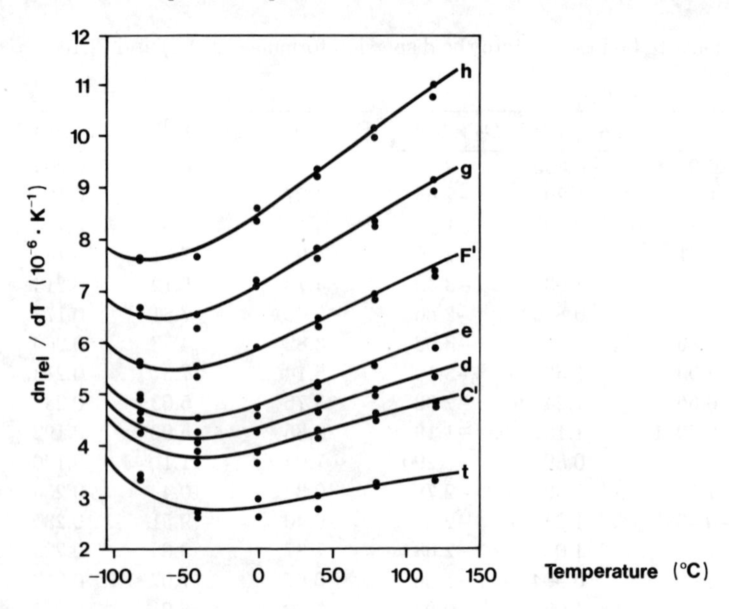

Fig. 2.48. Thermo-optic coefficient of the optical glass type LaSF N9 with respect to dry air as a function of temperature with the wavelength as a parameter. The full curves represent the results of calculations using (2.105), (2.109), (2.110), (2.89), and (2.90). Wavelengths: h = 404.66 nm, g = 435.83 nm, F' = 479.99 nm, e = 546.07 nm, d = 587.56 nm, C' = 643.85 nm, t = 1013.98 nm. $D_0 = -9.44 \times 10^{-7}/°\mathrm{C}$, $D_1 = 1.14 \times 10^{-8}/(°\mathrm{C})^2$, $D_2 = -1.87 \times 10^{-11}/(°\mathrm{C})^3$, $E_0 = 9.22 \times 10^{-7}/°\mathrm{C}$, $E_1 = 1.22 \times 10^{-9}/(°\mathrm{C})^2$, λ_0 = 255 nm; n (λ, 20 °C) $= 3.2994326 - 1.1680436 \times 10^{-2} \mu\mathrm{m}^{-2} \times \lambda^2 + 4.0133103 \times 10^{-2} \mu\mathrm{m}^2 \times \lambda^{-2} + 1.3263988 \times 10^{-3} \mu\mathrm{m}^4 \times \lambda^{-4} + 4.7438783 \times 10^{-4} \mu\mathrm{m}^6 \times \lambda^{-6} + 7.8507188 \times 10^{-5} \mu\mathrm{m}^8 \times \lambda^{-8}$

2.4.5 Photoelastic Properties of Glasses

In order to analyse stresses in optical glasses one has to know the stress-optical coefficient. Figure 2.51 shows a rectangular glass sample under external uniaxial stress σ (tensile stresses are assumed to be positive). The stress renders the glass anisotropic. Thus, the refractive index depends on the orientation of the plane of vibration if for example a linearly polarized electromagnetic wave penetrates the glass sample perpendicularly to the direction of the applied stress. For the plane of vibration parallel (index $\|$) or perpendicular (index $\perp$) to σ the refractive index n changes into

$$n_\| = n + \frac{dn_\|}{d\,\sigma}\sigma \tag{2.111}$$

and

$$n_\perp = n + \frac{dn_\perp}{d\,\sigma}\sigma \tag{2.112}$$

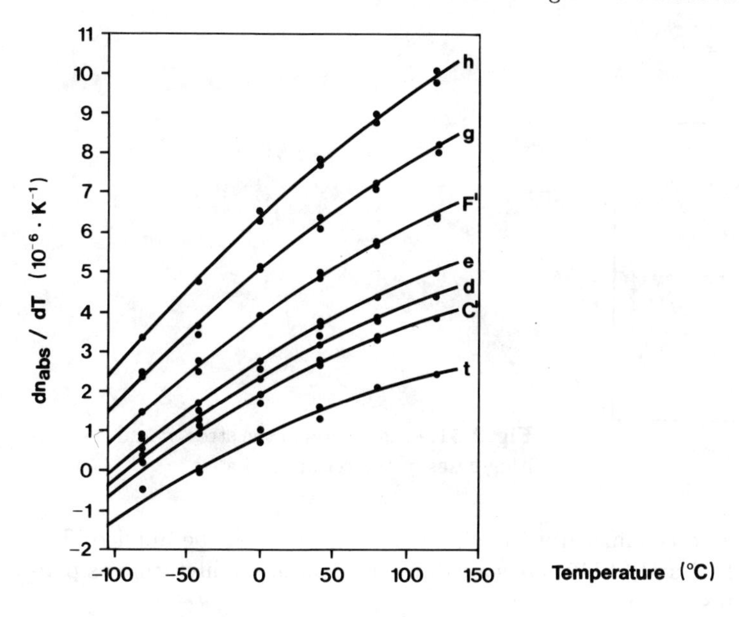

Fig. 2.49. Thermo-optic coefficient of the optical glass type LaSF N9 with respect to vacuum. The full curves represent the results of calculations using (2.105) with the same fitting parameters given in Fig. 2.48

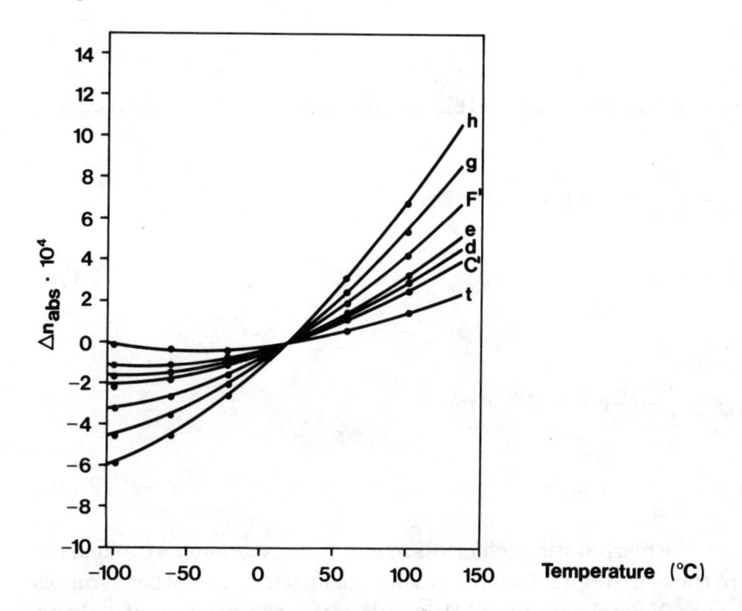

Fig. 2.50. Increments and decrements Δn_{abs} of the absolute refractive index of the optical glass type LaSF N9 with respect to $n(\lambda)$ at 20 °C as a function of temperature with the wavelength as a parameter. The full curves represent the results of calculations using (2.106) with the same fitting parameters given in Fig. 2.48

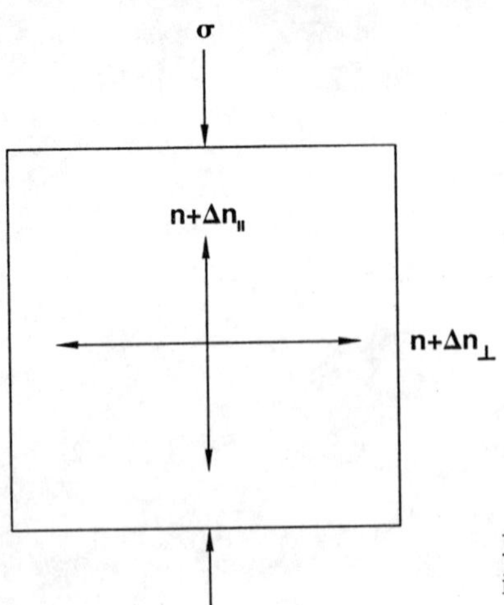

Fig. 2.51. Orientation of the stress σ and the limiting cases of the refractive index

as long as the stress is sufficiently small and Hooke's law can be applied. The derivatives in (2.111) and (2.112) are sometimes denoted in the literature as photoelastic coefficients [2.175]

$$K_{\parallel} = C_1 = \frac{dn_{\parallel}}{d\,\sigma} \tag{2.113}$$

and

$$K_{\perp} = C_2 = \frac{dn_{\perp}}{d\,\sigma}\ . \tag{2.114}$$

Other relationships (see Refs. [2.175–182] are also known for these derivatives, such as

$$\begin{aligned}\frac{dn_{\parallel}}{d\,\sigma} &= -\frac{n^3}{2}q_{11} = \frac{n^2}{E\,c}(q - 2\,\mu\,p)\\ &= -\frac{n^3}{2\,E}(p_{11} - 2\,\mu\,p_{12})\end{aligned} \tag{2.115}$$

and

$$\begin{aligned}\frac{dn_{\perp}}{d\,\sigma} &= -\frac{n^3}{2}q_{12} = \frac{n^2}{E\,c}(-\mu\,q + (1-\mu)p)\\ &= -\frac{n^3}{2\,E}(-\mu\,p_{11} + (1-\mu)p_{12})\end{aligned} \tag{2.116}$$

where q_{11} and q_{12} are the piezo-optic coefficients, q and p are Neumann's coefficients and p_{11} and p_{12} are the elasto-optic (strain-optical) coefficients. The other symbols E, μ, c, and n denote Young's modulus, Poisson's ratio, the velocity of light in vacuum and the refractive index without stress.

These coefficients can be determined experimentally by several methods, e.g. the four-point bending method. The experimental methods are described in [2.178–185]. Experimental data have been compiled by *Lindig* in [2.175]. A more recent collection of data for many optical glasses is given in [2.186]. In most cases, the data of the photoelastic coefficients have been determined for a single wavelength in the visible, very often for the sodium D lines at $\lambda = 589.3\,\mathrm{nm}$. According to *Pockels* [2.178], who has determined these coefficients as a function of wavelength for different glass compositions, their dispersion can in most cases be neglected. For the heavy flint glasses, however, the dispersion of these coefficients is considerable.

Mostly the refractive index increases with growing compressive stress for both directions of the plane of vibration, i.e. $K_{\parallel} = C_1 = dn_{\parallel}/d\sigma$ and $K_{\perp} = C_2 = dn_{\perp}/d\sigma$ are negative with $|K_{\parallel}| < |K_{\perp}|$ if a compressive (tensile) stress is considered to be negative (positive). In [2.186], very large values up to $9 \times 10^{-12}\,\mathrm{Pa}^{-1}$ are reported for the heavy flints. Furthermore, one observes that $|K_{\parallel}|$ increases with the PbO content stronger than $|K_{\perp}|$. Consequently, it is possible to melt a glass for which $K_{\parallel} = K_{\perp}$ at least for a given wavelength (because of the dispersion of $K_{\parallel}$ and $K_{\perp}$). Such a glass is nearly free of birefringence, since the changes of the refractive index for both directions of the plane of vibration are the same. A possible composition of a glass with this property has already been given by *Pockels* [2.178]. Therefore, this type of glass is called Pockels' glass. A commercial glass, which is nearly free of birefrigence under stress, is available under the mark SF 57 [2.187] (see Fig. 2.52b).

In the case of hydrostatic pressure, one has to consider that the refractive index of both the glass and the ambient medium can change. The derivative of n with respect to the hydrostatic pressure, P, is given simply by

$$\frac{dn}{dP} = K_{\parallel} + 2\,K_{\perp}\,. \tag{2.117}$$

The right-hand side can be transformed into a sum of other coefficients using (2.113) to (2.116).

Taking advantage of the changes of the refractive index under stress, glasses can be used in acousto-optical modulators [2.188, 189]. To characterize the efficiency for acousto-optic deflection, which depends on the design of the devices, one defines the following figures of merit (Sect. 2.5)

$$M_1 = \frac{n^7\,p_{ij}^2}{\rho\,v}$$

$$M_2 = \frac{n^6\,p_{ij}^2}{\rho\,v^3}$$

$$M_3 = \frac{n^7\,p_{ij}^2}{\rho\,v^2} \tag{2.118}$$

where v is the velocity of the acoustic wave, ρ is the mass density of the material and p_{ij} is the appropriate elasto-optic (strain-optical) coefficient depending on the orientation of the acoustic wave and the plane of vibration of the optical wave.

Data of M_1 and M_2 are collected in [2.186]. The glasses SF 58 and SF 59, which have the largest PbO contents in comparison with the other glasses, show the largest values of M_1 and M_2 among the data of [2.186].

For practical stress analysis of glasses it is necessary to know the stress-optical coefficient, K. In the simple case of Fig. 2.51, an electromagnetic wave penetrates the rectangular parallelepiped under normal incidence. The plane of vibration is oriented at an angle of 45° towards the direction of the uniaxial stress σ. The wave can be considered as being superposed by two coherent partial waves with planes of vibration parallel and perpendicular to σ. Since the refractive indices for both partial waves are different, a difference Δs in optical pathlength occurs. It is given by

$$\Delta s = (n + \frac{dn_{\parallel}}{d\,\sigma}\sigma - n - \frac{dn_{\perp}}{d\,\sigma}\sigma)\ell = (K_{\parallel} - K_{\perp})\sigma\,\ell \tag{2.119}$$

where ℓ is the geometrical path or the thickness of the sample. Δs transforms into a phase shift

$$\Delta\varphi = \frac{2\,\pi}{\lambda}(K_{\parallel} - K_{\perp})\sigma\,\ell = \frac{2\,\pi}{\lambda}K\,\sigma\,\ell \tag{2.120}$$

with the stress-optical coefficient

$$K = K_{\parallel} - K_{\perp}\,. \tag{2.121}$$

Thus K can be determined from the photoelastic coefficients. On the other hand, it can be determined directly from the relative phase shift of electromagnetic partial waves with the plane of vibration parallel and perpendicular to the uniaxial stress [2.178, 180, 183, 190–192].

There are numerous collections of data on the stress-optical coefficient of glasses in the literature. An arbitrary selection of references is [2.175, 178–181, 190–199]. There is also an abundance of work dealing with compositional dependence of this coefficient [2.175, 178, 190, 191, 194–199]. It is especially worth noting that K decreases with increasing PbO content until it reverses its sign for a very large lead content. Heavy flint glasses with about 70–75% PbO by weight do not show appreciable birefringence in the visible (Pockels' glasses, as mentioned above).

The temperature dependence of K has been investigated in [2.192] for soda-lime glasses and silicate glasses [2.193]. As long as the temperature is sufficiently below T_g, K varies only by some percent, whereas near T_g stronger variations are observed. This is to be expected, since in this case the viscoelastic properties of glasses become more important [2.143, 152].

The stress-optical as well as the photo-elastic coefficients depend on the wavelength [2.175, 178]. For most compositions of glasses, however, the dispersion is in the order of several percent in the visible spectral region [2.178, 199, 200]. In a recent investigation of K as a function of wavelength at 20 °C, this has been confirmed for a large number of optical glasses with different compositions [2.187]. The experimental data points for an arbitrary selection of optical glasses are shown in Figs. 2.52a and b.

Dispersion plays a major role for the high lead content glasses, such as SF 56 and SF 57. In order to fit the data points, the following dispersion formula has recently been derived [2.187, 200]

Table 2.12. Fitting parameters for the stress-optical coefficient of heavy flints using (2.122). Data from [2.187]

Glass type	A $10^{-6}\,\mathrm{mm^2N}$	B $10^{-14}\,\mathrm{mm^4/N}$	λ_{soc} µm
SF 1	3.4893	–9.9215	0.1447
SF 14	2.9517	–8.0661	0.1528
SF 18	3.2276	–8.2146	0.1452
SF 55	2.6038	–9.1004	0.1516
SF 57	0.3249	–9.6815	0.1617

$$K(\lambda) = \frac{n^2(\lambda) - 1}{2\,n(\lambda)} \left(A + \frac{B}{\lambda^2 - {\lambda^2}_{soc}} \right) \tag{2.122}$$

with the fitting parameters A, B and λ_{soc} and the refractive index $n(\lambda)$. If the dispersion is small, only the parameter A needs to be considered, because the parameter B can be assumed to be zero. Most of the full curves in Figs. 2.52a and b have been calculated under this assumption with only one fitting parameter A. A small dispersive effect is due to the prefactor $(n^2(\lambda) - 1)/2n$. For a fit of the data of the heavy flints, however, all of the three fitting parameters A, B and λ_{soc} have been determined. The fitting parameters are given in Table 2.12. In all of these cases, the experimental data can in fact be fitted reasonably well by the new dispersion formula (2.122), as can be seen by the full curves in Fig. 2.52b.

Neglecting the wavelength dependence of the prefactor $(n^2(\lambda) - 1)/2n$ in the visible spectral range, one can simplify (2.122) further into

$$K(\lambda) = A_s + \frac{B_s}{\lambda^2 - {\lambda^2}_{soc,s}} \,. \tag{2.123}$$

The dispersion of the stress-optical coefficient of several optical glasses has been investigated recently as a function of the temperature [2.201]. An example is given in Fig.2.53 for the optical glass BK 7. The data points show the experimental results for the temperature range from 21 °C to 537 °C. In order to fit the data points, (2.123) had to be extended. All three fitting parameters A_s, B_s, and $\lambda_{soc,s}$ can vary with the temperature. However, it was observed that the dominant temperature dependence in (2.123) is due to the parameter A_s. Using an expansion to second order for A_s, the dispersion of the stress-optical coefficient and its temperature dependence can be described by the dispersion formula

$$K(\lambda, T_c) = A_{s0} + A_{s1}T_c + A_{s2}T_c^2 + \frac{B_s}{\lambda^2 - \lambda^2_{soc,s}} \tag{2.124}$$

where T_c is the temperature difference with respect to some reference temperature (degree centigrade in Fig. 2.53). The data points in Fig. 2.53 have been fitted by (2.124). The result is shown by the full curves. Indeed, the fit of the data points is very satisfactory over the full temperature range between 21 °C and 537 °C, and the range of wavelength between 406.2 and 1021.6 nm.

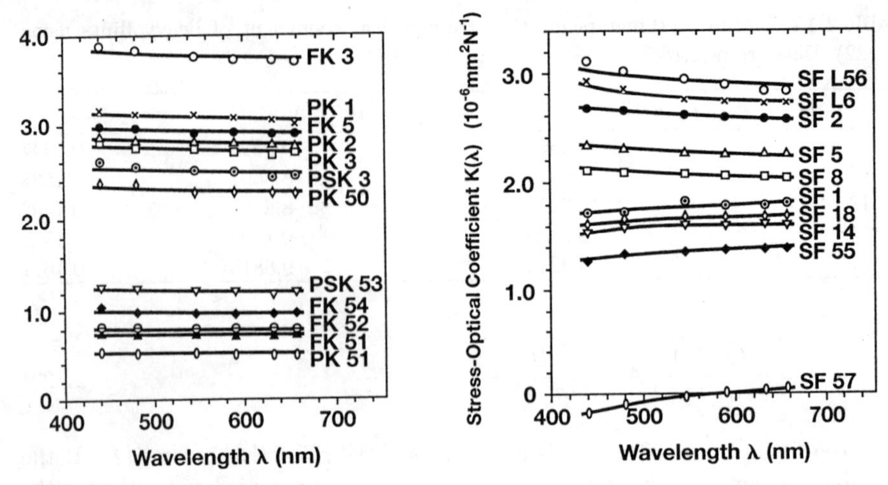

Fig. 2.52. Stress-optical coefficient as a function of wavelength λ at 20 °C for optical glasses; calculated with one fitting parameter A; except for SF1, SF14, SF18, SF55 and SF57, where all three fitting parameters of (2.122) have been used, see text and Table 2.12 (data from [2.187])

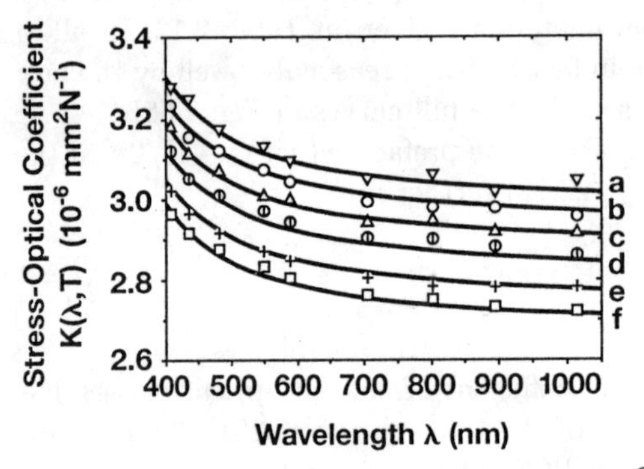

Fig. 2.53. The stress-optical coefficient of BK 7® as a function of wavelength for different temperatures. The full curves represent a fit of the experimental data points by the new dispersion formula (2.124) using the fitting parameters $A_{s0} = 2.67 \times 10^{-6}\,\mathrm{mm}^2/\mathrm{N}$, $A_{s1} = 8.19 \times 10^{-10}\,\mathrm{mm}^2/\mathrm{N}\,°\mathrm{C}$, $A_{s2} = -4 \times 10^{-13}\,\mathrm{mm}^2/\mathrm{N}\,°\mathrm{C}^2$, $B_s = 3.08 \times 10^{-14}\,\mathrm{mm}^4/\mathrm{N}$, $\lambda_{soc,s} = 237.2\,\mathrm{nm}$. Temperatures: a) 21 °C, b) 103 °C, c) 207 °C, d) 310 °C, e) 424 °C, f) 537 °C (according to [2.201])

2.4.6 The Electro-Optic Effect in Glasses (Kerr-Effect)

Applying a static electric field instead of mechanical stress to a glass in a similar configuration as in Fig. 2.51 causes an uniaxial anisotropy of the refractive index, which is known as Kerr-effect [2.202, 203]. The refractive index depends on the orientation of the plane of vibration of the electromagnetic wave towards the electric field. This anisotropy of the refractive index increases with the square of the electric field strength $\boldsymbol{E}$. (It must not be confused with the linear electro-optic effect occurring in crystals with unit cells that do not possess inversion symmetry. These effects are much larger than those dealt with in the present subsection). The difference of the refractive index $n_{\parallel} - n_{\perp}$ or the field-induced birefringence can be determined experimentally in a similar way as the stress-optical coefficient. For this purpose a linearly polarized electromagnetic wave penetrates a rectangular parallelepiped perpendicular to the electric field $\boldsymbol{E}$ (instead of stress σ in Fig. 2.51). The plane of vibration is oriented at an angle of 45° towards $\boldsymbol{E}$. The electromagnetic wave can be decomposed into two coherent partial waves with planes of vibrations parallel and perpendicular to $\boldsymbol{E}$. Due to the anisotropy $n_{\parallel} - n_{\perp} \neq 0$, a difference Δs occurs in the optical pathlength for both partial waves. It is given by

$$\Delta s = (n_{\parallel} - n_{\perp})\ell \tag{2.125}$$

where ℓ is the geometrical pathlength. This difference (2.125) of the optical pathlength transforms into a phase shift

$$\Delta\varphi = 2\pi\frac{\Delta s}{\lambda} = \frac{2\,\pi}{\lambda}(n_{\parallel} - n_{\perp})\ell \tag{2.126}$$

which can be measured by the application of polariscopic methods.

Since $n_{\parallel}$ and $n_{\perp}$ vary with the square of the electric field strength $\boldsymbol{E}$, $n_{\parallel} - n_{\perp}$ is written as

$$n_{\parallel} - n_{\perp} = \lambda\, B_{\mathrm{K}}\, E^2 \tag{2.127}$$

with the Kerr-constant B_{K}.

For oxide glass, B_{K} is of the order of 0.1 to $3 \times 10^{-12}\,\mathrm{cm/V^2}$, whereas for the chalcogenide glass As_2S_3 it is much larger, namely $8.7 \times 10^{-12}\,\mathrm{cm/V^2}$ [2.202]. Depending on the type of glass, both signs of B_{K} have been observed [2.203, 204].

With $B_{\mathrm{K}} \approx 10^{-12}\,\mathrm{cm/V^2}$, one estimates a field strength larger than 10^5 V/cm to induce a birefringence in the order of 10^{-6} in the visible spectral region. Thus, very large electric field strengths are necessary for glasses to render this effect useful for the modulation of polarized electromagnetic waves in the visible spectrum.

For a given field strength one can assume that both the induced birefringence, $n_{\parallel} - n_{\perp}$, and the changes of $n_{\parallel}$ or $n_{\perp}$ separately, are of the same order. Thus, the influence of an electric field on the refractive index can be neglected except for very large field strengths.

The Kerr-constant is known to show dispersion [2.203, 204]. Its value decreases considerably with increasing wavelength in the visible spectral range.

With respect to the *Kerr*-constant as a function of specific chemical constituents, it is well-known that its value increases with increasing PbO content in oxide glasses [2.203, 204].

2.4.7 Magneto-Optic Effects in Glasses

The plane of polarization of an electromagnetic wave which propagates in an optical medium can be rotated by a magnetic field. In the Faraday configuration (see Fig. 2.54), a linearly polarized electromagnetic wave travels in a transparent medium parallel to the magnetic field vector. The angle of rotation α is given by

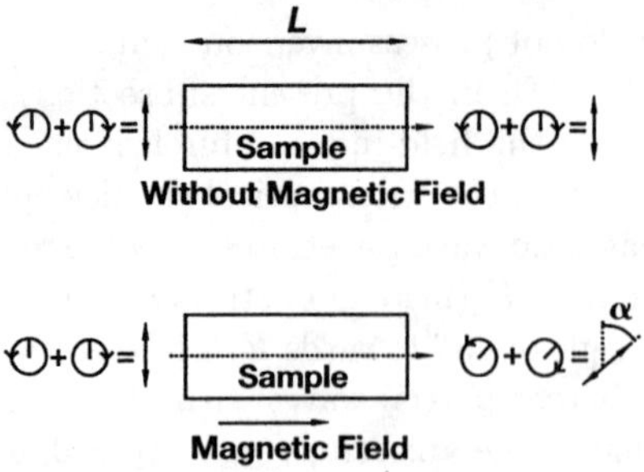

Fig. 2.54. Schematics of the Faraday rotation

$$\alpha = V \, \ell \, B \tag{2.128}$$

with the absolute value B of the magnetic flux density vector $\boldsymbol{B}$, the distance ℓ which the wave travels in the sample and the Verdet constant V, which is the factor of proportionality. The sign of V – and thus that of the rotation angle – is positive (negative) if the rotation is clockwise (counterclockwise) looking parallel to the vector of the magnetic flux density for both directions of the propagation of the electromagnetic wave, parallel as well as antiparallel to $\boldsymbol{B}$. Positive Verdet constants correspond to diamagnetic materials, whereas V is negative for paramagnetic samples. As the rotation is independent of the direction of the wave propagation with respect to $\boldsymbol{B}$, the Faraday rotation is used for non-reciprocal optical elements, i.e., optical isolators.

In conjunction with linear polarizers the Faraday rotation can be exploited to modulate the intensity of linearly polarized light and to construct optical isolators, which transmit electromagnetic waves in one direction only, whereas they block in the reverse direction. This effect is useful to suppress the interaction between the reflected laser beam and the laser itself or other components.

Since a linearly polarized electromagnetic wave can be represented by a superposition of a left-handed (index:$-$) and a right-handed (index:$+$) circularly polarized wave (see Fig. 2.54), the rotation angle α can be expressed for the standard Faraday configuration by

$$\alpha = \frac{\pi}{\lambda} \{n_+(B) - n_-(B)\} \, \ell \, . \tag{2.129}$$

Since $n_+(B) \neq n_-(B)$, the optical material is anisotropic with respect to circularly polarized electromagnetic waves, whereas stresses and electric fields cause anisotropies for linearly polarized electromagnetic waves. For reviews of magneto-optic effects see [2.205–207].

In general, one has to take into account that $n_+(B)$ and $n_-(B)$ are nonlinear functions of the magnetic flux density, B. Nonlinearities can be observed for paramagnetic materials especially at low temperatures and at large flux densities [2.208], which can result in a rotation angle saturating with B. In this case, the Verdet constant depends both on the temperature and on B. On the other hand, the Verdet constant of diamagnetic materials is expected to be nearly independent of the temperature and of B [2.206, 209].

Recently, the following dispersion formula of the Verdet constant has been developed on the basis of a single oscillator model [2.210, 211]

$$V = \frac{\pi}{\lambda}\left\{\frac{n^2 - 1}{2\,n}\right\}\left\{k_1 + \frac{k_2}{\lambda^2 - \lambda_0^2}\right\} \qquad (2.130)$$

with

$$k_1 = \frac{d(\ln f_+ - \ln f_-)}{dB} + 2\frac{d(\ln \lambda_{0+} - \ln \lambda_{0-})}{dB}$$

and

$$k_2 = 2\,\lambda_0^2 \frac{d(\ln \lambda_{0+} - \ln \lambda_{0-})}{dB} \, .$$

Here f and λ_0 are suitable average values of the strength and of the resonance wavelength of the oscillators.

Since the variation of the prefactor $(n^2 - 1)/2n$ with λ is small compared to the variation of $V(\lambda)$, (2.130) can be simplified further to

$$V = \frac{\pi}{\lambda}\left\{a + \frac{b}{\lambda^2 - \lambda_0^2}\right\} \qquad (2.131)$$

with the fitting parameters

$$a = k_1 \frac{n^2 - 1}{2\,n} \quad \text{and} \quad b = k_2 \frac{n^2 - 1}{2\,n} \, .$$

The data points in Figs. 2.55a, b represent experimental results of the Verdet constant of optical glasses in the visible spectral region. The curves result from a fit by (2.131). In fact, the dispersion of the Verdet constant can be described adequately by (2.131). The fitting parameters of a selection of optical glasses are given in Table 2.13.

The Verdet constant decreases monotonically with increasing wavelength λ in nearly all cases. For glasses of the SF type the Verdet constant increases monotonically with the PbO content. Very large values of V have been measured for the heavy flint glass type SF 59. This glass has an extremely large PbO content, namely about 80% by weight. Because of its large Verdet constant, this glass type is excellently suitable to be used as a Faraday rotator material. It is to be mentioned that almost all glasses of Table 2.13 show diamagnetic behaviour. On the other hand, special glasses with a large content of rare earth ions and large magnetic moments such as Tb^{3+} have been developed, providing larger Verdet constants [2.206–208, 213, 214]. Since these glasses are paramagnetic, the Verdet constant and consequently the Faraday rotation angle depend on temperature.

Taking advantage of the rotation in opposite directions, the diamagnetic and paramagnetic behaviour may compensate for appropriate compositions in yielding a Verdet constant close to zero. This has almost perfectly been achieved for the optical glasses SF L6 and SF L56 at room temperature. Thus, in these glasses, the plane of polarization of an electromagnetic wave is nearly unaffected by magnetic fields.

From a practical point of view, the dispersion formulae (2.130) or (2.131) can also be applied to paramagnetic materials as long as the Verdet constant

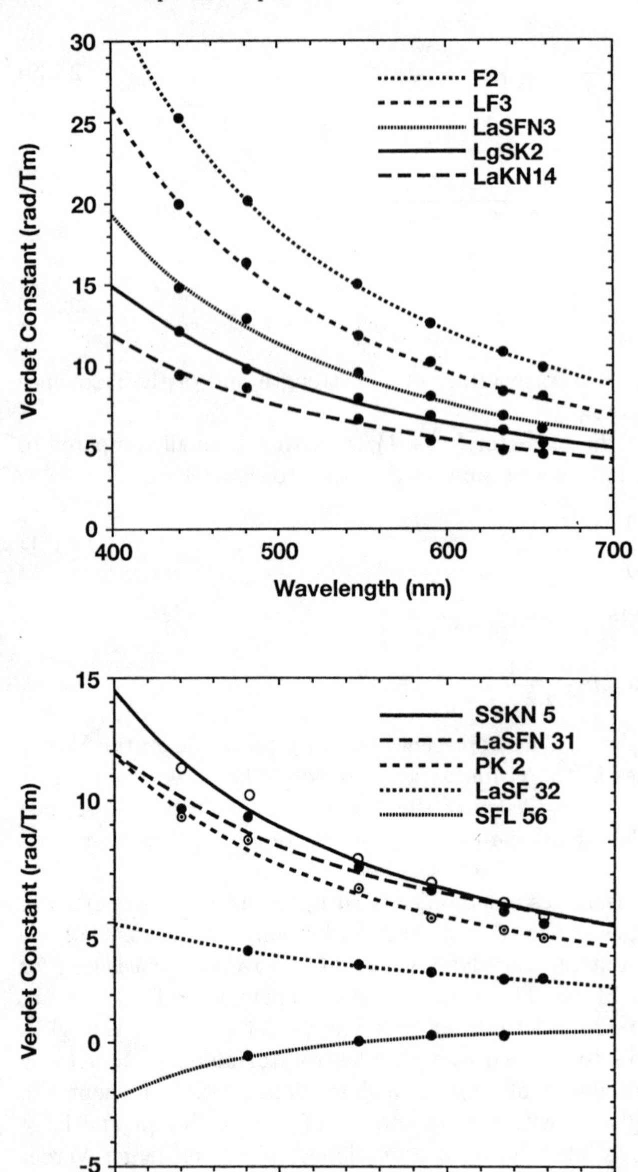

Fig. 2.55. (a, b) The Verdet constant V of different optical glasses as a function of wavelength in the visible spectral range. Experimental results are shown by the data points. The curves represent the results of fitting calculations according to (2.131)

Table 2.13. Parameters a, b and λ_0 for fitting the experimental data of the Verdet constant by the dispersion formula (2.131). Data from [2.212]

Type of glass	$a\ 10^{-9}/\mathrm{T}$	$b\ 10^{-20}\,\mathrm{m}^2/\mathrm{T}$	$\lambda_0\ 10^{-9}\,\mathrm{m}$
FK 3	584.86	10.1922	95.3
FK 5	590.95	10.7136	92.3
FK 51	449.47	10.3782	84.7
FK 52	324.24	14.0808	86.2
PK 2	620.50	13.2578	96.4
BK 3	558.73	13.1224	96.1
BK 7	445.56	19.0396	97.0
BaLK N3	499.68	20.0342	100.0
K 3	489.92	21.7512	101.0
BaK 50	630.79	19.3622	102.6
SK 16	519.45	22.6678	101.2
SSK N5	753.71	15.8852	110.6
LaK N12	714.57	20.8496	106.5
LaK N14	701.16	12.2336	106.5
LF 3	807.24	35.9836	120.4
F 2	1007.83	44.5356	129.7
F N11	515.34	0.7574	130.1
F 13	967.22	46.9302	130.4
LaSF N3	791.71	23.8658	125.1
LaSF N31	938.42	8.3116	125.4
LaSF 32	480.47	2.0898	143.9
SF 1	1297.31	68.6416	144.7
SF 2	575.35	64.9114	134.6
SF 5	1142.58	57.7910	138.2
SF 6	1682.15	88.7082	156.4
SF 8	1379.42	56.7350	140.5
SF 14	1261.61	65.9212	152.8
SF 18	1078.09	73.7398	145.2
SF 53	1041.75	74.4138	146.7
SF L56	273.20	-7.4974	154.3
SF 57	1867.83	98.1516	161.7
SF 58	2164.80	121.3394	170.5
SF 59	2843.20	112.5192	175.3
SF N64	367.16	-3.7126	142.8
TiK 1	697.09	12.6900	100.8
TiF 3	534.72	1.0448	119.9
TiF 6	371.12	2.4132	140.6
KzFS N4	822.45	30.2408	117.8

is independent of B. A possible dependence of the temperature can be taken into account by a suitable variation of the fitting parameters a, b and λ_0 with temperature.

If the electromagnetic wave propagates in a perpendicular — instead of a parallel — direction to the magnetic flux density vector, $\boldsymbol{B}$, we have the configuration of the Cotton-Mouton effect [2.215]. With respect to the orientation of the biasing field and to the direction of the propagating electromagnetic wave, this effect is analogous to the Kerr-effect. The material becomes anisotropic for linearly polarized electromagnetic waves. Corresponding to (2.127) the anisotropy can be formulated as

$$n_{\parallel} - n_{\perp} = \lambda\, K_{\mathrm{KM}} B^2 \ . \tag{2.132}$$

Data of the Cotton-Mouton constant K_{KM}, however, are not yet available for glasses but are for gases and liquids [2.216].

2.4.8 Further Possibilities

In this contribution various effects have been described which can change the refractive index. Some of the effects may be detrimental for applications, such as the variation of the refractive index due to the ambient temperature, to an annealing rate that is too fast, or to a changing environmental atmosphere. Such differential modifications of the refractive index may limit the performance of imaging optical systems. On the other hand, some of these effects, such as the Faraday effect, have advantageously been exploited to build measuring, sensoring or protecting devices.

An important positive aspect is the possibility that glasses can be melted in large volumes and can be tailored to meet the requirements of specific applications. The following examples illustrate some of the achievements:

1. The temperature coefficient with respect to standard air can be made approximately zero [2.164]. In this case, the deflection by a prism is insensitive to variations of temperature.

2. The temperature coefficient of the refractive index, dn_{abs}/dT, the refractive index, n, and the thermal expansion coefficient, α, can be matched in order to make the temperature coefficient of the optical pathlength within a glass sample nearly zero , i.e.,

$$\frac{d}{dT} n_{\mathrm{abs}} + (n_{\mathrm{abs}} - 1)\alpha \simeq 0 \ . \tag{2.133}$$

Such glasses are called "athermal glasses" [2.217]. They are useful, for example, for athermal Fabry-Perot etalons.

3. The parameters of laser glasses as active media can be adjusted by the chemical composition so that the thermal lensing of laser rods is a minimum [2.218]. The refractive power of a laser rod is given in the steady state by [2.219]

$$\frac{1}{f_r} = \frac{P}{kA} \left[\frac{1}{2} \frac{dn}{dT} + \alpha C_r n^3 + \frac{\alpha R(n-1)}{L} \right]$$

$$\frac{1}{f_\phi} = \frac{P}{kA}\left[\frac{1}{2}\frac{dn}{dT} + \alpha C_\phi n^3 + \frac{\alpha R(n-1)}{L}\right] \tag{2.134}$$

where f_r, f_ϕ, are the focal length of the radial and azimuthal component of an electromagnetic wave in the rod; P is the total heating power in the rod; k is the coefficient of heat conduction; A is the cross-section of the rod; dn/dT is the temperature coefficient of the absolute refractive index; n is the refractive index; α is the coefficient of linear thermal expansion; R and L are the radius and length of the rod; C_r and C_ϕ, are the radial and azimuthal photoelastic coefficients.

For a rod with $R/L = 1/30$ made of phosphate laser glass LG 750 the values of the sums in the bracket of (2.134) is very small in order to minimize the thermal lensing.

Beside the effects discussed above, there exist further possibilities to change the refractive index. These possibilities, however, would deserve more detailed considerations with respect to experimental techniques and practical applications than can be given in this survey. An important technical procedure is the variation of the refractive index due to ion exchange. This is the basis for integrated optics. Special optical glasses have been developed to meet the corresponding requirements [2.220]. Ion implantation is another possibility to vary the refractive index in thin layers, typically in the order of some tenths of a micrometer [2.221].

By leaching a glass the refractive index can be reduced in a surface layer. This has been used to produce antireflective surfaces [2.222, 223].

Furthermore, the refractive index of glass can be modified by irradiation with UV and γ-photons [2.224], α and β particles or neutrons [2.225]. This creates defects and disturbs the nearest neighbouring order. On the other hand, neutrons can be captured by nuclei which subsequently transmute into nuclei with the atomic number increased by unity, thus changing the isotopic or chemical composition of a glass.

Doping with suitable electron donors and acceptors enhances the possibility to change the refractive index locally. Photons may initiate the transfer of charge carriers from one site to another and change the refractive index. In glasses this effect plays only a minor role. Large effects are observed in electro-optic and photorefractive crystals (for a survey on photorefraction see [2.226]). The refractive index can also be changed by photodecomposition, which has been observed in PbI_2 films [2.227]. This effect may also play a role for surfaces of glasses which can be decomposed by the interaction of radiation.

In amorphous WO_3 films, the refractive index can be changed by injection of charge carriers [2.228]. Changes of the refractive index in the order of 0.01 have been observed in small semiconductor devices where free charge carriers have been injected by electric fields [2.229].

Furthermore, the refractive index can be changed by optical transfer of electrons into excited states, as has been discussed by *Baldwin* and *Riedel* [2.230]. For Eu-doped glasses, a permanent change of the refractive index has been observed under focused laser excitation [2.231].

Similar to the Kerr-effect, where the refractive index depends on the static or quasi-static electric field strength, it is possible that the electric field of the electromagnetic wave itself can modify n. Since the intensity I of an electromagnetic wave is proportional to the square of the time average of the electric field strength E,

$$I = \left[\frac{C}{4\pi}\right] n_0 |E|^2 \tag{2.135}$$

the refractive index increases proportionally to I of the electromagnetic waves [2.219]

$$\begin{aligned} n(I) &= n_0(\lambda) + \gamma\, I \\ &= n_0(\lambda) + 4.19 \times 10^{-3} \frac{n_2}{n_0} \left(\frac{\mathrm{cm}^2}{\mathrm{W}}\right) I \end{aligned} \tag{2.136}$$

with the nonlinear refractive index n_2. Typical values of γ are in the order of $4 \times 10^{-16}\ \mathrm{cm}^2/\mathrm{W}$. Thus, intensities I of several $10^9\ \mathrm{W/cm}^2$ are necessary to modify n in the sixth decimal place.

Comparing (2.127) with (2.136) and (2.135), the intensity dependent refractive index can be considered as a degenerate Kerr-effect since the frequencies of the applied electric field and of the electromagnetic wave are the same.

The intensity dependent refractive index belongs to a whole series of possible nonlinear effects. A survey of nonlinear effects occuring in glasses is given in Sect. 2.6 [2.232]. In glasses, mainly nonlinear absorptive effects can be exploited. Since glasses are statistically isotropic materials, phase matching of waves with different frequencies is not possible. Therefore, frequency mixing or multiplication cannot be used efficiently.

If the intensity, I, in (2.136) decreases monotonically as a function of the radius from the centre of an intense laser beam, we have a gradient index lens and the beam may be focused due to the self-induced inhomogeneity of the refractive index (self-focusing). With sufficiently large intensities this may cause the destruction of the material [2.233, 234].

Narrow focusing of intense laser beams can induce the destruction of any optical material. An overview of the damage threshold of optical glasses has been given by *Hack* and *Neuroth* [2.233]. These authors found correlations of the damage threshold for internal destruction of glasses with the nonlinear refractive index n_2, if very short pulses in the order of nanoseconds were used, and with the absorption constant at the laser wavelength (typically 1060 nm) if longer pulses in the order of milliseconds were applied. The interpretation of these correlations is the following: the larger n_2, the stronger is the focusing effect and, consequently, the intensity in the focus, which causes dielectric breakdown in the glass. If, on the other hand, the absorption constant is large, the heating power in the glass is increased, which may induce thermal breakdown (melting process). In addition to the internal destruction, damage can also occur on the surface of the sample. The threshold for this type of damage, however, is correlated with the surface quality and the polishing method.

Acknowledgements

I would like to thank Werner W. Jochs and his co-workers, Dipl.-Phys. Gerhard Westenberger, and Dipl.-Phys. Gudrun Przybilla for their help in determining many data in this article and for fruitful collaboration in the past years. I gratefully acknowledge the support and critical comments of Dr. N. Neuroth and critical reading of the manuscript by Dr. A. Marker, Schott Glass Technologies, Inc., Duryea, PA, USA.

2.5 Acousto-Optical Properties of Glasses

Hans-Jürgen Hoffmann

2.5.1 Acousto-Optical Effects and Applications

Acoustic waves propagate in gases and liquids normally as compressive waves, i.e. the acoustic wave causes periodic density variations in the direction of propagation. The length of the period is given by the wavelength

$$\Lambda_{\mathrm{ac}} = \frac{v_{\mathrm{ac}}}{f_{\mathrm{ac}}} \tag{2.137}$$

where v_{ac} and f_{ac} are the velocity and the frequency of the acoustic wave. Since the density and the refractive index of the gas or the liquid are modulated concomitantly, a phase grating is created. In a standard configuration shown in Fig. 2.56, a transducer excites in a tank plane acoustic waves. The planes of constant refractive index are indicated by dashed lines. Electromagnetic waves that pass the tank perpendicularly to the propagation vector of the acoustic waves become partially deflected or diffracted, depending on the wavelength of the acoustic wave (or the period of the phase grating) and the maximum differences of the optical pathlength in the layers of the phase grating [2.139–149]. As long as there are no additional anisotropies (e.g. dynamic anisotropies in liquids) the diffraction of the electromagnetic waves is independent of whether their plane of polarization is parallel or perpendicular to the vector of propagation of the acoustic waves. The diffraction angles Θ_m are given by the relation

$$\sin \Theta_m = m \frac{\lambda}{n\, \Lambda_{\mathrm{ac}}} , \tag{2.138}$$

where m is an integer, λ is the wavelength of the electromagnetic wave in vacuum and n is the refractive index of the medium. To achieve a deflection of at least 1° in the first order ($m = 1$) for an electromagnetic wave in the visible part of the spectrum ($\lambda = 550\,\mathrm{nm}$, refractive index $n = 1.5$), the wavelength Λ_{ac} of the acoustic wave must be about 20 μm corresponding to a frequency of about 200 MHz if the velocity of the acoustic wave is 4000 m/s. Unfortunately, the modulation depth of the refractive index is low in gases and the damping of the acoustic wave

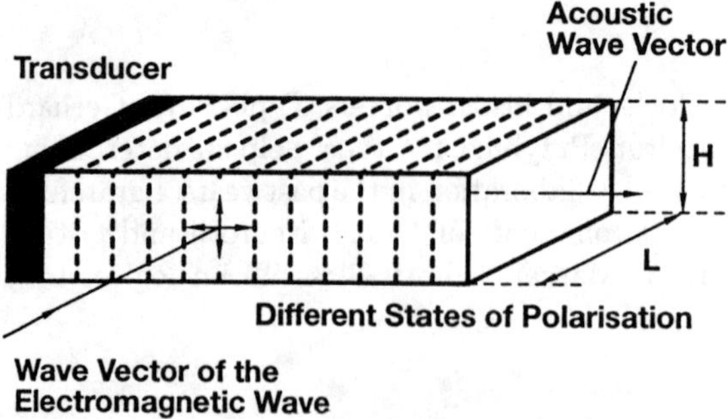

Fig. 2.56. Schematic configuration for an electromagnetic wave interacting with an acoustic wave. The different states of polarization of the electromagnetic wave are indicated by the vertical and horizontal arrows. The schematic device may extend further to the right for an acoustic absorber to avoid reflection of the acoustic wave. L: interaction length of the electromagnetic wave with the acoustic wave, H: lateral dimension of the acoustic field

is rather large in liquids, which makes it difficult to achieve effective and homogeneous diffraction of light beams using these materials. Since the damping of acoustic waves is as compared to gases and liquids at least one or two orders of magnitude smaller in crystalline and amorphous solids, one prefers these materials for applications in acousto-optic devices. In this case, however, both longitudinal compressive waves and elastic shear waves can be excited by a transducer. For simplicity, we consider the volume in Fig. 2.56 to be filled with an isotropic solid, e.g. homogeneous, statistically isotropic glass. The longitudinal compressive waves are excited by the transducer in the medium. Even in this simple case, one has to take into account that the refractive index n is modulated by the acoustic wave by different amounts depending on the direction of polarization of the electromagnetic wave. In this respect, consider Fig. 2.51 of the preceding Sect. 2.4, where the changes of the refractive index are indicated to be different whether the plane of vibration of the electromagnetic wave is parallel or perpendicular to the propagation vector of the (compressive) stress. This obviously occurs also in an acoustic wave.

An acousto-optic device operates in the Raman-Nath regime if $2\pi\lambda/n\Lambda_{\mathrm{ac}} \gg \Lambda_{\mathrm{ac}}/L$ (which corresponds to the Klein-Cook parameter $Q = 2\pi L\lambda/n\Lambda_{\mathrm{ac}}^2 \gg 1$ [2.240–244]. Since the interaction length L is relatively short in this regime, the diffraction efficiency is rather poor. In addition, multiple orders of diffracted beams are observed ($m = \pm1, \pm2, \pm3, \ldots$).

For practical use one prefers the Bragg regime, for which $2\pi\lambda/n\Lambda_{\mathrm{ac}} \ll \Lambda_{\mathrm{ac}}/L$ (corresponding to the Klein-Cook parameter $Q = 2\pi L\lambda/n\Lambda_{\mathrm{ac}}^2 \ll 1$ [2.240–244]). In this case, the device is usually tilted by the Bragg angle, so that the electromagnetic wave is incident obliquely with respect to the wave vector of the acoustic wave.

In the Bragg regime, the interaction length for the acoustic and the electromagnetic waves is long and the beam intensity of the electromagnetic wave deflected into the first order can be very large compared to the larger orders and to the zero order, provided that the refractive index is sufficiently strong modulated by the acoustic wave.

Both limiting cases, the Raman-Nath regime and the Bragg regime, are discussed in textbooks, of which Refs. [2.240–243] are just an arbitrary selection. The transition range from $Q \ll 1$ to $Q \gg 1$ has been treated by complete numerical calculations in [2.244].

Acoustic waves can be excited (and consequently propagate) not only in the bulk but also on the surface of solids. Thus, the propagating acoustic strain can be confined to a layer near the surface. This is the basis of surface acoustic wave devices. It is obvious that application of the interaction between acoustic and electromagnetic waves in surface layers is an interesting field of integrated optics. Surface acoustic wave interactions are treated for instance in [2.242, 243].

For applications in acousto-optical devices one varies the frequency and the intensity of the acoustic wave. In this connection, however, one has to point out that the frequency of the diffracted electromagnetic wave is modulated or shifted with respect to the incident wave by the frequency of the acoustic wave.

Though essentially only two parameters of the acoustic wave can be varied, there are numerous applications of acousto-optic devices. Just a few elements are to be mentioned starting from simple to more elaborate devices: switching elements, mode lockers and Q-switches in laser cavities, modulators, deflectors, scanners, wavelength tunable filters, spectrum analyzers, convolvers and correlators, optical computing matrix processors and Fourier transform processors [2.242, 243].

2.5.2 Characterization of Glasses

In the Bragg regime, the ratio of the beam intensities in the first and zero order is given approximately by [2.240]

$$\frac{I_1}{I_0} = \sin^2\left(\frac{\pi}{2}\frac{L}{H}\left(\frac{n^6 {p_{1i}}^2}{\rho\, v_{\mathrm{ac}}^3}\right)\frac{P_{\mathrm{ac}}}{\lambda^2}\right)^{\frac{1}{2}} = \sin^2\left(\frac{\pi}{2}\frac{L}{H}M_{2i}\frac{P_{\mathrm{ac}}}{\lambda^2}\right)^{\frac{1}{2}} \tag{2.139}$$

or for $\left(\frac{\pi}{2}\frac{L}{H}M_{2i}\frac{P_{\mathrm{ac}}}{\lambda^2}\right)^{\frac{1}{2}} \ll 1$

$$\frac{I_1}{I_0} \simeq \left(\frac{\pi}{2}\frac{L}{H}M_{2i}\frac{P_{\mathrm{ac}}}{\lambda^2}\right) . \tag{2.140}$$

In (2.139) and (2.140) $i = 1$ ($i = 2$) stands for the polarization of the electromagnetic wave parallel (perpendicular) to the propagation vector of the acoustic wave, L is the interaction length of the electromagnetic wave with the acoustic wave, H the lateral dimension of the acoustic field, λ the wavelength of the electromagnetic wave in vacuum, n the refractive index, p_{1i} the photo-elastic or strain-optical coefficients, ρ the mass density, P_{ac} the total acoustic power and v_{ac} the velocity of the acoustic wave.

The figure of merit [2.245–247]

$$M_{2i} = \frac{n^6 p_{1i}{}^2}{\rho\, v_{\mathrm{ac}}^3} \tag{2.141}$$

characterizes in (2.139) and (2.140) the diffraction efficiency which depends only on material parameters. The larger M_{2i}, the lower the power of the acoustic wave to achieve a large intensity of the deflected beam.

For acousto-optic scanning and switching elements, one has to take into account, in addition to the diffraction efficiency, the time constant corresponding to the bandwidth of the acousto-optic element. Therefore, one introduces in this case the figure of merit [2.246, 247]

$$M_{1i} = \frac{n^7 p_{1i}{}^2}{\rho\, v_{\mathrm{ac}}} \tag{2.142}$$

which characterizes the product of bandwidth and diffraction efficiency.

Considering the lateral dimension or the width H of the acoustic field, there is still a further figure of merit based on the following idea [2.246]: Since the required acoustic power decreases with the width H of the transducer (Fig. 2.56), H should be as small as possible. The lower limit of H corresponds to the waist of a (typically laser) beam focused in the acousto-optic interaction region. Then the modulation bandwidth Δf is about the reciprocal travel time of the acoustic wave across the beam waist $\Delta f = v_{\mathrm{ac}}/H$. This yields [2.246]

$$M_{3i} = \frac{n^7 p_{1i}{}^2}{\rho\, v_{\mathrm{ac}}^2} \,. \tag{2.143}$$

It characterizes an acousto-optical material with respect both to the diffraction efficiency and the bandwidth for minimum lateral dimensions under focusing.

The figures of merit M_{1i}, M_{2i} and M_{3i} have been calculated for a series of optical multicomponent glasses at 20 °C [2.248]. A selection of data is given in Table 2.14. For the velocity of the acoustic wave v_{ac} in (2.141–143)

$$v_{\mathrm{long}} = \sqrt{\frac{E(1-\mu)}{\rho(1+\mu)(1-2\mu)}}\,. \tag{2.144}$$

has been inserted, which is the velocity of longitudinal acoustic waves in isotropic solids extending to infinity. These values are larger than

$$v_d = \sqrt{\frac{E}{\rho}} \tag{2.145}$$

which is valid for longitudinal acoustic waves propagating along the axis of long, slender rods.

From the formulae (2.141–143) one concludes that the figures of merit are large, if the refractive index n and the strain-optical coefficients p_{1i} $(i = 1, 2)$ are large, whereas the mass density ρ and the velocity of the acoustic waves v_{ac} are small. However, these quantities are not all independent of each other. An analysis of these parameters has been made for crystalline materials in [2.249]. Inserting the relation (2.144) into (2.141–143), we obtain immediately the most important guideline for a large figure of merit. Since the dependence of the mass density is rather small in the denominator of (2.141–143), large figures of merit are expected if the refractive index n is large and Young's modulus E is small (in this case the strain-optical coefficients are also very often small). Therefore it is not surprising that TeO_2, $PbMoO_4$, α-HIO_3, As_2S_3, GaP, GaAs and Tl_3AsS_4 are among the materials with large figures of merit. The figures of merit M_{2i} of these materials are typically in the order of 20 to 500 $\times 10^{-18}\,s^3/g$ [2.246, 249, 250]. As compared to these large values most data of optical glasses are rather small. This can be understood by the fact that optical glasses have not been developed for a large acousto-optical figure of merit but for special requirements of the refractive index, dispersion and anomalous partial dispersion. Since, compared to crystals, the figures of merit of many different types of optical glasses are smaller by at least two orders of magnitude, one may generally conclude that glasses are not suitable for acousto-optical applications. Nevertheless, there exist optical glasses which possess a large figure of merit, namely the heavy flint glasses SF6, SF56, SF57, SF58, SF59 and SF61, for which the figures of merit M_{2i} are between 5 and $17 \times 10^{-18}\,s^3/g$. In this respect, the reader is referred to refs. [2.251, 252]. Glasses have the advantage over crystalline materials that they can be produced in large volume and very homogeneously. Furthermore, for the Pockels' glass (SF57, see Sect. 2.4) the piezo-optical coefficients q_{11} and q_{12} are approximately the same in the visible spectral range, since the stress-optical coefficient K is approximately zero [2.253]. The figures of merit and the diffraction of an acousto-optical device using this type of glass are consequently independent of the polarization of the incident electromagnetic wave with wavelength in the visible spectral range. By variation of the content of lead oxide one can shift that range towards the infrared spectral region. It seems to be unique for glasses to adjust the figure of merit for both planes of vibration by adjusting the composition. A remarkable success of materials science and engineering is to take advantage of this possibility.

One should point out that the data in Table 2.14 have been calculated for the wavelength $\lambda = 589.3\,\text{nm}$ of the electromagnetic wave. The specification of the wavelength is necessary, since the refractive index n and the strain-optical coefficients p_{1i} which enter the figures of merit depend on λ. Usually the dispersion of n is very well-known, whereas the dispersion of the strain-optical coefficients has not yet been investigated for many glasses.

Basically, dispersive effects in acousto-optical devices have two different origins. The first one arises from the diffraction. The diffraction angle Θ_m depends on the period of the modulated refractive index, which is given by the wavelength of the acoustic waves (2.138): The larger the wavelength λ of the electromagnetic wave, the larger the diffraction angle Θ_m. Thus, changing the wavelength λ of the electromagnetic wave implies a different angle of diffraction. This effect cannot be avoided except by adjusting the wavelength Λ_{ac} of the acoustic wave.

The second effect is related to the intensity of the diffracted beams and to the diffraction efficiency characterized by the figure of merit M_{2i} [see (2.139–141)]. Since the factor λ^2 occurs in the denominator of the argument in (2.139) and M_{2i} can show dispersion of the order of several ten percent, e.g. in the visible spectral

Table 2.14. Figures of merit of selected optical glasses for the wavelength $\lambda = 589.3\,\text{nm}$

Type of glass	p_{11}	p_{12}	v_{long} m/s	v_d m/s	M_{11} $10^{-7}\,\text{cm}^2\text{s/g}$	M_{12} $10^{-7}\,\text{cm}^2\text{s/g}$	M_{21} $10^{-18}\,\text{s}^3\text{/g}$	M_{22} $10^{-18}\,\text{s}^3\text{/g}$	M_{31} $10^{-12}\,\text{cm}\,\text{s}^2\text{/g}$	M_{32} $10^{-12}\,\text{cm}\,\text{s}^2\text{/g}$
FK 3	0.15	0.24	4920	4510	3	7	1	2	0.6	1.5
FK 5	0.14	0.23	5430	5030	2	6	1	1	0.4	1.2
FK 51	0.17	0.20	5440	4730	2	3	1	1	0.4	0.6
PSK 3	0.14	0.23	5760	5370	3	7	0	1	0.4	1.2
PSK 53A	0.17	0.20	5310	4650	5	6	1	1	0.8	1.2
BK 7	0.12	0.22	6030	5700	2	6	0	1	0.4	1.4
K 5	0.13	0.23	5630	5250	2	7	0	1	0.4	1.2
K 7	0.12	0.23	5540	5200	2	7	0	1	0.3	1.2
ZK N7	0.11	0.23	5660	5320	2	7	0	1	0.3	1.2
BaK 1	0.13	0.20	5250	4780	2	6	1	1	0.5	1.1
BaK 2	0.14	0.22	5360	4970	3	7	1	2	0.5	1.2
BaK 4	0.12	0.21	5400	4970	2	6	0	1	0.4	1.1
SK 2	0.14	0.20	5210	4700	3	6	1	1	0.6	1.2
SK 4	0.12	0.18	5360	4800	2	5	0	1	0.4	0.9
SK 5	0.14	0.21	5570	5060	3	6	1	1	0.5	1.1
SK 10	0.13	0.19	5320	4740	3	5	1	1	0.5	1.0
SK 14	0.13	0.19	5540	5000	3	5	1	1	0.4	0.9
SK 15	0.14	0.20	5410	4810	3	6	1	1	0.5	1.1
SK 16	0.16	0.22	5540	4980	4	7	1	1	0.7	1.3
BaLF 4	0.11	0.20	5250	4910	2	6	0	1	0.3	1.1
BaLF 5	0.14	0.23	5080	4690	3	7	1	2	0.5	1.5
SSK N5	0.11	0.17	5510	4880	2	5	0	1	0.4	0.9
SSK N8	0.13	0.21	5490	4990	3	7	1	1	0.5	1.3
LaK N7	0.11	0.16	5470	4850	2	4	0	1	0.4	0.7
LaK 8	0.10	0.16	6300	5510	2	5	0	1	0.3	0.7
LaK 9	0.11	0.17	6370	5600	2	5	0	1	0.3	0.8
LaK 10	0.10	0.16	6150	5390	2	5	0	1	0.3	0.8
LaK N12	0.13	0.17	5270	4620	3	5	1	1	0.6	1.0
LaK N13	0.15	0.19	5180	4530	4	6	1	1	0.8	1.3
LaK 21	0.16	0.22	5510	4920	4	7	1	1	0.7	1.4
LaK N22	0.13	0.18	5470	4900	3	5	1	1	0.5	1.0
LLF 6	0.15	0.25	4990	4730	3	8	1	2	0.6	1.8
BaF 4	0.14	0.21	4720	4330	3	7	1	2	0.7	1.6
BaF N10	0.14	0.20	5510	4860	3	7	1	1	0.6	1.3
BaF 50	0.14	0.19	5510	4950	3	7	1	1	0.6	1.2
BaF 51	0.09	0.16	5670	5090	2	5	0	1	0.2	0.8
BaF 52	0.12	0.19	5170	4700	3	6	1	1	0.5	1.1
LF 5	0.18	0.25	4600	4290	6	11	2	3	1.2	2.3
F 2	0.17	0.23	4300	4010	5	10	2	3	1.3	2.3
F 5	0.15	0.22	4370	4090	4	9	1	3	0.9	2.0
F N11	0.10	0.20	5990	5590	2	7	0	1	0.3	1.2

Table 2.14. (cont.)

Type of glass	p_{11}	p_{12}	v_{long} m/s	v_d m/s	M_{11}	M_{12} 10^{-7} cm^{2}s/g	M_{21}	M_{22} 10^{-18} s^3/g	M_{31}	M_{32} 10^{-12} cm s^2/g
BaSF 2	0.15	0.20	4490	4120	4	8	1	3	1.0	1.8
BaSF 51	0.12	0.17	4920	4310	3	6	1	1	0.6	1.3
LaF 2	0.11	0.15	5290	4630	3	5	1	1	0.5	0.9
LaF 3	0.11	0.15	5450	4780	2	5	0	1	0.4	0.8
LaF N7	0.13	0.17	4820	4260	4	7	1	2	0.8	1.4
LaF N21	0.07	0.12	6200	5390	1	3	0	0	0.2	0.5
LaF 22A	0.09	0.14	5530	4880	2	4	0	1	0.4	0.9
LaSF 3	0.08	0.14	6130	5360	2	5	0	1	0.3	0.8
LaSF N30	0.10	0.14	6070	5280	2	4	0	1	0.4	0.7
LaSF N31	0.10	0.14	5560	4790	3	5	1	1	0.5	1.0
SF 1	0.22	0.25	3860	3570	12	16	5	6	3.2	4.1
SF 2	0.19	0.25	4070	3780	8	13	3	5	1.9	3.2
SF 4	0.20	0.23	3710	3420	12	15	5	6	3.1	4.1
SF 5	0.18	0.22	4010	3710	7	11	3	4	1.8	2.7
SF 6	0.24	0.25	3580	3270	19	21	8	9	5.4	5.9
SF L6	0.09	0.16	5820	5260	3	8	0	1	0.4	1.4
SF 8	0.19	0.24	3920	3630	9	13	3	5	2.2	3.5
SF 10	0.20	0.24	4160	3860	10	15	3	5	2.5	3.6
SF 11	0.19	0.21	4020	3720	10	13	4	5	2.7	3.3
SF 14	0.20	0.23	4080	3780	11	15	4	5	2.8	3.7
SF 15	0.18	0.22	4160	3850	7	11	3	4	1.9	2.8
SF 53	0.19	0.22	3920	3620	10	13	4	5	2.4	3.3
SF 55	0.20	0.22	3760	3440	11	14	5	6	3.2	3.8
SF L56	0.07	0.15	5810	5280	2	6	0	1	0.3	1.2
SF 56	0.21	0.22	3740	3440	13	16	5	6	3.7	4.1
SF 57	0.23	0.23	3430	3130	20	20	9	9	6.0	6.0
SF 58	0.24	0.23	3290	2960	29	26	14	13	8.5	7.8
SF 59	0.25	0.24	3200	2870	34	30	17	15	10.5	9.7
SF 61	0.21	0.23	3810	3530	12	15	5	6	3.3	4.0
KzFS N4	0.13	0.20	4870	4320	3	7	1	2	0.6	1.5

region, the intensity of the diffracted beam clearly depends on the wavelength λ of the electromagnetic waves. In order to compensate this effect, the acoustic power has to be adjusted. On the other hand, the dispersion of the figure of merit M_{2i} in the numerator may compensate the factor λ^2 in the denominator. Therefore it is worth while investigating the figure of merit M_{2i} as a function of wavelength in different glass systems. If it is proportional to λ^2 in a certain range of wavelength, the intensity of the diffracted beams is independent of λ. There are indications that this occurs for glasses with a large content of lead oxide.

2.6 Nonlinear Optical Properties

Yuiko T. Hayden, Joseph S. Hayden

Nonlinear optics deals with the nonlinear changes in optical properties of materials induced by externally-applied light beams. In general, nonlinear optical effects are sufficiently small so that their observation requires the application of the intense light fields made available by lasers. The theoretical analysis of nonlinear optical effects was first reported in the early 1960s [2.254]. In the same time period, the available laser intensities from pulsed lasers reached a level sufficient that useful nonlinear phenomena such as the generation of a second harmonic laser beam became readily observable [2.255]. However, nonlinear effects were also responsible for undesirable thread-like damage observed in optical elements exposed to high intensity laser beams. It was quickly realized that this was due to the self-focusing of a propagating laser beam at high fluence locations from localized light-induced increases in refractive index. From this time to the early 1970s, research emphasized reducing nonlinear effects in order to maintain the beam quality of high power glass laser systems. Later, interest in nonlinear refraction broadened when optical bistability and phase conjugation were demonstrated in the latter 1970s.

Today, intensified research with nonlinear materials is currently directed at possible application areas such as all-optical switching, modulation, image processing, sensors, advanced targeting, optical limiting, adaptive optics and optical storage/ memory systems. In addition to these, more novel applications are still being discovered. There are numerous texts explaining the concepts and theory underlying nonlinear optics [2.256–258]. New technologies and potential commercial markets for nonlinear optical materials have been thoroughly reviewed in [2.259].

A wide variety of materials are known to possess nonlinear properties including gases, vapours, polymeric media, liquid crystals, biological systems, organic solutions, water, crystals and glasses [2.260]. Inorganic crystals and organic polymers are well-known for their large nonlinear effects, but they have several problems which often prevent practical application of these materials; e.g. cost, optical quality and laser damage threshold. Solid state nonlinear materials, on the other hand, are particularly attractive because of their easy handling; and they often possess a higher laser damage threshold. Of the solid state nonlinear materials, crystals exhibit more pronounced nonlinear behaviour than glassy materials. Indeed, several crystals have already been marketed for harmonic generation and parametric processes. However, glasses have several competitive advantages in comparison to crystals such as mechanical stability, easy handling, possible fabrication of large optics, excellent optical homogeneity, quicker production processes and potential lower cost.

Today, glasses are recognized as one class of usable nonlinear material. Numerous papers have appeared discussing their properties, the origin of their nonlinearity and actual device fabrication and performance. We will present a discussion of glass as a nonlinear material by classifying nonlinear glasses into homogeneous and heterogeneous types. Whereas homogeneous glasses are completely vitrified and isotropic on both a microscopic and macroscopic scale, heterogeneous glasses contain, on the other hand, additional zones, usually sub-microscopic of different chemical identity such as particles, crystals or other glassy phases. Most of the nonlinear examples discussed here are a result of the third order susceptibility, $\chi^{(3)}$, since glassy materials are, in general, isotropic and lack non-centrosymmetric structure, normally regarded as essential for the manifestation of $\chi^{(2)}$ processes [2.256–258]. Nevertheless, the best known $\chi^{(2)}$ process, second harmonic generation of light, has been observed in glasses as well [2.261]. Interested readers can also consult the several review papers which have appeared in the literature [2.262–267]. A short description of nonlinear terms and of related physical units is given in Appendix A.

2.6.1 Homogeneous Glasses

There have been several attempts to identify a relationship to predict the nonlinear response of a material including a purely empirical relationship known as *Miller's* Rule [2.268] and a semi-empirical approach using a perturbation treatment [2.269]. These early efforts all predict that glasses with high linear refractive indices will also possess a correspondingly high nonlinear refractive index. Today, the equation which is widely accepted to evaluate nonlinear refractive index is that proposed in [2.270],

$$n_2 = \frac{68(n_d - 1)\ (n_d^2 + 2)^2}{\nu_d \sqrt{1.52 + \frac{(n_d^2+2)\ (n_d+1)\nu_d}{6n_d}}}\,, \tag{2.146}$$

where n_2 is the nonlinear refractive index, n_d is the refractive index at 587.6 nm and ν_d is the Abbe number at 587.6 nm. According to (2.146), a high nonlinear refractive index is expected for glasses with a high linear refractive index and a low Abbe number (high dispersion). Figure 2.57 presents a schematic representation of lines of constant nonlinear refractive index, as calculated with (2.146), on the Abbe diagram. Current empirical laws thus suggest that a high $\chi^{(3)}$ value will be obtained by formulating a glass with heavy elements as the basic constituents.

Actual measurement of nonlinear refractive index values started in the 1970s when high intensity laser beam sources became available for laboratory use. The initial choice of test samples were active laser materials since self focusing due to high nonlinear index values was a problem in high power glass laser systems. A study by *Weber* et al., [2.271] utilized time-resolved interferometry in order to obtain nonlinear index values for optical glasses and crystals at 1064 nm. As a result of their work, the empirical expression above, (2.146), was shown to provide a good estimate of the nonlinear index of optical materials in the long-wavelength limit.

In search of higher nonlinear glass materials, glass developers have formulated experimental compositions including glasses with high contents of lead and bismuth oxides [2.272, 273], germanates [2.272], chalcogenides [2.272, 274, 275] and glasses with large amounts of titanium and niobium oxides [2.276]. Reported

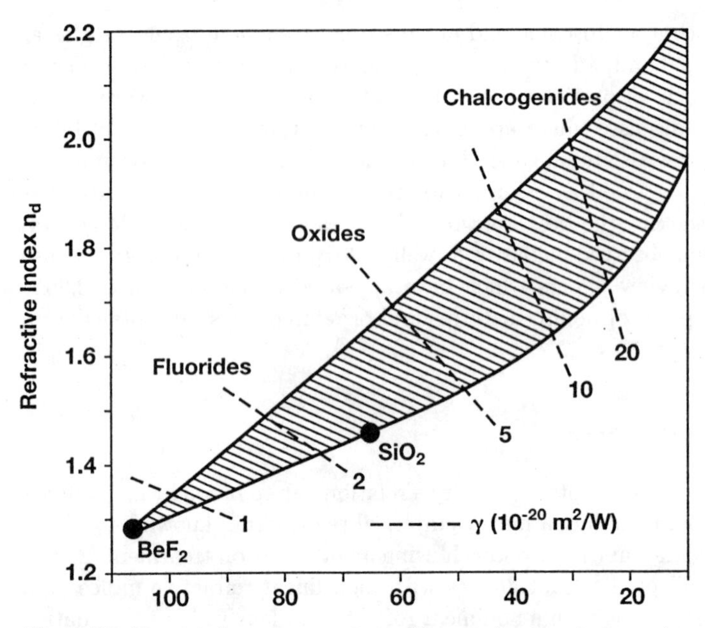

Fig. 2.57. Expected nonresonant nonlinear refractive index value, n_2, as a function of n_d and ν_d, as predicted by (2.146). The dashed lines correspond to lines of constant γ [2.263]. There are two different units for the nonlinear index n_2 and γ. For their relation to one another see Appendix A

results are summarized in Table 2.15 along with corresponding data for the commercial optical glass with the highest linear index, SF 59, for comparison. The highest $\chi^{(3)}$ value, in the range of 10^{-11} esu, was obtained in an As-S-Se composition [2.274, 275]. This is two orders of magnitude larger than that of conventional silicate type glasses and is comparable to values obtained with nonlinear organic compounds. The degree of nonlinearity in homogeneous glasses is felt to be sufficient to render them useful once available laser output powers become higher in the future.

One of the leading application pursuits in nonlinear optical research is the demonstration of an optical transistor or optical switch. Another contribution [2.278] has considered this application and defined a figure of merit for a functional material by considering the number of switching operations which can be performed within a thermal lifetime. They then investigated this figure of merit for a number of representative nonlinear materials: GaAs, a GaAs/GaAlAs multiple quantum well structure, the organic molecule PTS, a semiconductor-doped glass and Schott SF 59. They found that because of its extremely small absorption and inertness to thermal effects, the SF 59 glass had a larger figure of merit than any of the other nonlinear optical materials, despite the far larger nonlinear coefficients of the competitive semiconductor materials. This demonstrates how the advantages offered by glass can, under certain situations, make it the material of choice for a particular application [2.277, 278].

Consistent with material developments there has also been parallel research in fabricating guided wave structures of fibres and planar waveguides from homogeneous nonlinear glasses [2.279]. This work resulted in the observation of the self-induced confinement of a light beam referred to as a spacial soliton [2.280] as well as two newly discovered problems: two-photon absorption [2.281] and absorption by colour centres which mimics two-photon absorption [2.282]. Materials with a high nonlinear refractive index are necessary to accumulate a phase shift in an all-optical switching device. However, nonlinear absorption could cancel this effect before a switchable level is achieved. In [2.281] this limitation is investigated employing a lead-glass fibre consisting of an SF 6 glass core and an SF 56 cladding glass. It was reported that this system did not have usable properties at either 532 nm or 1060 nm, but that the mismatch at 1060 nm was nearly acceptable. Since the two photon transition probability drops significantly as the photon energy is decreased, we expect that adequate performance should be possible if such a device is operated further into the IR.

Although glassy materials are centrosymmetric and thus normally incapable of exhibiting a $\chi^{(2)}$ type of phenomena, observations of second harmonic generation (SHG) were first reported in germania-doped optical fibres by *Osterberg* in 1986 [2.261]. A number of models have been proposed to account for this phenomena which involve some kind of symmetry-breaking organizational process which gives a non-zero $\chi^{(2)}$ necessary for frequency doubling. Two leading theories to explain SHG in glass fibres are the creation of a DC polarization in the fibre through a third order interaction between the fundamental and second harmonic light beams [2.283] and the creation of electrostatic fields by a photogalvanic effect [2.284].

Table 2.15. Nonlinearity of high refractive index glasses

Glasses	$n_{623\,\mathrm{nm}}$	$n_{1060\,\mathrm{nm}}$	$n_2 \times 10^{-13}$ esu[f]
Silicate SF-59[a]		1.91	14.8
QR[a,b]		2.02	20.5
Germanate VIR-3[a]		1.84	9.8
QS[a,c]		1.94	15.6
Gallate RN[a,d]		2.30	68.8
Chalcogenide (As_2S_3)[e]		2.48	258.0
Oxides[g]			
$30TiO_2$-$20Na_2O$-$50SiO_2$	1.72		12.8
$15NbO_{5/2}$-$15TiO_2$-$20Na_2O$-$50SiO_2$	1.83		11.8
$30NbO_{5/2}$-$30TiO_2$-$20Na_2O$-$20SiO_2$	1.95		18.8

[a] Reference [2.272]
[b] QR: $60PbO$-$40SiO_2$
[c] QS: $60PbO$-$40GeO_2$
[d] RN: $40PbO$-$35BiO_{1.5}$-$25GaO_{1.5}$
[e] Reference [2.275]
[f] Reference [2.324]
[g] Reference [2.276]

Table 2.16. Nonlinearity of $CdSe_xS_{1-x}$ glasses

Glasses	Wavelength nm	Sulfur Fraction x	Measured $\chi(3)^a$ 10^{-9} esu	Calculated $n_2{}^b$ 10^{-8} esu	Estimated $n_2{}^c$ 10^{-8} esu
Corning #3484	532.0	0.9	13	32	7.4
Corning #2434	580.0	0.7	5	12	3.2
Schott RG 695	694.3	0.1	3	7	2.1

[a] Reference [2.287]
[b] Reference [2.324]
[c] Reference [2.288]

2.6.2 Heterogeneous Glasses – Semiconductor-Doped Glasses

Semiconductor-doped glasses are known as colloidally coloured, longpass filter glasses in the glass industry and they are readily available from glass filter catalogues [2.285]. The colourants in these glasses are, in most cases, due to small microcrystallites of semiconductor particles, principally of a mixed $CdSe_xS_{1-x}$ variety, which are nucleated and grown by a secondary heat treatment (known as a striking process) of the initially colourless glass. Commercial semiconductor-doped glasses present a yellow, orange, or red appearance with a very steep absorption edge at a wavelength which strongly depends upon the thermal history during the secondary heat treatment, as well as on the characteristics and the concentration of colourants and base glass constituents.

Although considerable research has been completed on the commercial filter glasses as supplied, there has been a number of disagreements in the findings among research groups. In addition to the fact that different companies may prepare these products with different base compositions and manufacturing schedules, these disagreements may also be partially in the fact that each filter glass supplier has the freedom, by selecting different glass formulations and by adjusting the employed striking schedule, to supply a given filter type with a variety of different glasses covering a range of base glass and crystalline phase compositions. Furthermore, use of this flexibility also leads to a corresponding variation in available crystal densities and size distributions.

The nonlinearity of these $CdSe_xS_{1-x}$ glasses was first noticed some time ago and the first application as a saturable absorber appears to have been in 1964 [2.286]. More active research on these materials began after the nonlinear refractive index had been measured for this type of glass in fourwave mixing experiments in 1983 [2.287]. These results are shown in Table 2.16 along with theoretically calculated values given in [2.288] where the near-resonant effect theory was employed.

Most efforts in this field have concentrated on revealing the mechanisms which cause the high nonlinearity in this type of glass, since the linear refractive indices of these glasses are fairly modest. Apparently the empirical correlation described in Sect. 2.6.1 does not explain such a high nonlinearity in this case. Two different nonlinear mechanisms have been proposed depending upon the microcrystalline particle sizes: a band-filling effect for particles larger than the effective radius of the exciton (a bound electron-hole pair), created in the semiconductor particles upon absorption of incident radiation, and a quantum confinement effect for particles smaller than this effective radius.

Band-Filling Effect

For the short-to-medium wavelength cut-on, commercial samples which are available in a ready-struck condition, the usual particle size is in the range of 100 Å. The observed nonlinearity for this size regime of semiconductor particles has been explained by many research groups as due to band-filling [2.289]. Correspondingly observed phenomena are saturation of absorption and a blue shifting of the absorption edge with increasing intensity of the incident probe light.

The response times reported for this case vary over a wide range: microseconds to nanoseconds to femtoseconds. An understanding of this wide range of response domains became possible with the observation given in [2.290, 291] of a photodarkening effect in which initially observed slow nonlinear response times (several tens of nanoseconds) changed into a much quicker (picoseconds) region. With this faster response, they also observed the disappearance of a broad luminescence band at higher wavelengths which they assigned to trapped-carrier recombination. In order to explain this faster response due to the photodarkening effect, a three-level system [2.292] was suggested which includes a conduction and valence band and an intermediate long-lived trapping level. The photodarkening phenomenon changes the carrier-recombination routes by population saturation of this low-lying trapping state. Later, a four-level system with a second, non-radiative trapping level was also proposed. The extremely slow millisecond regime response times have thus been assigned to recombination of this deep level trapped carrier, and the femtosecond order nonlinearity is attributed to intraband carrier decay [2.289].

Quantum Confinement Effect

When the microcrystallite becomes smaller, photo-excited carriers are effectively confined in volumes whose dimensions are all below a critical length. The optical behaviour of such systems is particularly affected by this quantum confinement effect. In particular, with decreasing particle size, the separation between individual energy states in the valence and conduction bands as well as the effective band gap of the semiconductor species increases. In order to exhibit distinct quantum confinement effects, the particle size must be smaller than its characteristic bulk exciton radius which depends upon the microcrystallite species [2.293–299].

Corresponding to the shifting of energy levels described above, quantum confinement effects are manifested by corresponding energy shifts of luminescence and absorption spectra and the appearance of discrete absorption structures attributable to transitions between discrete quantized levels. The response times in the quantum confined particle region are, as in the band filling regime above, also affected by photodarkening in room temperature experiments [2.300]. Apparently the discovery of a photodarkening effect in the $CdSe_xS_{1-x}$-doped glasses had been delayed since many of the initial experiments in the quantum confined region were carried out at lower temperatures.

Efforts have thus been made to develop particle-size control methods. These techniques include striking condition control [2.301] and doping of microcrystallites directly into porous glasses [2.302, 303]. Direct observation of CdSSe microcrystallites by transmission electron microscopy to determine particle size distributions and densities has indicated that these crystallites are hexagonal columns [2.304] and that they are spherical [2.305, 306]. The size distribution of the crystallites has also been reported [2.307].

Applications for $CdSe_xS_{1-x}$ Particle Doped Glasses

Fabrication of semiconductor-doped glass planar waveguides began in the mid-1980s. Ion-exchange is an effective method of increasing the refractive index. This technique has been applied to both bulk and thin-film glass samples. *Gabel* et al. presented the first effective generated four-wave mixing by utilizing the ion exchange technique [2.308]. Others have shown that phenomena observed in bulk semiconductor-doped glasses are also seen in waveguides, i.e. saturable absorption, photodarkening effect and resolution of thermal and electronic effects [2.309].

Successful fabrication of guided wave fibre structures has also been reported in [2.310]; however, attempts to demonstrate optical switching were unsuccessful due to saturation of the index change in the band gap energy region. They recently shifted their attention to the off-resonance regime; i.e., for photon energies well below the band gap. This was motivated by their discovery that certain semiconductor-doped glasses display substantial refractive nonlinearity (up to 40 times greater than silica) even at a wavelength of 1 μm. In order to utilize this off-resonance nonlinearity, the optical fibre must be prepared with sufficiently long pathlengths of a controlled microcrystallite size; a difficult process since the temperature regimes for fibre drawing usually lead to changes in the crystal size distribution.

Other Semiconductor Dopants

Semiconductor dopants other than $CdSe_xS_{1-x}$-mixed crystals have been studied in the search for systems of higher nonlinearity. These dopants include pure CdTe and CdSe, Bi_2S_3, PbS, HgSe, In_2Se_3 and AgI [2.311].

Researchers investigated the optical nonlinearity of pure CdTe quantum dots in glass, because CdTe has the largest Bohr radius of any semiconductor successfully grown as a quantum dot in glass [2.312]. Therefore, it was anticipated that quantum confinement effects would be observed in larger dots, making CdTe dots especially sensitive to electric fields. One observation of exceptional nonlinearity has been reported by another group who doped phosphate glasses with a concentration of CdSe two orders of magnitude higher than that found in conventional filter glasses [2.313]. Their degenerate four wave mixing experiments showed the $\chi^{(3)}$ of such a glass to be of the order of 10^{-6} esu near maximum resonance.

Table 2.17. Nonlinearity of metal particle dispersed glasses

Glasses	Measured $\chi(3)$ 10^{-9} esu[a]	Calculated $\chi(3)$ 10^{-9} esu[b]
Gold Particle Dispersed (530 nm)	15	0.87
Silver Particle Dispersed (400 nm)	2.4	0.18

[a] Reference [2.316]
[b] Reference [2.317]

An all-optical switching material figure of merit for this type of glass was presented and material figures of merit calculated for several semiconductors. ZnTe, CdTe and GaAs were identified as dopants which should exhibit higher nonlinear activity [2.314].

The next generation nonlinear optical material project in Japan disclosed CuCl and CuF-doped glasses as new materials which are expected to have a 1,000 times larger light sensitivity than existing $CdSe_xS_{1-x}$-doped glasses [2.315]. These systems have thus attracted considerable attention from the nonlinear glass research community.

2.6.3 Heterogeneous Glasses – Metallic Particle Doped Glasses

The nonlinearity of silver and gold particle dispersed glasses has been observed for several years. Successful optical phase conjugation utilizing such a system was reported in 1985 [2.316]. Representative nonlinear properties of these metallic particle doped glasses are presented in Table 2.17. The nonlinearity in these types of glasses is assigned to the local field effect enhanced near the surface plasma resonance. The observed nonlinear responses are in the picosecond region [2.317–319] even though a slower nonlinearity believed to be of thermal origin has also been reported for an Mn-doped glass [2.320]. Although the bulk of reported research has been in silver and gold colloid systems, modeling the effective nonlinearity of spheroidal metal particles embedded in a linear host medium suggests that aluminum can be exploited to obtain a larger enhancement of nonlinearity [2.321].

2.6.4 Heterogeneous Glasses – Other Systems

Theoretical calculations of nonlinear behaviour given in [2.322] have shown that model composites of nanospheres with a metallic core and nonlinear shell or with a nonlinear core and metallic shell can exhibit a 10^8 order of magnitude increase in nonlinear activity for photon energies where the optical field can match the surface-mediated plasmon resonance for this system. Such a nonlinear system has been demonstrated in solution, and there is strong motivation to develop techniques for fabricating such systems in a glass.

Lastly, there has also been research in the area of organically doped glasses. Here, research has focused on identifying suitable glass hosts which can be prepared with high optical quality. The difficulty is that these glasses must have the low viscosities required for a uniform blending of the organic dopant at temperatures low enough so that the employed organic molecules are not decomposed, vapourized, or experience combustion. One demonstrated example is lead-tin-fluorophosphate-based glasses which can be melted at or below 500 °C and which have glass transition temperatures as low as 95 °C [2.323]. Other methods to prepare a suitable glass host at low temperature, in many cases utilizing technology appropriate for the fabrication of integrated optic devices, are under development [2.262–264].

Today the field of nonlinear optics remains a field of research. The applications of this technology are in the early stage of development. Of particular current interest is the possibility of switching a light beam by the application of a second light beam in a very short time period. Such capability in the form of a commercial optoelectronic device is of extreme importance to the telecommunications and computer industry of the future.

Appendix A. Terms of Nonlinear Optics

If a light beam of low intensity is propagated through an isotropic medium such as glass, there is an induced polarization, P, in the material given by:

$$P = \chi E \tag{2.147}$$

where E is the magnitude of the applied electric field and the constant of proportionality, χ, is called the dielectric susceptibility. Note that for small values of E, the induced polarization depends linearly on the magnitude of the applied electric field. If the electric field strength of the applied light is very high, this linear relationship is no longer valid, and this function then has to be described in a series of powers of E:

$$P = \chi^{(1)} E + \chi^{(2)} E^2 + \chi^{(3)} E^3 + \dots \tag{2.148}$$

where $\chi^{(2)}$ and $\chi^{(3)}$ are the second- and third-order susceptibilities, respectively.

The refractive index can be expressed as:

$$n = n_0 + n_2 E^2 \tag{2.149}$$

where n_0 is the linear refractive index and n_2 the nonlinear refractive index. Both n_2 and E are often expressed in electro-static units (esu). Another equation also used (applying SI units) is:

$$n = n_0 + \gamma I \tag{2.150}$$

where γ is the nonlinear refractive index (in m^2W^{-1}) and I is the intensity of the light beam (in Wm^{-2}).

For the case of linear polarized, monochromatic light the relationship between n_2 and $\chi^{(3)}$ is given in (2.151):

$$n_2\,(\text{esu}) = \frac{12\pi}{n_0}\chi^{(3)}\,(\text{esu}) \tag{2.151}$$

and the relation between γ and $\chi^{(3)}$ is given in (2.152):

$$\gamma\,(\text{m}^2\text{W}^{-1}) = \frac{480\pi^2}{n_0^2 c}\chi^{(3)}\,(\text{esu}) \tag{2.152}$$

where c is the velocity of light ($3 \times 10^8\,\text{ms}^{-1}$).

If one has the value of $\chi^{(3)}$ in esu-units, the calculation in SI is done according to (2.153):

$$\chi^{(3)}\,(\text{m}^2\text{V}^{-2}) = 1.4 \times 10^{-8}\chi^{(3)}\,(\text{esu})\,. \tag{2.153}$$

The conversion between n_2 (esu) and γ (m^2W^{-1}) is made by (2.154):

$$n_2\,(\text{esu}) = \frac{c n_0}{40\pi}\gamma(\text{m}^2\text{W}^{-1})\,. \tag{2.154}$$

2.7 Resistance to Laser Radiation

Joseph S. Hayden

Optical glass is often utilized in transmissive or reflective applications in conjunction with laser radiation. In many cases, no change in the optical glass employed in these situations is observed. However, in some cases, generally those involving either high energy photons or high intensity light beams, changes are observed in the appearance and properties of the glass. Some of these changes are reversible, whereas others are not only permanent but can be serious enough in nature to render the optical element unusable.

We will discuss here the nature of changes which can take place in optical glass as a result of its use with laser radiation. Emphasis will be placed on a number of particular applications with specific information provided on the resistance of certain glasses to both reversible and irreversible changes in properties upon application with laser sources. Readers interested in a more complete treatment including, for example, discussions relating to crystalline materials, are referred to the literature [2.325–329].

2.7.1 Bulk Damage Resistance to Laser Radiation

Permanent internal damage in optical glass due to laser radiation takes several forms. For sufficiently strong applied electric field, dielectric breakdown of the optical material can take place. A related form of damage to be considered is thread-like damage due to self focusing of the laser beam which is correlated with the nonlinear refractive index of the sample. As described in Sect. 2.6, the refractive index of glass increases if it is under the influence of a strong electric field. The laser beam intensity profile is generally higher in the centre and decreases

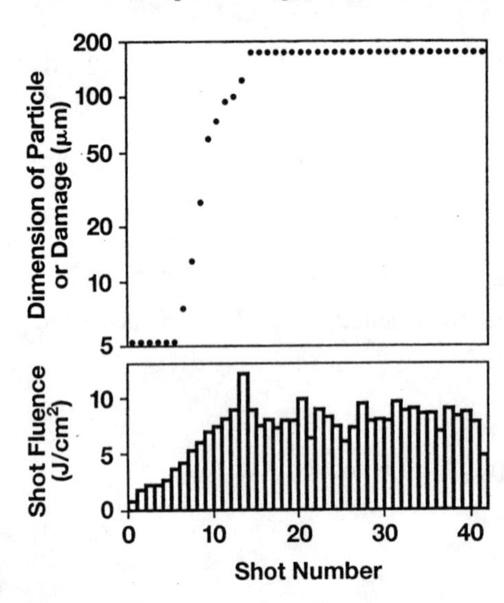

Fig. 2.58. Increasing of the damage dimension of a 4 µm inclusion in glass by irradiation with laser pulses of increasing energy density [2.333]

towards the outside region of the beam profile. The effective refractive index is altered according to the intensity distribution and this causes a lensing effect. By self focusing, the cross-section of the laser beam is confined so strongly that a dielectric breakdown is caused within the glass (forming thin dim threads).

There is a critical power, P_2, for self focusing given by

$$P_2 = \frac{3.77c\lambda^2}{32\pi^2 n_2} \tag{2.155}$$

where c is the speed of light, λ the laser wavelength and n_2 is the nonlinear refractive index (in esu). If the power of the laser beam is considerably less than P_2 (i.e. 25% of P_2), self-focusing does not influence the measurement of damage threshold [2.330].

Other forms of damage include that due to absorbing inclusions within the glass which, upon illumination, can heat up to a point that they explode within the glass; leaving behind point-like, individual damage sites. The absorbing inclusions can be undissolved particles of the glass batch, parts of the glass melting unit, or precipitated platinum particles. Another thermal effect occurs when the bulk glass is sufficiently absorbing that the irradiated portion of the optical element increases in temperature, ultimately leading to thermal deformation and even melting.

There have been numerous investigations of laser-induced damage over the last 30 years; for instance, see the Annual Reports of the Laser Program at Lawrence Livermore National Laboratory [2.331] and the series of Symposia on Laser Induced Damage in Optical Materials [2.332]. One consequence of this work is that transparent materials for use with laser beams must not have any absorbing particles larger than a few tenths of microns. Particles a few microns in size can lead to damage sites larger than 1 mm (see Fig. 2.58 [2.333]).

The short-pulse laser damage threshold for a wide range of optical glasses has been experimentally evaluated in [2.334] using a 3ns duration Nd:YAG laser at 1060 nm and in [2.335] with a 25 ns duration ruby laser at 694 nm. Analysis of

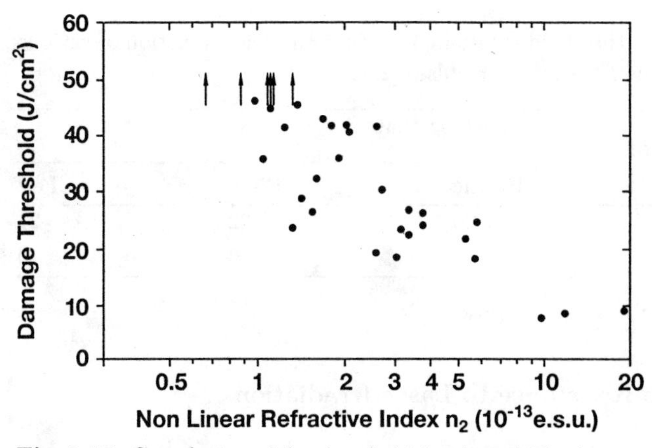

Fig. 2.59. Correlation of the threshold for thread-like damage caused by dielectric breakdown from self-focusing with the nonlinear refractive index [2.334]. Up to energy densities marked by the arrows no damage could be detected (higher energy densities were not measured)

these findings verifies the relationship of thread-like damage correlating with the nonlinear refractive indices of the materials tested. Additionally, self-focusing also scales with the thickness of the irradiated optic as given by the well-known B-Integral [2.329]. A summary of threshold levels for thread-like damage in optical glasses is presented in Fig. 2.59 along with computed values of nonlinear refractive index (2.146). In critical cases, it is advantageous to apply glass in the form of thin sheets because the damage by self-focusing occurs only for a sufficient accumulated pathlength. The thread-like damage starts first at the exit surface and grows with increasing beam intensity into the sample.

Point-like damage has been observed in nearly all of the twenty-two optical glasses investigated in Ref. [2.334], with the damage threshold varying from 7 to more than 47 J/cm^2 for the 3 ns pulse length Nd:YAG laser employed. Optical glasses of the highest optical quality are generally prepared in platinum melting units. Consequently, platinum particles in glass are of special importance in determining the resistance of glass to point-like laser damage. The aggression of platinum by the glass melt depends upon the chemical composition of the glass as well as the parameters of the melting process. Therefore, it is not possible to quantify a definite threshold energy for point-like damage of an individual glass type. The problem of a damaging particle in glass has been thoroughly modelled [2.336]; and, more recently, also for platinum particles in laser glass [2.337].

In 1986, as a result of manufacturing developments discussed further in Sect. 8.5, phosphate-based laser glasses became available in a form free of all damaging inclusions [2.338]. The bulk damage threshold for these materials is limited only by self-focusing. These glass compositions are already additionally characterized by low values of nonlinear refractive index. The complete absence of damaging inclusions is normally verified by the deliberate damage testing of the glass with a high fluence Nd:YAG laser beam [2.339].

Table 2.18. Surface damage threshold values in J/cm^2 for BK 7 as a function of various polishing parameters for 3 ns 1060 nm laser pulses [2.334]

	Polishing Plate		
Polishing Material	Plastic	Pitch	Felt
CeO_2	26	22	21
$0.8\ ZrO_2 + 0.2\ Fe_2O_3$	18	18	20

2.7.2 Surface Damage Resistance to Laser Irradiation

In practice it is the surface of an optical element which first experiences damage from laser radiation as the test conditions become more severe. Surface preparation plays an important role in the resistance of glass surfaces to short pulse laser radiation. The lowering of the damage threshold due to dust, stains and scratches on a glass surface has been considered in detail by *Bloembergen* [2.340]. So as not to play a role in the observed damage threshold of a material, such defects must generally be smaller than one-hundredth of the employed laser wavelength.

For sufficiently high laser fluences, surface damage becomes clearly visible upon inspection with a differential-interference microscope as circular elevations called pustules with diameters of about 0.01 to 0.1 mm. Other studies [2.334] indicate that the threshold for initial damage, when employing an Nd:YAG laser of nominal 3 ns pulse length, was nearly the same for all optical glasses if the polishing process of the samples was the same. The observed damage threshold depended only on the polishing technique employed and not on the chemical composition of the glass itself. The number of observed pustules increased with increasing energy density in the test beam and their size grew with increasing pulse intensity.

The damage mechanism has been assigned to the following: residue from the glass polishing material acts as the initial nucleation site for damage, especially when this residue is highly absorbing at the wavelength of interest, leading to localized strain points within the glass surface layer. Table 2.18 presents the findings given in [2.334] as a function of the polishing technique employed. Glass polished with cerium oxide showed a higher damage threshold than that which was polished with a mixture of ZrO_2 and Fe_2O_3. The threshold energy density for surface damage increases with the pulse length of the laser irradiation, τ, as $\tau^{0.4}$, see Fig. 2.60 [2.327] (with respect to the energy density the threshold is considerably smaller for smaller pulse length than for higher pulse length). In this figure it should be noted that the damage behaviour is very similar for a wide variety of glass types if they have been prepared with the same high quality polishing technique [2.331]. The measurements in Fig. 2.60 were done with a pulsed laser operating at a wavelength of 1060 nm; at shorter wavelengths the damage energy thresholds become smaller [2.341].

It is well established that application of an appropriate coating on the glass surface can reduce the damage threshold value considerably. The development of

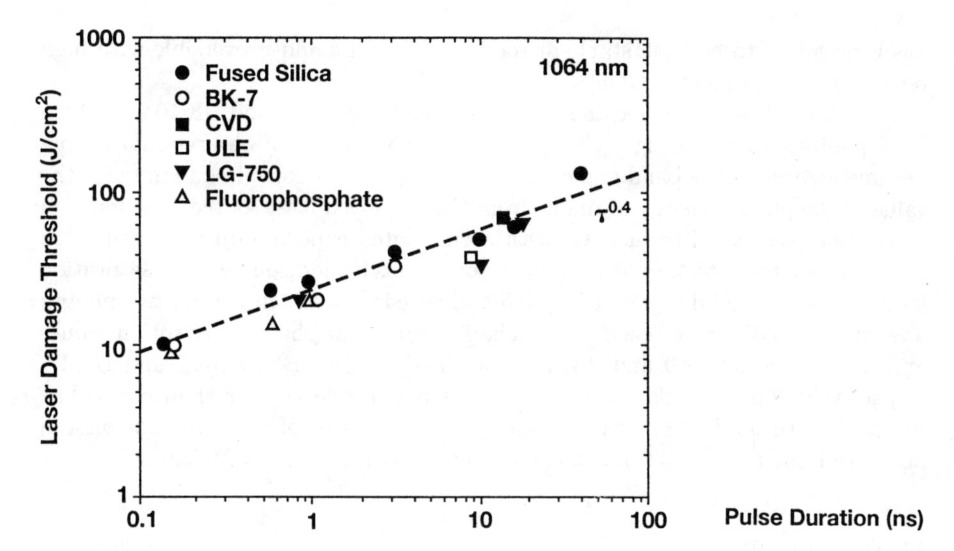

Fig. 2.60. Average threshold of surface damage of bare polished glasses as a function of laser pulse length, laser wavelength 1064 nm [2.327]

new coating techniques is still in process. Treatment of coatings by (moderate) laser irradiation increases their damage resistance.

2.7.3 Multiphoton Darkening of Optical Glass

Reversible damage to optical glass as a result of short pulse length laser illumination usually takes the form of darkening of the exposed glass. Although it is most commonly observed with high photon energies characteristic of UV light sources, this is often referred to as a solarization. This damage is termed reversible since, in many cases, the discolouration can be at least partially removed by application of a bake cycle at an appropriate elevated temperature. However, in some cases, this discolouration is correlated to defects which are sufficiently stable that the material's initial transparency cannot be recovered.

A discussion of UV-induced solarization and of optical glasses resistant to this and to other forms of high energy radiation can be found in Sects. 2.3 and 8.7. Here we will restrict our discussion to the laser-induced discolouration in optical glass which is assigned to multiphoton absorption processes.

The first reported solarization attributed to multiphoton absorption in glass was in 1968 [2.342]. Here discolouration was detected in a flint glass exposed to Nd:YAG radiation of 532 nm. Subsequently, others reported measurements of nonlinear absorption and of solarization in glasses exposed to laser light of UV and visible wavelengths at which the glasses had negligible single photon absorption [2.343, 344]. In general, the magnitude of induced solarization was undetectable for materials with band gap values more than twice the photon energies of the employed laser sources. However, unlike classic solarization, this form of damage has been found to be both stable at room temperature and irreversible with high temperature treatments [2.345].

White et al. have studied the optical glasses BK 3, BK 7 and BK 10 at 532 nm by a photothermal lensing technique [2.344]. The degree of observed darkening became greater as the band gap energy for the glasses became less than twice the value of the photon energy available from the employed laser source. The induced absorption was found to increase with accumulated exposure up to a saturation point proportional to the peak intensity of the excitation source. The saturation level of the induced discolouration also correlated with the measured two photon absorption coefficient of each glass. The reported two photon absorption values at 532 nm were 0.6, 2.9 and 0.4 cm^{-1} per TW/cm^2 for BK 3, BK 7 and BK 10, respectively. These levels are two orders of magnitude smaller than crystalline materials with similar band gap energies [2.346]. The level of induced single photon absorption for BK 7 at 532 nm was reported as 0.07 cm^{-1} per GW/cm^2.

References

2.1 W. Snell: *Risneri opticam cum annotationibus Willibrordi Snellii, Pars prima, librum primum continens* (I.A. Vollgraff, ed.) (Plantini, Ghent 1918)

2.2 R. Descartes: *Dioptrique, Météores* (anonymously published) (Leyden 1637)

2.3 R. Descartes: *Principia philosophiae* (Amsterdam 1644)

2.4 M. Herzberger: *Modern Geometrical Optics* (R.E. Krieger Publishing Company, Huntington 1980)

2.5 M. Born, E. Wolf: *Principles of Optics*, 6th ed. (Pergamon, Oxford 1980)

2.6 F.S. Jenkins, H.E. White: *Fundamentals of Optics*, 4th ed. (McGraw-Hill, New York 1981)

2.7 H.A. Lorentz: "Über die Beziehung zwischen der Fortpflanzungsgeschwindigkeit des Lichtes und der Körperdichte", Ann. Phys. **9**, 641–665 (1880)

2.8 L. Lorenz: "Über die Refractionsconstante", Wiedem. Ann. **11**, 70–103 (1881)

2.9 R. de L. Kronig, H.A. Kramers: "Zur Theorie der Absorption und Dispersion in den Röntgenspektren", Z. Physik **48**, 174–179 (1928)

2.10 H.R. Philipp: "Silicon dioxide (SiO_2) (glass)" in *Handbook of Optical Constants of Solids*, ed. by E.D. Palik (Academic Press, New York 1985) pp. 749–763

2.11 A.L. Cauchy: "Sur la réfraction et la réflexion de la lumière", Bull. des sc. math. **14**, 6–10 (1830)

2.12 Sellmeier: "Zur Erklärung der abnormen Farbenfolge im Spectrum einiger Substanzen", Pogg. Ann. **143**, 272–282 (1871)

2.13 Schott: *Optical Glass Catalog* (Mainz 1992)

2.14 C.F. Bohren, D.R. Huffman: *Absorption and Scattering of Light by Small Particles* (Wiley, New York 1983)

2.15 P. Drude: *The Theory of Optics* (Longmans Green, New York 1922)

2.16 Schott: *Optical Glass Catalog* (Mainz 1966)

2.17 M. Herzberger: "Color correction in optical systems and a new dispersion formula", Optica Acta **6**, 197–215 (1959)

2.18 W. Geffcken: "Bedeutung und Gültigkeitsbereich der Abbeschen Gleichung für die Teildispersion von Gläsern", in *Beiträge zur Angewandten Glasforschung*, ed. by E. Schott (Wissenschaftliche Verlagsgesellschaft, Stuttgart (1959) pp. 255–268

2.19 W. Geffcken: "Partial dispersions of glasses, part 2", Optica Acta **12**, 275–304 (1965)

2.20 W. Geffcken: "Partial dispersions of glasses, part 3", Applied Optics **12**, 2978–2990 (1973)

2.21 H.A. Buchdahl: *Optical Aberration Coefficients*, (Dover, New York 1968)

2.22 P.N. Robb, R.I. Mercado: "Calculation of refractive indices using Buchdahl's chromatic coordinate", Applied Optics **22**, 1198–1215 (1983)

2.23 P.N. Robb: "Selection of optical glasses", SPIE Vol. **554**, 60–75 (1985)

2.24 E.E. Bell: "Optical constants and their measurement", in *Handbuch der Physik*, ed. by S. Flügge, Vol. 25/2a (Springer, Berlin, Heidelberg 1967) pp. 1–58

2.25 L. Ward: *The Optical Constants of Bulk Material and Films* (Adam Hilger, Bristol 1988)

2.26 F. Kohlrausch: *Praktische Physik*, Vol. I, (Teubner, Stuttgart 1968) p. 408

2.27 B. Edlén: "The refractive index of air", Metrologia **2**, 71–80 (1966)

2.28 J.C. Owens: "Optical refractive index of air: dependence on pressure, temperature and composition", Appl. Opt. **6**, 51–59 (1967)

2.29 E. Abbe: *Gesammelte Abhandlungen*, ed. by S. Czapski, E. Wandersleb, M. von Rohr, Vol. 1–4 (Gustav Fischer, Jena 1904–1928)

2.30 G. Gliemeroth: "Light through optical glass", Schott Information **4** (1981)

2.31 H.-G. Zimmer: *Geometrische Optik* (Springer, Berlin, Heidelberg 1967)

2.32 T.S. Izumitani, K. Nakagawa: "Cause of the abnormal partial dispersion of optical glasses", VII. Internat. Congress on Glass (Paper I.1.5), Brussels (1965)

2.33 H. Rawson: *Inorganic Glass-Forming Systems* (Academic Press, London 1967)

2.34 A. Paul: *Chemistry of Glasses* (Chapman and Hall, New York 1982)

2.35 N.J. Kreidl: "Recent Highlights of Glass Science, Part 1", Glastechn. Ber. **60**, 249–260 (1987)

2.36 N.J. Kreidl: "Recent Highlights of Glass Science, Part 2", Glastechn. Ber. **63**, 277–287 (1990)

2.37 N.J. Kreidl: "Variability of Optical Properties and Structure of Glass", Glastechn. Ber. **62**, 213–218 (1989)

2.38 N.J. Kreidl: "Glass Forming Systems", in *Glass Science and Technology*, Vol. 1, ed. by D.R. Uhlmann, N.J. Kreidl, (Academic Press, New York 1983) pp. 1–49, 107–299

2.39 J.A. Savage: "Optical Properties of Chalcogenide Glass", J. Non Cryst. Sol. **47**, 101–106 (1982)

2.40 J.A. Savage: *Infrared Optical Materials and their Antireflection Coatings* (A. Hilger Ltd., Bristol 1985)

2.41 J.A. Savage: "Crystalline Optical Materials for Ultraviolet, Visible and Infrared Applications", in: *Optical Materials*, ed. by S. Musikant, B.J. Thomson, (M. Dekker Inc., New York 1990) pp. 323–419

2.42 M. Poulain: "Halide Glasses", J. Non Cryst. Sol. **56**, 1–14 (1983)

2.43 D.C. Tran, G.H. Sigel jr, B. Bendow: "Heavy Metals Flouride Glasses and Fibres: A Review", J. of Lightware Technology **LT-2**, 566–586 (1984)

2.44 J. Lucas, J.L. Adam: "Halide Glasses and Their Optical Properties", Glastechn. Ber. **62**, 422–440 (1989)

2.45 J.M. Parker, R.W. France: "Optical Properties of Halide Glasses" in *Glasses and Glass-Ceramics*, ed. by M. H. Lewis (Champman and Hall, London 1989) pp. 156–202

2.46 P.A. Tick, P.L. Bocko: "Optical Fiber Materials" in *Optical Materials*, ed. by S. Musikant, B.J. Thomson, (M. Dekker Inc., New York 1990) pp. 147–322

2.47 M. Poulain: "Glass Systems and Structures in Fluoride Glasses", in *Critical Reports on Applied Chemistry*, ed. by A.E.Comyns, Vol. 27 (J. Wiley and Sons, Chichester, 1989) pp. 11–48

2.48 T. Reuters, Schott Glaswerke, private communication

2.49 W.H. Zachariasen: "Die Struktur der Gläser", Glastechn. Ber. **11**, 120–123 (1933)

2.50 W.H. Zachariasen: "The atomic arrangement in glass", J. Amer. Chem. Soc. **54**, 3841–3851 (1932)

2.51 B.E. Warren, A.D. Loring: "X-Ray diffraction study of the structure of soda-silica glasses", J. Am. Ceram. Soc. **18**, 269–276 (1935)

2.52 W. Vogel: *Glaschemie* (VEB Deutscher Verlag für Grundstoffindustrie, Leipzig 1979)

2.53 H. Scholze: *Glas, Natur, Struktur und Eigenschaften* (Springer Berlin, Heidelberg 1988)

2.54 T. Okura, K. Yamashita, T. Kanazawa: "A Structural Explanation for the Phosphate Glass Anomaly", Phys. Chem. Glasses **29**, 13–17 (1988)

2.55 W. Geffcken: "Mehrstoffsysteme zum Aufbau Optischer Gläser", Glastechn. Ber. **34**, 91–101 (1961)

2.56 M. Faulstich, W. Jahn, G. Krolla, N. Neuroth: "Laserglas", German Patent 2717916 (1977)

2.57 M. Faulstich: "Mehrstoffsysteme zum Aufbau Optischer Gläser", Glastechn. Ber. **34**, 102–107 (1961)

2.58 W. Jahn: "Mehrstoffsysteme zum Aufbau Optischer Gläser", Glastechn. Ber. **34**, 107–120 (1961)

2.59 V.N. Polukhin: "A Review of Glass Forming Systems Used for the Synthesis of Various Types of Optical Glass", Fiz. Khim. Stekla **6**, 641–650 (1980)

2.60 N.J. Kreidl: "Optical properties", in *Handbook of Glass Manufacture,* ed. by F.V. Tooley, (Books for Industry, New York 1974) pp. 957–997

2.61 E. Zschimmer: "Die Glasindustrie in Jena" (E. Diedrichs Verlag, Jena 1909)

2.62 L.W. Eberlin: "Optical Glass", US Patent 2206081 (1936)

2.63 G.F. Brewster, N.J. Kreidl, T.G. Pett: "Lanthanum and barium in glass-forming systems", J. Soc. Glass Technol. **31**, 153–169 (1947)

2.64 G.W. Morey: "Optical Glass", US Patent 2150694 (1935)

2.65 D. Ehrt, W. Vogel: "Optische Gläser mit anomaler Teildispersion", Silikattechnik **31**, 147–151 (1982)

2.66 B.H. Slevogt: *Technische Optik* (Walter de Gruyter, Berlin 1974)

2.67 K. Fajans, N.J. Kreidl: "Stability of lead glasses and polarization of ions", J. Amer. Ceram. Soc. **31**, 105–114 (1948)

2.68 J.E. Stanworth: "On the structure of glass", J. Soc. Glass Technol. **32**, 154–172 (1948)

2.69 M.B. Volf: *Chemical Approach to Glass*, Glass Science and Technology 7 (Elsevier, Amsterdam 1984)

2.70 W. Vogel, W. Heindorf: "Color and structure of silicate glasses of high Pb content", Z. Chem. **3**, 394–395 (1963)

2.71 L. Roß, D. Grabowski, B. Speit: "Hoch PbO-haltige Gläser im System SiO_2-PbO-M_2O mit erhöhter chemischer Beständigkeit", German Patent Application 3404363 (1984)

2.72 R. Schmidt: *Die Rohstoffe zur Glaserzeugung* (Akademische Verlagsgesellschaft, Leipzig 1958)

2.73 D. Ehrt, C. Jäger, W. Vogel: "Reaktionsprozesse bei der Glasbildung und Kristallisation von Fluoroaluminatgläsern", Wiss. Zeitschr. Friedrich-Schiller-Univ. Jena, Naturwiss. Reihe **36**, 867–884 (1987)

2.74 In terms of workability at higher temperatures, a glass is called "short" if the gradient of the viscosity vs. the temperature ($d\eta/dT$) is high, i.e., the viscosity depends extremely on temperature. Similarly, glasses are called "long" if the gradient $d\eta/dT$ has a low value. Although the same terms are used for the dispersion characteristics, in this case the terms "short" and "long" are used to classify the thermal workability of glasses and not their dispersion properties. The use of the terms "short" and "long" is very common for the thermal workability as well as for the dispersion properties. In terms of workability, the glasses of this system are "short".

2.75 A.M. Bishay, P. Askalani: "Properties of antimony glasses in relation to structure", Comptes rendus du VIIe Congrès International du Verre, Bruxelles (1965) Part 1, Nr. 24, pp. 25–47

2.76 A.A. Appen: "Some anomalies in the properties of glass", Traveaux du IVth Congres International du Verre, Paris, (1956), pp. 36–40

2.77 A.A. Appen: "Attempt to classify components according to their effect on the surface tension of silicate melts", Zhur. Fiz. Khim. **26**, 1399–1404 (1952)

2.78 A.A. Appen: "Behavior and properties of the heavy metal oxides of zinc, cadmium and lead silicate glasses", J. Appl. Chem. USSR **25**, 1303–1309 (1952)

2.79 A.A. Appen: "Calculation of the physical properties of silicate glasses from their composition", Doklady Akad. Nauk SSSR **69**, 841–844 (1949)

2.80 A.A. Appen: "Relation between the properties and constitution of glasses. I. State and characteristics of silica in silicate glasses", Zhur. Priklad. Khim. **24**, 904–914 (1951)

2.81 A.A. Appen: "Relation between the properties and constitution of glasses. II. Behavior and properties of alkaline metal oxides in silicate glasses", J. Appl. Chem. USSR **24**, 1141–1148 (1951)

2.82 A.A. Appen: "Relation between the properties and constitution of glasses. III. Behavior and properties of alkaline earth metal oxides in silicate glasses", J. Appl. Chem. USSR **24**, 1273–1280 (1951)

2.83 A.A. Appen: "Relation between the properties and constitution of glasses. V. State and properties of Al_2O_3 and TiO_2 in silicate glasses", J. Appl. Chem. USSR **26**, 7–15 (1953)

2.84 A.A. Appen: "Relation between the properties and constitution of glasses. VI. The condition and properties of boron trioxide in silicate glasses", J. Appl. Chem. USSR **26**, 529–536 (1953)

2.85 A.A. Appen: "Relation between the properties and constitution of glasses. VIII. The structure of complex silicate glasses as a function of their properties", J. Appl. Chem USSR **27**, 113–117 (1954)

2.86 L.I. Demkina: *Investigation of the Properties of Glasses in Relation to their Composition* (Gosudarst. Izdatel. Oboronnoi Prom., Moscow 1958)

2.87 M.L. Huggins, K.H. Sun: "Calculation of density and optical constants of a glass from its composition in weight percentage", J. Am. Ceram. Soc. **26**, 4–11 (1943)

2.88 M.L. Huggins: "Density of silicate glasses as a function of composition", J. Optical Soc. Am. **30**, 420–430 (1940)

2.89 P. Gilard, L. Dubrul: "The calculation of physical properties of glass. III. The index of refraction", J. Soc. Glass Tech. **21**, 476–488 (1937)

2.90 F. Reitmayer: "Die Dispersion einiger optischer Gläser im Spektralgebiet von 0,33 bis 2,0 μ und ihre Abhängigkeit von den Eigenfrequenzen der Gläser", Glastechn. Ber. **34**, 122–130 (1961)

2.91 T. Izumitani, K. Nakagawa: "Cause of the abnormal partial dispersion of optical glasses", Comptes Rendus du VIIe Congrès International du Verre, Bruxelles (1965) part 1, Nr. 5 pp. 1–12

2.92 G.F. Brewster, J.R. Hensler, J.L. Rood, R.A. Weidel: "Partial dispersion ratios of some new borate and phosphate glasses", Appl. Optics **5**, 1891–1894 (1966)

2.93 H.H. Käs: "Änderung der Dispersion von Gläsern durch den Einbau zusätzlicher Absorptionszentren", Glastechn. Ber. **45**, 1–9 (1972)

2.94 T.S. Izumitani, *Optical Glass* (American Institute of Physics, New York 1986)

2.95 D. Grabowski, private communications

2.96 P. Brix: "Rechnergestützte Glasentwicklung", paper given at the meeting of the Fachausschuß III of the DGG in Würzburg at October 12, 1988, abstract published in: Glastechn. Ber. **62**, N21 (1989)

2.97 N.T. Huff, A.D. Call: "Computerized Prediction of Glass Compositions from Properties", J. Am. Ceram. Soc. **56**, 55–57 (1973)

2.98 J.A. Nelder, R. Mead, Computer Journal **7**, 308–313 (1965)

2.99 S.L. Morgan, S.N. Deming: "Simplex optimization of analytical methods", Analytical Chemistry **46**, 1170–1181 (1974)

2.100 The exact formula for P is

$$P = \frac{(1-r)^2}{1-r^2\tau_i^2}$$

For glasses with r-values in the size of 0.05 the approximation according to (2.78) can cause deviations from the exact value in the size of 1×10^{-3} at the most (depending on the τ_i-value). For r-values in the size of 0.1 there may be deviations in the size of 1×10^{-2} at the most (depending on the τ_i-value).

2.101 J. Derkosch: *Absorptionsspektralanalyse im ultravioletten, sichtbaren und infraroten Spektralgebiet* (Akademische Verlagsges., Frankfurt/Main, Germany 1967) pp. 24–196

2.102 H. Volkmann (ed.): *Handbuch der Infrarot-Spektroskopie* (Verlag Chemie, Weinheim 1972) pp. 23–126, 243–262

2.103 L. Ward: "The optical constants of bulk materials and films", in *Adam Hilger Series on Optics and Optoelectronics*, ed. by E.R. Pike and T.W. Welford (A. Hilger, Bristol, UK 1988)

2.104 C. Castellini, G. Emiliani, E. Masetti, P. Roggi, P. Polato: "Characterization and calibration of a variable-angle absolute reflectometer" Appl. Opt. **29**, 538–543 (1990)

2.105 G. Fuxi: *Optical and Spectroscopical Properties of Glass* (Springer, Berlin, Heidelberg 1992) pp. 62–77, 148–203

2.106 S. Hirota, T. Izumitani, R. Onaka: "Reflection spectra of various kinds of oxide glasses and fluoride glasses in the vacuum ultraviolet region", J. Non. Cryst. Sol. **72**, 39–50 (1985)

2.107 G.H. Sigel, Jr.: "Vacuum ultraviolet absorption in alkali doped fused silica and silicate glasses". J. Phys. Chem. Solids **32**, 2373–2383 (1971)

2.108 G.H. Sigel, Jr. "Optical absorption of glasses" in *Treatise on Material Science and Technology*, ed. by M. Tomozawa and R. H. Doremus (Academic Press, New York 1977) pp. 5–89

2.109 I. Simon: "Infrared Studies of Glass", in *Modern Aspects of Vitreous State*, ed. by J. D. Mackencie (Butterworths, London 1960) pp. 120–151

2.110 P.E. Jellyman, J.P. Procter: "Infrared reflection spectra of glasses", J. Glass Techn. **39**, 173T–179T (1955)

2.111 H. Bach, R. Haspel, N. Neuroth: "Ion beam milling of silica for IR transmission" J. Physics E **9**, 557–559 (1976)

2.112 Heraeus Quarzschmelze, catalogue of fused silica (1973)

2.113 T. Bates: "Ligand Field Theory and Absorption Spectra of Transition Metal Ions in Glasses", in *Modern Aspects of the Vitreous State*, ed. by J. D. Mackenzie, Vol. 2, Chap. 5, Butterworth, London (1962) pp. 195–254

2.114 D. Krause: "Aussagen der Absorptionsspektroskopie im Sichtbaren und UV zur Struktur von Glas", Deutsche Glastechnische Gesellschaft, Fachausschußbericht 70: Nahordnungsfelder in Gläsern (1974), pp. 188–218

2.115 C.R. Bamford: *Colour Generation and Control in Glass* (Elsevier Scient. Publ. Co., Amsterdam 1977)

2.116 T. Nassau: *The Physics and Chemistry of Color* (J. Wiley and Sons, New York 1983)

2.117 W.M. Yen: "Optical spectroscopy of ions in inorganic glasses", in *Optical Spectroscopy of Glasses*, ed. by I. Zschokke (D. Reidel Publ. Co., Dordrecht, The Netherlands 1986) pp. 23–64

2.118 R. Haspel: private communication

2.119 J. Schroeder: "Light Scattering of glass", in *Treatise on Material Science and Technology*, Vol. 12, ed. by M. Tomazowa and R. H. Doremus (Academic Press, New York, 1977) pp. 158–222

2.120 W. Jochs: private communication

2.121 M.J. Liepmann, A.J. Marker, J.M. Melvin: "Optical and physical properties of UV-transmitting fluorocrown glasses", SPIE Vol **1128**, 213–224 (1989)

2.122 H. Scholze: "Der Einbau des Wassers in Gläsern" part I Glastech. Ber. **32**, 81–88 (1959); part II ibid. 142–152, part III ibid. 271–281, part IV ibid. 314–320

2.123 F.L. Galeener, D.L. Griscom, M.J. Weber (eds.): "Defects in Glasses", Materials Research Society Symposia Proceedings Vol. 61, Material Research Society, Pittsburgh (1986) pp. 3–20, 177–222

2.124 J.J. Setta, R.J. Scheller, A.J. Marker, III: "Effects of UV solarisation on the transmission of cerium-doped optical glasses" in *Properties and Characterisaton of Optical Glass*, SPIE Vol. **970,** 179–191 (1988)

2.125 A.J. Marker III, J.S. Hayden, B. Speit: "Radiation resistant optical glasses", in *Reflective and Refractive Optical Materials for Earth and Space Applications*, SPIE Vol **1485**, 160–172 (1991)

2.126 M.J. Liepmann, A.J. Marker III, U. Sowada: "Ultraviolet radiation effect on UV-transmitting fluor crown glasses, SPIE Vol. **970**, 170–177 (1988)

2.127 B. Speit, S. Grün: "Irradition energy dependence of Discolouration in radiation shielding glasses", in *Properties and Characteristics of Optical Glass II*, SPIE Vol.**1327**, 92–99 (1990)

2.128 W. Jahn: "Einwirkung von radioaktiver Strahlung auf Glas", Glast. Ber. **31** 41–43 (1958)

2.129 K. Bermuth, A. Lenhart, H.A. Schaeffer, K. Blank: "Solarisationserscheinungen an Cer- und Arsenhaltigen Kalk-Natrongläsern" Glast. Ber. **58**, 52–58 (1985)

2.130 G.H. Sigel, Jr.: "Ultraviolet spectra of silicate glasses: A Review of some experimental evidence", J. Non Cryst. Sol. **13**, 372–398 (1973/74)

2.131 M.F. Bartusiak, J. Becher: "Proton-induced coloring of multicomponent glasses", Appl. Opt. **18**, 3343–3346 (1979)

2.132 D.L. Griscom, E.J. Friebele, K.J. Long, J.W. Fleming: "Fundamental defect centers in glass", J. Appl. Phys. **54**, 3743–3762 (1983)

2.133 D.G. Griscom, E.J. Friebele: "Effects of ionizing radiation on amorphous insulators", Radiation Effects **65**, 303–312 (1982)

2.134 F. Kohlrausch: *Praktische Physik*, ed. by Günter Lautz und Rolf Taubert (B. G. Teubner, Stuttgart 1968)

2.135 B. Edlén: "The Refractive Index of Air", Metrologia **2**, 71–80 (1965)

2.136 K.P. Birch: "Air refractometry and its application to precision length measurements", Proc. Electro-Optics, Laser International, ed. by H.G. Jerrard, pp. 230–243 (1984)

2.137 K.P. Birch and M.J. Downs: "Error sources in the determination of the refractive index of air", Appl. Opt. **28**, 825–826 (1989)

2.138 K.P. Birch, M.J. Downs, and R.E. Ward: "The measurement of humidity variations in gases resulting from the absorption of water on to surfaces", J. Phys. E: Sci. Instrum. **21**, 692–694 (1988)

2.139 P. Schellekens, G. Wilkening, F. Reinboth, M.J. Downs, K.P. Birch, and J. Spronck: "Measurements of the Refractive Index of Air Using Interference Refractometers", Metrologia **22**, 279–287 (1986)

2.140 J.H. Gladstone and T.P. Dale: "Researches on the Refraction, Dispersion and Sensitiveness of Liquids", Phil. Trans. **153**, 317–343 (1863)

2.141 Charles Kittel and Herbert Kroemer: *Thermal Physics*, 2nd edition, (W. H. Freeman and Co., San Francisco 1980) pp. 126 ff.

2.142 L.H. Adams and E.D. Williamson: "The Annealing of Glass", Journ. Franklin Institute **190**, 597–631 and 835–870 (1920)

2.143 E. Berger: "Gleichgewichtsverschiebungen im Glas und Einfluß der Wärmevorgeschichte auf seine physikalischen Eigenschaften", Glastechn. Ber. **8**, 339–367 (1930)

2.144 W.D. Kingery, H.K. Bowen, and D.R. Uhlmann: *Introduction to Ceramics*, 2nd ed., (Wiley, New York 1976)

2.145 F. Reitmayer and E. Schuster: "Homogeneity of Optical Glasses", Appl. Opt. **11**, 1107–1111 (1972)

2.146 J. Zarzycki: *Les verres et l'état vitreux* (Masson, Paris 1982), in English: J. Zarzycki: *Glasses and the vitreous state* (Cambridge University Press, Cambridge 1991)

2.147 W. Vogel: *Chemistry of Glass* (American Society, Inc., Columbus, Ohio, USA 1985)

2.148 N.M. Brandt: "Annealing of 517:645 Borosilicate Optical Glass: I, Refractive Index", J. Am Ceram. Soc. **34**, 332–338 (1951)

2.149 H.R. Lillie and H.N. Ritland: "Fine Annealing of Optical Glass", J. Am. Ceram. Soc. **37**, 466–473 (1954)

2.150 A.Q. Tool, L.W. Tilton, and J.B. Saunders: "Changes caused in refractivity and density of glass by annealing", J. Res. NBS **38**, 519–526 (1947)

2.151 S. Spinner and R.M. Waxler: "Relation between Refractive Index and Density of Glasses Resulting from Annealing Compared with Corresponding Relation Resulting from Compression", Appl. Opt. **12**, 1887–1889 (1966)

2.152 L. Prod'homme: "Certains aspects de l'évolution thermique et de la stabilisation des propriétés du verre", Verres et Réfractaires **11**, 351–369 (1957) and Verres et Réfractaires **12**, 3–22 (1958)

2.153 H.R. Lillie: "Viscosity time temperature relations in glass at annealing temperatures", J. Am. Ceram. Soc. **16**, 619–631 (1933)

2.154 H.R. Lillie: "Stress release in glass; phenomenon involving viscosity as variable with time", J. Am. Ceram. Soc. **19**, 45–54 (1936)

2.155 A.Q. Tool: "Relation Between Inelastic Deformability and Thermal Expansion of Glass in Its Annealing Range", J. Am. Ceram. Soc. **29**, 240–253 (1946)

2.156 H.N. Ritland: "Limitations of the Fictive Temperature Concept", J. Am. Ceram. Soc. **39**, 403–406 (1956)

2.157 O.S. Narayanaswamy: "A Model of Structural Relaxation in Glass", J. Am. Ceram. Soc. **54**, 491–498 (1971)

2.158 M.A. deBolt, A.J. Easteal, P.B. Macedo, and C.T. Moynihan: "Analysis of Structural Relaxation in Glass Using Rate Heating Data", J. Am. Ceram. Soc. **59**, 16–21 (1976)

2.159 G.W. Scherer: "Use of the Adam-Gibbs Equation in the Analysis of Structural Relaxation", J. Am. Ceram. Soc. **67**, 504–511 (1984)

2.160 H. Rötger and H. Besen: "Ein rationelles Kühlverhalten zur Toleranzeinigung des Brechungsverhältnisses optischer Gläser", Feingerätetechnik **10**, 547–554 (1961)

2.161 H. Rötger and H. Besen: "Brechzahlbeeinflussung durch Feinkühlung bei Jenaer optischer Gläser", Silikattechnik **13**, 424–427 (1962)

2.162 E.E. Danyushevskii: *The Basics of Linear Annealing of Optical Glass* (in Russian), Leningrad (1959)

2.163 V.S. Doladugina: "Changes in the refractive index of glasses when cooled at a steady state", Sov. J. Glass Phys. Chem. **10**, 391–396 (1984)

2.164 *Optical Glass Catalogue 10.000e*, Schott Glaswerke, Mainz, Fed. Rep. Germany (1992)

2.165 H.J. Hoffmann, W.W. Jochs, and N.M. Neuroth: "Relaxation phenomena of the refractive index caused by thermal treatment of optical glasses below T_g", SPIE **970**, 2–9 (1989)

2.166 G. Pulfrich: "Über den Einfluß der Temperatur auf die Lichtbrechung des Glases", Ann. Phys. u. Chem., N.F. **45**, 609–665 (1892)

2.167 "Dispersion Formula for the Temperature Coefficient of the Refractive Index of Glasses", *Technical Information No. 19*, Schott Glaswerke, Mainz, Fed. Rep. Germany (1988) 12 pages

2.168 H.J. Hoffmann, W.W. Jochs, and G. Westenberger: "A dispersion formula for the thermo-optic coefficient of optical glasses", SPIE **1327**, 219–231 (1990)

2.169 F.A. Molby: "Index of Refraction and Coefficients of Expansion of optical glasses at low temperatures", J. Opt. Soc. Am. **39**, 600–611 (1949)

2.170 R.M. Waxler and G.W. Cleek: "The Effect of Temperature and Pressure on the Refractive Index of Some Oxide Glasses", J. Res. NBS **77A**, 755–763 (1973)

2.171 T. Toyoda and M. Yabe: "The temperature dependence of the refractive index of fused silica and crystal quartz", J. Phys. D: Appl. Phys. **16**, L97–L100 (1983)

2.172 J.H. Wray and J.N. Neu: "Refractive Index of Several Glasses as a Function of Wavelength and Temperature", J. Opt. Soc. Am. **59**, 774–776 (1969)

2.173 T. Bååk: "Thermal Coefficient of Refractive Index of Optical Glasses", J. Opt. Soc. Am. **59**, 851–857 (1969)

2.174 G.N. Ramachandran: "Thermo-optic Behaviour of Crystals", *Progress in Crystal Physics Vol. 1*, ed. by R.S. Krishnan (Interscience Publishers, New York 1958) pp. 139–167

2.175 O. Lindig: Spannungsdoppelbrechung in Gläsern, in Landolt-Börnstein II/8, (Springer, Berlin, Heidelberg 1962) pp. 3–542

2.176 J.F. Nye: *Physical Properties of Crystals* (Clarendon Press, Oxford 1979)

2.177 F. Neumann: "Die Gesetze der Doppelbrechung des Lichts in komprimierten oder ungleichförmig erwärmten unkristallinischen Körpern", Pogg. Ann. Phys. Chem. **54**, 449–476 (1841)

2.178 F. Pockels: "Über die Aenderung des optischen Verhaltens verschiedener Gläser durch elastische Deformation", Ann. Phys. (Leipzig) **7**, 745–771 (1902)

2.179 Clemens Schaefer and Heinrich Nassenstein: "Bestimmung der photoelastischen Konstanten p und q optischer Gläser", Z. Naturforschung **8a**, 90–96 (1953)

2.180 Erich Schuster and Franz Reitmayer: "Die Änderung der Lichtbrechung von Gläsern bei eindimensionaler Druck- bzw. Zug-Beanspruchung", Glastechn. Ber. **34**, 130–133 (1961)

2.181 N.F. Borrelli and R.A. Miller: "Determination of the Individual Strain-Optic Coefficients of Glass by an Ultrasonic Technique", Appl. Opt. **7**, 745–750 (1968)

2.182 Hans Mueller: "Theory of Photoelasticity in Amorphous Solids", Physics **6**, 179–184 (1935)

2.183 Albert Feldman and William J. McKean: "Improved stressing apparatus for photoelasticity measurements", Rev. Sci. Instr. **46**, 1588–1589 (1975)

2.184 George Birnbaum, Earl Cory, and Kenneth Gow: "Interferometric Null Method for Measuring Stress-Induced Birefringence", Appl. Opt. **13**, 1660–1669 (1974)

2.185 Albert Feldman, Deanne Horowitz, and Roy M. Waxler: "Photoelastic constants of potassium chloride at 10.6 µm", Appl.Opt. **16**, 2925–2930 (1977)

2.186 "Modification of the Refractive Index of Optical Glasses by Tensile and Compressive Stresses", *Technical Information No. 20*, Schott Glaswerke, Mainz, Fed. Rep. Germany (1988) 10 pages

2.187 "Der spannungsoptische Koeffizient optischer Gläser", *Technical Information No. 15*, Schott Glaswerke, Mainz, Fed. Rep. Germany (1984) 9 pages

2.188 E.I. Gordon: "A Review of Acoustooptical Deflection and Modulation Devices", Appl. Opt. **5**, 1629–1639 (1966)

2.189 R.W. Dixon: "Photoelastic Properties of Selected Materials and Their Relevance for Application to Acoustic Light Modulators and Scanners", J. Appl. Phys. **38**, 5149–5153 (1967)

2.190 Roy M. Waxler and Albert Napolitano: "Relative Stress-Optical Coefficients of Some National Bureau of Standards Optical Glasses", J. Res. NBS **59**, 121–125 (1957)

2.191 Werner Schwiecker: "Komponentenabhängigkeit der spannungsoptischen Koeffizienten von Glas", Glastechn. Ber. **30**, 84–88 (1957)

2.192 Arthur F. van Zee and Henry M. Noritake: "Measurement of Stress-optical Coefficient and Rate of Stress Release in Commercial Soda-Lime Glasses", J. Am. Ceram. Soc. **41**, 164–175 (1958)

2.193 Peter Manns and Rolf Brückner: "Spannungsoptisches Verhalten einiger Silicatgläser im viskoelastischen Bereich", Glastechn. Ber. **54**, 319–331 (1981)

2.194 Kazumasa Matusita, Ryosuke Yokota, Tetsuo Kimijima, Takayuki Komatsu, and Chikashi Ihara: "Compositional Trends in Photoelastic Constants of Borate Glasses", J. Am. Ceram. Soc. **67**, 261–265 (1984)

2.195 Kazumasa Matusita, Chikashi Ihara, Takayuki Komatsu, and Ryosuke Yokoto: "Photoelastic Effects in Silicate Glasses", J. Am. Ceram. Soc. **67**, 700–704 (1984)

2.196 Kazumasa Matusita, Chikashi Ihara, Takayuki Komatsu, and Ryosuke Yokota: "Photoelastic Effects in Phosphate Glasses", J. Am. Ceram. Soc. **68**, 389–391 (1985)

2.197 Kazumasa Matusita, Hiroshi Kato, Takayuki Komatsu, Mamoru Yoshimoto, and Naohiro Soga: "Photoelastic Effects in some Fluoride Glasses based on the ZrF_4-BaF_2-BaF_2 System", J. Non-Crystalline Solids **112**, 341–346 (1989)

2.198 V.A. Levenberg and S.G. Lunter: "The stress-optical coefficient and the composition of a glass", Sov. J. Glass Phys. Chem. **2**, 61–66 (1976)

2.199 T.N. Vasudevan and R.S. Krishnan: "Dispersion of the stress-optic coefficient in glasses", J. Phys. D: Appl. Phys. **5**, 2283–2287 (1972)

2.200 H.J. Hoffmann, W.W. Jochs, G. Przybilla, and G. Westenberger: "The stress-optical coefficient and its dispersion in oxide glasses", Proceedings XVI International Congress on Glass, Madrid, October 4–9, 1992 Boletin de la Sociedad Española de Ceramica y Vidrio 31-C (1992) Vol. 4 (Glass Properties) pp. 187–192

2.201 S. Nzahumunyurwa, H.J. Hoffmann and G. Westenberger, to be published in Glastechn. Berichte 1995

2.202 N.F. Borrelli: "Electric field induced birefringence in glasses", Phys. Chem. Glasses **12**, 93–96 (1971)

2.203 O.D. Tauern: "Über das Auftreten des Kerrphänomens in Gläsern und über eine Bestimmung der Kerrkonstanten für Schwefelkohlenstoff", Ann. Physik, IV Folge, **32**, 1064–1084 (1910)

2.204 M. Paillette: "Dispersion de l'effet Kerr dans des verres aux silicates de plomb", Opt. Commun. **13**, 64–67 (1975)

2.205 M.J. Weber: "Faraday rotator materials for laser systems", SPIE **681**, 75–91 (1987)

2.206 M.J. Weber: Faraday Rotator Materials, Report M-103, Lawrence Livermore National Laboratory, Ca. (1982)

2.207 CRC Handbook of Laser Science and Technology, Vol. IV Optical Materials: Part 2; Marvin J. Weber, ed. (CRC Press, Inc., Boca Raton 1986)

2.208 Jeffrey A. Davis and Robert M. Bunch: "Temperature dependence of the Faraday rotation of Hoya FR-5 glass", Appl. Opt. **23**, 633–636 (1984)

2.209 Paul A. Williams, A.H. Rose, G.W. Day, T.E. Milner, and M.N. Deeter: "Temperature dependence of the Verdet constant in several diamagnetic glasses", Appl. Opt. **30**, 1176–1178 (1991)

2.210 H.J. Hoffmann, W.W. Jochs, and G. Przybilla: "The Verdet Constant of Optical Glasses", Natl. Bur. Stand. (USA), Spec. Publ. **697**, 266–269 (1985)

2.211 G. Westenberger, H.J. Hoffmann, W.W. Jochs, and G. Przybilla: "The Verdet constant and its dispersion in optical glasses", SPIE **1535**, 113–120 (1991)

2.212 "Faraday effect in optical glass. The wavelength dependence of the Verdet Constant", *Technical Information No. 17*, Schott Glaswerke, Mainz, Fed. Rep. Germany (1985) 6 pages

2.213 A. Balbin Villaverde and E.C.C. Vasconcelles: "Magnetooptical dispersion of Hoya glasses: AOT-5, AOT-44B, and FR-5", Appl. Opt. **21**, 1347–1348 (1982)

2.214 Heihachi Sato, Masotoshi Kawase, and Mitsunori Saito: "Temperature dependence of the Faraday effect in As-S glass fiber", Appl. Opt. **24**, 2300–2303 (1985)

2.215 K. Weber: "Polarisation und Doppelbrechung des Lichts", in Bergmann-Schäfer: *Lehrbuch der Experimentalphysik* Vol. IV Optik, 8th edition, Heinrich Gobrecht, ed. (Walter de Gruyter, Berlin, New York 1987) pp. 481–627

2.216 Landolt-Börnstein: II. Band: *Eigenschaften der Materie in ihren Aggregatzuständen*, 8. Teil: Optische Konstanten. (Springer, Berlin, Heidelberg 1962) pp. 826–844 and pp. 894–895

2.217 "Glasses for athermal optics", *Technical Information No. 12*, Schott Glaswerke, Mainz, Fed. Rep. Germany (1976) 5 pages

2.218 H.J. Hoffmann and J.S. Hayden: "Glasses as active materials for high power solid state lasers", SPIE **1021**, 42–50 (1988)

2.219 W. Koechner: *Solid-State Laser Engineering*, Springer Series in Optical Sciences, Vol. 1, 2nd ed. (Springer, New York 1989)

2.220 L. Ross: "Integrated optical components in substrate glasses", Glastechn. Ber. **62**, 285–297 (1989)

2.221 A.P. Webb and P.D. Townsend: "Refractive Index Profiles Induced by Ion Implantation into Silica", J. Phys. **D 9**, 1343–1354 (1976)

2.222 H. Schroeder: "A Method of Reducing Surface Reflection from Articles of Glass or Silicates", British Patent 698.831, filed on May 28, 1951

2.223 L.M. Cook and K.-H. Mader: Leached antireflection surfaces: Part 1: "Characterization of the neutral solution process", Glastechn. Ber. **60**, 234–238 (1986); L.M. Cook, A.J. Marker III, K.-H. Mader, H. Bach, and H. Müller: Part 2: "Characterization of surfaces produced by the neutral solution process", Glastechn. Ber. **60**, 302–311 (1987); L.M. Cook and K.-H. Mader: Part 3: "Chemical release data for leached BK-7 borosilicate glasses", Glastechn. Ber. **60**, 333–339 (1987); L.M. Cook: Part 4: "Proposed model for the neutral solution process", Glastechn. Ber. **60**, 368–375 (1987)

2.224 S. Gebala and I. Wilk: "Changes of refractive index in glass induced by UV-irradiation and a new possibility of their determination", Opt. Appl. **12**, 483–488 (1982)

2.225 G.V. Byurganovskaya, B.I. Kisin and N.F. Orlov: "Effect of Mixed Gamma-Neutron Radiation on Some Physicochemical Properties of Glasses", Sov. J. Opt. Technol. **34**, 167–173 (1967)

2.226 P. Günter (ed.): *Electro-optic and Photorefractive Materials*, Springer Proceedings in Physics Vol. 18, (Springer, Berlin, Heidelberg 1987)

2.227 A.B. Buckman and N.H. Hong: "On the origin of the large refractive index change in photolyzed PbI_2 films", J. Opt. Soc. Am. **67**, 1123–1125 (1977)

2.228 F.G.K. Baucke, J.A. Duffy, and P.R. Woodruff: "Optical properties of tungsten bronze surfaces", Thin Solid Films **148**, L 59–L 61 (1987)

2.229 A.F. Evans and D.G. Hall: "Measurement of the electrically induced refractive index change in silicon for wavelength $\lambda = 1.3$ μm using a Schottky diode", Appl. Phys. Lett. **56**, 212–214 (1990)

2.230 G.D. Baldwin and E.P. Riedel: "Measurements of Dynamic Optical Distortion in Nd-Doped Glass Laser Rods", J. Appl. Phys. **38**, 2726–2738 (1967)

2.231 Frederic M. Durville and Richard C. Powell: "Thermal lensing and permanent refractive index changes in rare-earth-doped glasses", J. Opt. Soc. Am. **B 4**, 1934–1937 (1987)

2.232 Yuiko T. Hayden and Alexander J. Marker III: "Glass as a nonlinear optical material", SPIE **1327**, 132–144 (1990)

2.233 H. Hack and N. Neuroth: "Resistance of optical and coloured glasses to 3-nsec laser pulses", Appl. Opt. **21**, 3239–3248 (1982)

2.234 M.J. Soileau, Eric W. Van Stryland, and William E. Williams: "Laser light induced bulk damage to optics", SPIE **541**, 110–122 (1985)

2.235 L. Brillouin: "Diffusion de la lumière et des rayons x par un corps transparent homogène influencé de l'agitations thermique", Annales de Physique **17**, 88–122 (1922)

2.236 P. Debye and F.W. Sears: "On the Scattering of Light by Supersonic Waves", Proc. National Academy of Science (U.S.) **18**, 409–414 (1932)

2.237 R. Lucas and P. Biquard: "Propriétés optiques des milieux solides et liquides soumis aux vibrations élastiques ultra sonores", J. Phys. Radium **3**, 464–477 (1932)

2.238 C.V. Raman and N.S.N. Nath: "Diffraction of Light by High Frequency Sound Waves", Part I, Proc. Indian Academy of Science **2**, 406–412 (1935); Part II, Proc. Indian Academy of Science **2**, 413–420 (1935); Part III, Proc. Indian Academy of Science **3**, 75–84 (1936); Part IV, Proc. Indian Academy of Science **3**, 119–125 (1936); Part V, Proc. Indian Academy of Science **3**, 459–465 (1936); Generalized Theory, Proc. Indian Academy of Science **4**, 222–242 (1936)

2.239 L. Bergmann: *Der Ultraschall* (Hirzel Verlag, Stuttgart 1954)

2.240 Milton Gottlieb, Clive L.M. Ireland, and John Martin Ley: *Electro-Optic and Acousto-Optic Scanning and Deflection*, Optical Engineering Vol. 3, Series editor: Brian J. Thompson (Marcel Dekker, New York 1983)

2.241 Adrian Korpel: *Acousto-Optics*, Optical Engineering Vol. 16, Series editor: Brian J. Thompson (Marcel Dekker, New York 1988)

2.242 Pankaj K. Das and Casimir M. DeCusatis: *Acousto-Optic Signal Processing: Fundamentals and Applications* (Artech House, Boston 1991)

2.243 Jieping Xu and Robert Stroud: *Acousto-Optic Devices: Principles, Design, and Applications*, Wiley Series in Pure and Applied Optics, ed. by Joseph W. Goodman (John Wiley, New York 1992)

2.244 W.R. Klein and B.D. Cook: "Unified Approach to Ultrasonic Light Diffraction", IEEE Trans. Sonics and Ultrasonics **SU-14**, 123–134 (1967)

2.245 T.M. Smith and A. Korpel: "Measurement of Light-Sound Interaction Efficiencies in Solids", IEEE J. Quantum Electronics (Correspondence) **QE-1**, 283–284 (1965)

2.246 R.W. Dixon: "Photoelastic Properties of Selected Materials and Their Relevance for Application to Acoustic Light Modulators and Scanners", J. Appl. Phys. **38**, 5149–5153 (1967)

2.247 E.I. Gordon: "A Review of Acoustooptical Deflection and Modulation Devices", Appl. Opt. **5**, 1629–1639 (1966)

2.248 "Modification of the Refractive Index of Optical Glasses by Tensile and Compressive Stresses", *Technical Information No. 20*, Schott Glaswerke, Mainz, Fed. Rep. Germany (1988) 10 pages

2.249 Douglas A. Pinnow: "Guide Lines for the Selection of Acoustooptic Materials", IEEE **QE-6**, 223–238 (1970)

2.250 A. Goutzoulis, M. Gottlieb, K. Davies, and Z. Kun: "Thallium arsenic sulfide acoustooptic Bragg cells", Appl. Opt. **24**, 4183–4188 (1985)

2.251 H. Eschler and F. Weidinger: "Acousto-optic properties of dense flint glasses", J. Appl. Phys. **46**, 65–70 (1975)

2.252 H. Eschler, R. Oberbacher, F. Weidinger, and K.-H. Zeitler: "Design of High-Resolution Acoustooptic Light Deflectors", Siemens Forsch.-u. Entwickl.-Ber. **4**, 174–184 (1975)

2.253 F. Pockels: "Über die Änderung des optischen Verhaltens verschiedener Gläser durch elastische Deformation", Ann. Phys. (Leipzig) **7**, 745–771 (1902)

2.254 R. Braunstein: "Nonlinear optical effects", Phys. Rev. **125**, 475–477 (1962)

2.255 P.A. Franken, A.E. Hill, C.W. Peters, and G. Weinreich: "Generation of Optical Harmonics", Phys. Rev. Lett. **7**, 118–119 (1961)

2.256 Y.R. Shen: *The Principles of Nonlinear Optics* (John Wiley & Sons 1984)

2.257 G.C. Baldwin: *An Introduction to Nonlinear Optics* (Plenum Press, New York 1969)

2.258 N. Bloembergen: *Nonlinear Optics* (W. A. Beniamin, Inc., New York 1969)

2.259 R.W. Bryant: *Nonlinear Optical Materials: New Technologies, Applications, Markets* (Business Communications Co. Inc., Norwalk 1989)

2.260 W.L. Smith: "Nonlinear Refractive Index" in *Handbook of Laser Science and Technology*, ed. by M. J. Weber, Volume III (CRC Press, Boca Raton 1986) pp. 259–265, 273, 274

2.261 U. Osterberg and W. Margulis: "Dye laser pumped by Nd:YAG laser pulses frequency doubled in glass optical fibers", Optics Lett. **11**, 516–518 (1986)

2.262 Y.T. Hayden and A.J. Marker III: "Glass as a nonlinear optical material", Current Overviews in Optical Science and Engineering II, ed. by Feinberg, SPIE Advent Technology Series **AT 2**, 406–418 (1990)

2.263 E.M. Vogel, M.J. Weber and D.M. Krol: "Nonlinear optical phenomena in glass", Physics and Chemistry of Glasses **32**, 231–254 (1991)

2.264 N.F. Borrelli and D.W. Hall: "Nonlinear Optical Properties of Glasses" in *Opt. Prop. Glass* ed. by D.R. Uhlmann and N.J. Kreidl, Am. Ceram. Soc., Westerville, Ohio, 87–124 (1991)

2.265 C. Flytzanis, J.L. Oudar: *Nonlinear optics: materials and devices* (Springer, Heidelberg, New York 1986)

2.266 C. Flytzanis, F. Hache, M.C. Klein, D. Ricard, Ph. Roussignol: "Nonlinear optics in composite materials" in *Progress in Optics* Vol. XXIX ed. by E. Wolf (North Holland Elsevier Science Publishers, Amsterdam, 1991)

2.267 C. Klingshirn: "Laser Spectroscopy of Crystalline Semiconductors" in *Laser Spectroscopy of Solids* II, ed. by W.M. Yen, (Springer Verlag Berlin, Heidelberg, New York)

2.268 R.C. Miller: "Optical Second Harmonic Generation in Piezoelectric Crystals", Appl. Phys. Lett. **5**, 17–19 (1964)

2.269 C.C. Wang: "Empirical relation between the linear and third-order nonlinear optical susceptibilities", Phys. Rev. **B2**, 2045–2048 (1970)

2.270 N.L. Boling, A.J. Glass and A. Owyoung: "Empirical relationships for predicting nonlinear refractive index change in optical solids", IEEE J. Quant. Electro. **QE14**, 601–608 (1978)

2.271 M.J. Weber, D. Milam and W.L. Smith: "Nonlinear refractive index of glasses and crystals", Opt. Eng. **17**, 463–469 (1978)

2.272 D.W. Hall, M.A. Newhouse, N.F. Borrelli, W.H. Dumbaugh, and D.L. Weidman: "Nonlinear optical susceptibilities of high-index glasses", Appl. Phys. Lett. **54**, 1293–1295 (1989)

2.273 Mir Akbar Ali and R.L. Ohlhaber: "High index infrared glasses for integrated optics", Glass Tech. **26**, 186–189 (1985)

2.274 H. Nasu, Y. Ibara and K. Kubodera: "Optical third-harmonic generation from some high-index glasses", J. NonCryst. Solids **110**, 229–234 (1989)

2.275 H. Nasu: New Glass **4(4)**, 13–20 (1990)

2.276 E.M. Vogel, S.G. Kosinski, D.M. Krol, J.L. Jackel, S.R. Fribert, M.K. Oliver and J.D. Powers: "Structure and optical study of silicate glasses for nonlinear optical devices", J. Non-Crys. Solids **107**, 244–250 (1989)

2.277 P.W. Smith: "Photonic Switching, present and future prospects", SPIE Proc. Vol. **881**, 30–37 (1988)

2.278 S.R. Friberg and P.W. Smith: "Nonlinear optical glasses for ultrafast optical switches", IEEE J. Quant. Electro. **QE23**, 2089–2094 (1987)

2.279 J.L. Jackel, E.M. Vogel, J.S. Aitchison: "Ion-exchanged optical waveguides for all-optical switching", Appl. Opt. **29**, 3126–3129 (1990)

2.280 J.S. Aitchison, A.M. Weiner, Y. Silverberg, M.K. Oliver, J.L. Jackel, D.E. Leaird, E.M. Vogel and P.W.E. Smith: "Observation of spatial optical solitons in a nonlinear waveguide", Opt. Lett. **15**, 471–473 (1990)

2.281 V. Mizrahi, K.W. DeLong, G.I. Stegeman, M.A. Saifi, and M.J. Andrejco: "Two-photon absorption as a limitation in all-optical switching", Opt. Lett. **14**, 1140–1142 (1989)

2.282 K.W. DeLong, V. Mizrahi, G.I. Stegeman, M.A. Saifi, and M.J. Andrejco: "Role of color center induced absorption in all-optical switching", Appl. Phys. Lett. **56**, 1394–1396 (1990)

2.283 R.H. Stolen and H.W.K. Tom: "Self-organized phasematched harmonic generation in optical fibers", Optics Lett. **12**, 585–587 (1987)

2.284 D.Z. Anderson, V. Mizahi, and J.E. Sipe: "A model for second harmonic generation in glass optical fibers based on asymmetric photoelectron emission from defect sites", Opt. Lett. **16**, 796–798 (1991)

2.285 See, for example, Schott Optical Glass Filters, Schott Glaswerke, Catalog 3555 (1984)

2.286 G. Bret and F. Gires: "Giant-pulse laser and light amplifier using variable transmission coefficient glasses as light switches", Appl. Phys. Lett. **4**, 175–176 (1964)

2.287 R.K. Jain and R.C. Lind: "Degenerate four wave mixing in semiconductor-doped glasses", J. Opt. Soc. Am. **73**, 647–653 (1983)

2.288 H. Nasu and J.D. Mackenzie: "Nonlinear optical properties of glasses and glasses based on sol-based composites", Opt. Eng. **26**, 102–106 (1987)

2.289 V.S. Williams, G.R. Olbright, B.D. Fluegel, S.W. Koch, and N. Peyghambarian: "Optical nonlinearities and ultrafast carrier dynamics in semiconductor-doped glasses", J. Mod. Opt. **35**, 1979–1993 (1988)

2.290 P. Roussignol, D. Ricard, J. Lukasik and C. Flytzanis: "New results on optical phase conjugation in semiconductor-doped glasses", J. Opt. Soc. Am. **B4**, 5–13 (1987)

2.291 P. Roussignol, D. Ricard, C. Flytzanis: "Nonlinear optical properties of commercial semiconductor-doped glasses", Appl. Physics **A44**, 285–292 (1987)

2.292 M. Tomita, T. Matsumoto and M. Matsuoka: "Nonlinear dynamic relaxation processes in semiconductor-doped glasses at liquid nitrogen temperature", J. Opt. Soc. Am. **B6**, 165–170 (1989)

2.293 A.I. Ekimov, A.L. Efros and A.A. Onuschchenko: "Quantum size effect in semiconductor microcrystallites", Solid State Comm. **56**, 921–924 (1985)

2.294 P. Roussignol, D. Ricard, C. Flytzanis, N. Neuroth: "Carrier recombination and quantum size effect in the nonlinear optical properties of semiconductor microcrystallites", SPIE Proc. Vol **1017**, 20–25 (1988)

2.295 A. Uhrig, D. Oberhauser, C. Coernfeld, C. Klingshirn, N. Neuroth: "Optical properties of CdS/Se microcrystallites in glasses", SPIE Proc. Vol. **1127**, 101–108 (1989)

2.296 A. Uhrig, L. Banyai, Y.Z. Hu, S.W. Koch, C. Klingshirn, N. Neuroth: "High-excitation photoluminescence studies of $CdS_{1-x}Se_x$ quantum dots", Z. Phys. B (Condensed Matter) **81,** 385–390 (1990)

2.297 A. Uhrig, A. Woerner, C. Clingshirn, L. Banyai, S. Gapenko, I. Lacis, N. Neuroth, B. Speit, K. Remitz: "Nonlinear optical properties of semiconductor quantum dots", J. Cryst. Growth **117**, 598–602 (1992)

2.298 K.E. Remitz, N. Neuroth, B. Speit: "Semiconductor-doped glass as a nonlinear material", Materials Science and Engineering **B9**, 413–416 (1991)

2.299 P. Roussignol, F. Hache, D. Ricard, C. Flytzanis: "Time scales and quantum-size effects in optical nonlinearities of semiconductor and metal microcrystallites in glasses", SPIE Proc. Vol. **1128**, 238–245 (1989)

2.300 P. Horan and W. Blau: "Photodarkening effect and the optical nonlinearity in a quantum-confined, semiconductor-doped glass", J. Opt. Soc. Am. **B7**, 304–308 (1990)

2.301 G. Potter and J.H. Simmons: "Quantum size effects in optical properties of semiconducting microcrystallites in glass", Phys. Rev. **B37**, 838–845 (1988)

2.302 N.F. Borelli and J.C. Luong: "Semiconductor microcrystals in porous glass", Proc. SPIE Vol. **866**, 104–109 (1987)

2.303 C.A. Huber and T.E. Huber: "A novel microstructure: semiconductor-impregnated porous Vycor glass", preprint, Journ. Appl. Phys. **64**, 6588–6590 (1988)

2.304 T. Yanagawa, Y. Sasaki and H. Nakano: "Quantum size effects and observation of microcrystallites in colored filter glasses", Appl. Phys. Lett. **54**, 1495–1497 (1989)

2.305 L.C. Liu and S.H. Risbud: "Analysis of TEM contrast of semiconductor crystals in glass", Phil. Mag. Lett. **61**, 327–332 (1990)

2.306 N.F. Borrelli, D.W. Hall, H.J. Holland and D.W. Smith: "Quantum confinement effects of semiconductor microcrystallites in glass", J. Appl. Phys. **61**, 5399–5409 (1987)

2.307 M. Allais, M. Gandais: "Structural study of Cd(S,Se) doped glasses. High resolution transmission electron microscopy (HRTEM) assisted by image treatment", J. Appl. Cryst. **23**, 418–429 (1990)

2.308 A. Gabel, K.W. DeLong, C.T. Seaton and G.I. Stegmann: "Efficient degenerate four-wave mixing in an ionexchanged semiconductor-doped glass waveguide", Appl. Phys. Lett. **51**, 1682–1684 (1987)

2.309 G.I. Stegeman, E.M. Wright, N. Finlayson, R. Zanoni, C.T. Seaton: "Third order nonlinear integrated optics", J. Light Tech. **6**, 953–970 (1988)

2.310 D. Cotter, C.N. Ironside, B.J. Ainslie and H.P. Girdlestone: "Picosecond pump-probe interferometric measurement of optical linearity in semiconductor-doped fibers", Opt. Lett. **14**, 317–319 (1989)

2.311 T. Rajh, M.I. Vucemilovic, N.M. Dimitrijevec, and O.I. Micic: "Size quantization of colloidal semiconductor particles in silicate glasses", Chem. Phys. Lett. **143**, 305–308 (1988)

2.312 V. Esch, B. Flugel, G. Khitrova, H.M. Gibbs, X. Jiajin, K. Kang, S.W. Koch, L.C. Liu, S.H. Risbud and N. Peyghambarian: "State filling, coulomb, and trapping effects in the optical nonlinearity of CdTe quantum dots in glass", Phys. Rev. **B42**, 7450–7453 (1990)

2.313 S. Omi, H. Hiraga, K. Uchida, C. Hata, Y. Asahara, A.J. Ikushima, T. Tokizaki and A. Nakamura: "Optical properties of CdSe microcrystallite-doped glasses", CLEO 1991 Tech. Digest Series **10**, 88–89, Optical Society of America (1991)

2.314 G.I. Stegeman and R.H. Stolen: "Waveguides and fibers for nonlinear optics", J. Opt. Soc. Am. **B6**, 652–662 (1989)

2.315 Japan New Mat. Rep., **V(2)**, 4 (1990)

2.316 D. Ricard, P. Roussignol and C. Flytzanis: "Surface mediated enhancement of optical phase conjugation in metal colloids", Opt. Lett. **10**, 511–513 (1985)

2.317 F. Hache, D. Ricard and C. Flytzanis: "Optical nonlinearities of small metal particles: surfacemediated resonance and quantum size effects", J. Opt. Soc. Am. **B3**, 1647–1655 (1986)

2.318 U. Kreibig: "Electronic properties of small silver particles: the optical constants and their temperature dependence", J. Phys. F: Metal Phys. **4**, 999–1014 (1974)

2.319 F. Hach, D. Ricard, C. Flytzanis, U. Kreibig: "The optical Kerr effect in small metal particles and metal-doped glasses: the case of gold", SPIE Proc. Vol. **1127**, 115–122 (1989)

2.320 M. Bertolotti, A. Ferrari, G. Gnappi, A. Montenero, C. Sibilia, and G. Suber: "The thermal nonlinearity of semiconductor-doped glasses", J. Phys. D: Appl. Phys. **21**, S17–S19 (1988)

2.321 J.W. Haus, N. Kalyaniwalla, R. Inguva, M. Bloemer, and C.M. Bowden: "Nonlinear-optical properties of conductive spheroidal particle composites", J. Opt. Soc. Am. **B6**, 797–807 (1989)

2.322 A.E. Neeves and M.H. Birnboim: "Composite structure for the enhancement of nonlinear-optical susceptibility", J. Opt. Soc. Am. **B6**, 787–796 (1989)

2.323 W.R. Tompkin, R.W. Boyd, D.W. Hall and P.A. Tick: "Nonlinear-optical properties of lead-tin fluorophosphate glass containing acridine dyes", J. Opt. Soc. Am. **B4**, 1030–1034 (1987)

2.324 Calculated using the equation, valid only for linear polarized light, $n_2\,(\text{esu}) = (12\pi/n_0)\chi^{(3)}\,(\text{esu})$

2.325 R. Wood: *Laser Damage in Optical Materials* (IOP Publishing Limited, Bristol 1986)

2.326 R.M. Wood: "Selected papers on laser damage in optical materials", SPIE Milestone Series **MS 24** (1990)

2.327 F. Rainer, R.M. Brusasco, J.H. Campbell, F.P. Demarco, R.P. Gonzales, M.R. Kozlowski, F.P. Milanovich, A.J. Morgan, M.S. Scrivener, M.C. Staggs, I.M. Thomas, S.P. Velsko, C.R. Wolfe: "Damage Measurements on Optical Materials for Use in high-peak power lasers", in *Laser Induced Damage in Optical Materials 1989*, NIST Spec. Publ. **801**, 74–85 (1990) also published by SPIE Proc. **1438** (1990)

2.328 F. Rainer, R.P. Gonzales, and A.J. Morgan: "Laser damage database at 1064 nm", in *Laser Induced Damage in Optical Materials*, National Institute of Standards and Technology Special Pub. **801**, Boulder, CO (1990) pp. 58–76

2.329 D.C. Brown: *High-Peak-Power Nd:Glass Laser Systems* (Springer, Berlin, Heidelberg 1981) pp. 170–183

2.330 M.J. Soileau, W.E. Williams, N. Mansour, E.W. Van Stryland: "Laser Induced Damage and the Role of Self-Focusing", Opt. Engin. **28**, 1133–1144 (1989)

2.331 J.H. Campbell, F. Rainer, M.R. Kozlowski, C.R. Wolfe, I.M. Thomas, and F.P. Milanovich: "Damage-resistance Optical Components", Lawrence Livermore National Laboratory 1990 Annual Report, 39–43 (1990)

2.332 *Laser Induced Damage in Optical Materials*, National Institute of Standards and Technology Special Pub. **462** (1976), **541** (1978), **568** (1979), **727** (1986), **746** (1985), **756**, (1987), Boulder, CO

2.333 R.P. Gonzales and D. Milam: "Evolution during multi-shot irradiation of damage surrounding isolated platinum inclusions in phosphate laser glasses", in *Proc. of the Boulder Damage Symposium, Oct. 1985, Boulder, CO*, National Institute of Standards and Technology Special Pub. **746**, Boulder, CO (1985) pp. 128–137

2.334 H. Hack and N. Neuroth: "Resistance of optical and colored glasses to 3-nsec laser pulses", Appl. Opt. **21**, 3239–3248 (1982)

2.335 H. Fuhrmann and C. Hoffmann: "Laserinduzierte Zerstörung optischer Medien (Teil II)", Feingerätetechnik **2**, 58–60 (1977)

2.336 R.W. Hopper and D.R. Uhlmann: "Mechanism of Inclusion Damage in Laser Glass", J. Appl. Phys. **41**, 4023–4037 (1970)

2.337 J.H. Pitts: "Modeling Laser Damage Caused by Platinum Inclusions in Laser Glass", Lawrence Livermore National Laboratory Rept. **UCRL-93249** (1985)

2.338 J.S. Hayden, D.L. Sapak and A.J. Marker: "Elimination of metallic platinum in phosphate laser glasses", Proc. SPIE **895**, 176–181 (1988)

2.339 C.L. Weinzapfel, G.J. Greiner, C.D. Walmer, J.F. Kimmons, E.P. Wallerstein, F.T. Marchi, J.H. Campbell, J.H. Hayden, K. Komiya and T. Kitayama: "Large Scale Damage Testing in a Production Environment", in *Proc. of the Boulder Damage Symposium, Oct. 26–28, 1987, Boulder, CO*, ed. by Philippe M. Fauchet and Karl H. Guenther, National Institute of Standards and Technology Special Pub. **756**, Boulder, CO, 112–122 (1987)

2.340 N. Bloembergen: "Role of Cracks, Pores and Absorbing Inclusions on Laser Induced Damage Threshold at Surfaces of Transparent Dielectrics", Appl. Opt. **12**, 661–664 (1973)

2.341 F. Rainer and T.F. Deaton: "Laser damage thresholds at short wavelengths", Appl. Opt. **10**, 1722–1724 (1982)

2.342 W.F. Hagen and E. Snitzer: "Nonlinear solarization in flint glasses by intense 0.53 μm light", IEEE J. Quant. Electron. **QE-4**, 361 (1968)

2.343 W.L. Smith, C.L. Vercimak, and W.E. Warren: "Excited-state-absorption", J. Opt. Soc. Am. **72**, 1782, Th E 1 (1982)

2.344 W.T. White III, M.A. Henesian, and M.J. Weber: "Photo-thermal-lensing measurements of two-photon absorption and two-photon-induced color centers in borosilicate glasses at 532 nm", J. Opt. Soc. Am. B **2**, 1402–1408 (1985)

2.345 M.J. Liepmann, A.J. Marker III, and U. Sowada: "Ultra-violet radiation effects on UV-transmitting fluor crown glasses", Proc. SPIE **970**, 170–179 (1988)

2.346 P. Liu, W.L. Smith, H. Lotem, J.H. Bechtel, N. Bloembergen, and R.S. Adhav: "Absolute two-photon absorption coefficient at 355 and 266 nm", Phys. Rev. B **17**, 4620–4632 (1978)

3. Optical Quality

Frank-Thomas Lentes, Norbert Neuroth

3.1 Tolerances of the Refractive Index and the Abbe Number

An optical glass type is defined by its refractive index and the Abbe number given in the catalogue. The melting process has to be controlled in such a way that the refractive index has a difference less than $\pm 1 \times 10^{-3}$ from the nominal catalogue value and the deviation of the Abbe number is less than $\pm 0.8\,\%$. It is possible to obtain glass with even better specifications, e.g., the deviation of the refractive index from the catalogue value is within $\pm 2 \times 10^{-4}$ and the deviation of the Abbe number is within $\pm 0.2\,\%$. During the production the conditions of the tank melting process can vary; for instance the mass flow may be changed because pieces of different sizes must be produced. Volatile components of the glass melt may vaporize or the glass melt itself attacks the tank wall; these are some reasons for variations of the chemical composition of the glass. Therefore, the refractive index and the Abbe number of the glass are continuously controlled so that both values are kept within the tolerances described above during the whole production process. Due to the large volume of a tank and the constant melting conditions, the properties of the glass vary only slowly with time, e.g., during a period of hours the variation of the glass properties is much smaller than the tolerances mentioned. So one can select parts of the production (batches) in which the refractive index varies by 1×10^{-4} or less (see classes of homogeneity in Subsect. 3.1.1).

Many aspects of optical quality (measuring method and tolerances) are described in the papers of national standards [3.1] and international standards [3.2].

3.1.1 Uniformity

Definitions

Optical glass must have a constant refractive index over the whole volume spatially and directionally. Deviations in the fourth and higher decimals – depending on the application – are already critical. Therefore, melting process conditions have to be kept very constant. That means raw materials (source and composition), melting temperatures and flow velocity must not be changed. Furthermore, the glass has to be annealed slowly in order to obtain small residual stress birefringence. Even if the production process is carried out very carefully, the refractive index of melts varies to a certain degree.

The following types of inhomogeneities of an optical glass can be found [3.3]:
a) **spatial** inhomogeneities in a glass block:

- micro-inhomogeneities with scale lengths less than or equal to the wavelengths of the optical range
- inhomogeneities of small widths (0.1 mm to a few millimetres) with a relative high refractive index gradient (called *striae*)
- large scale inhomogeneities with small refractive index gradients (usually designated by the term *homogeneity*)
- inclusions and bubbles

b) **directional** inhomogeneities in a glass block:

- stress birefringence due to temperature gradients during the annealing process

Micro-inhomogeneities cannot be detected macroscopically, yet they cause light scattering. If a larger quantity of optical glass with a very small variation of the refractive index is needed, glass from one and the same production batch should be taken.

Homogeneity

Optical glasses can be classified by their *homogeneities* depending on the large scale variations of the refractive index with small gradients. The smaller the variation, the higher the homogeneity. The influence of inhomogeneities on the image quality is discussed in [3.4] and [3.5].

Basically, there are two different approaches to evaluate inhomogeneities in optical glasses:

a) *Integrating method* by interferometry, preferably phase-measuring interferometry [3.6–11]. Today this is the preferred procedure since the deformation of the entering wavefront is directly measured. The homogeneity is evaluated by integrating over the light path in the glass sample. Therefore, a linear gradient of the refractive index (wedge) in the direction of the beam cannot be detected. In order to suppress surface irregularities the glass sample is put between high precision sandwich plates which are contacted by immersion oil. Some modifications of the immersion technique are found in [3.12] and [3.13].

b) *Statistical method.* From the upper and lower surfaces of the glass plate to be inspected several samples are cut from different places on the rim and in the centre. Each part from the edge is polished together with a piece from the centre. The optical path difference of a light beam passing through both samples is measured by double slit interferometry [3.3]. So the differences of the refractive index of all parts of the volume are measured.

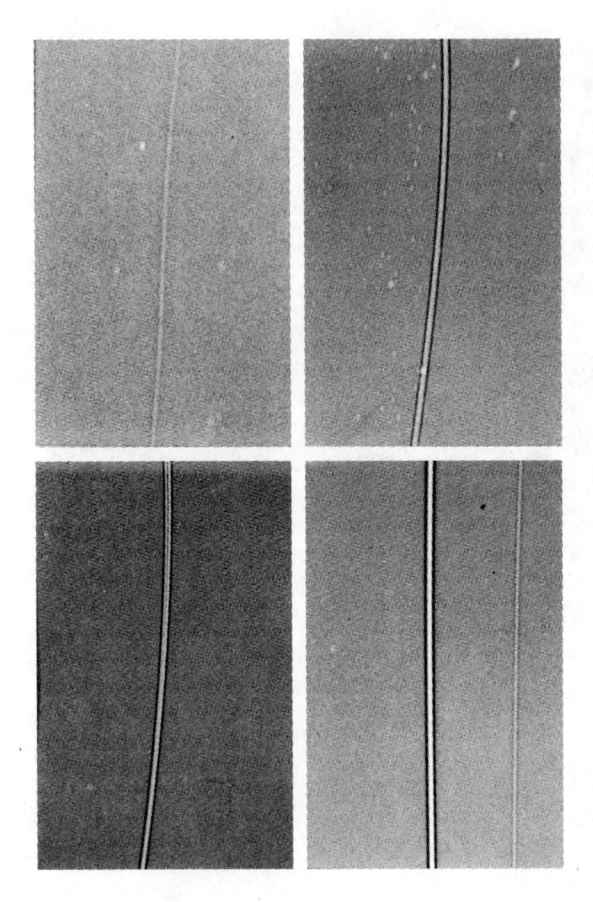

Fig. 3.1. Striae of glass melted in ceramic crucibles

Classes of homogeneity are defined as a variation of the refractive index n in the glass blank. The highest class of homogeneity has a variation of $n \leq 1 \times 10^{-6}$; lower grades have $\Delta n \leq 2 \times 10^{-6}$, $\leq 5 \times 10^{-6}$, or $\leq 2 \times 10^{-5}$.

Striae

Striae are abrupt discontinuities in the glass composition with short irregular local oscillations of the refractive index. They are caused by imperfect solution of raw materials in the melt, partial enrichment of the glass by parts of the wall materials of the tank and layers of varying density due to vaporization during the melting and casting process. If the melt is conducted in a ceramic pot, the form of striae looks like threads (Fig. 3.1). Striae due to a melting process in tanks resemble ribbons, strands, or show a veil without sharp local limitation indicating the structure of the flow process of the melted glass (Fig. 3.2). Since the striae patterns are irregularly distributed in space, the sample has to be rotated in the light beam. Sometimes the striae can only be detected in a special direction (parallel layer). It is difficult to measure the striae quantitatively.

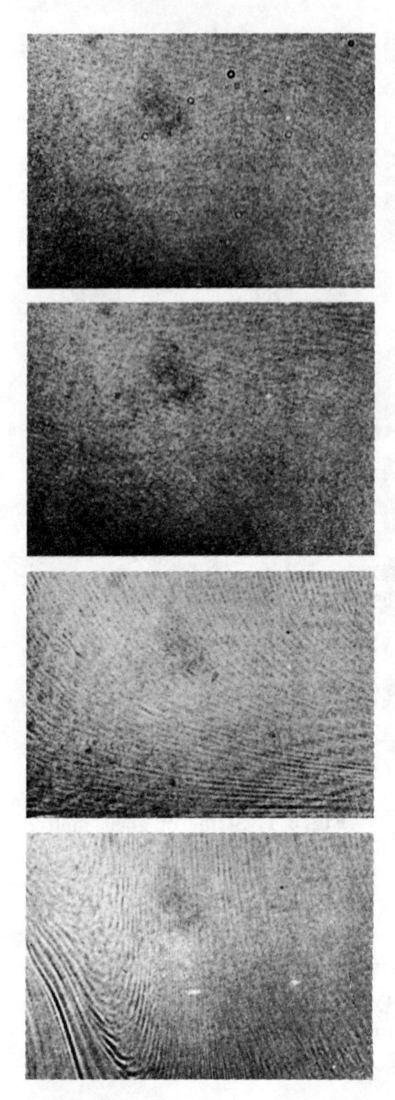

Fig. 3.2. Striae of glass melted in tanks

Striae cause a small scale deformation of the wavefront as well as scattering. The influence of a single well-defined stria on the image quality has been studied in [3.14–19]. The effects of a statistically deformed wavefront on the image quality are described in [3.20] and [3.5].

The most sensitive method to detect striae is the shadowgraph method [3.7, 21, 22]. Even very weak striae are detected and it is possible to compare the striae of the test sample with those of reference standards. A quantitative method is also possible by interferometric measurement; however, this requires an exceptional high quality sample surface. Today, the shadowgraph method is the only economic choice for measuring striae.

New standards have been proposed which are made synthetically. The optical path difference of these synthetic striae can be measured by interferometry. In this way this standard material is defined quantitatively (an ISO draft is in preparation).

A proposed variation of the shadowgraph method is the use of screens with different surface roughnesses (grain sizes from 0.3 to 12 μm). Weak striae are detected with the fine grain containing screen, but cannot be seen if a screen with a rough grain is used. Using screens with a consecutive greater grain size, it can be checked at which grain size the transition from the detectable to the undetectable striae occurs.

There are several national standard specifications for striae. The comparison between the classifications is difficult. The optical path differences caused by the striae are of the order of 10 to 100 nm, i.e. significantly smaller than the wavelength in the visual spectral range.

Inclusions and Bubbles

A material is homogeneous if the optical properties are constant over the whole volume and if there are no centres which deteriorate the light beam such as inclusions or bubbles.

Inclusions can be undissolved particles of the melt, like small crystals, stones, or metal particles. They are caused

a) by an imperfect melting process in special parts of the tank (e.g. regions with too low a temperature) or

b) by the attack of the wall material of the tank or crucible. Parts of the refractory material may not be totally solvable at the operating temperature of the tank.

Bubbles are also the result of an imperfect melting process if the gaseous reactions have not finished or if gases, dissolved in the melt at higher temperatures, appear at lower temperatures. The sources of the raw materials are selected thoroughly to get very clean and homogeneous substances. Furthermore, the melting process is guided very carefully. In spite of these efforts small inclusions and bubbles may occur in the glass.

The content of bubbles and inclusions of an optical glass is determined in the following way: a sample with a size of at least 100 × 50 × 20 mm is polished at the larger surfaces. With larger pieces of glass a part of the volume representative for the internal quality of the whole piece is selected. The plate or the block is illuminated from the small side and inspected through the larger surface. Bubbles and inclusions are then easily seen. The projected cross-sections of all defects greater than 50 μm are summarized. The content of bubbles and inclusions is classified depending on the projected area of inclusions calculated for a sample volume of 100 cm^3 (Table 3.1).

Small bubbles and inclusions cause some scattering if they are located in lenses near the pupils of an optical system. Depending on their size and number they

Table 3.1. Classification of bubbles and inclusions

Class	0	1	2	3	4
Projected area mm^2	0–0.03	0.03–0.10	0.11–0.25	0.26–0.50	0.51–1.00

may slightly reduce the image contrast. On the other hand, if they are found in optical elements near the image plane, they can be harmful (e.g. in reticles).

If the glass has to resist high laser fluences, the contents of inclusions (not bubbles) are critical. Absorbing particles even smaller than 50 µm are heated up by the laser beam; internal explosions occur which produce defects of much greater sizes than the original inclusions [3.23–27]. For such applications glass melts have to be taken which are free from absorbing particles. They are available on special request (see also Sect. 2.7).

Stress birefringence

Basically, optical glass is well annealed; nevertheless, it retains slight residual stresses. They cause birefringence which disturbs the optical homogeneity. The unit of the stress birefringence is nm/cm (see Sect. 2.4). The stress birefringence is measured near the edge of the plate. In the case of a circular plate the measuring spot is positioned at a distance of about 5 % of the plate diameter from the rim; in the case of a rectangular plate at a distance of about 5 % of the side length from the rim. The allowable birefringence depends on the application of the glass. The precision devices for measuring interference or polarization properties need the lowest degree of birefringence (< 2 nm/cm). For astro-optics a birefringence of ≤ 5 nm/cm may be allowed. Optical glass for cameras or microscopes could have 10 nm/cm. According to these values the classes of stress birefringence are defined by the manufacturers of optical glasses.

3.2 Colouration, Solarization and Fluorescence

Colouration

For many applications of optical glass the transmission in the visible as well as in the adjacent spectral regions is important, especially in the ultraviolet. The UV transmission rises with increasing wavelengths. The position and the slope of the absorption edge depend on the glass type, the purity of the raw materials, and the reactions of the glass melt with the walls of the crucible. In Fig. 3.3 the spectral transmission of plates of different thicknesses of one and the same glass type is shown. With an increasing thickness the transmission edge is shifted to longer wavelengths. The nearer the transmission edge is located to the visible region and the less the slope is, the poorer becomes the transmission in the blue region. Glasses – especially those with high refractive indices – may have a reduced transmission in the blue region if they are thicker (e.g., 30 mm). For applications with higher demands the knowledge of the complete spectrum of the internal

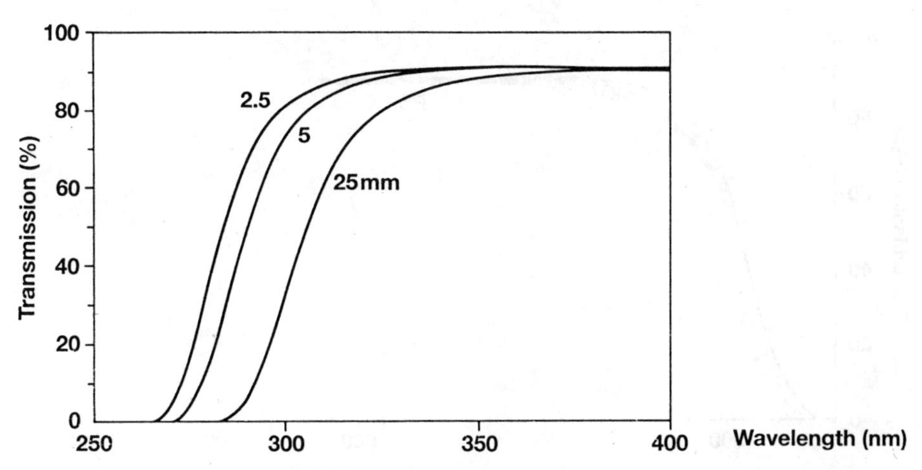

Fig. 3.3. Ultraviolet transmission edge of UK 50 glass with thicknesses of 2.5, 5, and 25 mm

transmittance τ_i is necessary. If only a rough classification of the short wavelength absorption is needed, there are the following definitions:

1) The sample thickness is fixed (e.g., 5, 10, or 25 mm) and the internal transmission is indicated at the wavelength 400 nm or at 400 and 365 nm (sometimes 370 nm is used).

2) From the transmission measurement of a 10 mm thick sample one obtains the two wavelengths at which the transmission is 80 % and 5 %, respectively (colour code according to the Japanese Optical Glass Industry Standard [3.1]). In this case information about both the spectral position and the slope of the absorption edge is provided.

Both methods are capable of indicating how much of the blue spectral region is cut off. This causes a yellowish colour of the glass. As mentioned in Sect. 2.3, iron oxide may increase the absorption in the spectral range around 400 nm. Normally the delivery specifications of optical glass refer to the refractive index and the Abbe number. The transmission curve of a glass varies in the short wavelength region from melt to melt. Therefore, the catalogue data should be used only for guidance. If one needs the transmission data as indicated in the catalogue or even more precisely, the glass manufacturer should be consulted.

Solarization

The influence of the electromagnetic radiation on the transmission of a glass depends on the glass type as well as on the wavelength of the radiation. Visible light (normal solar radiation on the ground) has a very small influence, whereas UV radiation has generally a stronger effect. The influence of visible and UV radiation on glass is called solarization. It is investigated by irradiation with a xenon or mercury lamp, or with UV-lasers. The xenon high pressure arc lamp shows a continuous spectrum in the UV and the visible spectral range without pronounced

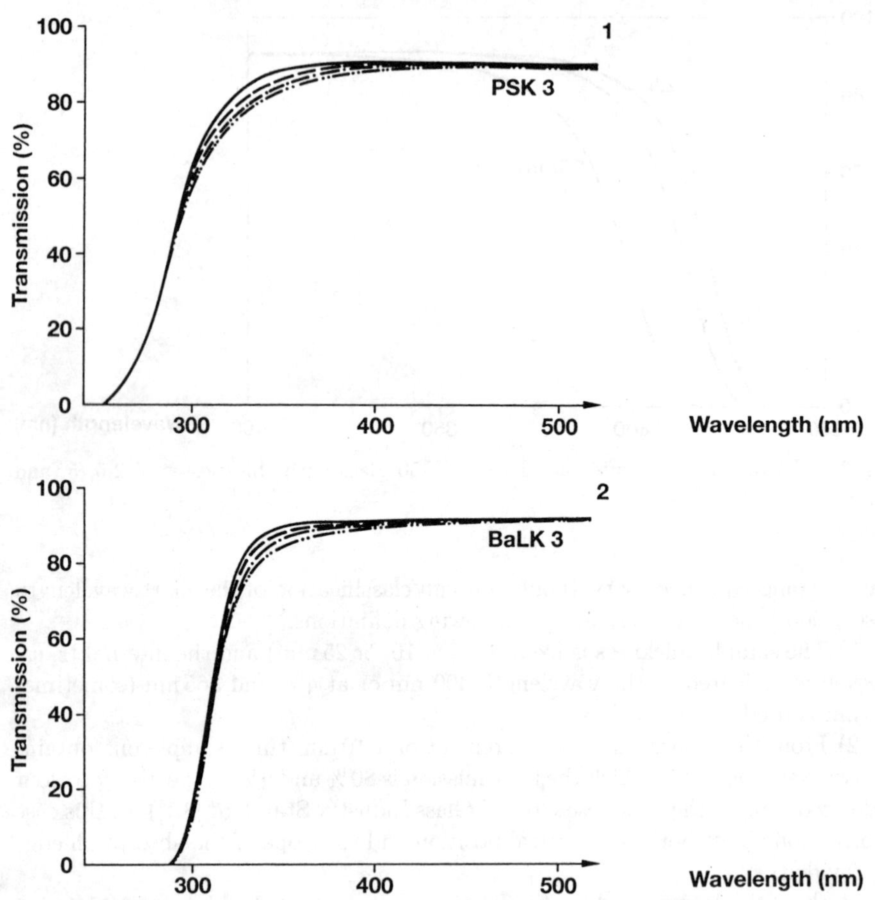

Fig. 3.4. Influence of ultraviolet radiation on the transmission of crown glasses that are 5 mm thick. Irradiation with a low pressure Hg-Lamp (main emission at 254 nm): 0, 75, 150, 300 hours; glass types: 1) PSK3, 2) BaLK3, 3) BaK1, 4) ZK1

emission lines. Mercury lamps have strong emission lines (mainly the 254 nm line if operated at low pressure, or mainly the 365 nm line if operated at high pressure). The solarization effect of these lamps can be different. Glasses with low UV transmission – e.g. with a high lead content (types F and SF) – normally have small solarization effects. Several crown glasses with a higher ultraviolet transmission change their UV transmission edge: PSK, BaLK, K, ZK, BaK, SK and LaK. In Fig. 3.4 some examples are given. The steepness of the transmission edge becomes smaller. This effect is permanent at room temperature, whereas at higher temperatures it can be reversed.

In Fig. 3.5 the influence of UV-laser radiation at the wavelength 248 nm on fused silica is demonstrated [3.28]. Synthetically produced materials (by the CVD process) and materials produced by melting natural rock crystals show different behaviour. In the synthetic material an absorption band occurs at 210 nm (caused

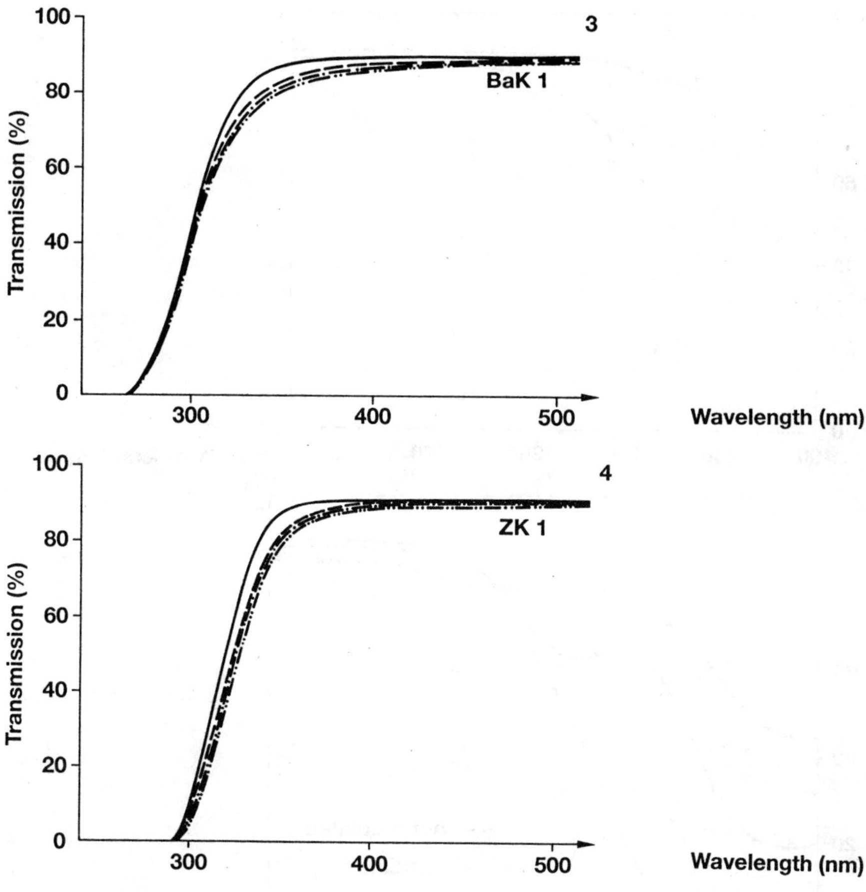

Fig. 3.4. (cont.)

by paramagnetic colour centres (E′ centres)). The intensity of this absorption band is weak if the OH content of the fused silica is high and vice versa. Fused silica produced by melting rock crystals has (even without irradiation) absorption bands at approximately 240 and 200 nm. In a material with a low OH content the 200 nm band is stronger than in a material with a higher OH content. Irradiation with a 248 nm laser diminishes the 240 nm band and makes the 200 nm band stronger and broader.

The influence of UV radiation can partly be suppressed by doping the glass with CeO_2. The doping shifts the UV absorption edge to longer wavelengths even without UV radiation. This edge is then stable. Also in the case of irradiation with sources of higher quantum energies (x-rays, γ-rays) CeO_2 doping proves protective (see also Sect. 8.7.2). In the case of neutron and electron radiation the protective effect of CeO_2 doping depends on the glass type. Therefore, the radiation load

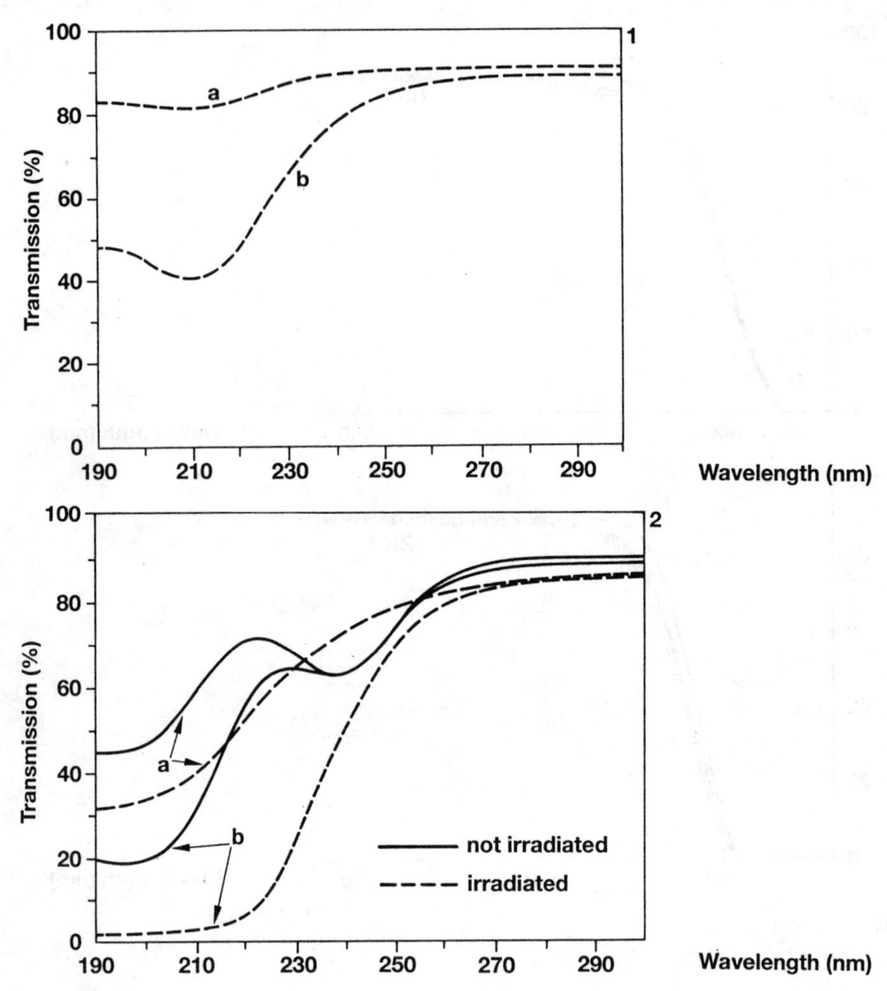

Fig. 3.5. Influence of laser irradiation (248 nm) on the transmission of fused silica: 1) synthetic material, 2) material melted from rock crystals a) with high OH content; b) with low OH content — before irradiation (not given in 1), ··· after irradiation [3.28]

to which the glass will be exposed must be considered when choosing a certain optical glass.

Fluorescence

Some transition elements (Cu, Mn, Cr, V, Ti in their lower oxidation states), several rare earth elements and uranium cause strong fluorescence in glass [3.29–33]. Sometimes only one emission line (or band) can be found, and sometimes several of them are located at different wavelengths in the visible or near infrared spectrum. The wavelengths of the maximum fluorescence are shown in Fig. 3.6.

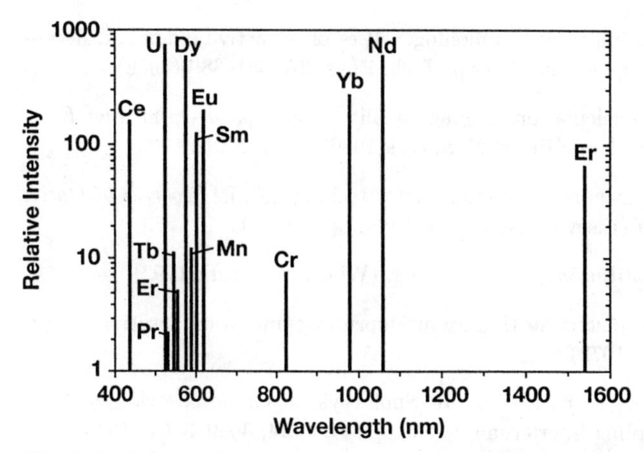

Fig. 3.6. Survey of the wavelength of maximum fluorescence with respect to activator ions in glass (the intensity depends on the glass composition and activator concentration)

The intensity depends on the glass composition, the melting process (e.g. the redox-condition) and the concentration of the activator ions. As these elements also cause absorption, they must consequently be avoided in order to get a colourless optical glass. They sometimes occur in glass as impurities.

Under certain conditions the elements Ag, Cu, Au, Sn, Pb, Bi, As, Sb, Te, Pt, Rh and Ir as atoms or particles may also induce fluorescence in glass [3.34, 35]. Another source of fluorescence is the colour centres which may be generated by oxygen vacancy or excess, or non-bridging oxygen hole centres [3.36–38]. The intensity of these types of fluorescence is weak. It cannot be detected in daylight but only in a dark room when ultraviolet light is used for excitation. In most of the applications of optical glasses this weak fluorescence does not cause any trouble, with the exception of lens systems e.g. for fluorescence microscopy. In this case the glass types and production batch has to be checked for fluorescence.

References

3.1 National standards: American Military Specification MIL-G-174B *Optical Glass* (1986); British Standards Institution BS 4301 (1982); French Standard NF S 10 – 001 to 004 (1978); DIN 3140, *Maß- und Toleranzangaben für Optikeinzelteile* (1978); DIN 58927, (1970); Standard of the GDR, TGL 21790, *Optisches Glas, Bestimmung der Schlierenhaltigkeit* (1980); Japan Optical Glass Industry Association (JOGIS) 11–75 (1975)

3.2 International Organisation for Standardization ISO 10110, *Optics and Optical Instruments* (1992), Part 2 "Material imperfections – stress birefringence", Part 3 "Bubbles and inclusions", Part 4 "Inhomogeneity and striae"

3.3 F. Reitmayer, E. Schuster: "Homogeneity of optical glasses", Appl. Opt. **11**, 1107–1111 (1972)

3.4 C. Hofmann: "On the influence of inhomogeneities of refractive index and decentering errors upon image forming", Exp. Tech. Phys. **26**, 381–389 (1978)

3.5 W.B. Wetherell: "The calculation of image quality", in *Applied Optics and Engineering*, Vol. VIII (Academic Press, New York 1980) pp. 172–315

3.6 K. Creath: "Phase-measurement interferometry techniques", in *Progress of Optics XXVI*, ed. by E. Wolf (Elsevier, New York 1988) pp. 350–393

3.7 D. Malacara (ed.): *Optical Shop Testing*, 2 ed. (Wiley, New York 1992)

3.8 J. Schwider: "Interferometrische Homogenitätsprüfung mit Kompensation", Opt. Commun. **6**, 106–110 (1972)

3.9 J. Schwider, R. Burow, K.-E. Elssner, R. Spolaczyk, J. Grzanna: "Homogeneity testing by phase sampling interferometry", Appl. Opt. **24**, 3059–3061 (1985)

3.10 D. Tentori: "Homogeneity testing of optical glass by holographic interferometry", Appl. Opt. **30**, 752–755 (1991)

3.11 F. Twyman, J.W. Perry: "Measuring small differences of refractive index", Proc. Phys. Soc. London **34**, 151–154 (1922)

3.12 P. Langenbeck: "Optical homogeneity measurement by a two angle method", Optik **28**, 592–601 (1968)

3.13 F.E. Roberts, P. Langenbeck: "Homogeneity evaluation of very large disks", Appl. Opt. **8**, 2311–2314 (1969)

3.14 R. Hild, S. Kessler, G. Nitzsche: "Influence of schlieren on imaging properties of an optical system: I. Point spread function", Optik **85**, 123–131 (1990)

3.15 R. Hild, G. Nitzsche, J. Hebenstreit: "Influence of schlieren on imaging properties of an optical system: II. Modulation transfer function (MTF)", Optik **85**, 177–179 (1990)

3.16 Hild, G. Nitzsche, S. Kessler: "Influence of schlieren on image quality of optical systems: III. Isoplanatic and nonisoplanatic imaging", Optik **86**, 1–6 (1990)

3.17 C. Hofmann, I. Reichardt: "On the variation of Strehl's definition by schlieren in optical systems", Exp. Tech. Phys. **23**, 513–523 (1975)

3.18 R. Keller: "Die Intensitätsverteilung im Bild eines punktförmigen Objektes bei einer mit Schlieren behafteten Abbildung", Optik **21**, 360–371 (1964)

3.19 H. Köhler: "Einfluß von Glasschlieren auf die optische Abbildung", Optik **21**, 339–359 (1964)

3.20 E. O'Neill: *Introduction to statistical optics* (Addison Wesley, Reading 1963)

3.21 H. Schardin: "Glastechnische Interferenz- und Schlierenaufnahmen", Glastechn. Ber. **7**, 1–12 (1954)

3.22 H. Wolter: "Schlieren-, Phasenkontrast- und Lichtschnittverfahren", in *Handbuch der Physik*, ed. by S. Flügge, Vol. 24 (Springer, Berlin 1956) pp. 555–645

3.23 H. Fuhrmann, C. Hofmann: "Laserinduzierte Zerstörung optischer Medien", Feingerätetechnik **26**, Teil I: 10–16, Teil II: 58–60 (1977)

3.24 H. Hack, N. Neuroth: "Resistance of optical glasses and colored glasses to 3-nsec laser pulses", Appl. Opt. **21**, 3239–3248 (1982)

3.25 J.S. Hayden, H.-J. Hoffmann: "Laser-Gläser ohne Platin-Teilchen", Werkstoff und Innovation **4**, 47–49 (1991)

3.26 D. Kitriotis, L.D. Merkle: "Multiple laser induced damage phenomena in silicates", Appl. Opt. **28**, 949–958 (1989)

3.27 W.H. Lowdermilk, D. Milam, F. Rainer: "Laser-induced damage in optical materials", Nat. Bur. Stand. (US) Spec. Publ. **541** (1978) and Spec. Publ. **568** (1979)

3.28 N. Leclerc, C. Pfleiderer, H. Hitzler, S. Thomas, R. Takke, W. Englisch, J. Wolfrum, K. O. Greulich: "KrF eximer laser induced absorption and fluorescence bands in fused silica related to the manufacturing process", in *Advanced Optical Materials*, ed. by A. Marker, SPIE Vol. **1327**, 60–68 (1990)

3.29 L.E. Ageeva,V. I. Abruzow, E. L. Raaben, M. N. Tolstoi, S. K. Shumilov : "Spectral-luminescence properties of glasses activated by rare-earth elements", English Translation of the Russian J. of Phys. and Chem. of Glasses **12**, 175–184 (1986)

3.30 R. Reisfeld: "Inorganic ions in glasses and polycrystalline pellets as fluorescence standard reference materials", J. Res. NBS Vol. **76 A**, 613–635 (1972)

3.31 R. Reisfeld: "Radiative and non radiative transitions of rare earth ions in glasses", in Structure and Bonding, Vol. **22** (Springer Verlag, Berlin 1975) pp. 123–125

3.32 G.E. Rindone: "Luminescence in the glassy state", in *Luminescence of inorganic solids*, ed. by P. Goldberg (Academic Press, New York 1966) pp. 419–464

3.33 T. Takahashi: *Introduction to luminescent Glasses*, RCA-Review **41**, March 1980

3.34 C. Bettinali, G. Ferraresso: "Luminescence centers in lead silicate glasses", J. Non-Cryst. Sol. **1**, 91–101 and 360–370 (1968)

3.35 S. Parke, R.S. Webb: "The optical properties of thallium, lead, and bismuth in oxide glasses", J. Phys. Chem. Sol. **34**, 85–95 (1973)

3.36 M. Kohketsu, H. Kawazoe, M. Yamane: "Luminescence centers in VAD SiO_2 glasses sintered under reducing or oxidizing atmospheres", Diffusion and Defect Data **53–54**, 127–133 (1987)

3.37 R. Tohmon, Y. Shimogaichi, S. Munekuni, Y. Ohki, K. Nagasawa, Y. Sakurai, Y. Hama: "The red luminescence in pure silica glasses", in *Proc. International Congress on Optical Science and Engineering*, SPIE Vol. **1128** (Paris, France 1989), paper No. 34, pp. 198–204

3.38 R. Tohmon, Y. Yamasaka, K. Nagasawa, Y. Ohki, Y. Hama: "Cause of the 5.0 eV absorption band in pure silica glasses", J. Non-Cryst. Sol. **95–96**, 671–678 (1987)

4. Mechanical Properties

Kurt Nattermann, Norbert Neuroth, Robert J. Scheller

4.1 Density

The density of the glass is mainly needed to calculate the weight of a lens or a lens system and the deformation of an optical component (e.g. a large mirror). Furthermore, it is used to calculate other properties, e.g. the mole volume or the mole refraction of the glass which are important quantities for the development of optical glasses.

In most cases glasses with higher densities also have higher refractive indices. In former times, lead oxide was chiefly used to obtain glasses with high refractive indices: the heavy flint glasses (SF-types) have very high refractive indices and also high densities (for instance, the glass type SF 59 with $n_d = 1.953$ and a density ρ of 6.26). If a lens system has many lenses, the weight becomes an important issue. Therefore, glasses with low densities, i.e., the so-called lightweight glasses, have been developed recently. Ophthalmic glasses, in particular, benefit from this development. Figure 4.1 gives a survey about the correlation of density and refractive index. It may be useful for selecting glass types of minimum weight. The density is mainly determined by the chemical composition. To a smaller degree (in the third decimal place), it can be influenced by the annealing process. The smaller the annealing rate in the vicinity of the transformation point, the higher the density, and vice versa.

As early as last century the first attempt was made to calculate the density from the chemical composition [4.1]. Several other approaches followed later [4.2–5].

Due to thermal expansion the densities of optical glasses decrease with increasing temperature.

4.2 Elastic Modulus, Poisson's Ratio, Specific Thermal Tension

Glass shows an almost perfect brittle-elastic behaviour at temperatures lower than the transformation point; i.e., the deformation is proportional to the stress

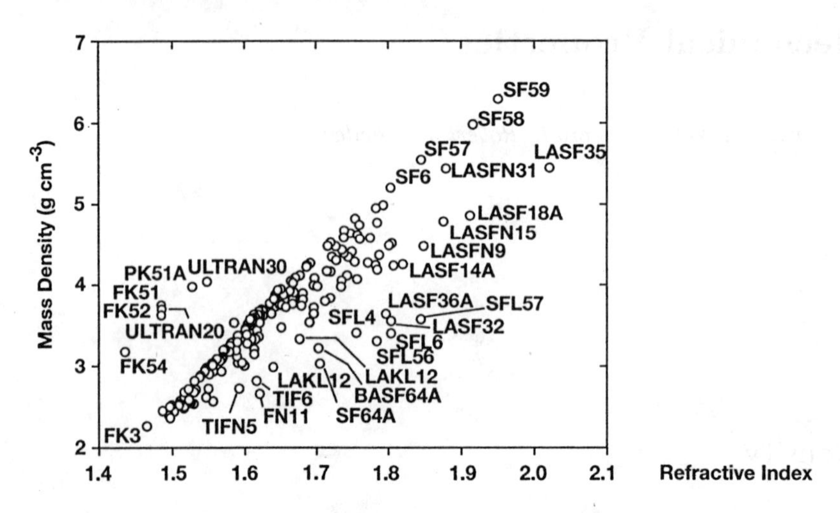

Fig. 4.1. Correlation between density and refractive index of optical glasses

(Hooke's law). If the ends of a rod are exposed to a stress σ, its relative elongation is given by

$$\frac{\Delta l}{l} = \frac{\sigma}{E} . \tag{4.1}$$

E is called the modulus of elasticity (Young's modulus). Due to the elongation of the rod its cross-section is decreased. The relation of the relative decrease of the thickness t to the relative elongation is called Poisson's ratio, defined as

$$\frac{\Delta t}{t} : \frac{\Delta l}{l} = \mu . \tag{4.2}$$

The torsion of a rod is characterized by the modulus of torsion (or shear modulus) G. The properties E, G and μ are characteristic constants of the individual glass type. They depend on the chemical composition of the glass (and to a lesser degree also on the annealing process).

The optical glasses vary in their compositions to a great extent. Correspondingly Young's modulus varies greatly: 40 to $140 \times 10^3\,\mathrm{N/mm^2}$. The Poisson's ratio changes from 0.17 to 0.3. The relation among these three constants reads:

$$E = 2(1 + \mu)G . \tag{4.3}$$

In Table 4.1 the glass types are sorted according to increasing Young's moduli.

If a temperature gradient exists in a glass piece, internal stress occurs. The elastic modulus E, the coefficient of thermal expansion α and the Poisson's ratio are needed to calculate the internal stress. If the temperature gradient in the sample is ΔT, the resulting stress can be estimated using

$$\sigma = \frac{E\alpha}{1 - \mu} \Delta T . \tag{4.4}$$

Table 4.1. Young's modulus of optical glasses (sorted according to increasing values)

Glass name	Young's modulus 10^3 N/mm^2	Glass name	Young's modulus 10^3 N/mm^2
FK3	46	LLF2	61
SF59	51	BaSF1	62
PSK54	52	FK5	62
SF58	52	LLF7	62
KzFN2	54	F9	63
SF57	54	LLF6	63
F13	55	BaF3	64
F7	55	SF10	64
SF16	55	SF13	64
SF2	55	BaLF5	65
SF6	55	K10	65
F1	56	KF6	65
F4	56	KzFSN5	65
KzFS1	56	SF14	65
SF1	56	TiF6	65
SF18	56	BaF4	66
SF3	56	BaSF2	66
SF4	56	BaSF56	66
SF5	56	KF3	66
SF55	56	KzFSN2	66
SF8	56	KzFSN9	66
F2	57	PK50	66
F6	57	SF11	66
SF56A	57	TiFN5	66
F14	58	BaSF10	67
F15	58	K11	67
F3	58	KF9	67
F5	58	LaF9	67
LF7	58	KzFS7A	68
SF19	58	ZK1	68
SF53	58	K7	69
SF54	58	KzFS8	70
SF63	58	ZKN7	70
SF9	58	BaK2	71
KzFS6	59	BaLKN3	71
LF5	59	BaSF54	71
F8	60	BK10	71
KzFSN4	60	K3	71
LF8	60	K4	71
LLF1	60	K5	71
SF12	60	LaF11A	71
SF15	60	BaK5	72
TiF3	60	K50	72
KzFN1	61	PSK50	72

Table 4.1. (cont.)

Glass name	Young's modulus $10^3\,N/mm^2$	Glass name	Young's modulus $10^3\,N/mm^2$
BaF8	73	LaFN23	82
BaK1	73	SK10	82
UK50	73	SK3	82
BaF52	74	SK4	82
BaSF12	74	FN11	83
BK1	74	SSKN8	83
BK3	74	BaF13	84
PK1	74	PK2	84
PK51A	74	PK3	84
SK51	75	PSK3	84
SSK3	75	SK15	84
BaLF4	76	SK5	84
FK54	76	SK7	84
ULTRAN30	76	SSK50	84
BaFN6	77	BaSF52	86
BaK4	77	LaK23	86
BaSF57	77	PSK2	86
BaF9	78	SK14	86
FK52	78	LaKN12	87
PSK52	78	LaKN13	87
PSK53A	78	BaFN11	88
SK12	78	LaKN6	88
SK13	78	SK18A	88
SK2	78	SSKN5	88
SSK2	78	BaF51	89
BaLF50	79	BaFN10	89
SK1	79	SF64A	89
SK11	79	SK16	89
SK6	79	LaK11	90
SK8	79	LaKN22	90
SSK1	79	LaKN7	90
SSK4A	79	SFL4	90
ULTRAN20	79	LaK21	91
BaSF51	80	SFL56	91
BaSF6	80	LaSF33	92
BK6	80	BaF50	93
BK8	80	LaF2	93
LaFN7	80	SFL6	93
SSK51	80	LaF20	94
BaK50	81	LaF13	95
BaSF13	81	LaF3	95
FK51	81	SK55	96
UBK7	81	SFL57	97
BK7	82	LaFN8	98

Table 4.1. (cont.)

Glass name	Young's modulus $10^3\,N/mm^2$	Glass name	Young's modulus $10^3\,N/mm^2$
LaF22A	100	LaFN10	117
BaSF64A	105	LaK16A	117
LaKL21	109	LaK28	117
LaSFN9	109	LaFN24	120
LaK9	110	LaSF3	121
LaKL12	110	LaFN28	123
LaK10	111	LaK33	124
LaKN14	111	LaSFN30	124
LaSF32	113	LaSFN31	124
LaK31	114	LaFN21	126
LaK8	115	LaSFN15	126
LaSF36A	115	LaSF18A	127
LaSF14A	116	LaSF35	132

The internal stress is proportional to the Young's modulus and to the coefficient of thermal expansion α. The expression

$$\varphi = \frac{E\alpha}{1-\mu} \tag{4.5}$$

is called specific thermal tension. In Table 4.2 the glass types are sorted according to increasing values of their specific thermal tension. The table gives an indication how sensitive the different glass types are to temperature shock.

The elastic moduli and the density are required to calculate the sound velocity, which is important when glasses are used as acousto-optical modulators.

4.3 Microhardness

The resistance of a material to indentation is characterized by the indentation hardness. Several testing methods can be used: scratching, abrasion, or penetration; however, the results are not exactly comparable. In the following, the last method will be looked at in greater detail. Two specific indenting geometries applying diamonds as tools are commonly used: Vickers hardness is determined by using a diamond in the form of a square pyramid with an included angle of 136° between the opposite faces. The procedure for measuring Knoop hardness differs by the use of a rhombus-shaped diamond so that the intersections between adjacent facets have included angles of 172.5° and 130.0°, respectively.

During the pressing of the diamond into the glass plate an *elastic* and a *plastic* deformation occurs. (If the movement of the diamond is very quick and the force is too great, cracks may also be generated). After the experiment a permanent indentation is left caused by the plastic flow of the glass. The indentation hardness is defined as the ratio of the force F to the area of the indentation which has the

Table 4.2. Specific thermal tension $\varphi = \frac{E\alpha}{(1-\mu)}$ of optical glasses (sorted according to increasing values)

Glass name	Specific thermal tension $Nmm^{-2}K^{-1}$	Glass name	Specific thermal tension $Nmm^{-2}K^{-1}$
KzFSN4	0.37	SF5	0.60
KzFSN9	0.37	SF57	0.60
KzFS1	0.39	SF8	0.60
BaK50	0.40	BaSF51	0.61
KzFSN5	0.40	LLF1	0.61
ZKN7	0.40	LLF2	0.61
KzFSN2	0.41	SF12	0.61
KzFN2	0.42	SF55	0.61
KzFS6	0.42	SF9	0.61
KzFS7A	0.46	BaSF52	0.62
BK3	0.49	F13	0.62
FK3	0.50	F6	0.62
KzFS8	0.51	F9	0.62
BK10	0.52	SF15	0.62
K10	0.52	SF3	0.62
LaF11A	0.53	SF53	0.62
SF11	0.53	SF63	0.62
K11	0.54	SK5	0.62
KF6	0.56	F1	0.63
KzFN1	0.56	F8	0.63
SF14	0.56	SF10	0.63
K59	0.57	SF58	0.63
LLF7	0.57	BaLF4	0.64
F14	0.58	SK2	0.64
LF7	0.58	BaF3	0.65
PK1	0.58	K50	0.65
SF19	0.58	LF8	0.65
LaFN7	0.59	SK1	0.65
LLF6	0.59	SK8	0.65
SF1	0.59	SSK2	0.65
SF13	0.59	K4	0.66
SF4	0.59	LaF9	0.66
SF54	0.59	SSK4A	0.66
SF56A	0.59	UK50	0.66
SF6	0.59	PSK3	0.67
F15	0.60	SK12	0.67
F2	0.60	SK6	0.67
F3	0.60	SSK3	0.67
F4	0.60	ZK1	0.67
F5	0.60	KF3	0.68
SF16	0.60	SF59	0.68
SF18	0.60	SK11	0.68
SF2	0.60	SSK1	0.68

Table 4.2. (cont.)

Glass name	Specific thermal tension $Nmm^{-2}K^{-1}$	Glass name	Specific thermal tension $Nmm^{-2}K^{-1}$
BaF4	0.69	SSKN8	0.79
BaF8	0.69	BaSF6	0.80
BaF9	0.69	BK6	0.80
BaLF5	0.69	FN11	0.80
LF5	0.69	SK15	0.80
BaSF1	0.70	TiF3	0.80
SK14	0.70	LaKN22	0.81
SK3	0.70	SSK51	0.81
BaK4	0.71	BaF52	0.83
BaLKN3	0.71	BaFN11	0.83
BaSF54	0.71	SSKN5	0.83
BaSF56	0.71	BaFN10	0.84
F7	0.71	PSK50	0.84
BaSF2	0.72	LaK21	0.85
BK1	0.72	LaKN14	0.85
PSK2	0.72	LaKN6	0.85
SK13	0.72	SSK50	0.85
SK4	0.72	BaF13	0.86
UBK7	0.72	BaLF50	0.87
BaSF57	0.73	PSK54	0.87
BK7	0.73	LaKN7	0.88
PK2	0.73	LaK10	0.89
SK7	0.73	LaFN24	0.91
BaK1	0.74	LaK11	0.91
BaK2	0.74	LaK16A	0.91
BaK5	0.74	LaK31	0.91
FK5	0.74	LaK8	0.91
K7	0.74	LaFN8	0.92
BK8	0.75	LaKL21	0.92
K5	0.75	PSK52	0.92
PK3	0.75	LaFN23	0.93
K3	0.76	LaKN12	0.93
LaF13	0.76	LaFN10	0.94
PK50	0.76	LaK28	0.94
SK16	0.76	LaSF3	0.94
BaSF10	0.77	SK51	0.94
SK18A	0.77	LaK23	0.95
BaFN6	0.78	LaF20	0.96
SK55	0.78	LaF22A	0.96
TiFN5	0.78	LaSF14A	0.96
BaSF12	0.79	LaK9	0.97
BaSF13	0.79	LaF3	1.01
SK10	0.79	LaFN28	1.01

Table 4.2. (cont.)

Glass name	Specific thermal tension $Nmm^{-2}K^{-1}$	Glass name	Specific thermal tension $Nmm^{-2}K^{-1}$
SF64A	1.01	LaSFN9	1.13
BaF51	1.02	SFL6	1.13
LaKN13	1.03	SFL57	1.14
PSK53A	1.03	LaSFN15	1.16
BaSF64A	1.04	LaSF36A	1.17
LaKL12	1.04	LaSF32	1.20
BaF50	1.05	LaSFN31	1.20
LaK33	1.05	TiF6	1.22
LaF2	1.06	ULTRAN30	1.29
SFL56	1.06	PK51A	1.33
LaSF18A	1.08	LaSF35	1.40
LaSFN30	1.09	ULTRAN20	1.46
LaFN21	1.11	FK51	1.52
SFL4	1.12	FK54	1.55
LaSF33	1.13	FK52	1.58

form of a parallelogram. If the length of the greater diagonal is d, the hardness is defined by (4.6) and (4.7), respectively:

Vickers hardness

$$HV = 0.1855\frac{F}{d^2} , \tag{4.6}$$

Knoop hardness

$$HK = 1.4233\frac{F}{d^2} . \tag{4.7}$$

The length of the diagonal d has the dimension of a few microns and is measured with the aid of a microscope. There may be an error caused by the diffraction of light but this effect can be eliminated by a correction. The corrected values are smaller (in the range of 4 to 8 %). The indentation hardness depends on the load applied. (With increasing load the hardness values become smaller). Therefore, the applied load has to be specified together with the hardness value. The details of the measurement of the Knoop hardness have been stated in the ISO paper 9385 [4.6]. For newer methods of measuring the microhardness see [4.7–9].

The indentation hardness depends on the chemical composition of the glass [4.10–12]. Glasses with a high content of network formers – e.g. those with a high content of silica and/or boron oxide – have quite large hardnesses; barium-lanthanum-borate glasses have the highest hardnesses: LaK 8, LaK 16A, LaK 28, LaK 31, LaSF 3, LaSF N15, LaSF 18A, LaSF N30, LaSF N31, LaSF 35. An increasing alkaline and/or lead content decreases the indentation hardness (see Fig. 4.2).

The hardnesses of freshly broken glass samples are about 1–9 % greater than those of samples which have been ground and polished. This effect is caused by

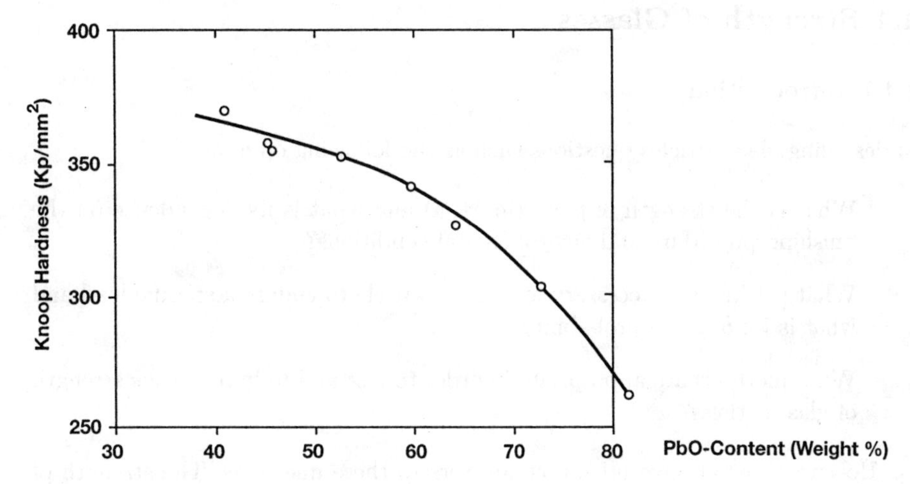

Fig. 4.2. Knoop hardness of lead-alkali-silicate glasses as a function of the PbO content

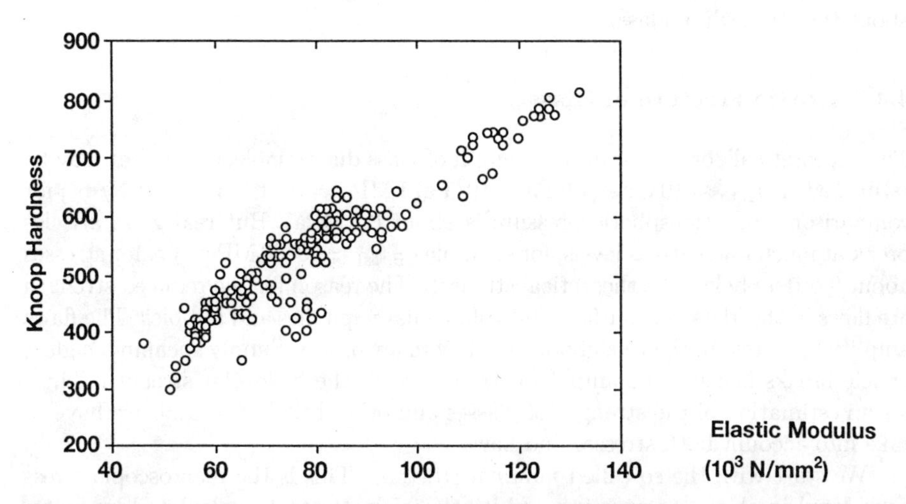

Fig. 4.3. Knoop hardness in relation to the elastic modulus of optical glasses

the hydrolysed coating which is generated during the machining process. Furthermore, after the attack of aqueous solutions the indentation hardness is reduced. There is no simple correlation between the grinding hardness and the indentation hardness since the chemical resistance and mechanical properties also influence the grinding behaviour (see Chap. 7). But there is a rough correlation between the lapping hardness and the indentation hardness [4.13–15]. Since the hardness of a solid material and its elastic modulus depend on the molecular binding forces, a correlation between these two properties is expected. This is shown in Fig. 4.3.

4.4 Strength of Glasses

4.4.1 Introduction

In designing glass articles questions such as the following often arise:

- What is the strength of glass (in MPa) and what is its dependence on the finishing procedure and environmental conditions?
- What thickness is necessary for a glass article to endure a specific load and what is its fracture probability?
- What methods are appropriate in order to test and to improve the strength of glass articles?

However, we can give no simple answers to these questions. The strength of inorganic glasses depends strongly on surface defects and the humidity under which the strength is measured. In the following we will give some general information about the strength of glass.

4.4.2 Brittle Fracture of Glass

The "theoretical" or "structural" strength of glass due to molecular forces may be estimated as $\sigma_{\mathrm{B,th}} \approx 10\,\mathrm{GPa}$ ($1\,\mathrm{GPa} = 10^9\,\mathrm{Pa}$, $1\,\mathrm{MPa} = 10^6\,\mathrm{Pa}$, $1\,\mathrm{Pa} = 1\,\mathrm{N/m^2}$; for comparison, the atmospheric pressure is about 0.1 MPa). But real glass articles break at much lower stress levels, for example $\sigma_{\mathrm{B,real}} \approx 20$–$30\,\mathrm{MPa}$ – i.e. for stresses about 3 orders below the theoretical strength. The reason for the reduced strength are flaws in the glass, e.g. surface and volume flaws, pitches and bubbles: The flaws amplify the stress in their neighbourhood by many orders. Simply speaking, a glass article breaks because the amplified stress exceeds the molecular strength. Thus, for an estimation of the strength of glasses and other brittle materials we have to take into account both stresses and flaws.

We start with the so-called *nominal stress* σ_0. This is the macroscopic stress considered in classic engineering and it can, for instance, be calculated by finite-element methods. There exist several contributions to the nominal stress, e.g. stress from external forces acting on the article, due to an inhomogeneous temperature profile, structural relaxations generated by the annealing process, or due to chemical or thermal strengthening of the article. In a region around the tip of a crack we find the *local stress* σ_{loc}. Typically, it depends on the distance from the tip as follows:

$$\sigma_{\mathrm{loc}} \propto \frac{K_i}{\sqrt{r}}, \quad \text{with } K_i \propto \sigma_0 \tag{4.8}$$

where r denotes the distance from the tip of crack, $i = \mathrm{I}$, II, or III refers to the crack mode (see below). The K_i are the so called *stress intensity factors* (for crack mode I, II, or III). Note that the stress intensity factors are proportional to the

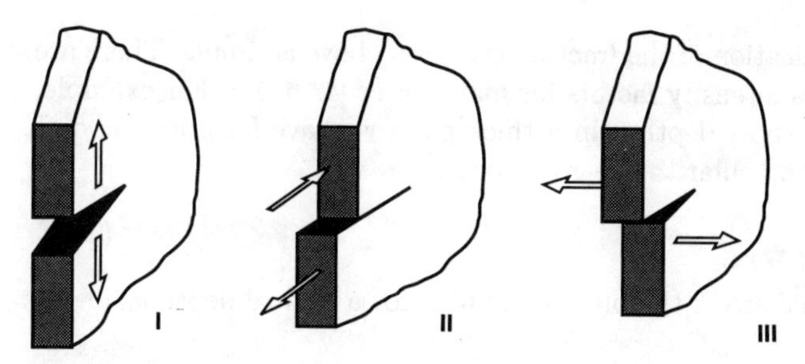

Fig. 4.4. Fundamental crack modes

nominal stress. For $r \to 0$ we find $\sigma_{\mathrm{loc}} \to \infty$: The local stresses diverge at the tip of the crack!

The crack modes or fracture modes take into account the orientation of the crack in the stress field. The three main fracture modes are (see Fig. 4.4):

mode I: tensile forces normal to the crack plane, crack propagation perpendicular to the forces, with crack opening;

mode II: forces acting parallel to the crack plane, crack propagation parallel to the forces, no crack opening;

mode III: forces parallel to the crack plane, but now crack propagation perpendicular to the forces, no crack opening.

Mode I, i.e. the cracking due to tensile forces acting perpendicular to the plane of a flaw, is of utmost importance for the fracture of glass. Note that for mode I only tensile forces open the crack – compressive forces would close a crack without its growing.

For crack mode I the following fracture criterion holds for brittle materials: A flaw will result in a fracture if

$$K_{\mathrm{I}} \geq K_{\mathrm{Ic}} \,, \tag{4.9}$$

where K_{Ic} denotes the *critical stress intensity factor* for crack mode I, the so-called *fracture toughness.* K_{Ic} is a material constant. Stress will lead to a fracture in mode I if the stress intensity factor exceeds the fracture toughness. Then it fractures fast and uncontrollably. Table 4.3 shows experimental values for K_{Ic} for some glasses:

We see that all these glasses have similar values for K_{Ic}. The values for K_{Ic} for silicate glasses differ from one another only by a few dozens per cent.

Table 4.3. Fracture toughness of some glasses

Glass	K_{Ic} MPa$\sqrt{m}$
BK7	1.08
F5	0.86
SF6	0.74
K50	0.77
Duran®	0.85

In practice, the application of the fracture criterion above is simple: There are several tables with stress intensity factors for many cases [4.16, 17]. For example, in case of a flaw with a short depth a in a thick plate we have for crack mode I (nominal stress σ_0 perpendicular to the crack-plane):

$$K_\mathrm{I} \approx 2\sigma_0\sqrt{a}\ . \tag{4.10}$$

Thus, for a given nominal stress the plate will break for a critical depth a_c of

$$a_c \approx \left(\frac{K_\mathrm{Ic}}{2\sigma_0}\right)^2\ . \tag{4.11}$$

Numerical example: For $\sigma_0 \approx 50\,\mathrm{MPa}$ and $K_\mathrm{Ic} \approx 1\,\mathrm{MPa}\sqrt{\mathrm{m}}$ we find $a_\mathrm{c} \approx 100\,\mu\mathrm{m}$. Thus, for glass flaws with very small depths ($\ll 1\,\mathrm{mm}$) can give rise to a crack at comparatively low stress levels.

Note two remarks on the stress intensity factors:

a) Since glass is a material that is elastic in a very wide range, for a combined loading, we can add the stresses from all contributions and take only their resultant into account. Thus, if we have a thermally strengthened glass with internal compressive stress σ_int, an additional stress σ_ext due to external forces will result in a mode I fracture if it is tensile and overcomes the internal stress.

b) Unfortunately, we have sufficient material data on the critical stress intensity factor only for crack mode I (the fracture toughness $\mathrm{K_{Ic}}$) and not for the higher or combined modes. In such cases we can work with an effective stress intensity factor which may be compared with the fracture toughness to give a fracture criterion [4.18]:

$$K_\mathrm{I,eff} = \sqrt{K_\mathrm{I}^2 + K_\mathrm{II}^2 + 1.25 \cdot K_\mathrm{III}^2} \geq K_{Ic}\ . \tag{4.12}$$

4.4.3 Fractography

When glass cracks, the study and characteristics of the crack itself can reveal why, how and where the crack occurred: Examination of the cracks will usually show features such as mirror areas, Wallner lines, or hackle marks [4.19].

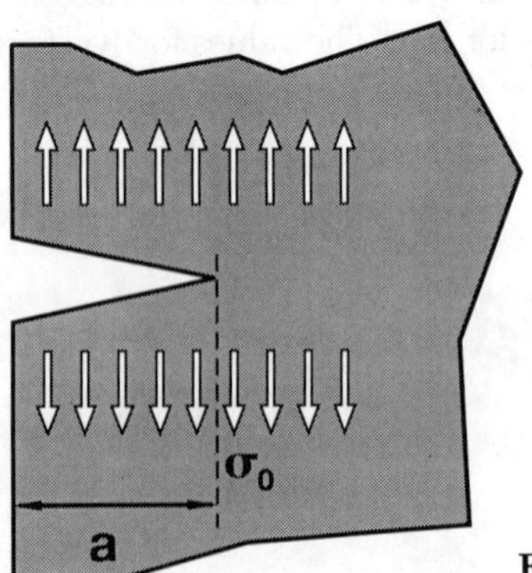

Fig. 4.5. Fracture due to mode I

Mirror areas have an extreme flatness and are nearly circular in shape (or semicircular for cracks starting from surface flaws). At the centre of the mirror areas we usually find easily identifiable flaws. Mirror areas are usually associated with comparatively slow speed cracks [4.20].

Wallner lines are usually associated with cracks of mechanical nature. They appear as small ridges in the surface of the crack. Since Wallner lines are always concave toward the origin of the crack, the ridges can be traced back to the initial flaw [4.20, 21].

Hackle marks are tiny fractures or ridges which appear on the edge of the fractured surface. The direction of these marks shows the direction of crack propagation. They usually appear when the fracture is of violent nature. Hackle marks frequently appear at the edge of a well defined mirror area where the crack changes its direction drastically, or along the surface where the crack breaks through [4.20, 21].

Fracture conditions are seldom constant and there are a wide range of crack conditions, where no tell-tale sign occurs. But somewhere in the fracture, one of the three signs – mirror areas, Wallner lines, or hackle marks – is visible and useful for determining the source of a failure [4.22].

4.4.4 Sub-critical Crack Growth

Most glasses, however, also exhibit slow crack growth for a stress intensity factor well below the critical value. The most important sub-critical crack growth occurs in the presence of water: water combined with stress may dissolve the glass. The amount of water needed for *stress-corrosion* is very low, less than 10 mg per m^2 new surface. Under normal atmospheric conditions at any level of humidity there will be enough water for stress-corrosion! Note that the water must penetrate the crack to its tip; thus, crack-corrosion will only occur for mode I with crack opening. *Wiederhorn* [4.23] considered the details of stress-corrosion of glasses. It has been found that the velocity of a crack can be described (with some modifications) by

$$\frac{da}{dt} = \nu_{\text{lim}} \left(\frac{K_{\text{I}}(a)}{K_{\text{Ic}}} \right)^n \quad \text{for } K_{\text{I}}(a) < K_{\text{Ic}} \tag{4.13}$$

where a denotes the depth of the crack, K_{Ic} the fracture toughness, ν_{lim} the limiting value for the velocity of the crack and the exponent, n, the so-called *stress-corrosion constant.*

Thus a delayed failure may result from slow (sub-critical) propagation of the crack until the fracture condition holds. Furthermore, glass is subject to fatigue; i.e. it is stronger under instantaneous loading than under prolonged stresses [4.24].

Table 4.4 shows some values for the stress-corrosion constant:

Numerical Example

Consider a long flaw with an initial depth a_0 well below the critical depth for an instantaneous fracture in a thick plate under static tensile stress (crack mode I with stress σ_0): With $K_{\text{I}}(a) \approx 2\sigma_0\sqrt{a(t)}$ and $n \gg 1$ we estimate for the breakage time:

Table 4.4. Stress-corrosion constants of some glasses

Glass	n
Duran®	22.4
Zerodur®	33
Soda-Lime-Glass	18.1

$$t_b \approx \frac{2}{n}\frac{a_0}{\nu_{\text{lim}}}\left(\frac{K_{\text{Ic}}}{2\sigma_0\sqrt{a_0}}\right)^n . \tag{4.14}$$

Numerical example: For $a_0 = 10\,\mu\text{m}$, $\sigma_0 = 50\,\text{MPa}$, $K_{\text{Ic}} = 1\,\text{MPa}\sqrt{\text{m}}$, $n = 20$, and $\nu_{\text{lim}} \approx 1500\,\text{m/s}$, we find $t_b \approx 7\,\text{s}$.

To calculate the effect of stress-corrosion we need the stress-corrosion constant n. It must be determined experimentally. This can be done by a direct observation of the crack growth, or, more practically, by a dynamic strength testing (e.g. with the ring-on-ring test, see below). For example, in the case of measuring the breakage stress σ_{B} with a linearly increasing stress $\sigma(t) = \dot{\sigma}t$ (with constant stress rate $\dot{\sigma} > 0$) we find the following relationship of the breakage stress to the stress rate:

$$\ln(\sigma_{\text{B}}) \propto \frac{1}{n+1}\ln(\dot{\sigma}) . \tag{4.15}$$

Thus we can calculate the stress-corrosion constant n from the slope of the $\ln(\sigma_{\text{B}})$-vs-$\ln(\dot{\sigma})$ plots. However, since $n \approx 20$ we have to change the stress rate by several orders (at least by more than three orders) to achieve sufficient results. But, with the help of modern automatic testing apparatus this does not present any problem.

4.4.5 Fracture Probability

So far we have considered the fracture from a deterministic point of view: With knowledge of the data on a flaw (depth, orientation) and the stress we can calculate whether or not the flaw will result in a fracture. However, in real life we do not know all the data on the flaws. Usually, we have many flaws, scratches and notches with a statistical distribution on the surface of a glass article and, therefore, it is practically impossible to determine the data for all cracks.

Instead of calculating we can measure the fracture strength of a glass article. For example, we take N apparent identical specimens of an article and test their fracture strengths by the experimental procedures discussed below. However, in doing this we will find N different values for the fracture strengths $\{\sigma_1, \sigma_2, \ldots, \sigma_{\text{N}}\}$. Obviously, there is some statistics in the failure stress which has its reason at least in the statistics of the initial surface flaws.

In general, the statistics for the fracture stress for glasses and ceramics may be well described by the Weibull distribution; i.e., the cumulative fracture probability for a given stress σ is given by the formula

$$F(\sigma) = 1 - \exp\left(-(\sigma/\sigma_{\text{c}})^m\right) \tag{4.16}$$

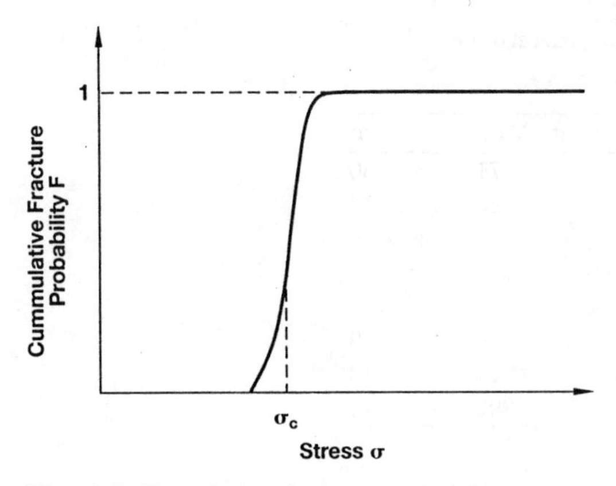

Fig. 4.6. Cumulative fracture probability vs. stress (Weibull distribution), σ_0: characteristic value

($0 \leq F \leq 1$), where σ_c denotes the characteristic value (approximately the mean value of the distribution) and m the so-called *Weibull* modulus (determining the standard deviation) of the distribution. The meaning of the cumulative fracture probability is simple: $F(\sigma)$ is the probability for a fracture of the specimens if the applied stress is less than σ.

The characteristic value and the modulus of the Weibull distribution depend on the testing procedure (see below) as well as on the treatment of the surfaces of the glass article (grinding, polishing, etching, etc.): as a rule, grinding with narrow grain size distributions will lead to comparatively narrow strength distributions (i.e. larger m) for the ground surfaces. Decreasing grain sizes result in smaller depths of the surface flaws and, consequently, in higher strengths (higher σ_c).

Polished surfaces may have the highest strength. However, this refers only to cases where the cracks induced by the previous machining have been eliminated. This may be achieved by a subsequent grinding of the surfaces with decreasing grain sizes, each grinding process taking off a material layer with a thickness of at least three to four times the maximum depth of the cracks from the preceding process. By these procedures we can increase σ_c. However, the strength distributions of glasses with polished surfaces vary more significantly than those of glasses with only ground surfaces: Polishing of glass decreases the Weibull modulus m. In general we find for a better surface quality that σ_c increases and m decreases.

There is another way to increase the strength of a glass surface, namely, to round the tips of the cracks by etching or to remove the damaged surface layers completely by etching with acid (HF, HCl, HNO_3) or alkaline solutions (KOH, NaOH).

The characteristic value σ_c of the Weibull distribution is not a material parameter as the Young's modulus or the fracture toughness. It depends on the size of the article since deeper cracks occur more frequently when the size increases. For example, we have the following dependence of the characteristic value σ_c on the size A of the surface:

Table 4.5. Values for σ_c and m (determined by a ring-on-ring test)

Glass		σ_c MPa	m
BK7	SiC 600	71	30
BK7	D 64	50	13
BK7	D 64 + etched	236	4
Zerodur®	SiC 600	108	
Zerodur®	SiC 320	71	12
Zerodur®	SiC 100	54	19
Zerodur®	D 64	64	13
Zerodur®	D 64 + etched	292	4

$$\sigma_c \propto 1/A^{1/m} \,. \tag{4.17}$$

From this we can measure the strength of a glass article using samples with small surfaces and estimate the strength of articles with large surfaces.

4.4.6 Testing of Glasses

Glass users must ultimately have confidence that glass articles will have a reliable minimum strength. Thus we have to test the strength. There are a lot of testing procedures adequate for the special articles, several kinds of *flexure tests* and *impact tests*. Impact tests, for example, have a definite place in fracture research. However, for these tests the true fracture energy corresponds to the elastic energy stored in the specimen at the moment of failure and depends on the specimen used, the location tested and the testing conditions. Whatever method of testing is used it should be applied to the articles under conditions that correspond as exactly as possible to those met in service [4.24].

The strength testing can be applied to representatives of an article (testing until fracture occurs) or, individually, to every article (100 % testing). In general, if glass articles are to be proof-tested individually before being placed in service, precautions should be taken to reduce the probability that they are significantly weakened by the test.

In addition to the testing of the strength of special glass articles, we are interested in the question of how to measure the strength of glass itself (more precisely: the strength with respect to the state of its surface). The main methods of testing the strength of glass are: three- and four-point bending and ring-on-ring test (all *flexure tests*). We will discuss some features of these testing methods in the following.

4.4.7 Three-Point Bending

The simplest case is to use three-point bending with a symmetric arrangement (Fig. 4.7): Three parallel wedges are arranged at equal distance, $L/2$, at opposite sides of the specimen. A force F applied to the upper (bending) wedge induces a

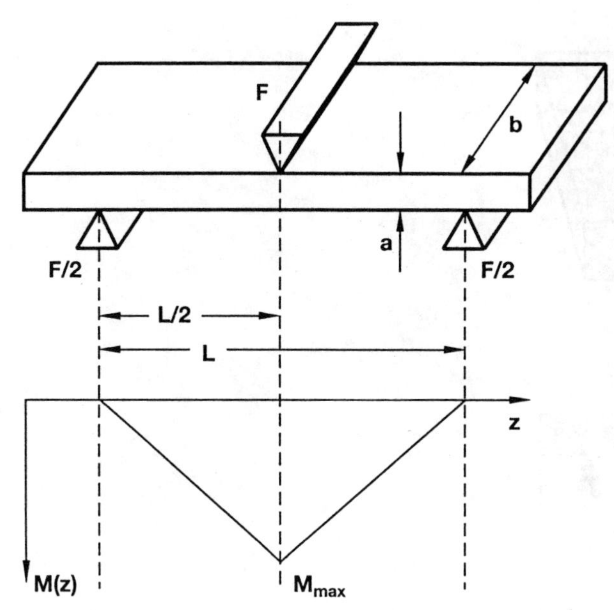

Fig. 4.7. Three-point bending

bending moment: the bending moment is maximum just under the upper roller and drops linearly to zero at the support rollers. The tensile stress is nearly uniaxial parallel to the z-axis and appears on the specimen's lower side. Its maximum value is

$$\sigma_{\max} = 3\frac{FL}{a^2 b} \ . \tag{4.18}$$

However, this arrangement has three disadvantages:

1. The stress is inhomogeneous. Thus the question arises what stress we have to take into account for the calculation of the data for the fracture statistics. (For this reason, four-point bending is preferred to three-point bending [4.25]).

2. The stress is uniaxial. Thus, the result of the strength testing depends on the orientation of the initial flaw with respect to the axes of the sample.

3. Usually the fracture does not start from the sample's surface – instead it starts from one of the edges, from small flaws at the edges (even if it has been manufactured very carefully). Thus we actually do not test the "strength of the surface" – we test the "strength of the edges"!

4.4.8 Ring-on-Ring Test

This method overcomes the disadvantages of three- and four-point bending. Here the specimen (circular or square size) lies on an "outer ring" with a larger radius, and a force F is applied to a central "inner ring" with a smaller radius (Fig. 4.8). Using this arrangement we induce a nearly homogeneous and isotropic stress in the area in the inner ring:

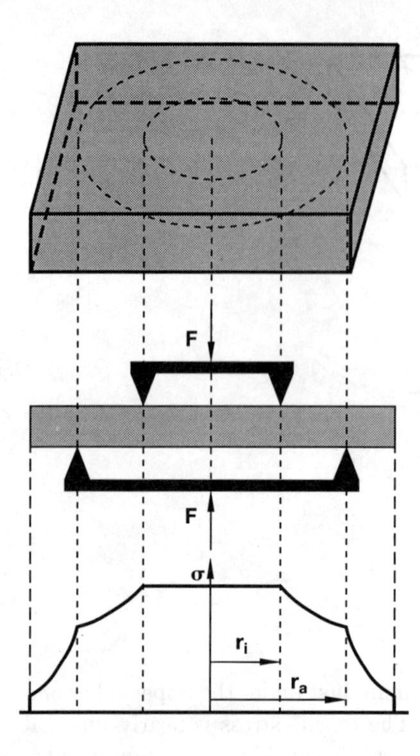

Fig. 4.8. Ring-on-ring test

$$\sigma_r = \sigma_t \propto \frac{F}{s^2} \tag{4.19}$$

(s: thickness of the specimen), with some correction factors of order one with respect to the radii of the rings and the specimen's shape (circular or square specimens). The stress in the inner ring also depends on the Poisson number of the material, but since we have only small variations of μ for glasses (mostly $\mu \approx 0.2$–0.3) we can neglect this dependence.

Outside the inner ring the stress becomes anisotropic and decreases. Unfortunately, it does not decrease to zero, and we have stresses at the edges of the sample. Thus we have to make the specimen sufficiently large compared to the diameter of the rings.

4.5 Strengthening of Glass

In spite of its brittle nature, glass can be strengthened by several processes. The most common strengthening methods applied in industry are the so-called *tempering* or *thermal strengthening* (or *toughening*) and *chemical strengthening*.

4.5.1 Thermal Strengthening

Thermal strengthening has traditionally been performed by heating a glass article to a temperature between the transformation temperature and the softening point and then cooling it rapidly. The surfaces become rigid and when the interior cools and contracts, it pulls on the surfaces, causing residual compressive stress on the surfaces and tension in the interior. Normally, glass is not strong under tension, but this is only because cracks spreading from the surface propagate rapidly across the sample. But the interior region of toughened glass which is under tension has no surface from which cracks may spread. Since the surface is under compression a crack cannot penetrate unless tensile forces from outside compensate this surface compression. As a result, commercially toughened glass will withstand much stronger impacts than annealed glass of the same composition.

This method can be applied quite easily and inexpensively during processing, but it is hard to control and often results in non-uniform surface stress. For a rough estimation, the maximum compressive stress at the surface of a glass plate is given by [4.26]

$$\sigma_{\max} = \frac{E\alpha}{1-\mu} \frac{\alpha' \frac{d}{2}}{\lambda + \alpha' \frac{d}{2}} (T_{\mathrm{h}} - T_{\mathrm{g}}), \tag{4.20}$$

where α denotes the coefficient of thermal expansion, E the Young's modulus, μ the Poisson number [the first factor is the specific thermal tension; see Table 4.2], α' the coefficient of heat exchange (from the glass surface to the surrounding cooling medium), λ the thermal conductivity of the glass, d the thickness of the glass plate, T_{g} the transition temperature of the glass, T_{h} the starting temperature for the strengthening process (well above the transition temperature but below the softening point).

Firstly, a high compressive surface stress requires fast cooling and a high coefficient of heat exchange α'. In Table 4.6 some typical values are listed.

Furthermore, a high figure of merit for the specific thermal stress $E\alpha/(1-\mu)$ is required. For example, we have $E\alpha/(1-\mu) \approx 0.80$ MPa/K for soda-lime-glass leading to a possible maximum stress of about 200 MPa, and $E \cdot \alpha/(1-\mu) \approx 0.25$ MPa/K for a borosilicate glass with maximum stress of about 75 MPa. Finally, thermal strengthening may not be suitable when the glass is extremely thin – i.e. for glass articles with a thickness d less than 1 mm.

Thermal strengthening can result in high stresses (well above 100 MPa) in glass articles with a comparatively large depth of the layer under compression (the stress profile as a function of the depth is nearly parabolic): The depth of

Table 4.6. Heat exchange coefficients

Method	α' W/m^{2}K
Simple blowing with air	100–300
Spraying with an air-water mixture	500–1500
Dipping in water (with an oil-layer)	up to 5000

Table 4.7. Ionic radii

	r_0 10^{-10} m
Li	0.78
Na	0.98
K	1.33
Cs	1.65

the compressive layer is about 10...15 % of the total thickness of the article. Thus, the stored elastic energy in the glass article is relatively high and when strong thermally strengthened glass fractures, it will shatter into small squares or cubes. In principle, we find for the number of the pieces N and the surface stress σ, a dependence $N \propto \sigma^4$; this relationship has been obtained from damage tests.

4.5.2 Chemical Strengthening

Chemical strengthening is another method of strengthening glass by developing a compressive stress at its surfaces with an ion exchange process. In this process a glass article is immersed in a salt bath at a temperature at which alkali ions diffuse out of the glass and are replaced by larger alkali ions from a salt bath. For example, a glass containing sodium (Na) is immersed in a bath containing highly concentrated potassium nitrate (KNO_3). The larger potassium ions from the salt bath replace the smaller sodium ions in the glass, which results in compressive stresses in the diffusion layer at the surface. Table 4.7 gives the ionic radii r_0 of some alkali atoms:

The compressive stress and the thickness of the diffusion layer can vary depending on the composition of the glass and the particular strengthening cycle incorporated (usually the diffusion layer has a thickness of less than 200 µm because of the low diffusion coefficient of the ions). Chemical strengthening of glasses optimized for this processing can result in compressive stresses up to 700 MPa (!). This process can be employed on thinner glass articles than the thermal tempering. But while the surface compression can be higher than that produced by thermal strengthening, the compression layer is much thinner than when thermal strengthening is applied. Chemical strengthening is widely used in the ophthalmic industry for strengthening thin eyeglass lenses.

References

4.1 A. Winkelmann, O. Schott: "Über die thermischen Widerstandskoeffizienten verschiedener Gläser in ihrer Abhängigkeit von der chemischen Zusammensetzung", Ann. Phys. **51**, 730–746 (1894)

4.2 G.W. Morey: *The properties of glass* (Reinold Publishing Corporation, New York 1954) pp. 221–262

4.3 E.B. Shand: *Glass engineering handbook* (McGraw-Hill, New York 1958) pp. 22–23

4.4 H. Scholze: *Glas, Natur, Struktur und Eigenschaften* (Springer Verlag, Berlin 1988) pp. 181–199

4.5 W. Vogel: *Glaschemie*, 3rd ed., (Springer, Berlin 1992) p. 467

4.6 International Standard Organisation (ISO) paper No. **9385** or Deutsche Industrie-Norm (DIN) paper No. **52**, p.333, American Society for Testing and Materials, paper No. **C** (1989) pp. 730–785

4.7 F. Fröhlich, P. Grau, W. Grellmann: "Performance and analysis of recording microhardness tests", Phys. stat. sol. **42**, 79–89 (1977)

4.8 F. Fröhlich, P. Grau, W. Grellmann: "Untersuchung mechanischer Eigenschaften von Glas mit Hilfe moderner Härtemeßverfahren", Wiss. Zeitschr. Fr.-Schiller- Universität Jena, Math.-Nat. R. **28**, 449–463 (1979)

4.9 G.H. Frischat, E. Özmen, T. Richter, B.D. Michels: "Lastunabhängige Makrohärte an verschiedenen Gläsern", Glastech. Ber. **55**, 119–125 (1982)

4.10 N.A. Ghoneim, H.A. El Batal, Ali M.A. Nassar: "Microhardness and softening point of some alumo-borate glasses as flow dependant properties", J. Noncryst. Sol. **55**, 343–351 (1983)

4.11 L.I. Demkina, E.V. Smirnova: "Hardness of lead silicate glasses", Sov. J. Opt. Tech. **53**, 711–714 (1986)

4.12 V.V. Akimov: "Microhardness of alkali borosilicate glasses", Phys. and Chem. of Glasses **15**, 869–873 (1988)

4.13 T.S. Izumitani, I. Suzuki: "Lapping hardness and Knoop hardness of optical glasses", in Proceed. Intern. Congress Glass, Versailles, Vol. **I** (1971) pp. 605–611

4.14 T.S. Izumitani, I. Suzuki: "Indentation hardness and lapping hardness", Glass Technology **14**, 35–41 (1973)

4.15 T.S. Izumitani: "Polishing, lapping, and diamond grinding of optical glasses", in *Treatise on materials science and technology*, ed. by M. Tomozawa, R.H. Doremus, Vol. 22 (Academic Press, New York 1982) pp. 115–171

4.16 D.P. Rooke and D.J. Cartwright: Compendium of stress intensity factors (Her Majesty's Stationary Office, London 1974)

4.17 H. Tada, P.C. Paris, G.R. Irwin: *The stress analysis of cracks handbook*, 2nd ed. (Paris Productions, Inc. 1985)

4.18 H.A. Richard: "Bruchvorhersagen bei überlagerter Normal- und Schubbeanspruchung sowie reiner Schubbeanspruchung von Rissen"; (in German); Habilitationsschrift, Universität Kaiserslautern (1984)

4.19 A.K. Varshneya: *Fundamentals of Inorganic Glasses* (Academic Press, Inc., Boston 1993), pp. 409–452

4.20 R. Matheson: "Reading Shattered Glass" in *The Glass Industry* (1977) p. 27

4.21 H.M. Bateson: "Some Views on the Breaking of Glass, Derived from the Examination of Fracture Surfaces", J. Soc. Glass Technol. **34**, 114–118 (1950)

4.22 F. Kerkhof: "Bruchentstehung und Bruchausdehnung von Glas" (in German) in *Glastechnische Fabrikationsfehler*, 3rd ed., ed. by H. Jebsen-Marwedel and R. Brückner (Springer, Berlin 1980), pp. 526–587

4.23 S.M. Wiederhorn and L.H. Bolz: "Stress Corrosion and Static Fatigue of Glass", J. Am. Ceram. Soc. **53**, 543–548 (1970)

4.24 G.W. McLellan and E.B. Shand: "Strength and Strength Testing", in *Glass Engineering Handbook* (McGraw-Hill, New York, NY 1984)

4.25 K. Blank, D. Dürkop, M. Durchholz, H. Gruters, G. Helmich, W. Senger: "Strength tests of flat glass by means of four-point-bending", Glastechn. Ber. **67**, 9–15 (1994)

4.26 W. Kiefer: "Thermisches Vorspannen von Gläsern mit niedriger Wärmeausdehnung" (in German), Glastechn. Ber. **57**, 221–228 (1984)

Additional References

R.H. Doremus: "Fracture and Fatigue of Glass" in *Treatise on Materials Science and Technology*, Vol. 22 (Academic Press New York, NY 1982) pp. 170–225

G. Exner: "Erlaubte Biegspannung in Glasbauteilen im Dauerlastfall" (in German), Glastechn. Ber. **56**, 299–312 (1983)

G. Exner: "Abschätzung der erlaubten Biegespannung in vorgespannten Glasbauteilen" (in German), Glastechn. Ber. **59**, 259–271 (1986)

V.D. Frechette: "Fractology" in *Introduction to Glass Science*, ed. by D. Pye, H.J. Stevens, and W.C. LaCourse (Plenum Press, NY 1972)

V.D. Frechette: "Fracture of glass in the presence of H_2O", Glastechn. Ber. **58**, 125–129 (1985)

C. Gurney and S. Pearson: "Effect of Surrounding Atmosphere on Delayed Fracture of Glass", Proc. Phys. Soc. (London) **62**, 469–476 (1979)

R.H. Koopshot and R.P. Mikesell: "Strength and Fatigue of Glass at Very Low Temperatures", J. Appl. Phys. **28**, 610–614 (1957)

W.C. LaCourse: "The Strength of Glass" in *Introduction to Glass Science*, ed. by D. Pye, H.J. Stevens, and W.C. LaCourse (Plenum Press, NY 1972)

T.A. Michalske and B.C. Bunker: "Slow Fracture Model Based on Strained Silicate Structures", J. Am. Ceram. Soc. **56**, 2686–2693 (1984)

H. Rawson: "Why do we make glass so weak? A review of research on damage mechanisms", Glastechn. Ber. **61**, 231–246 (1988)

J.E. Ritter jr. and C.L. Sherburne: "Dynamic and Static Fatigue of Silicate Glasses", J. Am. Ceram. Soc. **54**, 601–605 (1971)

E.B. Shand: "Experimental Study of the Fracture of Glass I", J. Am. Ceram. Soc. **37**, 52–60 (1954)

B. Vonnegut and J.L. Glathart: "The Effect of Temperature on the Strength and Fatigue of Glass Rods" J. Appl. Phys. **17**, 1084–1100 (1946)

5. Thermal Properties of Glass

Ulrich Fotheringham

The term "thermal properties" refers to those physical properties whose temperature dependence is the crucial aspect either for the handling alone or for both the handling and the functioning of the material.

Almost all of these Physical Properties have characteristic features at the glass transition, i.e. in the temperature range where the change from the liquid to the glassy state or vice versa takes place. The measurement of these features reveals a lot about the nature of the glassy state; thus, the thermal properties are a key to the understanding of it.

5.1 Heat Capacity

The heat capacity or the specific heat c of a substance is defined as the increase of its specific thermal energy e (i.e. the thermal energy per gram) per degree of temperature increase [5.1]:

$$c = \frac{de}{dT} . \tag{5.1}$$

where T is the absolute temperature. As the sample volume V has to be kept constant for the measurement (otherwise the energy exchange measured will be the sum of the thermal energy increase and the work linked with the volume change), the specific heat defined by (5.1) is called c_V, the *specific heat at constant volume.*

As it is experimentally difficult to keep a sample volume constant, usually the energy exchange which accompanies a temperature increase at constant pressure is measured. This gives de/dT plus $p\,dv/dT$. The quantity $p\,dv$ is the specific work linked with the change dv of the specific volume v at constant pressure p (specific = per gram, see above). Introducing $h \equiv e + pv$, the specific enthalpy, $de/dT + p\,dv/dT$ can be replaced by dh/dT ($dh = de + p\,dv$ at constant pressure). Analogously to $c_V = de/dT$, the *specific heat at constant pressure* c_p is defined by:

$$c_p = \frac{dh}{dT} . \tag{5.2}$$

From the latter, c_V can be calculated using a fundamental thermodynamical relation [5.2]:

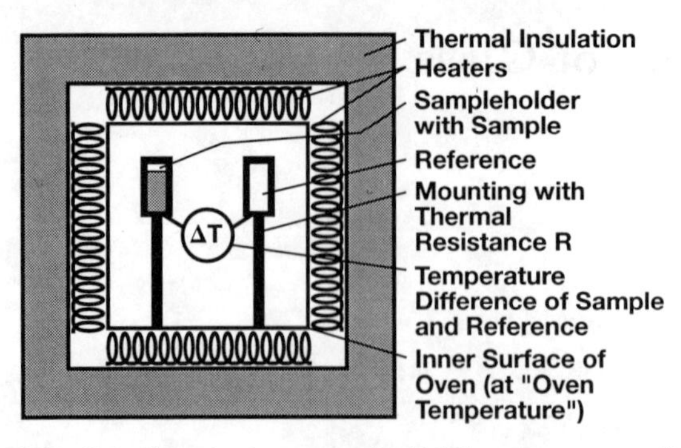

Fig. 5.1. Principal set-up of a DSC-measurement (heat flux type). The overall effect of all thermal bridges between oven and sample (or reference) is represented by the thermal resistance "R" which is attributed to the mounting

$$c_p - c_V = 9\alpha^2 K \frac{T}{\rho} \tag{5.3}$$

with α being the thermal expansion coefficient, K the compression modulus, and ρ the density.

A suitable device for the measurement of the specific heat at constant pressure is a differential scanning calorimeter (DSC [5.3], Fig. 5.1). A sample is introduced into an oven whose temperature is raised at a constant rate. Accordingly, there is a steady heat flux to the sample which amounts to $C_{\text{total}} dT/dt$ with dT/dt being the imposed heating rate and C_{total} being the heat capacity of the sample with its holder. This heat flux is equal to the temperature difference $T_{\text{O}} - T_{\text{S}}$ of the oven body and the sample, divided by the thermal resistance R between them. With the latter being known from a calibration, C_{total} can be determined by measuring $T_{\text{O}} - T_{\text{S}}$.

If the experimental set-up does not allow this calibration, the measurement is called DTA (differential thermal analysis).

To allow a measurement of the sample heat capacity alone, the DSC is laid out as a twin system with a second (reference) sample holder. If the latter is left empty during the measurement, the temperature difference between it and the oven is proportional to $C_{\text{sampleholder}}$ alone. Consequently, the temperature difference ΔT between the sample and the reference is a measure for C_{sample}:

$$C_{\text{sample}} = \frac{\Delta T}{R dT/dt} \,. \tag{5.4}$$

Equation (5.4) is only valid if neither exothermic nor endothermic processes occur in the sample. If an enthalpy Q is absorbed in a certain temperature range, the following equation holds:

$$\frac{dQ}{dt} + C_{\text{sample}} \frac{dT}{dt} = \frac{\Delta T}{R} \,. \tag{5.5}$$

As already mentioned, R is determined through an advance measurement of a standard with known specific heat, e.g. sapphire. Figure 5.2 shows the specific heat of an optical glass, BK 7® from Schott, in the temperature range 100–800 °C.

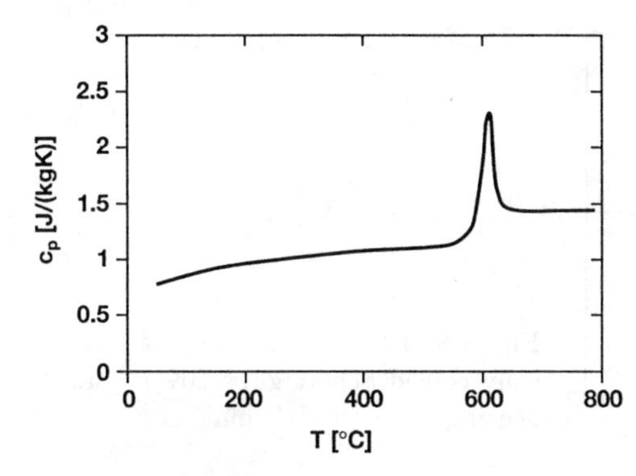

Fig. 5.2. DSC-measured c_p of optically cooled Schott glass BK 7® at a heating rate of 10 °C/min

For an interpretation, the principles of solid state thermodynamics and a special feature of glass have to be considered. According to the microscopic picture, the thermal energy of solids is stored in atomic vibrations. At high temperatures, each atom may oscillate freely and independently from the others in all three dimensions of space, with kT being the mean value for the energy associated with each oscillation. T is the absolute temperature and k is Boltzmann's constant [5.4]. As this energy changes periodically from being kinetic to being potential, $1/2kT$ is the average value for either state.

Consequently, the heat capacity per atom, i.e. the derivative of the total vibrational energy of one atom with respect to the temperature, is given by $3k$. So for one mole of an element, the heat capacity is given by $3N_{\mathrm{L}}k = 3R$ (Dulong-Petit law [5.4]), where N_{L} is the number of atoms in a mole and $R \equiv N_{\mathrm{L}}k$ the gas constant. With n_i being the mass fraction of each element in a solid and m_i being the corresponding molar mass, the resulting heat capacity per unit mass or specific heat of the solid is (Neumann-Kopp rule [5.5]):

$$c_V = \sum_i n_i \frac{3R}{m_i} \,. \tag{5.6}$$

Of course, glass is not a crystalline solid but a supercooled liquid. For the Dulong-Petit law to be valid, however, it is only necessary that the motional modes of the material are harmonic oscillations and that configurational effects may be neglected [5.7].

At low temperatures, the assumption that all atoms may be treated as independent oscillators with the mean energy kT for each dimension is not true. First it has to be taken into account that the quantum-mechanical calculation gives $h\nu/(\mathrm{e}^{h\nu/kT} - 1)$ as the exact value for the mean energy per oscillator, with ν being the frequency of the oscillation. This expression is equal to kT at high temperatures only. At low temperatures, it is equal to zero; the oscillation is said to be frozen-in ($\mathrm{e}^{h\nu/kT} \sim 1 + h\nu/kT$ at high temperatures where $kT \ll h\nu$; $\mathrm{e}^{h\nu/kT} \sim 0$ at low temperatures) [5.4].

Second, the coupling of the atomic oscillations has to be considered. This brings about that the vibrational modes of the solid – the "phonons" – are not identical to the individual atomic oscillations. Rather, they are collective excitations. Some of them imply the simultaneous counter-swinging of all neighbouring

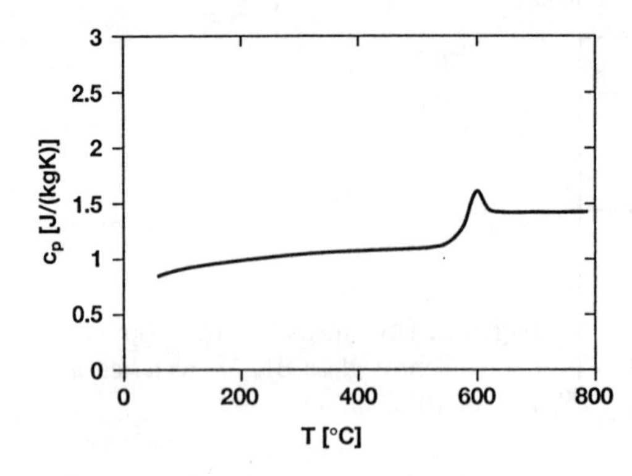

Fig. 5.3. DSC-measured c_p of technically cooled Schott glass BK 7® at a heating rate of 10 °C/min

atoms. They come close to the individual atomic oscillations or linear combinations of them, respectively. Others imply the counter-swinging of whole areas of atoms. Both of them are phenomena with a spatial periodicity, which has the order of magnitude of one interatomic distance in the first case and of several ones in the second case.

These two facts, the freezing-in of the oscillations and the various spatial periodicities of them, determine the behaviour of the specific heat at low temperatures. Due to the freezing-in effect, the specific heat decreases with decreasing temperature to become zero at zero Kelvin. Due to the broad distribution of spatial periodicities and, consequently, oscillation frequencies, this decrease is smooth [5.4]. (There is an approximate proportionality between the inverse spatial periodicity and the oscillation frequency ν: With the velocity of sound v_s assumed to be constant, the relation $\lambda\nu = v_s$ is valid. λ is the spatial proportionality – or better wavelength. So the oscillations with big spatial periodicities have low frequencies and freeze in at very low temperature therefore.)

This principal behaviour is found in the BK 7® measurement (Figs. 5.2, 5.3). In the high temperature range, the specific heat is given by the Dulong-Petit value plus $c_p - c_V$. The Dulong-Petit value follows from the composition of BK 7® (approximate fractions of the major components: 70% SiO_2, 11.5% B_2O_3, 1.5% BaO, 9.5% Na_2O, 7.5% K_2O) to be 1257 J/kgK.

With the high temperature thermal expansion coefficient $\alpha = 45 \times 10^{-6}$/K, $\rho = 2250\,\mathrm{kg/m^3}$ at 800 °C, and an estimated 15.5 GPa for the high temperature compression modulus K (this value follows from the rule of thumb that the low temperature compression modulus, 46.5 GPa, is three times the high temperature compression modulus [5.6]), $c_p - c_V = 9\alpha^2 KT/\rho$ amounts to 135 J/kgK at 800 °C. So at 800 °C, the theoretical value for c_p is 1395 J/kgK, which is in reasonable agreement with the experimental one, $c_p = 1420$ J/kgK (Fig. 5.3).

Between the high temperature region and the low temperature region, a striking feature of glass can be observed, the *glass transition*. It is characterized by a step change of the specific heat plus an endothermal peak similar to the one observed during a melting process.

This effect is due to the configurational degrees off freedom of glass. The fact that glass is not a crystalline substance but a supercooled liquid means that the atoms have no fixed positions but can choose among almost a continuum of locations.

In general, these locations correspond to different energy levels. Therefore, the equilibrium distribution of the atoms depends on the temperature. A temperature change will always be followed by an attempt of the glass to adapt to the new temperature through a redistribution of the atoms, i.e. a configurational change. Because of the entropy gain, an upward step in temperature will evoke the hopping of several atoms from the more densely populated low energy positions to the less densely populated high energy positions and vice versa. So the configurational degrees of freedom give rise to a calorimetric effect as well as the oscillatory ones.

There are two circumstances which restrict the occurrence of a measurable calorimetric effect to the medium temperature range. As the hopping of an atom from one position to another requires *thermal activation*, the adaptation of the glass to a temperature change is not possible at low temperatures, i.e. the configurational degrees of freedom are *effectively frozen in*. In particular, this brings about that glass is never at equilibrium at low temperatures.

In the DSC-measurement, the onset of the thermal activation of the configurational degrees of freedom is registered as a significant rise of the *specific heat*, i.e. the *step change* observed. (The accompanying *endothermal peak* will be discussed in detail below.)

At high temperatures, the equilibrium distribution of the atoms comes close to an equal distribution over all possible locations. This is independent from the exact value of the temperature so that there is almost no configurational change and thus no calorimetric effect attached to a temperature change. It is the reason why configurational effects may be neglected at high temperatures and one finds the Dulong-Petit law to be valid.

The turnover from the medium temperature range to the high temperature range is not accompanied, however, by a striking feature in the DSC measurement. There are two reasons for this. First, the decrease of the configurational contribution to the specific heat is smooth. Second, it is compensated by the increase of the vibrational contribution which approaches its high temperature limit from below. Because of this mutual compensation, the specific heat takes the Dulong-Petit value plus $c_\mathrm{p} - c_\mathrm{V}$ immediately above the endothermal peak.

The configurational change for the purpose of taking the equilibrium structure at the temperature given is called *structural relaxation.* For a quantitative description, a measure for the deviation of the glass structure from the equilibrium structure is needed. If the glass structure is characterized by the so-called *fictive temperature* T_f, i.e. the temperature at which a glass with such a structure would be at equilibrium, the difference between it and the actual temperature can be taken as such a measure.

The freezing-in effect of the configurational degrees of freedom implies that during a cooling process, the glass structure deviates more and more from the equilibrium structure. In terms of the fictive temperature, this means that at the start of the cooling process, at high temperatures, the fictive temperature is equal to the environmental temperature. During the cooling process, it lags behind the environmental temperature, being always higher than the latter. The final value of the fictive temperature can be used for the characterization of the glass structure at the end of the cooling process and is called *glass temperature* T_g.

In first order, one may take the glass temperature as the temperature value above which the configurational degrees of freedom are thermally activated and below which they are frozen in. So the glass temperature is a characteristic value for the temperature range of the glass transition which can be interpreted as an equilibrium to non-equilibrium transition during a cooling process and vice versa during a heating process such as in a DSC.

The exact value of the glass temperature depends on the cooling rate. This is a consequence of the structural relaxation kinetics. Very much simplified, one can imagine a structural relaxation process as follows. The atoms are captured in potential wells in which they are oscillating back and forth. Each time they touch the potential barrier separating their potential well from a neighbouring one, those atoms which have enough thermal energy will jump over.

With this picture in mind one expects that the structural relaxation is a time consuming process that can be characterized by a time constant which is proportional to the period of an atomic oscillation and inversely proportional to the part of atoms that have enough energy to jump over the potential well. As a consequence, the ability of the glass to equilibrate at a certain temperature during the cooling process depends on the time spend at this temperature. So the lower the cooling rate is, the lower the glass temperature will be.

The effect of different cooling rates on the glass temperature can be detected analysing the *endothermal peak*. As it has been explained above, the structural change of the glass following a temperature change dT is linked with an enthalpy effect dh. This enthalpy effect is given by $c_{p,\text{structure}} dT$ with $c_{p,\text{structure}}$ being the height of the specific heat step at the glass transition.

(It is assumed that around the glass transition, both the vibrational and the configurational contribution have constant values. So for a temperature change dT leading from one equilibrium state to the next, the structural enthalpy effect is $dh = c_{p,\text{structure}} dT$.)

If the glass is cooled slowly, the glass structure is frozen-in at a comparatively low glass temperature. During a re-heating at a high rate, the onset of the structural relaxation takes place at a higher temperature. So at this temperature, the difference of this temperature and the glass temperature, multiplied by $c_{p,\text{structure}}$, has to be provided additionally. The slower the previous cooling has been, the bigger the resulting endothermal peak in the DSC-curve is (Figs. 5.2, 5.3, 5.4).

So in contrast to a crystalline solid, where – unless there are solid-solid phase transitions – the structure does not change below the melting point, the details of the glass structure are determined as late as during the glass transition. This affects the low temperature glass properties and has a crucial meaning for the glass manufacturing therefore. Via the dependence of the phonon spectrum on the glass structure, the specific heat at low temperatures depends on the thermal history also.

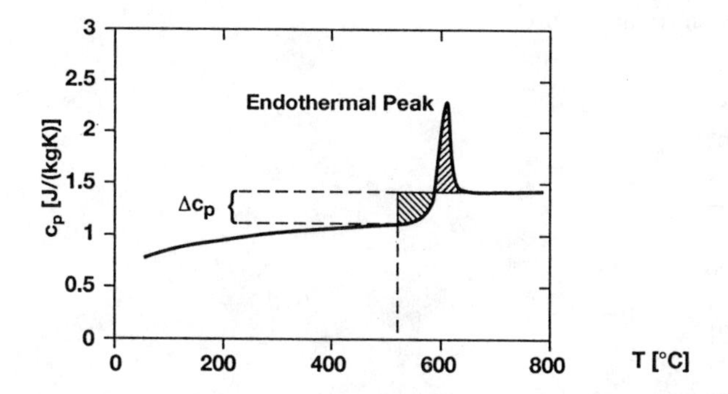

Fig. 5.4. The derivation of the glass temperature T_g from the DSC-curve. At the thawing temperature, the difference of this temperature and the glass temperature, multiplied by Δc_p, has to be provided, additionally (endothermal peak; the peak temperature can be identified with the "thawing temperature", T_t). So T_g is the temperature for which $\Delta c_p(T_t - T_g)$ is equal to the area of the endothermal peak and can be identified by a graphical construction (the two hatched areas have to be of the same size).

For the quantitative description of the time dependence of the fictive temperature, more sophisticated models than the one above are used. They give that the elementary process, i.e. the response of a glass that has been equilibrated at the temperature T to a infinitesimal temperature change from T to $T + \Delta T$, can be described by the Kohlrausch law [5.7]. It says:

$$T_f(t) = T + \Delta T(1 - e^{-(t/\tau)^\beta}) . \tag{5.7}$$

τ is the relaxation time and $0 < \beta < 1$ the Kohlrausch parameter. Over a small temperature range, the Kohlrausch parameter β may be treated as a constant and the Arrhenius law is applicable for the relaxation time: $\tau \propto e^{H/(RT)}$, with H being the activation enthalpy. R is the gas constant.

For a broad temperature range representation of τ, the Vogel-Fulcher-Tammann equation is better suited:

$$\tau = \tau_0 e^{\frac{Q}{R}\frac{1}{T-T_0}} . \tag{5.8}$$

This is the same formula as for the temperature dependence of the viscosity which will be discussed in detail below. Usually one even finds the same parameters Q and T_0 [5.8].

A suitable ansatz for a broad temperature range representation of β seems to be $\beta(T) = (1 - T_0/T)/(1 - T_0/T^*)$ with T^* being a critical temperature above the glass transition range [5.9].

In general, it may not be assumed that the glass is always close to equilibrium so that (5.8) holds. For the non-equilibrium case, the equation $\tau = \tau_0 \exp((Q/R)/T(1 - T_0/T_f))$ has been proposed [5.10]. With this and the further assumption that all temperature changes in the thermal history may be linearly superponed (Boltzmann's superposition principle), the general case of an arbitrary history can be treated.

Table 5.1. Specific heat at room-temperature

Oxide	As crystal J/gK	As glass component J/gK
Li_2O	1.11	1.5
Na_2O	1.06	1.08
K_2O	0.83	0.84
MgO	0.96	1.01
CaO	0.78	0.77
BaO	0.28	0.28
B_2O_3	0.88	0.92
Al_2O_3	0.8	0.85
SiO_2	0.77	0.81
ZnO	0.49	0.52
PbO	0.2	0.2

As it has been said above, the exact values of the specific heat *below the glass transition* depend on the structure which has been frozen in. In first order (and for a limited temperature range), however, they are linear combinations of the specific heats of the oxydes involved. For the latter, one can take data derived from different glass compositions [5.11] or, usually at a slight loss of precision, the values of the crystalline oxides themselves [5.12] (Table 5.1).

At room temperature, a mixture of them with the same composition as BK 7® would have a specific heat of 0.8 J/gK. If the specific heat values of the crystals are replaced by the data derived from glasses with different compositions, the value is 0.84 J/gK. From the measurement, the value 0.79 J/gK follows.

The astonishing fact that the specific heat of glass at room temperature may be derived from crystal data can be explained considering the mutual order of the atoms. According to the X-ray diffraction pattern, the short-range order of a glass is very much like the one in a crystal, whereas the long-range order reflects the amorphous state [5.13]. So at temperatures where, first, the short-range order is fixed, and, second, those atomic vibrations are dominant the wavelengths of which have the same dimension as the short-range order, the specific heat should be similar to the one of a corresponding mixture of crystals. At very low temperatures where only the atomic vibrations with very long wavelengths are not frozen in, significant deviations occur [5.14].

5.2 Thermal Conductivity

Usually the thermal conductivity is defined as the physical quantity measured by the following experiment (Fig. 5.5): Two opposite sides of a sample are held at different temperatures, at T and at $T + \Delta T$. Lateral thermal insulation ensures

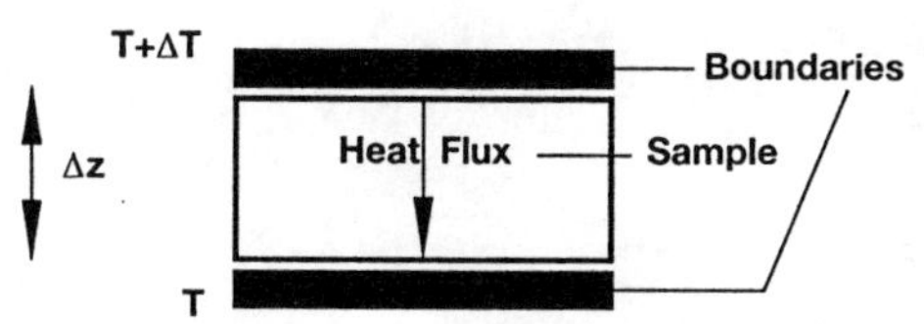

Fig. 5.5. Principal set-up of a thermal conductivity measurement

that the resulting temperature gradient in the sample is uniform over each cross-section. The resulting heat flux is measured.

For a technical realization, the test specimen is sandwiched between two identical reference samples whose thermal conductivity is known (comparative heat flux technique). This stack is placed between two heating elements with different temperatures. At thermal equilibrium, there is a constant heat flux through the sandwich (there are *two* reference samples to make sure that this is really the case) and a corresponding temperature distribution which is measured with thermocouples. The heat flux is calculated from the temperature gradient along the reference samples and their thermal conductivity.

With most materials, one finds a linear proportionality between the temperature gradient $\Delta T/\Delta z$, Δz being the sample thickness, and the heat flux density j. The constant of proportionality is called the thermal conductivity κ.

$$j = -\kappa \frac{\Delta T}{\Delta z} \,. \tag{5.9}$$

In general, the relation is $j = -\kappa \nabla T$. With the continuity equation $\nabla j = \rho c_p \partial T/\partial t$, this leads to the heat transfer equation $\rho c_p \partial T/\partial t = -\nabla(\kappa \nabla T)$.

In glass, there are two mechanisms contributing to the total heat transfer, the phononic thermal conduction and the heat transfer via thermal radiation.

5.2.1 The Phononic Contribution to the Thermal Conductivity

For the first contribution – which is the major one at temperatures below about 400 °C –, the microscopic theory of Debye gives [5.15]:

$$\kappa_{\text{phononic}} = \frac{1}{3} c_V \rho v_s l \,, \tag{5.10}$$

where c_V is the heat capacity at constant volume. ρ is the density. l and v_s are the mean free pathlength and the mean velocity of the phonons.

To arrive at this equation, one has to consider that the thermal energy of the solid is carried by vibrational excitations or phonons, which are capable of a wave-like motion through the glass volume. Due to the regular scattering and other interaction processes the phonons are subjected to, this motion is a drift process.

(The interaction processes bring about a re-arrangement of the phonon distribution. Any deviation from the equal distribution of the thermal energy among all types of phonons is thus corrected. So these processes are responsible for the *local thermal equilibrium.*)

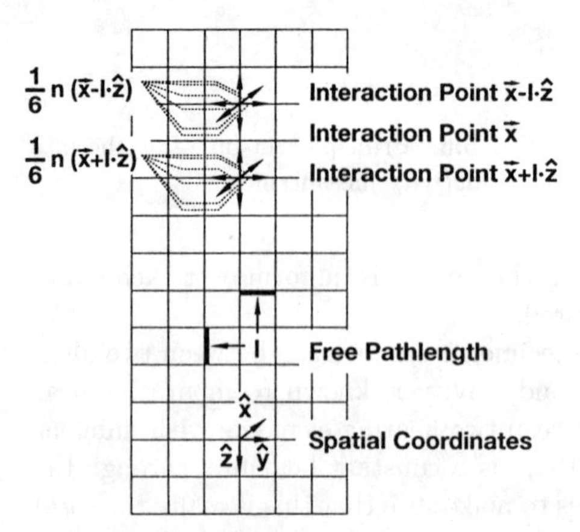

Fig. 5.6. Microscopic mechanism of thermal conduction. It is assumed that the energy from each interaction point is equally distributed among the six directions indicated.

The distance a phonon may cover before it is subjected to the next interaction is called the free pathlength l. In the following, it is assumed that l is a constant number and that the interaction processes take place at fixed points in space.

To calculate the resulting heat flux density in the direction of the temperature gradient, one has to consider the net phonon flux density passing one interaction point $\boldsymbol{x}$ (Fig. 5.6). It is the difference of the flux densities coming from the two neighbouring interaction points $\boldsymbol{x} - l\hat{\boldsymbol{z}}$ and $\boldsymbol{x} + l\hat{\boldsymbol{z}}$. $\hat{\boldsymbol{z}}$ is the unit vector with the direction of the temperature gradient. As all directions are equal at an interaction point, it can be assumed that both from $\boldsymbol{x} - l\hat{z}$ and from $\boldsymbol{x} + l\hat{z}$, one sixth of the phonons moves towards $\hat{z}$. With $n(\boldsymbol{x} - l\hat{z})$ and $n(\boldsymbol{x} + l\hat{z})$ being the corresponding phonon densities, the resulting net flux density is

$$\left(\frac{1}{6}n(z+l) - \frac{1}{6}n(z-l)\right) v_s \sim \frac{1}{3}\frac{dn}{dz} l v_s \; . \tag{5.11}$$

As n is proportional to the local energy density, $e\rho$ ($e\rho$ denotes the energy density, i.e., the energy per volume, whereas e alone denotes the specific energy, i.e., the energy per mass), this is equivalent to an energy flux density

$$j = \frac{1}{3}\rho\frac{de}{dz} l v_s = \frac{1}{3}\rho\frac{de}{dT}\frac{dT}{dz} l v_s \; . \tag{5.12}$$

With $c_V =: de/dT$, the resulting expression for j is:

$$j = -\frac{1}{3}\left(c_V \rho l \frac{v_s}{\rho}\right)\frac{dT}{dz} \; . \tag{5.13}$$

This corresponds to the above equation

$$\kappa_{\text{phononic}} = \frac{c_V \rho l v_s}{3} \; . \tag{5.14}$$

5.2.2 The Radiative Heat Transfer in Opaque Glass

As stated above, the radiative heat transfer has to be taken into account at temperatures above approximately 400 °C. In an opaque glass, it is characterized by a single radiative thermal conductivity. The thermal conductivity measured in the prototype experiment (Fig. 5.5) – the *effective* thermal conductivity – is the sum of the phononic and the radiative contribution:

$$\kappa_{\text{effective}} = \kappa_{\text{phononic}} + \kappa_{\text{radiation}} \,. \tag{5.15}$$

With the same argument as for the phononic thermal conductivity, the following expression can be obtained for the radiative thermal conductivity [5.16]:

$$\kappa_{\text{radiation}} = \frac{16}{3} n^2 \frac{\sigma}{k_{\text{R}}} T^3 \,. \tag{5.16}$$

n is the refractive index, σ is the Stefan-Boltzmann constant and k_{R} is the Rosseland mean for the absorption coefficient (see (5.21) below).

To derive this expression, one starts considering the energy flux density which is given by a similar expression as above,

$$j = -\frac{1}{3}\frac{de}{dz} l \frac{v_{\text{l}}}{n} \,. \tag{5.17}$$

This time, e is the energy density stored in the radiation field – as photons –, l is the inverse of the absorption coefficient k and v_{l}/n is the velocity of light (vacuum velocity of light divided by the refractive index). (Note that as "e" is defined as the energy density of the radiation field, "e" from here corresponds to "$e\rho$" from above.)

Usually the absorption coefficient is wavelength dependent. Then the energy flux density has to be calculated separately for each wavelength λ:

$$j_\lambda = -\frac{1}{3}\frac{de_\lambda}{dz} l(\lambda) \frac{v_{\text{l}}}{n} \equiv -\frac{1}{3}\frac{de_\lambda}{dz}\frac{1}{k(\lambda)}\frac{v_{\text{l}}}{n} \tag{5.18}$$

(the wavelength dependence of n is suppressed). The integration over all wavelengths leads to the total radiative energy flux density $j = \int_0^\infty j_\lambda \, d\lambda$.

Due to the intense exchange of energy through absorption and emission in opaque glasses, there is strong coupling of the atomic vibrations – the phonons – and the radiation field – the photons – and, thus *local thermal equilibrium* of both. Therefore e_λ is given by 4π times Planck's expression [5.17] for the equilibrium spectral intensity over the velocity of light, i.e., $4\pi B(\lambda, T)/(v_{\text{l}}/n)$.

(The spectral intensity I indicates the amount of radiation energy moving in a certain direction, per unit wavelength interval, per unit solid angle, per unit area normal to this direction, per time. The energy density e_λ is obtained from this through division by the velocity of light and integration over the whole solid angle. Because of the *local thermal equilibrium*, I is identical to Planck's function B, $I(\ldots, x, y, z) = B(\lambda, T(z))$ and consequently, $e_\lambda(z) = 4\pi B(\lambda, T(z))/(v_{\text{l}}/n)$ is valid.)

Because of the coupling of temperature and radiation, a temperature gradient gives rise to a gradient of I, which results in a preferred direction of the energy transfer through alternating photon emission and absorption.

With the relation $e_\lambda(z) = 4\pi B(\lambda, T(z))/(v_l/n)$ and

$$\frac{\partial B}{\partial z} = \frac{\partial B}{\partial T}\frac{dT}{dz}, \tag{5.19}$$

the total radiative energy flux density j is:

$$j = -\frac{4\pi}{3}\frac{v_l}{n}\left(\int_0^\infty \frac{\partial B(\lambda, T)}{\partial T}\frac{1}{v_l/n}\frac{1}{k(\lambda)}\,d\lambda\right)\frac{dT}{dz}. \tag{5.20}$$

With the Rosseland mean

$$k_R = \frac{\int_0^\infty \frac{\partial B}{\partial T}\,d\lambda}{\int_0^\infty \frac{\partial y}{\partial T}\frac{1}{k(\lambda)}\,d\lambda} \tag{5.21}$$

and the identity

$$\pi \cdot \int_0^\infty B(\lambda, T)\,d\lambda = n^2\sigma T^4, \tag{5.22}$$

the expression for j can be rewritten as:

$$j = -\frac{16}{3}n^2\frac{\sigma}{k_R}T^3\frac{dT}{dz}. \tag{5.23}$$

This corresponds to the above expression for the radiative thermal conductivity.

As it has been mentioned above, the effective thermal conductivity is the phononic plus the radiative term:

$$\kappa_{\text{effective}} = \frac{1}{3}c_V\rho v_s l + \frac{16}{3}n^2\frac{\sigma}{k_R}T^3. \tag{5.24}$$

5.2.3 The Radiative Heat Transfer in Transparent Glass

Considering glass which is transparent for a part of the electromagnetic spectrum (Fig. 5.7), i.e. glass where the free pathlength of those photons is bigger than the dimension of the glass volume, one has to distinguish between the *active* and the *passive* thermal conductivity.

The active thermal conductivity determines the heat flux which is actively transferred via phonons or photons with a short free pathlength. The passive thermal conductivity indicates the flux of the photons with a long free pathlength which, therefore, may pass through the glass without interaction. It is both generated and deleted outside the glass volume. The sum of the *active* and the *passive* contribution is called *apparent thermal conductivity* [5.18].

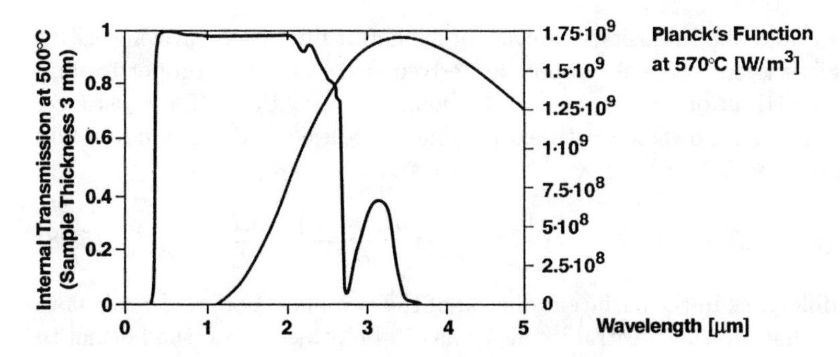

Fig. 5.7. Net spectral transmission of Schott glass Duran® at an intermediate temperature (500 °C) of the cooling experiment described below. To indicate the spectral range of interest, Planck's function at the initial temperature (570 °C) of this experiment is also given.

(To match the above nomenclature, the term *effective* thermal conductivity could always be used when the sum of the phononic part and a radiative contribution is addressed. This, however, would lead to long expressions like *apparent effective thermal conductivity.*)

It is the latter quantity which comes out of the prototypical thermal conductivity measurement (Fig. 5.5) if the data are evaluated as usual. *The temperature distribution in a cooling piece of glass, however, is determined only by the active thermal conductivity.*

To determine this quantity also from the prototype experiment, measurements on samples with different thicknesses Δz, $\Delta z \pm dz$ are necessary. The active thermal conductivity is then obtained from the different heat fluxes j, $j \pm dj$ by [5.19]:

$$\kappa_{\text{active}} = -\frac{1}{\Delta T}\frac{dj}{d\left(\frac{1}{\Delta z}\right)}\,, \tag{5.25}$$

whereas, as has been said, the apparent thermal conductivity is determined as usual:

$$\kappa_{\text{apparent}} = -\frac{j}{\Delta T/\Delta z}\,. \tag{5.26}$$

Equation (5.25) can be derived by considering the calculation of the various components of the heat flux in the prototypical thermal conductivity measurement. For reasons of simplicity, it is assumed that with respect to the sample thickness Δz, the absorption coefficient $k(\lambda)$ is either close to zero or almost infinite; i.e. either $k(\lambda)\Delta z \ll 1$ or $k(\lambda)\Delta z \gg 1$ is valid. The wavelengths for which the first inequation is true will be indexed with "t" (λ_t, as *transparent*), the others with "o" (λ_o, as *opaque*).

For the latter, the consideration leading to the above expression for the flux density is again valid:

$$j_{\lambda_o} = -\frac{4\pi}{3}\frac{dB(\lambda_o,T)}{dz}\frac{l}{k(\lambda_o)}\,. \tag{5.27}$$

The existence and the intensity, however, of a photon flux with wavelengths for which the glass is transparent cannot be derived from the glass properties and temperature distribution but depend on the boundary conditions. If, for instance, the glass sample is sandwiched between two black bodies, the corresponding flux density is given by [5.20]

$$j_{\lambda_t} = -(\pi B(\lambda_t, T + \Delta T) - \pi B(\lambda_t, T)) \sim -\pi \frac{\partial B(\lambda_t, T)}{\partial T} \Delta z \frac{dT}{dz} . \tag{5.28}$$

This is the difference in the surface radiation of the two black bodies. To calculate the surface radiation, the spectral intensity has to be projected on the normal to the surface (cosine law) and integrated over the half sphere. Reflection effects at the surface are suppressed. In the second (approximate) equation, $\Delta T/\Delta z$ has been replaced by dT/dz on the right side. This is exactly true for a linear temperature profile $T(z)$, only. Remarkably, the expression for j_{λ_t} almost equals the expression for j_{λ_o} if the sample thickness Δz is inserted as free pathlength.

So the active thermal conductivity is given by:

$$\kappa_{\text{active}} = \frac{1}{3} c_V \rho v_s l + \frac{4\pi}{3} \int_{\lambda_o} \frac{dB(\lambda, T)}{dT} \frac{1}{k(\lambda)} d\lambda . \tag{5.29}$$

The integral over λ_o means the integration over all wavelengths for which the glass is opaque.

For black bodies as boundaries, the passive thermal conductivity amounts to

$$\kappa_{\text{passive}} = \pi \Delta z \int_{\lambda_t} \frac{dB(\lambda, T)}{dT} d\lambda , \tag{5.30}$$

where the integral over λ_t means the integration over all wavelengths for which the glass is transparent.

The sum of both the active and the passive contribution is the apparent thermal conductivity:

$$\kappa_{\text{apparent}} = \frac{1}{3} c_V \rho v_s l + \frac{4\pi}{3} \int_{\lambda_o} \frac{dB(\lambda, T)}{dT} \frac{1}{k(\lambda)} d\lambda + \pi \Delta z \int_{\lambda_t} \frac{dB(\lambda, T)}{dT} d\lambda . \tag{5.31}$$

As stated above, it is this apparent thermal conductivity which is measured by the prototype experiment if the latter is evaluated according to (5.26). *To obtain the active thermal conductivity, the measurement data have to be evaluated according to the differentiation rule* (5.25). This follows from the relation of the heat flow j and the sample thickness Δz:

$$j = -\frac{1}{3} c_V \rho v_s l \frac{\Delta T}{\Delta z} - \frac{4\pi}{3} \left[\int_{\lambda_o} \frac{dB(\lambda, T)}{dT} \frac{1}{k(\lambda)} d\lambda \right] \frac{\Delta T}{\Delta z} - \pi \left[\int_{\lambda_t} \frac{dB(\lambda, T)}{dT} d\lambda \right] \Delta T \tag{5.32}$$

$$\Rightarrow -\frac{1}{\Delta T} \frac{dj}{d\left(\frac{1}{\Delta z}\right)} = \frac{1}{3} c_V \rho v_s l + \frac{4\pi}{3} \int_{\lambda_o} \frac{dB(\lambda, T)}{dT} \frac{1}{k(\lambda)} d\lambda \equiv \kappa_{\text{active}} . \tag{5.33}$$

The different significances of the apparent thermal conductivity and the active thermal conductivity can be illustrated best by considering two typical heat transfer problems. Firstly, the heat flux through a fire protective glass *in action* is to be estimated. Secondly, the temperature profile inside a piece of glass during the cooling process is to be determined.

In the first case, the quantity of interest is the total heat flux coming from the fire, including both the actively guided and the passively transmitted radiation. This is indicated by the apparent thermal conductivity.

(It is assumed that both the furnace and the opposite side can be regarded as black bodies. For a "standard fire", the furnace temperature follows from the standard time-temperature curve $T(t) - T(0) = 345\,^\circ\mathrm{C}\cdot \log_{10}(1 + 8t/\mathrm{min})$ [5.21].)

In the second case, the temperature distribution in a piece of glass which is gradually cooled down is asked for. The resulting temperature gradient from the maximum temperature in the centre of the glass volume to the surface temperature is especially interesting because it determines the stress which arises during the cooling process. The amount of this gradient is determined by the *active* capability of the glass to transport the stored vibrational energy outside; i.e., the *active thermal conductivity*. (The energy transport outside can either be achieved by the motion of the vibrational modes or by the emission of radiation. As there is no emission at wavelengths where there is no absorption – because of the identity of the emissive spectrum and the absorptive spectrum –, the latter is limited to the wavelengths for which the glass is opaque.) The passively transmitted radiation is not of any interest. If the enclosure of the glass volume has a uniform temperature, the corresponding net flux even amounts to zero.

Under the above condition that, with respect to the sample thickness Δz, the absorption coefficient $k(\lambda)$ is either close to zero or almost infinite, i.e. either $k(\lambda)\Delta z \ll 1$ or $k(\lambda)\Delta z \gg 1$ is valid, the active heat transfer is exactly characterized by the active thermal conductivity. Likewise, the overall heat transfer is indicated by the apparent thermal conductivity which in the case of a linear temperature profile is given by (5.26).

Unfortunately, the first statement is not true if radiation is involved whose inverse absorption coefficient has the order of magnitude as the inverse sample thickness, i.e., $k(\lambda)\Delta z \sim 1$ is valid. The coupling of this radiation and the vibrational modes is too weak to provide a thermal equilibrium of both, so that the equation

$$j_\lambda = -\frac{4\pi}{3}\frac{dB(\lambda, T)}{dz}\frac{1}{k(\lambda)} \tag{5.34}$$

is not applicable for the corresponding wavelengths. On the other hand, it is too strong to allow this radiation to be neglected in the calculation of the actively transferred heat flux.

For an exact solution of the heat transfer problem, the transfer of this radiation has to be calculated separately. In both the corresponding transfer equations – one for each spectral band; for calculating the radiative heat transfer, the electromagnetic spectrum is divided into bands in such a way that for each band the absorption coefficient can be regarded as constant – and the regular heat trans-

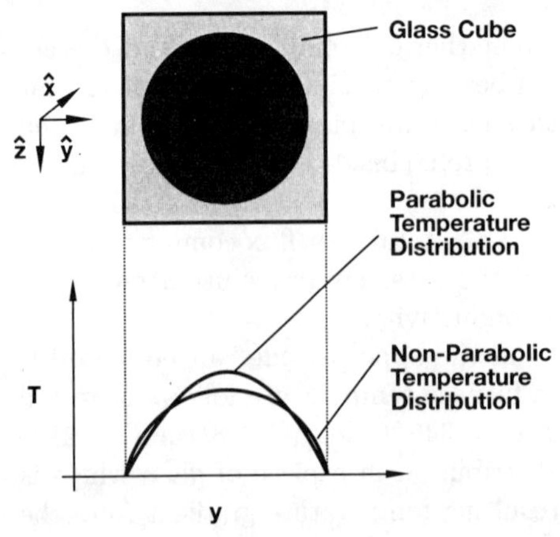

Fig. 5.8. Temperature distribution in a cooling glass cube (the shading intensity inside is proportional to the temperature). If the heat transfer inside is exactly characterized by the active thermal conductivity, the temperature distribution is parabolic. If not, the temperature distribution has "temperature jumps" at the surface and a flat shape in the centre, both typical features of radiative heat transfer if photons are involved the freepath length of which has the same dimension as the cube geometry.

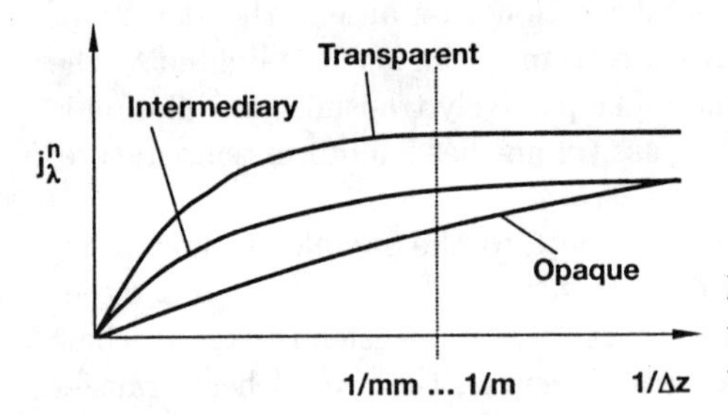

Fig. 5.9. Flux versus inverse sample thickness for the different contributions to the heat transfer

fer equation, sources and sinks have to be introduced which quantify the energy exchange via emission and absorption.

By these terms, the solutions of the heat transfer equation are significantly affected. In the case of an opaque glass or a glass fulfilling the above condition that either $k(\lambda)\Delta z \ll 1$ or $k(\lambda)\Delta z \gg 1$, the temperature profile of a cooling glass plate, for instance, has a parabolic shape. If a significant amount of radiation with $k(\lambda)\Delta z \sim 1$ has to be taken into account, it is closer to a higher-order power law profile z^{2n}, $n > 1$ [5.22], (Fig. 5.8).

The simultaneous calculation of a system of transfer equations, however, requires a lot of computer power and yields a result that is not always easy to comprehend. So often the characterization of the active and the passive heat transfer by a *single* number for each is preferred.

According to its definition, the apparent thermal conductivity as always characterizes the overall heat transfer. Likewise, the differentiation rule (5.25) yields a suitable indicator for the actively transported heat flux. The latter is not clear a priori and, therefore, a remarkable fact. It can be made plausible considering the way how the various parts of the heat flux depend on the inverse sample thickness $1/\Delta z$ (Fig. 5.9).

As it has been explained above, both the phononic heat flux, j_{phononic}, and the radiative heat flux at wavelengths for which the glass is opaque, $j_{\lambda\text{o}}$, are proportional to $\Delta T/\Delta z$. With the temperature difference ΔT having a constant value, this is equivalent to a linear relation between these two contributions and $1/\Delta z$. In contrast to this behaviour, the radiative heat flux at wavelengths for which the glass is transparent, j_{λ_t}, is independent of either Δz or $1/\Delta z$.

The latter, however, is true only as long as the above condition $k(\lambda)\Delta z \ll 1$ is valid. For samples with the thickness Δz being bigger than $1/k(\lambda)$ – provided that $k(\lambda)$ is not exactly equal to zero, which will never be the case in real glass – almost the whole corresponding radiation will be completely absorbed at least once on its way through the sample. Even if the whole energy involved were re-emitted at the original wavelengths (in reality, the absorbed energy is distributed among the vibrational modes and, according to Planck's law, the whole radiation field), only one half would move on in the original direction. Because of the re-emission being isotropic, the other would be sent back to where it came from. Therefore, j_{λ_t} is not independent of the sample thickness for very large values of the latter. (The classification as opaque or transparent refers to technically relevant values of Δz.) In the limit of infinite sample thickness, there is always a similar linear relationship to $1/\Delta z$ as for $j_{\lambda\text{o}}$. In the diagram, this results in a saturation curve with the turning point at $\Delta z(\lambda) = 1/k(\lambda)$.

In the intermediate case, the turning point of the saturation curve lies in the region of the technically relevant values of Δz, i.e. $k(\lambda)\Delta z \sim 1$ is valid for them. The corresponding flux density is indexed with "i" for intermediate: $j_{\lambda i}$.

According to the differentiation rule (5.25), the slope of each of these curves is proportional to the corresponding contribution to the active thermal conductivity. The constant of proportionality is the temperature difference ΔT which according to the above presupposition has a constant value for all experiments. It can be eliminated by introducing the normalized flux density j^n, which is the quotient of the flux density j and the temperature difference ΔT. So the active thermal conductivity is given by:

$$\kappa_{\text{active}} = -\frac{dj^n}{d\left(\frac{1}{\Delta z}\right)} . \tag{5.35}$$

For low values of $1/\Delta z$ or high values of Δz, respectively, the differentiation rule yields the sum of all zero point slopes as active thermal conductivity. (The sum of all zero point slopes is equal to the radiative thermal conductivity – as has been said, the latter is characteristic of all glasses if the thickness is infinite.) This corresponds to the fact that above a certain thickness the glass is opaque for all wavelengths. In the opposite limit, the differentiation rule sifts the phononic contribution from all the radiative ones. This, in return, corresponds to the fact that below a certain thickness, the glass is transparent for all wavelengths.

For the intermediate values of Δz – the technically important ones –, the differentiation rule counts the full zero point slope of both j^n_{phononic} and $j^n_{\lambda_\text{o}}$. As $j^n_{\lambda_\text{t}}$

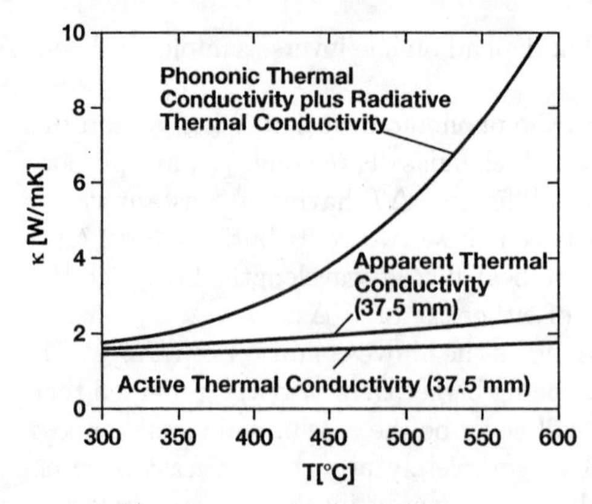

Fig. 5.10. Thermal conductivity of Schott glass Duran® versus temperature for sample thicknesses 37.5 mm and infinity

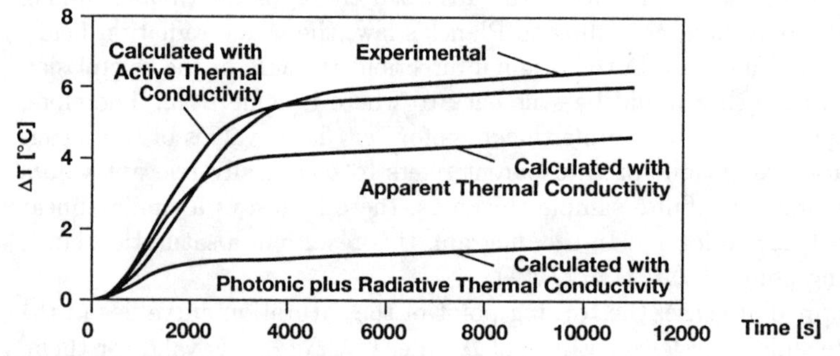

Fig. 5.11. Difference between maximum and minimum temperature of a cooling glass plate, thickness 75 mm

has a constant amount, it is not taken into consideration. For $j^n_{\lambda_i}$, the differentiation rule yields an intermediate value.

Obviously the latter is more reasonable than any alternative. Calculating the contribution to the thermal conductivity from the difference quotient $j^n_{\lambda_i}/(1/\Delta z)$ instead of the differential quotient $dj^n_{\lambda_i}/d(1/\Delta z)$ means proceeding according to the definition of the apparent thermal conductivity and will lead to an over-estimation of the effect of $j^n_{\lambda_i}$. Calculating the contribution of j_{λ_i} from the equation

$$j_\lambda = -\frac{4\pi}{3}\frac{dB(\lambda,T)}{dz}\frac{1}{k(\lambda)} \tag{5.36}$$

(which comes to calculating with the radiative thermal conductivity as defined by (5.16)) will even lead to a higher value if $1/k(\lambda) > \Delta z$ is valid. On the other hand, it is not justified to neglect this radiation completely.

Figure 5.10 shows the active and the apparent thermal conductivity of Schott glass Duran® for sample thicknesses 37.5 mm and infinity (where both are equal and identical to the phononic thermal conductivity plus the radiative thermal conductivity as defined by (5.16)).

The concept of the active thermal conductivity is verified by a cooling experiment. A glass plate (Schott glass Duran®, 75 mm thickness) is equilibrated at

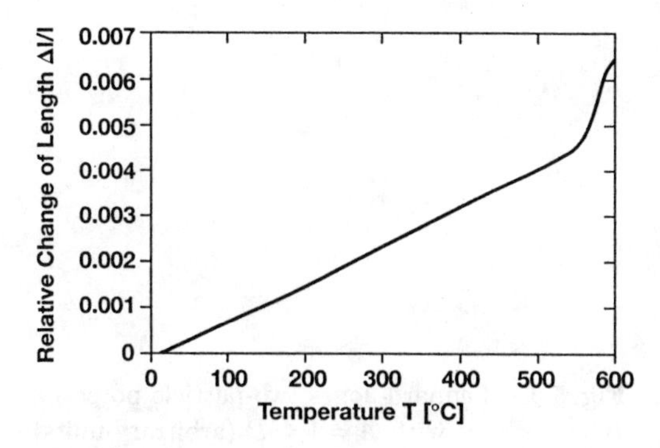

Fig. 5.12. Dilatometric measurement of technically cooled Schott glass BK 7® (heating rate 2 K/min)

570 °C and subsequently cooled down at 20 °C/h. Figure 5.11 shows the difference between the maximum and the minimum temperature in the plate as a function of the time after the start of the cooling. The three theoretical curves have been calculated on the basis of the different thermal conductivities from Fig. 5.10, i.e. the active thermal conductivity, the apparent thermal conductivity and the sum of the phononic thermal conductivity and the radiative thermal conductivity as defined in (5.16).

The experimental curve has been determined with a spectral radiometer. The radiation emitted by the glass plate is measured at both a wavelength for which the glass is opaque (3.6 μm) and a wavelength for which the glass is half-transparent (2.5 μm). If an assumption about the principal shape of the temperature distribution is made (a parabolic one, for instance), both the surface temperature and the difference between the maximum and the minimum temperature can be extracted from these data.

As expected, the correspondence between the experimental curve and the theoretical curve calculated with the sum of the phononic thermal conductivity and the radiative thermal conductivity is the least. Neither is the correspondence with the theoretical curve calculated with the apparent thermal conductivity satisfactory. The best fitting theoretical curve is derived with the active thermal conductivity.

5.3 Thermal Expansion

The thermal expansion is defined as the relative change of size as a response to a temperature change. By linear thermal expansion, the relative change of length is addressed. It is usually measured with a dilatometer on a rod-shaped sample.

Figure 5.12 shows the dilatometric measurement of BK 7® from Schott. The curve itself indicates the relative change of length $\Delta l/l$; the thermal expansion coefficient α is given by the slope:

$$\alpha = \frac{1}{l}\frac{dl}{dT} \tag{5.37}$$

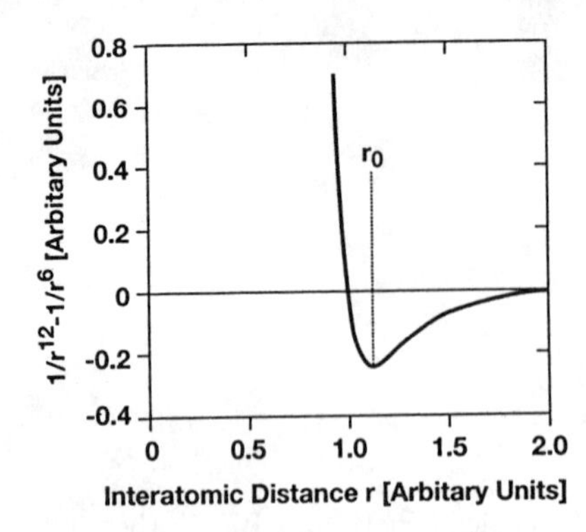

Fig. 5.13. Lennard-Jones two-particle potential $B/r^{12} - A/r^6$ with $A = 1 = B$ (arbitrary units)

A microscopic interpretation of the thermal expansion can be found considering the atomic interaction. To first order, the latter can be described by a two-particle potential such as the one introduced by Lennard and Jones (Fig. 5.13).

One particle is assumed to be fixed, the other may move in the potential well, bouncing from one point of return to the other. In this picture, the energy of the interatomic motion (the thermal energy per particle) is represented by the height above the potential minimum at which the second particle moves. *According to the asymmetry of the interatomic potential* in the neighbourhood of the potential minimum, the mean distance of the second particle to the first one is shifted to higher values with rising energy.

With the interatomic potential $U(r - r_0) \equiv U(x)$, which can be represented by

$$U(x) = ax^2 - bx^3 \tag{5.38}$$

in the neighbourhood of the potential minimum r_0, the thermal expansion coefficient α is:

$$\alpha = \frac{4\pi^2 r_0^2}{3} \frac{3b}{4a^2} \rho c_p \; . \tag{5.39}$$

So besides identifying the asymmetry of the interatomic potential as being the cause of thermal expansion, one finds a relation between the thermal expansion coefficient and the specific heat.

A quick derivation of (5.39) can be obtained with the following argument. Let the arithmetic mean of $x_{1,2}$, the two points of return, be the average localization of the second particle. With H being the enthalpy of the two-particle interaction, $ax_{1,2}^2 - bx_{1,2}^3 = H$ is valid. A change ΔH of the latter will cause a shift $\Delta x_{1,2}$ of each of them. The amount of the shift follows from $\Delta H \approx U'(x_{1,2})\Delta x_{1,2} = (2ax_{1,2} - 3bx_{1,2}^2)\Delta x_{1,2}$. The resulting shift of the average localization is

$$\Delta \left\{ \frac{[x_1 + x_2]}{2} \right\} = \frac{\Delta x_1 + \Delta x_2}{2} = \frac{1}{2} \left\{ \frac{\Delta H}{(2ax_1 - 3bx_1^2)} + \frac{\Delta H}{(2ax_2 - 3bx_2^2)} \right\} . \tag{5.40}$$

With $|bx_{1,2}^3| \ll ax_{1,2}^2$ and consequently $x_1 \approx -x_2$ in first order, $\Delta\,[x_1 + x_2]\,/2 = (3b/4a^2)\Delta H$ holds. If (with some imagination, e.g. by smearing the second atom around the first) the problem is made rotationally symmetric, an interatomic distance change of $(3b/4a^2)\Delta H$ means a volume change of $4\pi^2 r_0^2(3b/4a^2)\Delta H$. Consequently the *relative* volume change is $4\pi^2 r_0^2(3b/4a^2)\Delta(h\rho)$ with $h\rho$ the volume density of the enthalpy or, with $\Delta(h\rho)$ replaced by $\rho c_p \Delta T$, it is $4\pi^2 r_0^2(3b/4a^2)\rho c_p \Delta T$. The linear expansion coefficient follows as $(4\pi^2 r_0^2/3)(3b/4a^2)\rho c_p$.

Replacing $6\pi^2 r_0/a$ by the compressibility χ and introducing the dimensionless Grüneisen parameter γ for $br_0/(2a)$, one arrives at the Grüneisen rule [5.23]:

$$\alpha = \gamma\chi\rho c_p/3 \ . \tag{5.41}$$

A quick derivation of the expression for χ, i.e. $\chi = 6\pi^2 r_0/a$, can be obtained with the following argument. Exerting a pressure p is equivalent to adding the linear term $p4\pi^2 r_0^2 x$+const. to the two-particle potential (pressure = energy/volume). As the zero level of this term is arbitrary, "const." is set equal to zero. With the third-order term $bx_{1,2}^3$ of the potential being suppressed, $H = ax_{1,2}^2 + p4\pi^2 r_0^2 x_{1,2}$ is valid at the two points of return. As H does not change with a pressure change, the equation

$$0 = (2ax_{1,2} + p4\pi^2 r_0^2)\Delta x_{1,2} + \Delta p 4\pi^2 r_0^2 x_{1,2} \tag{5.42}$$

holds in first order. With the initial pressure $p \equiv 0$, $\Delta x_{1,2}/\Delta p = -4\pi^2 r_0^2/2a$ is valid. So a pressure change Δp results in an average localization shift of $-\Delta p 4\pi^2 r_0^2/2a$, which, in turn, corresponds to a relative volume change of $-6\pi^2 r_0 \Delta p/a$. Consequently, the compressibility amounts to $6\pi^2 r_0/a$.

As γ, χ are given by atomic constants and ρ is only weakly temperature dependent, the Grüneisen rule predicts a proportionality of the thermal expansion coefficient and the specific heat.

The Grüneisen rule is not strictly valid for all temperatures. This is because instead of just one triple (a, b, r_0) there are various triples of atomic constants which are differently affected by the effective freezing-in of a number of (motional) degrees of freedom at the glass transition.

Above and below the glass transition, however, where for (a, b, r_0) average values may be used that do not change in the temperature range under consideration, the Grüneisen rule is applicable. The thermal expansion coefficient particular can be expected to have an almost constant value in the high temperature limit (if $c_p \sim c_V$ is valid; the latter is constant according to the Dulong-Petit law). With $\alpha = 45 \times 10^{-6}$/K, $\rho = 2250\,\mathrm{kg/m^3}$, $c_p = 1420$ J/kgK, $\chi = (1/16)\,\mathrm{GPa^{-1}}$ at 800 °C, the high temperature Grüneisen parameter of BK 7® is $\gamma = 0.65$.

By the effective freezing-in of a significant number of degrees of freedom at the glass transition, the thermal expansion coefficient is affected much stronger than c_p (Fig. 5.14). Interpreting this result with the Grüneisen rule one finds that this is "due" to a change of the compressibility by a factor of three and a change of the Grüneisen parameter by a factor of 1.3. With $\alpha = 7.7 \times 10^{-6}$/K, $\rho = 2500\,\mathrm{kg/m^3}$, $c_p = 840$ J/kgK, $\chi = (1/46.5)\,\mathrm{GPa^{-1}}$, the room temperature Grüneisen parameter of BK 7® is $\gamma = 0.49$. (As for c_p, the data measured at technically cooled BK 7® are referred to this time.)

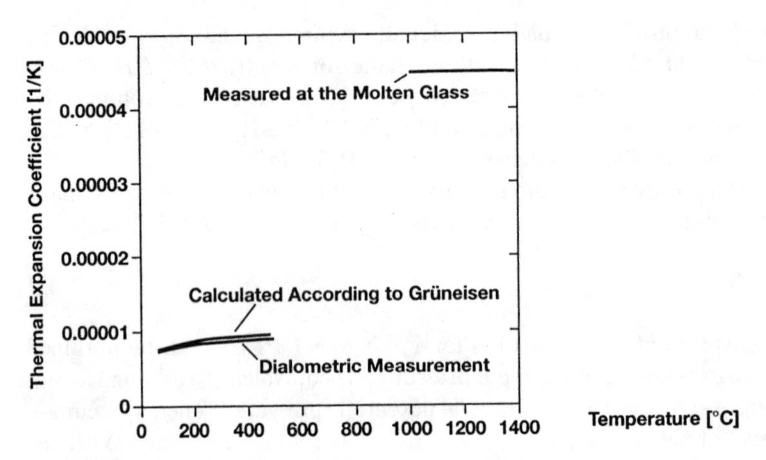

Fig. 5.14. Thermal expansion coefficient α of technically cooled Schott glass BK 7®, measured with a dilatometer (below T_g) and with the buoyancy method (at the molten glass). For a comparison, the values calculated according to Grüneisen from the specific heat are also given (below T_g). The two curves are almost indistinguishable.

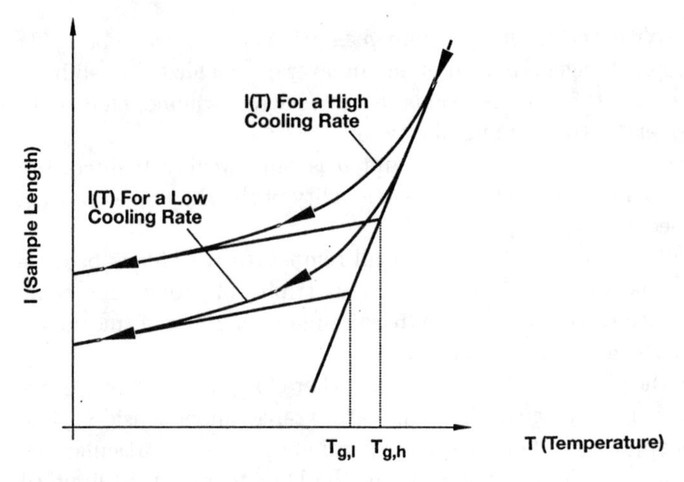

Fig. 5.15. Effect of the cooling rate on the position of the glass transition. The smaller the cooling rate is, the lower is the temperature T_g at which the dilatometer curve changes its slope (high cooling rate → $T_{g,h}$, low cooling rate → $T_{g,l}$).

The details of the thermal expansion curve at the glass transition are determined by the kinetics of the structural relaxation. As mentioned above, the exact position of the glass temperature depends on the cooling rate (Fig. 5.15) [5.24].

5.4 Viscosity

When two layers of a fluid move parallel in the same direction but at different velocities, a mutual force is exerted between them. Those fluids for which a proportionality is found between this force and the velocity gradient perpendicular to the direction of the fluid flow are called Newtonian fluids. The constant of proportionality is called the (shear) viscosity η:

$$f = \eta \nabla \times v \, . \tag{5.43}$$

f is the friction per unit area, v is the velocity.

Like BK 7,® the viscosity–temperature curve of which is given in Fig. 5.16, glass is always characterized by an extraordinary increase of the viscosity with decreasing temperature (which is a result of the effective freezing-in of the configurational degrees of freedom and, therefore, a typical feature of the glassy state). During glass processing, several decades of Pas are covered. Usually, the temperatures at which the viscosity amounts to 10^{12} Pas (upper annealing point), $10^{6.6}$ Pas (softening point), 10^{3} Pas (working point) are given as representative points of the viscosity–temperature curve.

It is not possible to measure the entire viscosity-temperature curve with one universal apparatus. In the temperature range of the working point, the friction of a rotating cylinder in a crucible with melted glass is measured (Fig. 5.17a).

If the gap between the crucible wall and the cylinder surface is small, the viscosity follows from $\eta = MD/(4\pi^2 R^3 H\omega)$ with M: angular moment, D: gap thickness, R: cylinder radius, H: height, ω: angular frequency.

In the temperature range of the softening point, the viscosity is usually determined via the elongation of a fibre which is loaded with a constant weight (Fig. 5.17b). The viscosity follows from $\eta = Fl^2/3Vv$ with F: load, l: fibre length, V: fibre volume, v: elongation speed [5.25].

Within the temperature range of the annealing point, the three-point bending method is used (Fig. 5.17c). The sample rests on two edges. In the middle, a load is applied. The viscosity follows from $\eta = (F + g\rho Sl/1.6)l^3/(144Iv)$ with F: load, ρ: density of glass, S: area of cross-section, l: distance between the two edges, I:

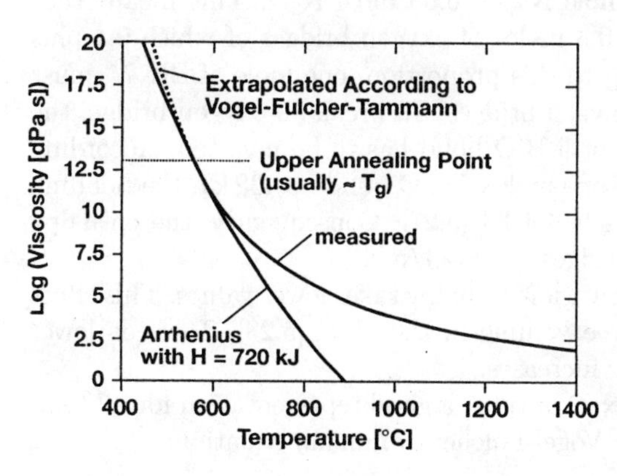

Fig. 5.16. Viscosity–temperature curve of Schott glass BK 7®

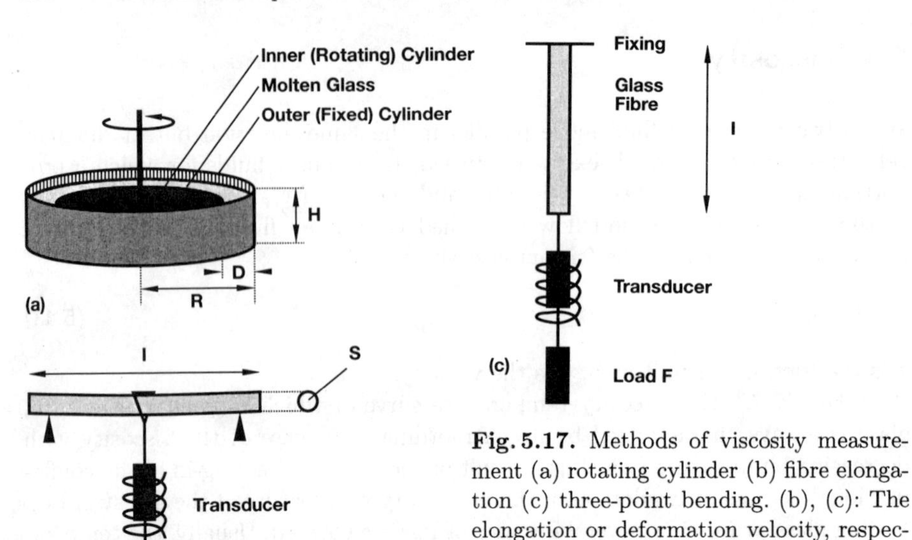

Fig. 5.17. Methods of viscosity measurement (a) rotating cylinder (b) fibre elongation (c) three-point bending. (b), (c): The elongation or deformation velocity, respectively, is monitored via an impedance transducer.

geometric moment of inertia ($= \pi r^4/4$, r: radius, for a circular cross-section), v: deformation speed [5.25].

A first guide for measuring the temperature dependence of the viscosity is the Arrhenius law for thermally activated processes with the enthalpy H necessary for breaking all oxygen bridges as activation enthalpy (as long as the shear force is small compared with the interatomic forces, the shearing is a drift process of which each step needs thermal energy; R is the gas constant) [5.26]:

$$\eta = \eta_0 \mathrm{e}^{\frac{H}{R}\frac{1}{T}} . \tag{5.44}$$

Indeed, one finds such a relation in the annealing region (Fig. 5.16). For a quick analysis, the argument is: One mole of SiO_2 means two moles of oxygen bridges, one mole of B_2O_3 means three moles of oxygen bridges. One mole of either Na_2O or K_2O means the destruction of one mole of oxygen bridges. The composition of BK 7® approximately corresponds to one mole of BK 7® consisting of 0.75 mole SiO_2, 0.10 mole B_2O_3, 0.10 mole Na_2O, 0.05 mole K_2O. This means that 1 mole of BK 7® contains 1.5 plus 0.3 moles of oxygen bridges of which 0.1 plus 0.05 moles are destroyed. According to this proportion, one mole of BK 7® must contain 1.65 moles of functioning oxygen bridges. To break an oxygen bridge, the bonding enthalpy of either an Si-O or a B-O bond has to be invested. According to Pauling, the bonding enthalpy of one mole of Si-O bonds is 432 kJ, the bonding enthalpy of one mole of B-O bonds is 460 kJ [5.27]. Consequently, the enthalpy necessary for breaking all oxygen bridges is 720 kJ/mol.

For higher temperatures, the activation enthalpy takes lower values. This effect is connected with the increasing free volume in the glass [5.28]. Towards lower temperatures, the activation energy increases.

As above for the structural relaxation time, a good representation for all temperatures can be obtained with the Vogel-Fulcher-Tammann equation:

$$\eta = \eta_0 e^{\frac{Q}{R}\frac{1}{T-T_0}}. \quad (5.45)$$

For the activation enthalpy H to be inserted in the Arrhenius equation which is to yield the same viscosity and have the same slope as the the Vogel-Fulcher-Tammann equation at a given temperature T, $H = QT^2/(T - T_0)^2$ is valid. So the activation enthalpy of the "best-fitting Arrhenius" increases with decreasing temperature and becomes infinite at $T = T_0$.

The physical nature of the Vogel-Fulcher-Tammann equation has been discussed a lot, especially with respect to the introduction of a characteristic temperature T_0 at which the viscosity becomes infinite. The parallel between the temperature dependence of the natural logarithm of η, $\ln \eta \propto 1/(T - T_0)$, to the paramagnetic susceptibility of ferromagnetic materials above the Curie temperature indicates that there is a second-order phase transition at T_0.

There is a thermodynamic argument in favour of this [5.29]. The entropy of the glass phase decreases stronger with decreasing temperature than the one of the corresponding crystal phase(s). Consequently, there is a crossover temperature, the Kauzmann temperature T_K, at which the entropy of the glass becomes smaller than the entropy of the crystal phase(s). This so-called Kauzmann paradox is solved by the assumption that there is a second order phase transition at T_K. As a consequence, T_0 is the lowest temperature at which an equilibrated glass may exist. Usually, the Kauzmann temperature T_K and the Vogel-Fulcher-Tammann temperature T_0 are of the same order of magnitude, but not exactly the same.

This point of view is supported by the above-mentioned argumemts about the kinetics of the restructuring or *relaxation* processes [5.30]; this time the kinetics by which the glass tries to adapt to the new shape imposed by the shearing force.

During constant rate shearing, there is a steady state of adaption to the new shape – which results in a decrease of the stress imposed by the shearing force – and renewed deformation. The viscosity is a measure of the steady state adaptation or *stress relaxation* capability. The stress response $\sigma(t)$ to a suddenly imposed shear by an angle γ is again described by a Kohlrausch law: $\sigma(t) = \sigma(0)e^{-(\frac{t}{\tau})^b}$; b Kohlrausch parameter; $\sigma(0) = G\gamma$; G instantaneous shear modulus.

As well as the structural relaxation, the overall stress relaxation is the sum of the statistically happening elementary processes such as the breaking and closing of single chemical bonds at high temperatures. At lower temperatures, again complex relaxation steps involving the simultaneous occurrence of several of such elementary processes are necessary in addition.

Because of this phenomenon of *cooperativity*, the temperature dependence of the overall steady state relaxation capability, i.e. the viscosity, is larger than that of the single elementary process. This explains that the activation energy exceeds the value estimated from the bonding strength at temperatures below the (upper) annealing point. The number of elementary processes and the size of the *correlation volume* involved increases with decreasing temperature. It becomes infinite at T_0 which is typical of a second-order phase transition.

It has to be emphasized that the Vogel-Fulcher-Tammann equation which gives the equilibrium viscosity is not applicable for a non-equilibrium glass. Usually, the glass transition occurs at temperatures far above T_0 so that the second-order phase transition problem is circumvented. Below the glass transition, in the non-equilibrium, the following modification of Hodge's equation [5.10] holds:

$$\eta = \eta_0 e^{\frac{Q/R}{T(1-T_0/T_g)}} \ . \tag{5.46}$$

In general, Hodge's equation says: $\eta = \eta_0 \exp((Q/R)/(T(1-T_0/T_f))$. So if the fictive temperature is above T, the viscosity is less than the equilibrium viscosity at T and vice versa.

As has been said, both the structural and the stress relaxation time are proportional to the viscosity. Therefore

$$\tau_{\text{structural}} \propto \tau_{\text{stress}} \sim \exp\left(\frac{Q/R}{T(1-T_0/T_f)}\right) \ . \tag{5.47}$$

Acknowledgements

I am grateful to Mr. Harald Krümmel, Mr. Thomas Korb, and to the coworkers of the laboratory of Dr. Rolf Müller, especially Mrs. Klaas, Mrs. Leicht, Mrs. Diehl, Mrs. Kalden, Mrs. Leopold-Adler, Mrs. Weitzel, and Mr. Wernst for carrying out the measurements.

References

5.1 F. Reif: *Statistical Physics. Berkeley Physics Course*, Vol. 5 (Educational Services Incorporated, Library of Congress Catalog Card Number: 64-66016, Orlando, Florida1967) p. 216

5.2 G. Burns: *Solid State Physics* (Academic Press, Inc., Orlando, Florida 1985), United Kingdom edition: (Academic Press, Inc. (London) Ltd., Library of Congress Catalog Card Number: 85-70104) p. 356

5.3 W. Hemminger, G. Höhne: Grundlagen der Kalorimetrie (Verlag Chemie, Weinheim, Deerfield, Basel 1979) p. 195

5.4 G. Burns: *Solid State Physics* (Academic Press, Inc., Orlando, Florida 1985), United Kingdom edition: (Academic Press, Inc. (London) Ltd., Library of Congress Catalog Card Number: 85-70104) p. 360–365

5.5 C. Weimantel, C. Hamann: *Grundlagen der Festkörperphysik* (VEB Deutscher Verlag der Wissenschaften, Berlin 1979) Licensed edition: (Springer, Berlin, Heidelberg, New York) p. 294

5.6 G.W. Scherer: *Relaxation in Glass and Composites* (John Wiley & Sons, Inc., New York, Chichester, Brisbane, Toronto, Singapore 1986) Library of Congress Catalog Card Number: 85-17871, p. 268

5.7 G.W. Scherer: *Relaxation in Glass and Composites* (John Wiley & Sons, Inc., New York, Chichester, Brisbane, Toronto, Singapore 1986) Library of Congress Catalog Card Number: 85-17871, pp. 130–144

5.8 S.M. Rekhson, A.V. Bulaeva, O.V. Mazurin: Sov. J. Inorg. Mater. **7** (4), 622–623 (1971), quoted after: George W. Scherer: *Relaxation in Glass and Composites* (John Wiley & Sons, Inc., New York, Chichester, Brisbane, Toronto, Singapore 1986) Library of Congress Catalog Card Number: 85-17871, p. 143

5.9 S. Rekhson, J.-P. Ducroux: "Treatment of Non-Linearity in Structural Relaxation in Glass", in: L. David Pye, W.C. LaCourse, H.J. Stevens: *The Physics of Non-Crystallines Solids* (Taylor & Francis, London, Washington DC 1992) pp. 315–320

5.10 G.W. Scherer: *Relaxation in Glass and Composites* (John Wiley & Sons, Inc., New York, Chichester, Brisbane, Toronto, Singapore 1986) Library of Congress Catalog Card Number: 85-17871, p. 157

5.11 A. Makishima, J.D. Mackenzie: "Calculation of Thermal Expansion Coefficient of Glasses", Journal of Non-Crystalline Solids **22**, 305–313 (1976)

5.12 A. Winkelmann: "Über die spezifischen Wärmen verschieden zusammengesetzter Gläser", Ann. Physik **49**, 401–420 (1893), quoted after: H. Scholze: *Glas* (Springer, Berlin, Heidelberg, New York, London, Paris, Tokyo 1988) p. 332

5.13 W. Vogel: *Glaschemie* (Springer, Berlin, Heidelberg, New York, London, Paris, Tokyo 1992) pp. 49–65

5.14 C. Bergmann, L. Schäfer: *Lehrbuch der Experimentalphysik*, Vol. IV, Part 1 (Walter de Gruyter, Berlin, New York 1981) pp. 882–887

5.15 C. Kittel: *Einführung in die Festkörperphysik* (R. Oldenbourg, München, Wien 1968) pp. 231–243

5.16 W. Geffcken: "Zur Fortleitung der Wärme in Glas bei hohen Temperaturen", Part 1, Glastechnische Berichte **25** (12) 392–396 (1952)

5.17 R. Siegel, J.R. Howell: *Thermal Radiation Heat Transfer* (Hemisphere Publishing Corporation, Washington, Philadelphia, London 1992) p. 22

5.18 G.K. Chui, R. Gardon: "Interaction of Radiation and Conduction in Glass", J. Am. Ceram. Soc. **52** (10), 548–553 (1969)

5.19 U. Fotheringham, F.-T. Lentes: Die aktive Wärmeleitfähigkeit heißen Glases, Proceedings of the "68. Glastechnische Tagung", edit. by "Deutsche Glastechnische Gesellschaft" (Frankfurt) pp. 21–27

5.20 R. Siegel, J.R. Howell: *Thermal Radiation Heat Transfer* (Hemisphere Publishing Corporation, Washington, Philadelphia, London 1992) p. 21

5.21 BS 476, Part 8, (1972) ("BS": British Standard), Fire Tests on Building Materials and Structures, Part 8, Tests Methods and criteria for the fire resistance of elements of building construction, edit. by the British Standards Institution

5.22 R.E. Field, R. Viskanta: "Measurement and Prediction of the Dynamic Temperature Distribution in Soda-Lime Glass Plates", J. Am. Ceram. Soc. **73**, 2047–2053 (1990)

5.23 C. Weimantel, C. Hamann: *Grundlagen der Festkörperphysik* (VEB Deutscher Verlag der Wissenschaften, Berlin 1979), licensed edition by (Springer, Berlin, Heidelberg, New York) p. 321

5.24 G.W. Scherer: *Relaxation in Glass and Composites* (John Wiley & Sons, Inc., New York, Chichester, Brisbane, Toronto, Singapore 1986) Library of Congress Catalog Card Number: 85-17871, p. 120

5.25 DIN 52312, Part 1, September 1983. ("DIN": German Industrial Norm), Messung der Viskosität, edit. by "Deutsches Institut für Normung" (Berlin)

5.26 R.W. Douglas: J. Soc. Glass Technol. **33**, 108 (1949), quoted after: J.M. Stevels: "The Structure and the Physical Properties of Glass", in: S. Flügge: *Handbuch der Physik*, Vol XIII, *Thermodynamik der Flüssigkeiten und Festkörper* (Springer, Berlin, Heidelberg, New York 1962) p. 584

5.27 L. Pauling: *General Chemistry* (W.H. Freeman and Company, San Francisco 1970), licensed edition by (Verlag Chemie, Weinheim 1973), (German Title: *Grundlagen der Chemie*) Library of Congress Catalog Card Number: 72-89596, pp. 795–798

5.28 A. Feltz: Amorphe und glasartige anorganische Festkörper (Akademie-Verlag, Berlin 1983) pp. 41–45

5.29 G.W. Scherer: *Relaxation in Glass and Composites* (John Wiley & Sons, Inc., New York, Chichester, Brisbane, Toronto, Singapore 1986) Library of Congress Catalog Card Number: 85-17871, pp. 9–10

5.30 K.L. Ngai, R.W. Rendell: "Fundamental Issues Confronting Models of Non-Linear Structural Relaxation", in: L. David Pye, W.C. LaCourse, H.J. Stevens: *The Physics of Non-Crystallines Solids* (Taylor & Francis London, Washington DC 1992) pp. 309–314

6. Chemical Durability of Optical Glass: Testing Methods – Basic Information, Comparison of Methods Applied by Different Manufacturers and International Testing Standards

Wilfried Heimerl, Arnd Peters

6.1 Introduction

Optical glasses have primarily been developed to obtain certain desired optical characteristics. In order to achieve the optical performance demanded in practice, chemical compositions are required which sometimes result in glasses with low chemical resistance. In view of the diversity of chemical components it is not possible to specify any *single* universal testing procedure which takes into consideration all possible changes produced by chemical influences. The different chemical reactions which may occur between a glass surface and its surrounding environment have been described and explained in the literature [6.1–7].

The main effects which can be observed as a consequence of slight surface deterioration are dimming (white spots or a cloudy film), staining (interference colours), latent scratching "developed" to visible scratches, or even heavier deterioration like holes, crusts, etc. It is well-known that tests of the chemical resistance of the basic glass mass (bulk glass) commonly provide information which in the case of optical glass is of little practical value. The behaviour of the outermost layers is of much greater interest because damage due to slight corrosion of these surface layers makes the glass body useless for the envisaged optical purpose. It is thus highly important for the user of optical glass in the field of manufacturing of optical instruments to obtain the most comprehensive information that is possible concerning the expected chemical behaviour of the surfaces of optical glasses. Testing methods serve to support him.

The extensive group of optical glasses covers an extremely wide range from chemically highly inert glasses to chemically very easily attackable glasses. The various testing procedures, which are described later, yield information concerning the differing chemical behaviour.

Since there exists no universal method covering the examination of all possible influences, each actual testing method aims to determine the behaviour of the glass with respect to certain specific influences. Due to the reasons mentioned above it is desirable that these testing procedures should concentrate on testing the surface.

In practice, however, there is a great problem: Analytical surface analysis methods need highly complicated, expensive apparatus, such as ESCA, AES, or SIMS, for instance, and also highly experienced analytical experts. Of course, these methods are of great importance to analyze a certain, given damage [6.8–14], but because of their complicity they are not generally to be applied as testing methods. *Bach* and *Baucke* described results of surface investigations on glass electrodes by an ion sputtering method [6.15, 16] which of course is also applicable on optical glasses [6.9, 10].

To escape from this dilemma, the International Organization for Standardization ISO agreed in Technical Committee 172, Subcommittee 3 and its Working Group 1 "Raw Optical Glass" on a compromise: after discussing which information is highly needed, it was decided that the working group should establish International Standards which are simple, quick and inexpensive. They should give usable information about the principal chemical durability, especially for the aspects of cleaning optical glasses after polishing and – recently – for surface reactions under very small masses of water ("water in deficiency", like humidity and climate conditions in general). Additionally, as qualitative information, the results contain observations of possible changes on the glass surface after the test which can be seen with the naked eye. To understand what is going on chemically on the glass surface, some explanations may be helpful.

6.2 General Chemical Reactions with Water (Neutral Aqueous Solutions) on the Glass Surface

In all cases in which glass surfaces react chemically, water is the basic condition. Water may be present in excess as a liquid [6.17], or in deficiency as humidity, droplets, or suchlike [6.18, 19]

When aqueous solutions in excess react with a surface of a low resistance glass, two mechanisms take place: One in dissolving the glass network (dissolution of the glass) and the other presenting an interdiffusion reaction (ion exchange with special elements in the surface layer). The first reaction is determined by the dissolution velocity, the second by the diffusion velocity [6.20]. For the visible results of these reactions it is essential which velocity is the quickest. When it is the dissolution velocity the glass surface is quickly removed. This process is often called "chemical polishing", and takes place especially with alkaline solutions at higher temperatures, but also to a certain degree with acids. The resulting surfaces are mostly blank without visible changes. Because there is usually an overlapping of the two mechanisms, however, it needs great experience to render the new surface really blank. If the diffusion velocity is dominant, which is often the case in neutral aqueous solutions, water molecules diffuse into the surface and can react with cations. In optical glasses these can be very numerous, and many of them can form insoluble or poorly soluble new components like hydroxides or, in the presence of air, carbonates [6.21, 22]. *Bach* et al. reported about analysis and chemical properties of hydrolyzed surface layers [6.9, 23, 24]. Reactions with water in deficiency are often summarized as environmental influences. Additionally, traces of acids, like sulphur or carbon dioxide from the atmosphere, acids in human sweat, or of other compounds, can make the attack more acidic and, therefore, more severe.

After local ion exchange reactions the reaction products often cannot migrate out of the surface even if they are soluble (because of the deficiency of water) and, thus, they cannot be taken away. So new compounds can form in the surface layer. They first appear in the interference colours from blueish-green-yellow-brown as a function of thickness, then turn to cloudy hazes (whitish) and, in the worst cases, to thicker white crusts. So trace amounts of water can be most dangerous because they are often overlooked, but they can easily damage unprotected glass surfaces. Because these ion-exchange reactions often occur during storage of unprotected glass, the producers of optical apparatus mostly cover the surfaces of sensitive glasses as soon as possible after the polishing process with thin protective films [6.2, 25, 26] or with an optical coating which is mostly at the same time highly resistant and protects the glass [6.2].

6.3 Processing and Chemical Reactions

Processing comprises all mechanical treatments such as lapping, grinding, polishing, etc. A liquid phase must always be present during such treatments, but it is desirable to work as far as possible in the absence of water, especially during the polishing stage.

The polishing agents, finely ground oxides, such as cerium or zirconium oxide, are preferably used with plastics polish, often even in a slurry of water. During the polishing process the slurry is progressively enriched by the fine abrasive particles, which can easily be leached by water. Especially the alkali-ions from usual glasses are extracted very quickly and increase the original pH-value of a neutral slurry (pH of about 7) intensively, as said before. Depending on the glass type, pH-values of even 10 or 11 are reached more or less quickly and the first aggressive attack by water is overwhelmed by the different and much more severe alkaline attack mechanism.

The pH-value can readily be measured with glass electrodes or litmus paper and can serve as a guide for the condition of the solution. Unproblematic processing of sensitive optical glasses generally demands a pH-value in the vicinity of the neutral point, i.e. an actual value lying between 6 and 8. The pH-value should be checked regularly and kept within these limits in order to ensure optimum (safe) processing conditions. The frequency of the pH-check measurements will depend on the actual conditions such as type of abrasive, nature of the liquid phase, glass type and specific experience. Measurements may be necessary several times daily.

Alkaline attack commences at a very significant rate when the pH-value of a polishing medium has risen to 9. Major damage may then be caused. For many optical glasses, a pH-value of 8 already represents the safe limit which should not be exceeded. Because of these dangers the slurry is often not made with water, but with water-free or nearly water-free liquids like petroleum or ethanol.

There is an exception however: glasses with a phosphate basis form acids with water after dissolution, so that the pH-value can decrease to values lower than 5. Hence in the polishing process the reactions described in Sect. 6.2 can take place, and what actually happens depends on the reaction mechanism. For a more detailed description of the polishing of optical glasses see Chap. 7. To avoid problems with very sensitive glasses, the final polishing step should be carried out in the absence of water (dry polishing). Thereafter, the processed glasses should not be touched again under any circumstances with naked fingers.

When the reactive amount of water is small, but not too small, the pH-value of this originally neutral water (e.g., droplet) may increase because of the released, quickly soluble and easily migrating alkali-ions, and the attack mechanism changes from the weak hydrolytical attack to the more severe strong alkali attack.

6.4 Cleaning and Its Possible Effects

The polished optical glasses must be cleaned very carefully, and, if possible, at once. As the danger of chemical damage by water is well-known by experienced customers, they have developed automatic cleaning apparatus, in which – in circular processing – the completely polished glasses are in contact with cleaning solutions only for a short time. These are mostly aqueous, phosphate containing detergent solutions, which are alkaline. Here a quick dissolution of the upper surface takes place, supported by a slightly increased temperature of 50 °C. With this dissolution of the upper surface, adherent polish particles, dust and other impurities are "swept off", so that a new clean surface is obtained. These short washing centres are followed by water-free solutions, mostly 2-propanol [6.17, 27]. The cleaning process is supported by ultrasonic equipment, which is a good means of cleaning even a smaller number of polished items [6.27]. It is offered, for instance, by Roag Fabrikationstechnik, Switzerland. In special cases it may be recommendable to study the chemical reactions of a special glass by surface analysis after polishing – also taking into consideration the type of polishing and cleaning agent and storage conditions in order to derive the most suitable processing parameters.

6.5 Testing Methods

In order to predict the chemical behaviour of an optical glass, different testing methods have been developed. Due to the variety of optical glasses and the great number of parameters, which influence the chemical attack, a single test cannot

Table 6.1. List of testing methods for the chemical durability of optical glasses according to recent catalogue data of glass producers

Code		Meaning
Corning, France/USA		
– Acids	=	Acid Resistance (predecessor-method of ISO 8424)
Hoya, Japan		
– D_{W}	=	water durability by the powdered glass method (resistivity to dimming)
– D_{A}	=	acid durability by the powdered glass method
– T_{blue}	=	staining resistivity by the surface method
– D_{NaOH}	=	latent scratch resistivity
– D_{STPP}	=	latent scratch resistivity to polymerized phosphoric ions
– D_{O}	=	intrinsic chemical durability to water
Ohara, Japan		
– $\mathrm{RW}_{(\mathrm{p})}$	=	Water Resistance (Powder)
– $\mathrm{RA}_{(\mathrm{p})}$	=	Acid Resistance (ISO 8424)
– $\mathrm{W}_{(\mathrm{s})}$	=	Weathering Resistance (Surface)
– PR	=	Phosphate Resistance (ISO 9689)
Pilkington, Great Britain		
– Clim. R.	=	Climate Resistance
– Acid R.	=	Acid Resistance (ISO 8424)
– Alk. R.	=	Alkali Resistance (ISO DIS 10 629)
Schott Glaswerke, Germany		
– CR	=	Climate Resistance
– FR	=	Resistance to Staining
– SR	=	Acid Resistance (ISO 8424)
– AR	=	Alkali Resistance (ISO DIS 10 629)

cover the whole range of cases. Even for a given optical glass several different tests have normally to be carried out to get an answer to its reaction to chemical attack.

In the past, testing methods have been developed in different countries individually, Table 6.1. Depending on the pH-value of the attacking medium these tests can roughly be divided into three categories:

1. Acid resistance tests
2. Alkali resistance tests including detergent tests
3. Hydrolytic resistance tests (concerning climate, weather, water).

In most of these procedures solid glass plates are used; only seldom does one uses glass powder. They have been individually created and are therefore not comparable with each other, even if the typical type of test seems to be the same. Closer information is given in Subsect. 6.5.2.

To make the data in the catalogues from the different producers of optical glasses comparable, standard procedures have been requested that are internationally agreed upon Table 6.2. This task was accepted by WG1 in ISO/TC 172/SC3.

Table 6.2. List of testing methods used by glass producers arranged according to the chemical attack

1. Acid resistance	
SR (ISO 8424)	Corning France/USA
	Schott Glaswerke
	Pilkington
	Ohara
DA	Hoya
FR	Schott Glaswerke
$RA_{(p)}$	Ohara
2. Alkali resistance, incl. Phosphate resistance	
AR (ISO 10 629)	Schott Glaswerke AR
	Pilkington
D_{NaOH}	Hoya
D_{STPP}	Hoya
PR (ISO 9689)	Ohara
3. Hydrolytic resistance	
T_{blue}	Hoya
D_O	Hoya
D_W	Hoya
$RW_{(p)}$	Ohara
$W_{(s)}$	Ohara
CR	Pilkington
	Schott Glaswerke

6.5.1 International Standard Procedures

At present, there are four International Standards established or on trial; one as a "pure" alkaline test, another applying an alkaline phosphate containing solution, one acid test in which two aqueous solutions of different pH-values are used, and finally – on trial – the test with water in deficiency as a climate test. The principle in all cases is to use solid polished samples. The principles of the first three standards (Tables 6.3–6) are the use of liquid in excess and an identical evaluation following a classification according to the loss in weight and information about surface changes after the test. Classification is carried out by a calculation of the time needed to corrode a layer being 0.1 μm thick (calculated from the loss in weight and the glass density).

Phosphate-Containing Alkali Test (ISO 9689/1990, "PR-Test") "Raw optical glass – resistance to attack by aqueous alkaline phosphate-containing detergent solutions at 50 °C – testing and classification"

This International Standard was published in December 1990. The attacking medium is a solution of penta-sodium-triphosphate ($Na_5P_3O_{10}$) in a concentration of 0.01 mol/l. As the liquid has an alkaline reaction, this can be regarded as

Table 6.3. Testing of phosphate resistance

Abbreviation:	PR
Status of standardization:	ISO 9689 (Dec.1990)
Title:	Raw optical glass – Testing of the resistance to attack by aqueous alkaline phosphate containing solutions at 50 °C Testing method and classification
Attacking medium:	aqueous solution of penta sodium triphosphate ($Na_5P_3O_{10}$), $c = 0.01$ mol/l), pH: 10, in excess (2 litres)
Temperature:	50 °C = 0.2 °C
Duration of the test:	0.25–16 h
Samples:	glass plates polished on all sides 30 × 30 × 2 mm, edges bevelled
Further conditions:	– vessel of silver or stainless steel – stirring (100 rpm) – samples fastened with a platinum wire are hanging in the solution
Measuring process:	– weight of the sample before and after treatment – visible surface changes
Classification:	according to the time which is necessary to etch a surface layer of a 0.1 μm (derived from the weight loss per cm^2 and the density)
Decimal:	grade of visible changes on the surface

the main alkali-test because it gives information about the chemical resistance of optical glasses to those cleaning solutions which are mostly applied by the consumers in the optical industry.

The features of this test are given in Table 6.3, the classification is shown in Table 6.5.

“Pure” Alkali Test (ISO DIS 10629, “AR-Test”)
“Raw optical glass – resistance to attack by aqueous alkaline solutions at 50 °C – testing and classification”

This test is very similar to the PR-test, but the attacking agent is different: a sodium hydroxide solution, $c\,(\mathrm{NaOH}) = 0.01$ mol/l. The alkaline attack is also mostly covered by the PR-test, but some special glasses are attacked by the phosphate ions to a significantly higher degree, e.g. the lathanum-containing glasses. Table 6.4 gives details of this test. In Table 6.5 the classification is shown. The third test uses two different acids (Table 6.6).

Table 6.4. Testing of alkaline resistance ISO DIS 10 629

Abbreviation:	AR
Status of standardization:	ISO DIS 10 629
Title:	Raw optical glass – Testing of the resistance to attack by aqueous alkaline solutions at 50 °C Testing method and classification
Attacking medium:	aqueous sodium hydroxide solution, $c_{NaOH} = 0.01$ mol/l, pH = 12 in excess (2 litres)
Temperature:	50 °C ± 0.2 °C
Duration of the test:	0.25–16 h
Samples:	glass plates polished on all sides, 30 × 30 × 2 mm, edges bevelled
Further conditions:	– vessel of silver or stainless steel – stirring (100 rpm) – samples fastened with a platinum wire are hanging in the solution
Measuring process:	– weight of the sample before and after treatment – visible surface changes
Classification:	according to the time which is necessary to etch a surface layer of a 0.1 μm (derived from the weight loss per cm^2 and the density)
Decimal:	grade of visible changes on the surface

Acid Test (ISO 8424/1987, "SR-Test")
"Raw optical glass – resistance to attack by aqueous acidic solutions at 25 °C – testing and classification"

Two reagents are used:

- the stronger nitric acid ($c_{HNO_3} = 0.5$ mol/l) for the more resistant glass types, and
- a weaker buffer solution of sodium acetate (pH = 4.6) for the very low resistant glasses.

More details about this test are given in Table 6.6; the classification is shown in Table 6.5.

For the fourth test, standardization is on the way:

Climate Test ("CR-Test", ISO working draft introduced in 1993)

At present, climate resistance tests for optical glasses are carried out by Ohara, Pilkington and Schott Glaswerke. The conditions of these tests are slightly different; they are summarized in Table 6.7.

Ohara applies a test with constant temperature and constant exposure time. At Pilkington and Schott Glaswerke, temperatures alternating between 40 and

Table 6.5. Classification of phosphate, alkali- and acid-resistance

a) Phosphate resistance (ISO 9686) and alkali resistance (ISO DIS 10 629)

Classes of phosphate and alkali resistance	Time in hours needed to etch a depth of 0.1 μm
1	> 4
2	1–4
3	0.25–1
4	< 0.25

b) Acid resistance (ISO 8424)

Acid resistance class	pH	Time in hours needed to etch a depth of 0.1 μm
1	0.3	> 100
2	0.3	10–100
3	0.3	1–10
4	0.3	0.1–1
5	0.3	< 0.1
	4.6	> 10
51	4.6	1–10
52	4.6	0.25–1
53	4.6	< 0.1

c) Classification according to the visible surface changes (for all three testing methods given as decimal figure)

First decimal	Optical variation
.0	no visible changes
.1	clear, but irregular surface (wavy, pockmarked, pitted)
.2	staining and/or interference colours (slight selective leaching)
.3	tenacious thin whitish layer (stronger selective leaching, a cloudy/hazy/dullish surface)
.4	loosely adhering thick layer, such as insoluble friable surface deposit (may be a cracked and/or peelable surface crust, or cracked surface, strong attack)

50 °C are used. Exposure times vary between 30 and 180 hours at Pilkington, whereas at Schott Glaswerke the exposure time is limited to 30 hours.

Table 6.6. Testing of acid resistance with two different acids

Abbreviation:	SR
Status of standardization:	ISO 8424 (Sept. 87)
Title:	Raw optical glass – Testing of the resistance to attack by aqueous acidic solutions at 25 °C and classification
Attacking medium:	a) nitric acid, c = 0.5 mol/l, pH = 0.3 (for high resistance glass types) b) sodium acetate buffer, pH = 4.6 (for glasses with lower acid resistance) in each case in excess (2 litres)
Temperature:	25 °C ± 0.2 °C
Duration of the test:	0.2–100 h
Samples:	glass plates polished on all sides, 30 × 30 × 2 mm, edges bevelled
Further conditions:	– vessel borosilicate glass – stirring (100 rpm) – samples fastened with a platinum wire are hanging in the solution
Measuring process:	– weight of the sample before and after treatment – visible surface changes
Classification:	according to the time which is necessary to etch a surface layer of a 0.1 μm (derived from the weight loss per cm^2 and the density)
Decimal:	grade of visible changes on the surface

The test used by Schott Glaswerke has been accepted as a basic proposal for an international standard and is currently treated by ISO/TC172/SC3/WG1.

According to this method polished glass samples are exposed in a climate chamber to a water-vapour-saturated atmosphere. Temperatures are alternated between 40 °C and 50 °C. That produces alternating condensation on the surface with very small water droplets, and again their evaporation in a one hour cycle. The result of the water attack is evaluated by a measurement of a possibly developed cloudiness by means of a sphere transmission haze meter.

6.5.2 Comparison of Methods for Testing the Chemical Durability of Optical Glasses, Using Solid Polished Glass Plates as Samples

In the following the methods of testing the chemical durability of optical glasses, used by the leading manufacturers in Germany, Japan, Great Britain and France/USA, are compared. In Tables 6.1 and 6.2 these methods are listed. Details taken from the latest catalogues of optical glass from Schott Glaswerke, Ohara, Hoya, Pilkington and Corning France, (available until autumn 1992) served as a basis for

Table 6.7. Climate Resistance Test

	Samples	Medium	Temp. °C	Exposure time h	Classification	
					Method	Classes
Ohara	polished glass plates	moist air (58% humidity	50 const.	24 (6)	microscopic inspection	4
Pilkington	polished glass plates	saturated water vapour	alternating between 40 to 50 in cycles of 45 min.	varying 30–180	measuring of percentage light scatter	4
Schott Glaswerke	polished glass plates	saturated water vapour	alternating between 40 to 50 in cycles of 60 min.	30	measuring of percentage light scatter	4

this comparison. The comparison is carried out stepwise following the different test groups (categories) related to the different types of chemical attack.

Phosphate-Containing Alkali Test (e.g. PR)

Ohara makes use of this method and has presented the data in its latest catalogue. Schott Glaswerke has also used the test (PR) and presented respective data for the first time in the 1992 edition of its catalogue. Hoya uses a test called “Latent Scratch Resistivity (DSTPP)”. The polished glass sample is immersed in a solution of 0.01 N $Na_5P_3O_{10}$, equal to the ISO PR-Test. To compare the data from this test with ISO see also the paragraph titled “Pure” Alkali Test (e.g. AR).

The currently catalogues from Corning France and Pilkington do not contain phosphate resistance data. Since the solution of the phosphate testing method is an alkaline one, the reaction of the phosphate is overlapped by the reaction of the hydroxide ions. Therefore, in most cases the phosphate test is sufficient for an examination of the glass resistance to both types of attack. There are, however, exceptions where the phosphate attack can be far more severe (see the above section on “ ‘Pure’ Alkali Test (ISO DIS 10629, ‘AR-Test’ ”, lanthanum-containing glasses).

“Pure” Alkali Test (e.g. AR)

In the new catalogue from Schott Glaswerke the alkali resistance data are already based on the final version of this test. Data from Pilkington are from an earlier test version which is close to the one described, but with a temperature of 90 °C instead of 50 °C, and a NaOH concentration of 10^{-4} mol/l instead of 10^{-2} mol/l. Most of the results received by these two methods are comparable (belong to the same class).

Hoya applies an alkali resistance test also called "Latent Scratch Resistivity (D_{NaOH})". A polished glass sample is immersed in a NaOH solution (c_{NaOH} = 0.01 mol/l) at 50 °C for 15 h under stirring. The loss in weight in mg/(cm^2 · 15 h) is used to classify the glasses. The conditions are very close to those of the ISO draft AR, but the exposure time is fixed (15 h) and the expression of the results is different (in the ISO draft, identical for all ISO procedures, the time necessary for the corrosion of a 0.1 μm thick layer is used for classification).

With this value a link can be made between the data of the two methods. Ohara and Corning France do not refer to alkali resistance in their actual catalogues for optical glass.

Acid Test (e.g. SR)

Although in practice optical glasses are rarely exposed to acid attacks this testing method gives some valuable information. It was the first one to be internationally accepted.

Tests using polished glass plates:

In ISO 8424 "Acid Test-SR" polished glass samples are used, for the conditions see the paragraph on "Acid-Test" and Tables 6.5 and 6.6. This testing method is used by Corning, Ohara, Pilkington and Schott Glaswerke; i.e. the majority of all glass manufacturers is included in this comparison.

Another test which uses weakly acidic reagents in deficiency is carried out at Schott Glaswerke. This test simulates the attack by humidity from, e.g. breath precipitation, fingerprints, etc. Reagents are sodium acetate-buffer solutions with a pH-value of 4.6 or 5.6, respectively. Contact is made between two or three droplets and a polished glass surface at a temperature of 25 °C for a period of 0.2 to 100 h. The surface is then visually controlled and compared with attacked reference plates to evaluate the formation of staining causing a specific change of colour from brown to blue (the interference layer being 0.137 μm thick). Depending on the time needed for the formation of such a layer 6 classes are formed. In class 0 the glass does not show any staining even after a duration of exposure of 100 h at a pH-value of 4.6. In class 5 staining already occurs within 12 minutes at a pH-value of 5.6 [6.28]. Attention must be paid to the fact that there are glasses which are severely attacked (by dissolution!) without staining (e.g., PSK glasses).

The methods of Hoya and Ohara are equivalent. Due to the very different conditions (especially the different temperatures and the different classifications) the results of these two methods cannot be compared with the results of the test according to ISO 8424.

Hydrolytic Resistance (to Climate, Weather, Water)

Hitherto testing methods of the hydrolytic resistance have not been standardized on an international level. Currently, there is a great variety of different methods. A new piece of work, however, has been internationally agreed upon. A proposal has been made and the work was started in spring 1993 – see the section entitled "Climate Test (New Working Draft 1993, 'CR-Test')".*

Climate resistance tests on polished plates *(water in deficiency)* are to predict the chemical behaviour of optical glasses under the influence of a moist atmosphere. Data on climate resistance have been given in the catalogues from Schott Glaswerke, Ohara and Pilkington.

The main feature of these tests is that polished glass plates are exposed to an atmosphere with high relative humidity at a temperature of 40 °C to 50 °C. There is a deficiency of the reagent (water, only in droplets).

The methods used by Pilkington and Schott Glaswerke (and now under consideration in ISO) are similar and, consequently, the results are comparable. One characteristic is the alternating condensation and evaporation of water on the plate (1-hour period).

The tests done by Ohara are carried out at a constant temperature of 50 °C. Due to the different testing conditions one cannot compare the results of this test with those of the tests applied by Pilkington and Schott Glaswerke.

Another test on hydrolytic resistance using *water in excess* has been carried out by Hoya. Hoya also applies a test of water resistance with glass plates as samples. A plate polished on both sides of diameter 43.7 mm and with a thickness of 5.5 mm is used. The plate is immersed in water at 50 °C. The water is kept clean from reaction products by a continous pumping process through layers of ion-exchange resin.

There are two ways to evaluate the results of this test:

1. Determination of the loss in weight of the sample in 10^{-3} mg/(cm^2 h). This test gives the "intrinsic chemical durability to water (D_O)". There are five classes ranging from a loss in weight of $\leq 3 \times 10^{-3}$ mg/(cm^{2}h) (class 1) to $> 15.1 \times 10^{-3}$ mg/(cm^{2}h) (class 5).

2. Determination of the reaction time that is necessary until a blueish stained layer is formed which can be observed visually. This test is called "Staining Resistivity by the Surface Method (T_{blue})". 5 classes are formed: glasses for which the formation of a stained layer needs more than 45 h belong to class 1. The glasses of class 5 only need 5 h for the very same effect.

6.5.3 Comparison of Methods for Testing the Chemical Durability of Optical Glasses Using Powdered Glass as Sample

Acid Resistance

Hoya provides data on acid durability that are based on the leaching of powdered glass with diluted nitric acid. Ohara uses a similar test in addition to the test of ISO 8424. In contrast to ISO 8424 we have in this case freshly broken glass and it is not a polished glass surface which is exposed. In Table 6.8 the conditions of

*see the part of the preceeding section entitled "Climate Test" ("CR-Test", ISO working draft introduced in 1993)

Table 6.8. Acid Resistance Test with Powdered Glass

	Acid	Conc. mol/l	Temp. °C	Duration min.	Grain size	Weight-loss	Classes
Hoya, Ohara	HNO_3	0.01	100	60	420–590	(%)	6

the test are summarized.

Hydrolytic Resistance

Tests of this type are used by Hoya and Ohara. Several grammes of freshly prepared glass grains (grain size 420–590 µm) are placed in a platinum basket and exposed to an excess of boiling water for 60 minutes. By weighing before and after the test the loss in weight is determined. Depending on the loss in weight, expressed in %, 6 classes of water resistance can be determined (class 1: loss in weight $< 0.05\,\%$, class 6: loss in weight 1.10 %). This method is equivalent to the acid resistance method.

6.6 Discussion

Steps in the direction of international standardization have been made by developing the tests for phosphate resistance (ISO 9689 PR), acid resistance (ISO 8424 SR) and alkali resistance (ISO DIS 10629 AR). These procedures are valid International Standards or (ISO DIS) in the stage of a Draft International Standard. They have already been introduced in many catalogues of the large producers of optical glasses.

The common feature of the three testing methods is the excess of reagent. Therefore, they are useful to get information on the behaviour of optical glasses coming into contact with solutions, e.g. with detergent solutions after polishing during cleaning procedures (especially phosphate resistance, alkali resistance). Another similarity of these tests is the use of polished glass discs as samples, what is also true for the fourth method, the climate resistance test of which the first working draft is under discussion. This is reasonable because normally optical glasses with polished surfaces are used in technology. Therefore, the corrosion resistance of such surfaces and not that of glass grains is of interest. This means, that tests with glass grains do not provide sufficient information on the final behaviour of the real surface, but on the general chemical durability. This information can be important for scientific investigations if one has to know which ions are released in which amount.

Beside the attack of reagents in excess, it is important to know the response of glasses towards reagents in deficiency (moisture from the atmosphere, fingerprints, etc.) In this field there is still a lack of internationally standardized testing methods. This work has just been started with the climate test CR regarding the alternating condensation and evaporation of water on a polished sample in the temperature range from 40 °C to 50 °C (already introduced in two of the examined catalogues).

Another type of this test with reagent in deficiency should simulate the attack of traces of weakly acidic moisture, caused by fingerprints, breath precipitation, etc. Such a test is applied by Schott Glaswerke (see the paragraph titled "Acid Test (e.g., SR)".

Finally, it can be concluded that at present still some of the data on the chemical resistance of optical glasses given in the catalogues of different manufacturers cannot be compared due to the application of different testing methods. But important steps have been taken to improve the situation by an international standardization of such methods. These efforts, however, have to be continued in order to achieve a complete comparability.

References

6.1 R. Wendler, H. Eiche, C. Hofmann: "Empfindlichkeit optischer Gläser gegenüber chemischer Einwirkung, Feingerätetechnik **24,** 436-439 (1975)

6.2 T.S. Izumitani: *Optical Glass* (American Institute of Physics, Translation Series, New York 1986) pp. 42 ff.

6.3 R.H. Doremus: "Reactions of Glasses with aqueous and non-aqueous environments", Mat. Res. Symp. Proc. **125**, 177-188 (1988)

6.4 L.L. Hench: "Corrosion of silicate glasses: an overview", Mat. Res. Symp. Proc. **125**, 189-200 (1988)

6.5 L.M. Cook: "Chemical processes in glass polishing", J. Non-Cryst. Solids **120**, 152-171 (1990)

6.6 W. Wagner, F. Rauch, H. Bach: "Chemical properties of hydrolized surface layers on SiO_2-BaO-B_2O_3 glass", Glastechn. Ber. **63**, 351-362 (1990)

6.7 G. Gliemeroth and A. Peters: "Optical Glass, the pragmatic view of chemical durability", J. Non-Cryst.Sol. **38-39,** 625-630 (1980)

6.8 K. Neumann: "Ellipsometrische Bestimmung von Oberflächenschichten auf polierten optischen Gläsern", Optica Acta **30**, 967-980 (1983)

6.9 H. Bach: "Analysis of surface layers" in *Optical Surface Technology*, ed. by H. Walter, SPIE Vol. **381**, 113-128 (1983)

6.10 H. Bach: "Problems with the analysis of glass and glass ceramic surfaces and coatings", Fresenius Z. Anal. Chem. **333**, 373-382 (1989)

6.11 D. Sprenger, H. Bach, W. Meisel, P. Gütlich: "XPS study of leached glass surfaces", J. Non-Cryst. Sol. **126**, 111-129 (1990)

6.12 F.W. Oertmann: "Zur Optimierung des Polierprozesses optischer Gläser", Optica Acta **30**, 243-251 (1983)

6.13 J. Kross and H. Gerloff: “Optical Investigations on optical surfaces”, SPIE Vol. **381**, 138-149 (1983)

6.14 I.F. Stowers: “Advances in cleaning metal and glass surfaces to micron level cleanliness”, (California Univ., Livermore, USA, Lawrence Livermore Lab. 1977)

6.15 H. Bach, F.G.K. Baucke: “Measurement of ion concentration profiles in surface layers of leached (“swollen”) glass electrode membranes by means of luminiscence excited by ion sputtering”, Electroch. Acta **16**, 1311-1319 (1971)

6.16 F.G.K. Baucke: “The Glass Electrode - Applied Electrochemistry of Glass Surfaces”, J. Non-Cryst. Sol. **73**, 215-231 (1985)

6.17 A. Peters: “Chemische Korrosion ausgewählter optischer Gläser durch handelsübliche Reinigungsmittel”, Glastech. Ber. **57**, 21-29 (1984)

6.18 U.S. Milit. Standard 810 C: *Humidity Resistance of Optical glasses*

6.19 H.V. Walters and P.B. Adams: Effects of humidity on the weathering of glass, J. Non-Cryst. Sol. **19**, 183-199 (1975)

6.20 J. Schäfer, H.A. Schaeffer: “Leaching of alkali silicate glasses - formation of hydrated layers, surface - and diffusional-controlled kinetics”, Rivista della Staz. Spez. Vetro **5**, 79-82 (1984)

6.21 H. Bach, F.G.K. Baucke: “Investigations of reactions between glasses and gaseous phases by means of photon emission induced during ion beam etching”, Phys. Chem. Glasses **15**, 123-129 (1974)

6.22 H.H. Dunken: “Glass Surfaces” in *Treatise on Materials Science and Technology*, Vol. 22, Glass III, ed. by R.H. Doremus, M. Tomozawa (Academic Press, New York 1982) pp. 1-74

6.23 H. Bach: “The use of surface analysis to optimize the production parameters of the surface of glasses and glass ceramics”, Glass Technology **30**, 75-82 (1989)

6.24 H. Bach, K. Großkopf, P. March, F. Rauch: “In-depth analysis of elements and properties of hydrated subsurface layers on optical surfaces of a SiO_2-BaO-B_2O_3 glass with SIMS, IBSCA, RBS, and NRA, Part 1: Experimental procedures and results”, Glastech. Ber. **60**, 21-46 (1987); Part 2: Discussion of results, ibid. 33-46 (1987)

6.25 H. Pforte: *Feinoptiker, Teil 1 - Werkstoffe für Fein- und Brillenoptik* (VEB Verlag Technik Berlin 1985)

6.26 J.M. Bennett, Lars Mattson, M.P. Keane, L. Karlsson: “Test of strip coating materials for protecting optics”, Appl. Optics **28**, 1018-1026 (1989)

6.27 K.H. Günther and H. Enssle: “Ultrasonic precision cleaning of optical components”, Optical World, 8-14 (Nov./Dec. 1985)

6.28 “The Chemical Resistance of Optical Glasses in the Acid-Proof Class 5c”, in *Schott Technical Information No. 5* (12/72)

7. Processing (Grinding and Polishing)

Knut Holger Fiedler

7.1 Introduction

7.1.1 Grinding and Polishing

Firstly, the nature and the goal of the three traditional manufacturing steps of optics, and the techniques applied there, should be pointed out:

The initial rough shape of the optics is provided by *generating.* The next step is a *fine grinding* operation which provides a surface shape as close as possible to the final geometry. Additionally, the roughness and sub-surface damage left from generating have to be reduced. Finally, the optical surface is obtained by *polishing.* Sometimes, there is an additional polishing step called figuring which means corrective polishing to achieve the final surface shape.

The manufacturing steps introduced above can be performed by means of different material removal techniques. Generating is usually done by fixed abrasive grinding (Sect. 7.2.3). Until the 1970s, fine grinding was typically performed by loose abrasive grinding, sometimes also called lapping (Sect. 7.2.2). Then, especially in high volume production, the fixed abrasive grinding tools and machines that were fairly expensive became common, too. Loose abrasive grinding is similar to the standard polishing process (Sect. 7.3.1). Nevertheless, there is an essential difference: loose abrasive grinding is dominated by mechanical material removal effects whereas polishing is characterized by a combination of mechanical and chemical processes. In this context, *Brown* [7.1] created the model of the "chemical tooth".

The need for the distinction between the manufacturing steps of optics, and the techniques applied may be surprising because polished surfaces are glossy. On the other hand, the appearance of a typical ground surface is a mat finish due to the brittle fracture removal process. However, recent progress has been made in grinding as a finishing process. Now ground surfaces are obtainable whose appearance is similar to that of polished ones, although their character might be different; see the paragraph on "Shear Mode Grinding" in Sect. 7.2.3 below.

7.1.2 Material Removal Rate

In order to be able to communicate about "removal rates" a figure which is *system independent* is required. It seems to be a fairly difficult and uncertain task to reasonably compare material removal data on grinding or polishing regimes that work with quite different machine configurations, tool sizes and shapes. Nevertheless, with the *Preston* equation, see [7.1] and its constant, there is a tool which has proven quite helpful to overcome such constraints.

The theoretical basis of material removal prediction was established already in the 1920s by *Preston* [7.2], and was originally verified by loose abrasive grinding experiments. Later it has successfully been applied to polishing and even fixed abrasive grinding [7.3]. The resulting equation

$$dh/dt = c\,p\,v \tag{7.1}$$

simply states that for a given process

- the rate of removal dh/dt of a certain spot

is the product of

- the pressure p at that spot,
- the relative velocity v between lap and workpiece at that spot, and
- a process constant **c** which is a function of the workpiece material, lap material and slurry.

The Preston equation can be used to verify the proper functioning of a material removal process by monitoring the Preston constant, or in establishing advanced polishing machines [7.4]. It has to be pointed out that a high Preston constant does not automatically mean a high removal rate: when inherently different processes are compared, the pressures and velocities applied may differ tremendously. This becomes obvious when the Preston constant is compared with the *specific material removal rate* Q'_{w}, which is widely used in mechanics to characterize machining efficiency without beeing concerned about the intrinsic material removal mechanism. The *specific material removal rate* Q'_{w} is defined as

$$Q'_{\mathrm{w}} = a_{\mathrm{e}}\, v_{\mathrm{f}} \tag{7.2}$$

with the factors:

- depth of cut a_{e}; and
- feed rate v_{f}.

An example of typical Preston coefficients and removal rates is given in Table 7.1.

7.1.3 Historical Background

According to *Parks* [7.7] the tradition of the fabrication of glass optics has existed for thousands of years. In those days, the philosophers of the cultures were quite aware of the lensing effect of spherical surfaces. Shiny surfaces were obtained by

Table 7.1. Preston coefficients and specific material removal rates of different BK 7® machining processes utilizing diamond tools with grit sizes of a few microns for precision grinding and cerium oxide slurry for polishing [7.3, 5, 6 and internal results]

Machining Process	Preston coefficient c 10^{-13} Pa^{-1} internal results [7.3, 6]		Removal Rate Q'_w mm^3/mm g Miyashita [7.5]
Fixed abrasive brittle mode grinding	2–10	[internal]	$> 10^{-1}$
Loose abrasive brittle mode grinding	30–101	[7.3]	10^{-4}–10^{-1}
Fixed abrasive shear mode grinding	0.05–0.15 0.1–1.2	[internal] [7.6]	$> 10^{-4}$
Loose abrasive shear mode grinding	4–23	[7.3]	
Full lap polishing (pitch of polyurethane lap)	6–10	[internal]	$< 10^{-4}$

using loose abrasives containing corundum. The Romans were expert craftsmen in glass blowing and molding. Spectacles have been mentioned in the literature from 1300 A.D. onward.

Later, progress in the fabrication of optics was furthered by astronomers. Generally, the Dutchman *Hans Lippershey* is given credit for making the first telescope in 1609. Because polishing was the final manufacturing step for the improvement of both slope and finish it has been paid more attention than grinding. *Hooke* in his "Micrographia" (1665) and *Newton* in his "Opticks" (1704) were the first to regard pitch polishing not just as an act of burnishing but as a kind of material removal process [7.8]. About a century later, *Hershel* was the first to write of cutting a pitch lap into facets to allow the pitch to "flow" when a lap is pressed against a workpiece during lap preparation [7.7], thus assuming its shape. Around 1900, extensive polishing experiments performed by *Lord Rayleigh* brought new points of view into discussion.

7.2 Glass Grinding

7.2.1 Indentation and Scratching

Several indentation techniques have been developed to characterize the hardness of ductile materials. If hardness measurements are performed on brittle materials, care should be taken in order to avoid fracturing. It has turned out that the measured hardness of brittle materials is not proportional to the yield stress as it is with ductile materials. *Grau* [7.9] therefore proposed a new formula to develop an appropriate hardness value for glasses. This formula takes into account fracturing effects, and elastic and plastic material displacement.

Izumitani [7.10] proved that Knoop hardness values listed in glass catalogues do not correlate with the grindability of glasses. For practitioners in glass grinding it is more helpful to apply a grinding hardness value which also takes into account the intrinsic mechanical strength of the material and not just the indentation hardness. The results of grindability tests performed by *Golini* [7.11] showed reasonably good correlations with the Schott abrasion hardness of optical glasses, when the data of M-pellets were used for comparison [7.12]. Some updated grindability tests are right now under discussion to be standardized internationally (ISO/TC 172/SC 3).

In comparison to ductile materials, such as copper, the brittleness of glasses is governed by its minuscule level of plasticity and extremely low fracture energy. Therefore, fracturing occurs before the flow limit is reached, which means that the critical depth of penetration by an indentor is fairly small [7.13]. Nevertheless, *Busch* [7.14] and *Schinker* [7.15] proved that crack-free glass machining, "shaving", can be realized under a limited set of conditions. Later, *Puttick* [7.16] observed that after ductile shaving, additional spiral (shaped) swarf is delaminated from the bottom of the machined grooves. It is assumed that the the spiral swarf is due to longitudinal compressive stress along the tool path.

Takata [7.17] found that the crack initiation threshold increases with increasing water content, up to 12 wt%, in experimental glasses. He assumed that, due to a dissolution of large amounts of water in glass, deformation by plastic flow is promoted. This finding seems to contradict most crack growth experiments and grinding experience, see the paragraph on "Brittle Fracture Grinding" (Sect. 7.2.3), which prove that the speed of crack growth is increased in the presence of water.

The transition of the material removal regime from viscoplastic to brittle is extremely crucial with respect to the depth of the cut and to the introduced forces. Therefore the technical and commercial realization of diamond turning of glass as a production process is very unlikely. The limitations imposed by a critical depth of cut of some tens of nanometers [7.13] have been overcome by machining glass with multi-point diamond tools: fixed abrasive grinding has become a favourite method of gentle glass removal, see the paragraph on "Shear Mode Grinding" (Sect. 7.2.3). The overview given by *Fawcett* [7.18] is helpful in practice because it shows how the chip geometry, due to different grinding arrangements, influences the obtainable surface quality.

7.2.2 Loose Abrasive Grinding

Loose abrasive grinding is the traditional way of fine grinding of optical surfaces as a preparation for polishing. Both processes, loose abrasive grinding and polishing, usually have a "three-body wear" character, meaning that an abrasive containing liquid is supplied into the gap between workpiece and tool, which is called a lap. The force between lap and workpiece is transmitted by the abrasive grains, which are tumbling through the gap producing a random wear pattern. Sometimes combinations of lap material, grain type and process parameters are chosen in a way that fractions of the grains are temporarily embedded into the lap surface. This results in different structures of the workpiece surfaces due to the process change from loose to fixed abrasive grinding (Sect. 7.2.3).

The machines used in loose abrasive grinding are fairly simple and inexpensive. Usually the laps are either made of metals such as cast iron, aluminum, brass, or of glass and ceramics. For special purposes, even wood and composite materials (plastics with embedded metal particles) are used. The abrasive used initially with this type of grinding was alumina. Later, the harder abrasives, such as silicon carbide and boron carbide were introduced. Nowadays, advanced aluminas with special grain forms and structures have improved wear properties, and are widely used again. To obtain ultra-smooth surfaces, even expensive diamond powders are applied. The lubricants used with the loose abrasives have to be adapted to the lap properties and the workpiece material. Oil or water based systems are widely used and for special applications other organics such as polyhydric alcohols are useful.

Utilizing diamond as the microcutting tooth, material removal in the ductile regime is possible when the diamond type and the grain sizes used in the sequence of the manufacturing steps are chosen carefully. *Golini* [7.19] has recently reported on a dramatic change in the material removal regime just by a methodical variation of the lubricant. Ductile grinding of glass with loose abrasives is more of academic importance, and is applied only to a few special manufacturing tasks. *Brown* [7.3] reported on the tremendous change of the workpiece roughness and the wear coefficient depending on the mode of grinding. Proceeding from brittle fracture to shear mode removal, the surface roughness of BK 7® specimens dropped from 19 nm rms (mat surface) to 2.8 nm rms (glossy surface), and the Preston coefficient went from $30 \times 10^{-13}\,\mathrm{Pa}^{-1}$ to $3.9 \times 10^{-13}\,\mathrm{Pa}^{-1}$ at the same time. This means that the advantage of a smoother surface is countered by the disadvantage of a slower process. For comparison: polished surfaces have roughnesses below 1 nm. The Preston coefficient for both pitch and polyurethane polishing of glass is between $6 \times 10^{-13}\,\mathrm{Pa}^{-1}$ and $10 \times 10^{-13}\,\mathrm{Pa}^{-1}$ (internal results).

In loose abrasive grinding as well as in traditional full lap polishing, the localized wear of the workpiece surface is governed by the relative tool velocity and the acting pressure. A comprehensive overview of the application of these rules to surface curvature control in loose abrasive grinding has been given by *Rupp* [7.20].

7.2.3 Fixed Abrasive Grinding

Fixed abrasive grinding is a "two-body wear" process during which the orientation of the abrasive grains towards the workpiece surface is constant. The grinding fluid contains on purpose no abrasives, and has just the function of a coolant for the tool and the workpiece. The coolant has also to flush away the debris.

As in loose abrasive grinding, there are two basically different material removal processes in glass grinding with fixed abrasives, but in this case both processes are of economic importance. In all conventional grinding processes, the material near the surface is removed in the so-called *brittle fracture mode*. The material is crushed and removed as macroscopic chips. The remaining surface has a mat finish. It is characterized by lateral, or even median cracks [7.21]. The depth of the layer that is sub-surface damaged may range from about 3 µm to 100 µm. The results in shear mode or ductile mode grinding are quite different: a thin layer of material is removed without any fracturing. There is no significant sub-surface damage. For cosmetic reasons, 1 µm to 3 µm have to be removed by polishing before surfaces ground in shear mode meet tight imperfection tolerances [7.22].

Brittle Fracture Grinding

Before 1940, the generating of glass was performed by loose abrasive grinding. During World War II it started to be replaced by fixed abrasive grinding. Afterwards, in Europe, progress in the manufacturing of optics was driven by the booming post-war market for binoculars, photo lenses and eyeglasses. Therefore, from the middle of the 1950s onward, generating by fixed abrasive grinding, followed by the labour intensive loose abrasive grinding, was reorganized and replaced by multi-step, fixed abrasive grinding. At Carl Zeiss, new machines had to be built especially for the fine grinding of eyeglasses. At that time, no adequate machines were commercially available.

The shape of the grinding tools applied in fine grinding depends firstly on the shapes which have to be ground, and secondly on the machine type used. Spherical surfaces, for example, can be machined either with cup wheel tools [7.23], crown tools or semi-spherical tools [7.24] mounted with grinding pellets, or grinding disks. The grains used in the bond-filler-abrasive matrix are basically the same as those applied in loose abrasive grinding. A very common tool is the combination of a bronze bond with diamonds. For special purposes vitrified bonds or plastic bonds are used.

Over the years, diamond tool manufacturers have developed combinations of matrix properties and diamond abrasives to suit the variety of machining situations. Beside natural diamond, quite a number of synthetic, "cool grinding" diamond abrasives have been introduced [7.25] which turned out to be a prerequisite for special machining tasks, such as shear mode grinding; see the paragraph on "Shear Mode Grinding" below.

In the 1950s, petroleum was used very successfully as a grinding fluid. It offered the advantage of providing low and constant grinding forces because it helped to maintain the so-called self-sharpening of the tool. Petroleum was primarily discarded because of its constant danger of combustion. Nowadays coolants consist of a small quantitiy of a bio-degradable additive and of about 98 % water. These grinding fluids firstly have the disadvantage that they do not allow grinding to occur under a continuous self sharpening condition, which requires intermittent dressing of the tool. Secondly, during the grinding action, a glass surface is subject to enhanced stress. In such an environment, acqueous coolants lower the strength of the silicon-oxygen bond of silicate tetrahedrons by a factor up to twenty [7.26]. The resulting dissociative chemisorption may raise the crack growth speed by several orders of magnitude, up to a factor of 10^{14}! Therefore, the fracturing of the glass is encouraged which means that the average chip size will increase [7.27], and the surfaces obtained will be coarser than those similarly ground with oily lubricants.

Recently, a Japanese group has revitalized an old dressing idea in order to overcome the poor self-dressing properties of metal bonds, especially cast iron bonds, in the presense of aqueous lubricants. *Ohmori* [7.28] called the method electrolytic in Process dressing (ELID) in ductile mode "mirror grinding". With ELID, a pulsed voltage is applied across a sub-mm gap between an electrode and the active face of the grinding tool. This voltage source is utilized throughout the grinding process. The conductive grinding tool bond is then oxidized by electrolysis. Eventually, when an insulating oxide layer has been built up, the current between the electrode and the tool decreases, whereas the voltage increases. The active grinding grains of the outermost layer become loose in the weaker oxide layer and preferentially break out during the grinding action. The conductive bond is once again exposed and the process continues, in a steady state manner with continuous selective erosion of the bond. The method works for all grinding and trueing operations. It is especially successful when fine grain tools, with mesh sizes up to #120.000, or 0.1 μm, respectively, are used.

Some years ago, a new finishing process for brittle and ductile materials in high volume productions was introduced by the application of diamond lapping films. Either polymer films [7.29], or cloth [7.30] are used as carriers for the abrasive grains. The commercialized grit sizes span from 0.25 to 30 μm. The stock removal rate and its reproducibility is higher than in conventional loose abrasive grinding when an appropriate grinding foil is used with grits of a narrow size distribution.

Shear Mode Grinding

As mentioned above Sect. (7.2.1), an exact control of the depth of the cut of each diamond grain ploughing through the glass surface is the prerequisite for shear mode grinding. Therefore, it is obvious that an extremely high performance is needed in machine construction, particularly with respect to a precise but smooth feed control, low vibration level, high dynamic stiffness and high thermal stability. *Yoshioka* [7.31] was among the first to report on the efforts to stabilize "micro-grinding". Heavy machine elements and special carriages were introduced in order to absorb vibration energy and to raise the dynamic stiffness to 40 N/μm. Later, *Fiedler* [7.32] described an aspheric generator capable of 100 % "ductile mode grinding" with a loop stiffness of above 18 N/μm and 100 N/μm, depending on the direction. *Namba* [7.33] succeeded in stabilizing his grinding process by using a spindle rotor made from zero-expansion glass ceramics. Recently, as a rule of thumb, *McKeown* et al. have asked for a static "loop-stiffness" between tool and workpiece of at least 300 N/μm in order to establish safe conditions for shear mode grinding [7.34].

There are still several reasons for choosing a material whose results are similar to those of the single point diamond turning of a ductile material where sub-μm figure accuracy and nm-level rms-microroughness are typically reached. The most important one is that the shiny surfaces ground in ductile mode can easily be tested by interferometry. Time consuming and less stable processes – such as polishing in the production of aspherics – can be avoided [7.32]. It has also been found that the modulus of rupture of a workpiece can be increased when its surface is ground in a ductile mode instead of having a standard fine grind, or even instead of being polished.

One disadvantage of grinding glass in the ductile mode is the low material removal rate. This type of machining efficiency can be expressed, in a normalized manner, independent of the machining details, by means of the *Preston* coefficient [7.2]. An increase of the coefficient and an essential change of the obtainable surface quality may be achieved either by chemical modifications of the lubricant [7.19], or by adding other removal processes additional to ductile grinding, loose abrasive grinding of the grinding wheel bond and polishing of the glass surface at the same time. The grinding fluid can be modified by adding a loose abrasive, such as submicron size ceria. In the case of machining BK 7®, the Preston coefficient increases by an order of magnitude (from $10^{-14}\,Pa^{-1}$ to $12{\times}10^{-14}\,Pa^{-1}$) [7.6].

7.3 Glass Polishing

The material removal process in glass polishing is described either by wear theory, flow theory, chemical theory or combinations of these. *Izumitani* [7.35] proved by numerous experiments that the polishing rate of glasses depends on the chemical durability of the glasses, and that there is no relationship to the microhardness, or the softening point of glasses. That is true because polishing does not remove the bulk material itself, but just the soft hydrated layer. This is a unique property of glass which is the source of ultra-smooth surfaces (Sect. 7.3.3).

In practice, the composition of this hydrated layer and its reactivity are decisive for the determination of production costs and the number of rejects. Fairly often, the acid resistance value (SR) indicated in a glass catalogue gives a reasonable hint of how a glass type might behave during polishing and ultrasonic cleaning (Chap. 6).

7.3.1 Full Lap Polishing Techniques

Pitch Polishing

Until the 1950s, only pitch polishing techniques were used to manufacture glass optics. Skilled craftsmen created the shape of the work by loose abrasive grinding. The final shape and finish were obtained by pitch polishing. In the grinding process, corundum was still used as the abrasive whereas iron oxide had become the dominant polishing powder. There were many different types of "rouge" [7.36], and it was the secret of the skilled optician to choose the proper combination of pitch and rouge to make the miracle happen: production of surfaces - thanks to the unique properties of glass - with sub-nm smoothness and sub-μm shape accuracy. At that time, these small numbers expressed surface qualities that were unattainable in other production areas. However, it has to be clearly pointed out that these nominally outstanding qualities could only be obtained for plano, or more general, spherical optics, but not, for example, aspherics. This is due to the relative motion between tool and workpiece which is necessary in the manufacturing process. After being ground, the almost spherical workpiece is fitted into an already nearly spherical, visco-plastic pitch-lap. The tangential component of any force put on the workpiece will force it to slide along the lap surface, which is wetted with a slurry containing the abrasive. Beside the speed of the motion, the normal force is reponsible for the amount of material removed from the workpiece surface as well as the amount of pitch displaced. The formerly rough surface of the workpiece will be smoothened by this process. At the same time, high spots of the lap and the workpiece surface undergo higher pressures than the average parts, and, therefore, they are removed preferentially. Thus, a nearly ideal spherical workpiece surface is produced almost automatically. The proper choice of the pitch viscosity and the temperature control of the environment are decisive for the shape accuracy [7.37] and the obtainable smoothness. Then it depends on the optician's skills to govern the pressure and the motion by hand or with a machine in a way that not just a sphere is produced, but that the sphere with the desired radius is achieved just at the end of the smoothening process.

A quite comprehensive discussion of the aspects of pitch polishing has been given by *Brown* [7.38].

Polyurethane Polishing

Progress on the grinding side of the manufacturing process of optics, see the paragraph on "Brittle Fracture Grinding", has consequently stimulated the advance on the finishing side. At Carl Zeiss, the big demand for TESSAR® lenses required the introduction of high output techniques, especially for all the time consuming processing steps. Therefore, at Carl Zeiss and elsewhere, new fast polishing systems, featuring the "copying principle", were introduced in the late 1950s. In order to reach lower polishing cycle times, the polishing speed had to be increased by one or two orders of magnitude. As a consequence increased polishing pressures were also necessary so that the workpiece did not merely float hydrodynamically on the polishing slurry film – without any material removal. Therefore, in the late 1960s, a new generation of stiffer polishing machines was designed and built in our firm. At Carl Zeiss, pitch, cloth and felt laps were replaced by wear resistant, stuctured polyurethane pads. Other companies have used different thermoplastic pads, fairly often foamed and filled with abrasives.

It has to be pointed out that, according to our own experiments, the Preston coefficients for all kinds of standard glass polishing techniques are almost the same as long as a similar chemistry is applied, see Table 7.2. The differences of the material removal rate are merely due to the differing speeds and pressures see (7.1). Data from our own work, and those estimated in the literature, show

Table 7.2. Preston coefficients of some glass polishing regimes utilizing cerium oxide slurries

Polishing Technique	Glass Type	Lap	Preston Coefficient c $10^{-13}\,\mathrm{Pa}^{-1}$
High speed polishing of eyeglasses	Crown glass	Polyurethane pad	10
High speed polishing of spherical precision optics	BK 7	Polyurethane pad with filler	6–10
Pitch polishing of spherical precision optics	BK 7	Pitch	8
Pitch polishing of spherical precision optics	Fused silica	Pitch	2
Full aperture pitch polishing of aspherical optics	ZERODUR	Pitch with cloth	2–8
Sub-aperture pitch polishing of aspherical optics	ZERODUR	Pitch with felt	6
Elastic emission machining	BK 7	Polyurethane	
[7.50]		ZrO_2 grains	0.013
internal results		CeO_2 grains	0.3
Float polishing [7.46]	Sapphire	Tin/SiO_2	0.02

that the Preston coefficients of new polishing regimes, such as elastic emission machining and float polishing, are different.

In contrast to visco-plastic pitch laps, these almost inelastic polyurethane pads have to be trued to the desired shape prior to operation. During the polishing process this master shape is then copied at high speed to the ground workpiece surface. The fairly rigid lap surface does not flow as is the case with pitch polishing. As a result, quite a number of workpieces can be polished using one tool, and there is no interruption due to a necessary refurbishing of the instruments.

In the 1970s the traditional “rouge” was finally replaced step by step by the “faster”, more efficient rare earth polishing compounds containing cerium oxide. Evaluating potential polishing compounds, it has, for a long time, not clearly been understood why some of the oxide forms of iron, cerium, zirconium, and thorium polish glass well, whereas the related oxides of titanium [7.39] and tin do not [7.40]. Polishing tests performed by *Kaller* [7.41] showed that beside the nominal chemical composition, the internal structure of polishing grains determines their chemical reactivity and therefore their effectiveness in glass polishing. *Silvernail* [7.42] proved that in an aqueous environment cerium hydroxide is an accelerator of glass removal. *Cook* [7.43] showed the strong correlation of the polishing removal rate and the isoelectric point of the polishing compound. All these results confirm the early experimental findings of *Izumitani* [7.44].

Teflon Polishing

Leistner [7.45] established a TEFLON® *polishing* system for the production of ultra precise flats. Using a teflon polisher, the friction forces between lap and workpiece were reduced dramatically in comparison to pitch polishing. This "slow" machine polishing process is able to produce plano surfaces repeatably, with flatnesses of $\lambda/200$ which are needed for Fabry-Perot interferometers.

Float Polishing

In the 1970s and 1980s, two polishing processes were introduced in which structured tin laps were used. Both processes have the common feature that areas where the worked surface sticks out are removed preferentially and a smoothening effect occurs:

Namba [7.46] introduced the *float polishing* technique which was later applied to the gentle polishing of ceramics, cermets and glasses [7.47], too. The technological and commercial importance of the process was gained because of its superiority in polishing the contacting faces of ferrites used in VCR (Video Camera Recorder) heads. Float polishing uses a rapidly rotating tin lap with an aqueous polishing slurry of colloidal silicon oxide. The size of these polishing particles is in the range of just 4–7 nm, whereas the particles in conventional polishing are in the order of 1 μm. The inventor of the process claimed that the removal mechanism is an elastic bombardment of the workpiece surface by the polishing grains, leading to contact bonding and removal of weakly bonded atoms from the surface.

Hader's [7.48] explanation of their *hydrothermal wear* process in superpolishing sapphire is different. Microcavities are thought to form for a short time at the sapphire–tin interface. The internal high pressure and temperature allow hydrothermal conditions to be established so that alumina groups of the sapphire surface are dissolved. Under proper conditions, the alumina, together with the colloidal silica of the slurry, forms stable kaolinite which is important in order to prevent redeposition effects (Sect. 7.3.3).

Fixed Abrasive Polishing

In the 1980s polishing with fixed abrasive was introduced as another even faster polishing system. It was especially suitable for the high volume production of lower quality lenses with diameters up to about 20 mm. Plastic laps, highly charged with sub-micron ceria abrasive, are used with pure water as a lubricant for only one use. Beside the higher removal rate, the advantage of this system over polyurethane polishing is that no chemically unreliable slurry is needed (Sect. 7.3.3).

7.3.2 Sub-Aperture Polishing

Polishing of glass optics by using laps which have a much smaller area than that of the workpiece (for example, less than 10 % of full aperture) is a widely used process for generating weak aspheres with small departures from the best fit sphere, or for figuring during a correction step on high quality spherical or aspherical surfaces. Once again the field of astronomy is the driving force behind this time-consuming and therefore, relatively expensive process because of its need of odd, but highly accurate, optical surfaces. The polishing technique itself is not new. It has been practised for decades already by skilled opticians using spot-like or ring-shaped polishing tools.

Since the tool motion itself does not automatically generate the desired surface slope, as in polishing of spheres, see the paragraph on "Pitch Polishing" (Sect. 7.3.1), the success of a sub aperture polishing technique depends very much on the ability to measure surface slopes. Therefore, this polishing technique was revitalized when advanced measuring systems, such as interferometric methods, were standardized. Finally, with increased reproducibility, optical surfacing became less expensive and was more widely used - a development that was due to the application of computer control [7.49].

A new polishing head, which could be used in sub-aperture polishing, was invented by *Mori*. It was introduced to the western world by *Tsuwa* [7.50]. In this process called "elastic emission machining" (EEM) a polyurethane ball is spun at high speed close to the workpiece surface that is placed in a bowl filled with polishing slurry. In a small area, a hydrodynamically induced material removal takes place by the use of polishing grains which are squeezed through the gap between ball and workpiece.

7.3.3 Influences of Chemistry in Glass Polishing

As proven by *Izumitani* [7.51], polishing does not actually remove glass itself, but rather the hydrous leached layer of glass. As already stated in the paragraph on "Polyurethane Polishing" (Sect. 7.3.1) chemical reactivity between the glass, sometimes the lap, and always, however, the polishing compound – containing water, polishing powder and additives – is desired. This reactivity is the fundamental reason for the unique smoothness that is obtained with polished glass surfaces. Nevertheless, there is undesired, uncontrolled chemical activity which is responsible for a substantial amount of refuse in the production of optics.

Problems associated with the chemical part of the glass removal mechanism became apparent when the revolutionary change from pitch to polyurethane polishing took place. Whereas in pitch polishing a small amount of the polishing slurry was brought onto the lap with a brush, ample slurry is flooded onto the workpiece continuously in polyurethane polishing. The recirculated polishing slurry contains additives to prevent settling and is supplied by pipelines. The same slurry may be used for months to polish many different glass types. This means that the aging polishing slurry is being loaded with the debris of the polished glasses that is partially dissolved and changes the chemical activity of the slurry in a manner that is not reproducible. The choice of adequate slurry additives is essential to keep the behaviour of the slurry as stable and as "forgiving" as possible, especially for the more sensitive glasses.

After hundreds of hours of use, an aged and loaded slurry will eventually cause "stain" on the surface of the optics. It is either due to a thick leached layer, or to hydrated glass redeposited from the slurry onto the glass. Neither of the effects can easily be distiguished directly after polishing. Unfortunately, they are usually not obvious before an AR (anti-reflection) coating; afterwards they are generally called "haze".

According to *Tesar* [7.52], the redeposition of hydrated glass onto the work surface represents a general danger when the polishing slurry is recirculated. The redeposited material may behave plastically for a short time before condensation. On the other hand, opticians report that locally redeposited material behaves like canine teeth and destroys the polyurethane lap immediately after redeposition. In any case, the redeposited material cannot be removed even by harsh cleaning procedures.

The freshly leached layer is soft and can easily be damaged by standard handling procedures. At last, after coating, when the defect is visible, the resulting nonuniformity is described as "texture". The appearance of leaching effects or tarnishing strongly depends on the glass composition [7.53]. The material layers near the surface are generally called "leached layers", "hydrolyzed layers", or "Beilby-layers". Their composition and their properties are actually determined by the ion exchange between the bulk glass, the near surface material layer(s), the coating – if present – and the environment [7.54]. Even worse, using quite different methods of analysis, *Bach* [7.55] by means of spectral reflectance curves, *Wagner* [7.54] using NRA (nuclear reaction analysis) and RBS (Rutherford backscattering spectroscopy), and *Sprenger* [7.56] by XPS (X-ray photoelectron spectroscopy), have proved that the leached layer varies within time. Depending on the glass type and the storage conditions, the leached layer may, after months of storage, condense to a thin but solid layer of glass with a composition of its own. Latest EXAFS (extended x-ray absorption fine structure) and REFLEXAFS (reflection fluorescence extended x-ray absorption fine structure) experiments performed on special glass types such as SK 16 and LaK N12 [7.57] showed that there is not just one but a stack of leached layers with varying composition. During the last years, *Cook* [7.43] put some effort into the science of polishing. He wanted to obtain a material removal rate, a surface finish and an integrity of the material layers forming the surface that were predictable and stable.

7.4 Advanced Material Removal Techniques

A fairly recent development is the use of an ion beam as a non-contact tool for local slope correction. The first commercial benefits of this method, sometimes called "ion polishing", have been obtained by *Wilson* [7.58]. The advantage of ion beam figuring – in comparison to chemo-mechanical sub-aperture polishing, Sect. 7.3.2 – is that it does not produce "roll off" at the edges of the workpiece. Today at Carl Zeiss, the surfaces of fairly unstable lenses for use in microlithography lens systems are efficiently slope corrected by ion beam figuring to accuracies better than $\lambda/20$ [7.59]. A seeming disadvantage of the method is that the workpiece has to be transferred into a vacuum chamber to be machined. Nevertheless, this might turn out to be helpful when the ion machined virgin surface is immediately transferred into an adjacent coating facility. By this means it should be possible to obtain coatings with a reproducibility which can otherwise only be reached by extreme effort in the cleaning procedure.

A plasma based radio frequency method has recently been described [7.60], which is principally suited to remove chemically and physically ill defined surface layers whose production could not be avoided during polishing and cleaning procedures. This method minimizes the disturbance of the depth distribution of the glass elements by collisional interactions and eliminates any disturbing surface charging.

7.5 Conclusion and Future Work

In spite of all the advances achieved in modern times, the understanding of glass polishing has a long way to go. Only very specialized types of mass production using a very limited spectrum of glass types guarantee a satisfactory and uninterrupted working process without daily surprises. In a typical optics polishing shop, things often seem to be miraculous. This is especially so because the relationship between a change in processing and the variation of a complex glass surface is not obvious. The deterioration is often not visible until a coating is applied, perhaps only weeks later.

It seems that grinding has now been fairly well understood. About a decade ago, the introduction of shear mode grinding brought some new accents. Shear mode grinding seemed to be correlated to super stiff, ultra precise machine tools. Nevertheless, "Mirror Grinding", see the paragraph on "Brittle Fracture Grinding", Sect. 7.2.3, can be performed with standard, fairly compliant machines as well. A few years ago, a new potential chapter in glass grinding was announced when *Golini* [7.19] reported on the influence of the type of lubricant chosen. The zeta potential between the lubricant and the surface of the substrate, which may be characterized by an isoelectric point, seems to trigger whether the material removal process happens at a high level in a brittle mode, or in shear mode with a low rate of material removal. The importance of the isoelectric point is that it links the processes of grinding and polishing grains, see the paragraph on "Polyurethane Polishing", Sect. 7.3.1.

In the future, the mass production of many simple types of glass optics will most likely be performed by the application of finished-lens-molding techniques, rather than by grinding and polishing. New, mold-friendly glass types will push that development. Polishing tools with fixed abrasive will be developed further and their usage may become the standard polishing technique for most applications in the future. Not so much gradient index optics, but replication techniques of diffractive optics for the visible, or Si-based lithographic processes will presumably make feasible the realisation of ultra-compact and inexpensive optical devices. The use of liquid lenses should help to reach a greater freedom in the design of mass production optics. High-precision end optics will probably be fine formed and cleaned by ion beam figuring prior to on-line coating. One has only begun to use UV lasers for the shaping of hard and brittle materials such as glass. This may play an important role in the future of predictable high volume material removal.

References

7.1 N. Brown, L. Cook: "Role of Abrasion in the Optical Polishing of Metals and Glass", in *The Science of Polishing* (Techn. Digest, Monterey, Optical Society of America 1984) TuB-A3-1 to TuB-A3-3

7.2 F.W. Preston: "The Theory and Design of Plate Finishing Machines", J. Soc. Glass Tech. **11**, 214–256 (1927)

7.3 N.J. Brown: "Brittle to Shear Grinding Mode Transition for Loose Abrasive Grinding", in *Optical Fabrication and Testing* (Techn. Digest **9**, St. Clara, Optical Society of America 1988) pp. 23–26

7.4 G. Reiter, H. Carl: "Optimiert Polieren in der optischen Fertigung", Feinwerktechnik u. Messtechnik **98**, 29–32 (1990)

7.5 M. Miyashita: "Review on Ultraprecision Grinding Technology", in *Ultraprecision in Manufacturing Engineering*, ed. by M. Weck, R. Hartel (Springer, Berlin, Heidelberg 1988) pp. 41–57

7.6 P.J. Davis, K.L. Blaedel, J.S. Taylor, N.J. Brown, T.T. Saito, I.F. Stowers: *Experiments to Increase the Efficiency of Ductile Grinding* (to be published)

7.7 R.E. Parks: "The Traditions of Optical Fabrication", Proc. SPIE **315**, 56–64 (1981)

7.8 W. Klemm: "Das Polieren Optischer Flächen", Zeiss-Opton Werkzeitung **6**, 67–68 (1952)

7.9 P. Grau: "Die Härte von Gläsern und Keramiken", in *Strukturabhängiges Mechanisches Verhalten von Festkörpern*, ed. by D. Schulze (Akademie Verlag, Berlin 1979) pp. 278–287

7.10 T. Izumitani: "Polishing, Lapping, and Diamond Grinding of Optical Glass", in *Treatise on Materials Science and Technology*, ed. by M. Tomozawa, R.H. Doremus, Vol.17 (Academic Press, New York 1979) pp. 156–164

7.11 D. Golini, W. Czajkowski: Center for Optics Manufacturing Process Development (Status Report 1/15/92, Rochester, Center for Optics Manufacturing 1992)

7.12 "Abrasion Hardness and Knoop Hardness of Optical Glasses" (Technische Information – Optisches Glas No. 4, Mainz, Schott Glaswerke, November 1972)

7.13 M.G. Schinker, W. Döll: "Untersuchung der Abtragungsvorgänge und -mechanismen bei der Bearbeitung Optischer Gläser mit Diamantwerkzeugen", Industrie Diamanten Rundschau **18** (4), 3–11 (1984)

7.14 D.M. Busch: "Ritz- und Verschleißuntersuchungen an Spröden Werkstoffen mit Einzelkornbestückten Hartstoffwerkzeugen"; Ph.D. Thesis, Technische Universität Hannover (1968)

7.15 M.G. Schinker, W. Döll: "Turning of Optical Glass at Room Temperatures", Proc. SPIE **802**, 70–81 (1987)

7.16 K.E. Puttick, M.R. Rudman, K.J. Smith, A. Franks, K. Lindsey: "Single point diamond machining of glasses", Proc. R. Soc. Lond. A**426**, 19–30 (1989)

7.17 M. Takata, M. Tomozawa, E.B. Watson: "Effect of Water Content on Mechanical Properties of Na_2O-SiO_2 Glasses", Comm. Am. Cer. Soc., Sept. 1982, C156–C157 (1982)

7.18 S.C. Fawcett, T.A. Dow: "Influence of Wheel Speed on Surface Finish and Chip Geometry in Precision Contour Grinding", Prec. Engin. **14** (3), 160–167 (1992)

7.19 D. Golini, S.D. Jacobs: "Chemo-Mechanical Effects in Loose Abrasive Grinding of ULE", in *Science of Optical Finishing* (Techn. Digest, Monterey, Optical Society of America 1990) pp. 25–28

7.20 W. Rupp: "Loose Abrasive Grinding of Optical Surfaces", Appl. Opt. **11** (12), 2797–2810 (1972)

7.21 M.G. Schinker, W. Döll: "Basic Investigations into the High Speed Processing of Optical Glasses with Diamond Tools", Proc. SPIE **381**, 32–38 (1983)

7.22 K.H. Fiedler: "The Importance of Shear Mode Grinding for Fabrication of Asphercis", Proc. 5. Intl. Prec. Engin. Seminar, Monterey/CA, USA, September 18–22, 1989, pp. 137–146

7.23 W. König, Cl. Schmitz-Justen, N. Koch: "Grinding Optical Glasses with Fine Diamond", Industrial Diamond Review **50** (536), 20–25 (1990)

7.24 H.R. Meyer: "Feinstschleifen Optischer Gläser", Feinwerktechnik & Meßtechnik **93**, 5–8 (1985)

7.25 H. Wapler, H.O. Juchem: "Diamond Abrasives for Machining Glass", Industrial Diamond Review **4/87**, 159–162 (1987)

7.26 T.A. Michalske, B.C. Bunker: "The Fracturing of Glass", Sci. Am., December 1987, 122–129 (1987)

7.27 E.M. Fetisova, V.P.Korovkin, V.M. Al'tshuller, Y.V. Ashkerov: "Modeling the Abrasive Breakup of the Surface of Optical Glass During Fine Diamond Grinding", Sov. J. Opt. Technol. **52** (8), 481–483 (1985)

7.28 H. Ohmori, T. Nakagawa: "Mirror Surface Grinding of Silicon Wafers with Electrolytic In-Process Dressing", Ann. CIRP **39** (1), 329–332 (1990)

7.29 N. Kawashima, T. Hattori, K. Orii, S. Tochihara: "Properties of Diamond Lapping Films", Industrial Diamond Review **5/91**, 244–247 (1991)

7.30 N.N.: "Diamond Polishing Cloths for Super-Finishing", Industrial Diamond Review **3/92**, 115–117 (1992)

7.31 J. Yoshioka, K. Koizumi, M. Miyashita, A. Kanai, M. Shimizu, H. Yoshikawa: "Surface Grinding with a Newly Developed Ultra Precision Grinding Machine", Techn. Paper Precision Machining Workshop, St. Paul/MN, USA, June 8–10, 1982, Soc. of Manufacturing Engineers, P.O. Box 930, Dearborn, Michigan 48128, 20 pages

7.32 K.H. Fiedler: "Precision Grinding of Brittle Materials", in *Ultraprecision in Manufacturing Engineering*, ed. by M. Weck, R. Hartel (Springer, Berlin, Heidelberg 1988) pp. 72–76

7.33 Y. Namba, R. Wada, K. Unno, A. Tsuboi: "Ultra-Precision Surface Grinder Having a Glass-Ceramic Spindle of Zero-Thermal Expansion", Ann. CIRP **28** (1), 331–334 (1989)

7.34 P.A. McKeown, K. Carlisle, P. Shore, and R.F.J. Read: "Ultra-Precision, High stiffness CNC Grinding Machines for Ductile Mode Grinding of Brittle Materials", Proc. SPIE **1320**, 301–313 (1990)

7.35 T. Izumitani: "Polishing Rate of Optical Glass", in *Optical Fabrication and Testing* (Techn. Digest, Palo Alto, Optical Society of America 1982) Ma1-1

7.36 W. Rebentisch, F. Dinkelacker: "Elektronenmikroskopische Untersuchungen an Polierrot", Glastech. Ber. **32**, 321–327 (1959)

7.37 N.J. Brown: "Flat Lapping Revisited", in *Optical Fabrication and Testing* (Techn. Digest, Tucson, Optical Society of America 1979) pp. 13–16

7.38 N.J. Brown: "Optical Polishing Pitch", University of California Preprint UCRL-80301, November 10, 1977, paper prepared for the Optical Society of America Workshop on Optical Fabrication and Testing, Nov. 10–12, 1977, San Mateo, CA

7.39 J.A. Seckold: "Optical Polishing with Titanium Oxide", Appl. Opt. **15** (7), 1693 (1976)

7.40 W.L. Silvernail, N.J. Goetzinger: "The Mechanism of Glass Polishing", The Glass Industry, May 1971, 172–175 (1971)

7.41 A. Kaller: "Einfluß des Äußeren und Inneren Aufbaus der Poliermitteloxide auf deren Griffigkeit beim Polieren von Glas", Silikattechnik **35** (5) 134–138 (1984)

7.42 W.L. Silvernail: "The Role of Cerium Oxide in Glass Polishing", in *The Science of Polishing* (Techn. Digest, Monterey, Optical Society of America 1984) TuB-B1-1 to B1-4

7.43 L.M. Cook: "Chemical Processes in Glass Polishing", J. Non-Cryst. Solids **120**, 152–171 (1990)

7.44 T. Izumitani: "Polishing, Lapping, and Diamond Grinding of Optical Glass", in *Treatise on Materials Science and Technology*, ed. By M. Tomozawa, R.H. Doremus, Vol. 17 (Academic Press, New York 1979) pp. 131–140

7.45 A.J. Leistner: "Teflon Polishers: Their Manufacture and Use", Appl. Opt. **15** (2), 293–298 (1976)

7.46 Y. Namba, H. Tsuwa: "Ultra-Fine Finishing of Sapphire Single Crystal", Ann. CIRP **26**, 325–329 (1977)

7.47 J.M. Bennet, J.J. Schaffer, Y. Shibano, Y. Namba: "Float Polishing of Optical Materials", Appl. Opt. **26** (4), 696–703 (1987)

7.48 B. Hader, O. Weis: "Precision Measurements of Material-Removal Rates in Superpolishing Sapphire", Proc. SPIE **1015**, 114–122 (1989)

7.49 R.A. Jones: "Optimization of Computer Controlled Polishing", Appl. Opt. **16**, 218–224 (1977)

7.50 H. Tsuwa, N. Ikawa, Y. Mori, K. Sugiyama: "Numerically Controlled Elastic Emission Machining", Ann. CIRP **28** (1), 193–197 (1979)

7.51 T. Izumitani: "Polishing, Lapping, and Diamond Grinding of Optical Glass", in *Treatise on Materials Science and Technology,* ed. By M. Tomozawa, R.H. Doremus, Vol. 17 (Academic Press, New York 1979) pp. 116–140

7.52 A. Tesar, W. Eickelberg: "Polished Substrate Surface and Cleaning Study for Coated Optic Quality", in *Optical Fabrication and Testing* (Techn. Digest **24**, Boston, Optical Society of America 1992) pp. 145–147

7.53 K.H. Fiedler: "Problems with Polishing and Its Quality Assessment", in *Optical Fabrication and Testing* (Techn. Digest **13**, St. Clara, Optical Society of America (1988) pp. 42–47

7.54 W. Wagner, F. Rauch, H. Bach: "Chemical Properties of Hydrolyzed Surface Layers on SiO_2-BaO-B_2O_3", Glastech. Ber. **63** (12), 351–362 (1990)

7.55 H. Bach: "The Use of Surface Analysis to Optimize the Production Parameters of the Surface of Glasses and Ceramics", Glass Technology **30**, 75–82 (1989)

7.56 D. Sprenger, H. Bach, W. Meisel, P. Gütlich: "XPS Study of Leached Glass Surfaces", J. Non-Cryst. Solids **126**, 111–129 (1990)

7.57 A.J. Dent, B.R. Dobson, G.N. Greaves: "An Investigation of the Surface Structures in a Number of Silicate Glasses by X-ray Spectroscopy and Reflectivity" (Report of SERC Daresbury Laboratory, Daresbury 1991) unpublished

7.58 S.R. Wilson, D.W.Reicher, J.R.McNeil: "Surface Figuring Using Neutral Ion Beams", Proc. SPIE **966**, 74–81 (1988)

7.59 K. H. Fiedler, I. W. Roblee, Y. L. Chen, M. Weiser: *Manufacturing Technology for Soft X-Ray Lithography Optics* (Top. Meeting Digests Series: Vol. 4, European Optical Soc., Engelberg, Switzerland 1994) pp. 58–59

7.60 D. Martin, H. Oechsner: "Ein plasmagestütztes Hochfrequenzverfahren zum niederenergetischen Ionenbeschuß von Isolatorwerkstoffen, insbesondere von Gläsern", Vakuum in der Praxis **4**, 263–268 (1993)

8. Selected Applications

8.1 Ophthalmic Glasses

David Krashkevich, Susan R. Loehr

8.1.1 History of Ophthalmic Lenses

Since ancient times, lenses fashioned from sections of clear glass or crystalline spheres have been used for the converging or concentrating of the sun's rays to ignite fires. These so-called burning glasses as well as other hand-held magnifying lenses were known to the ancient Greeks as well as to the Chinese. Records describing the wearing of lenses before the eyes, in some sort of crude frame fastened to the head, to modify the vision of the wearer, did not appear until the latter decades of the thirteenth century in northern Italy [8.1, 2]. By the fifteenth century spectacles had become more widely accepted; however, only the wealthy could afford them. The frames became more elaborate and fashionable, but not much in the way of advancement in the correction of vision was realized. In the mid seventeenth century, the German astronomer Johannes Kepler described the refraction of incoming light by the cornea and the crystalline lens of the eye and the formation of an inverted image of the retina. He also demonstrated that the eye could be farsighted or nearsighted and that these conditions could be improved with the aid of convex or concave ophthalmic lenses. As the understanding of the physiology of vision improved, advances in spectacle design appeared. Notable inventions of the eighteenth century included achromatic lenses in 1758 by Dollan and bifocals in 1784 by the famous American scientist and statesman, Benjamin Franklin.

8.1.2 The Optics of Vision and Ophthalmic Lenses

One must understand a bit of the biology of human vision and elementary optics in order to fully appreciate the need for ophthalmic lenses. Many excellent texts provide detailed descriptions of the mechanics of human vision, far beyond the scope of this text [8.3, 4]. A basic overview of the formation of images by the visual system will be given. The details of the perception of colour will be omitted here for the sake of brevity.

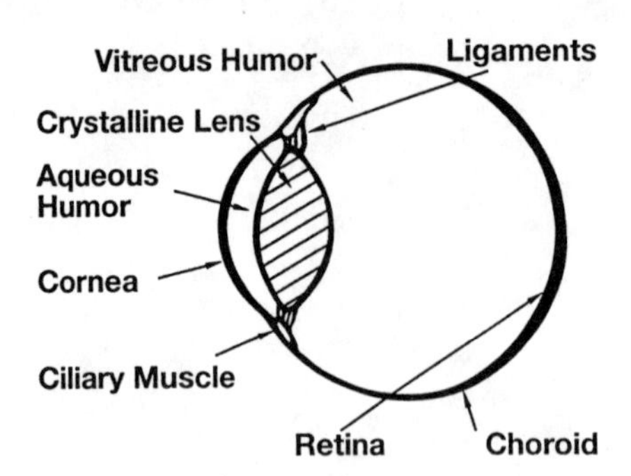

Fig. 8.1. Schematic view of the human eye

One can describe the human eye as a positive lens system, containing several refractive elements, which gathers light and forms a real image on a light sensitive surface. Figure 8.1 illustrates a simplified version of its basic components. The eye is a nearly spherical body surrounded by a tough membrane called the sclera. With the exception of the transparent front portion of the cornea, the sclera is white and opaque. The air-cornea interface provides the greatest refraction of light within the human eye due to the large difference in refractive index between the air and the cornea. The chamber behind the cornea is filled with a clear watery liquid called the aqueous humor. Little refraction of incoming light will occur at this interface due to the similarities in refractive index between the cornea and aqueous humor. This chamber also contains a diaphragm known as the iris which controls the amount of light entering the eye through the circular pupil at its centre. The pupil contracts or dilates in response to changes in the intensity of ambient light. The pupil will also contract to increase image sharpness when doing close work, thereby enhancing the depth of focus to the eye. The crystalline lens, a complex layered fibrous structure surrounded by an elastic membrane, is located just beyond the iris. Its laminar structure varies in refractive index from 1.406 at the centre to approximately 1.386 at the surface. It is attached to the ciliary muscles by a series of radially arranged ligaments around the periphery of the lens. The crystalline lens provides the visual system with the necessary means of adjusting fine focusing through changes in its shape which vary focal length and, therefore, the power of the lens. Following the lens is another chamber filled with a transparent, gelatinous material known as the vitreous humor. The inner surface of the sclera is covered by an inner shell called the choroid, whose main function is to absorb stray light entering the eye. The final component of the eye's optical system is the retina which is composed of a thin layer of light-sensitive cells that covers a large part of the choroid. These receptor cells absorb the light impinging upon the retina and convert the image via photochemical reactions to an electrical signal which is transmitted via the optic nerve to the brain. It is interesting to note that while the image focused on the retina is both inverted and reversed with respect to the true orientation of the object, the brain has learned to interpret this signal as normal [8.3].

Fine focusing or accommodation is performed by the crystalline lens by the contraction of the ciliary muscles. When viewing objects at a distance, the muscles will be fully relaxed. In this state the ligaments exert tension on the periphery of the lens drawing it into a flat configuration which in turn increases its focal length. Conversely, as an object moves closer to the eye, contraction of the ciliary muscles in sphincter-like fashion relieves some of the tension on the ligaments and allows the lens to assume a more spherical shape, thereby increasing the curvature of the lens and decreasing the focal length.

A normal or emmetropic eye is one which can focus parallel rays of light on the retina while the ciliary muscles are fully relaxed. When this is not possible, the eye is said to be ametropic. Common forms of refractive errors of vision are myopia (nearsightedness), hyperopia (farsightedness) and astigmatism. These conditions frequently occur due to defects in the shape of the cornea, lens, or the eye itself. The crystalline lens may also lose the ability to accommodate as a person ages and the lens loses some of its elasticity. This condition, know as presbyopia, is common in people over forty. Compensation for these deficiencies may be made through the use of suitably tailored spectacle lenses in either single vision or multifocal designs.

Basic spectacle prescriptions are specified in terms of the dioptric power of a lens. The power of a lens is the reciprocal of its focal length, f. If the focal length of the lens is expressed in meters, then the unit of lens power is the inverse meter or diopter. For our purposes, we shall consider spectacle lenses to be thin lenses. Thus, we will neglect the contribution of thickness to the calculation of the total power of a lens. Readers interested in a more thorough treatment of this subject are referred to Morgan ([8.1] p. 33–60). The lensmaker's formula gives the power, D, of a thin lens in air:

$$D = \frac{1}{f} = (n-1)\left[\frac{1}{R_1} - \frac{1}{R_2}\right] \tag{8.1}$$

where n is the refractive index of the lens material and R_1 and R_2 are the radii of curvature of the lens surfaces. Lenses of positive power cause parallel rays of light to converge, while those of negative power cause them to diverge.

Myopia or nearsightedness is a condition where parallel light rays are brought into focus in front of the retina in the unaccommodated eye. This condition may result when the combined power of the cornea and crystalline lens is too large and the focal length is shorter than the axial length of the eye or when the eye itself is elongated along the optical axis. Whatever the cause, the far point (defined as the most distant point that can be brought into focus) will be closer than infinity and all objects beyond it will appear blurred; hence the term nearsightedness. One can correct the symptoms of this condition by placing a negative or diverging lens in front of the eye such that the combined lens system of the eye and spectacle now has its second focal point on the retina. Hyperopia or farsightedness occurs when the focal point of the unaccommodated eye lies behind the retina. A converging positive power spectacle lens is needed in this case to move the second focal point of the visual system back to the retina. Another common defect of the visual system is astigmatism. Astigmatism usually results from an asymmetric cornea in which the curvature and, therefore, the refractive power is unequal in different meridians. If the meridians or maximum and minimum refractive power are perpendicular, the

Table 8.1. Properties of typical Schott ophthalmic glasses

Glass Type	UV W76	HC weiß 0389	High Lite®	BaSF 64	LaSF 36A	LaSF 39[1]
n_e	1.5251	1.6040	1.7064	1.7052	1.8000	1.8936
n_d	1.5230	1.6006	1.7010	1.7010	1.7947	1.8868
v_e	58.3	41.8	30.5	39.3	35.4	30.3
v_d	58.6	42.1	30.8	37.7	35.7	30.5
Softening point °C	710	753	688	712	799	
Annealing point °C	536	621	570	581	680	639
Transformation temperature °C	530	615	578	580	683	651
Density g/cm^3	2.55	2.67	3.00	3.20	3.62	3.99
Coefficient of therm. expansion α_{20-300} 10^{-6}/K	9.3	8.0	9.7	8.7	8.7	8.3

® High Lite is a registrated trademark of SCHOTT GLASS TECHNOLOGIES, Inc., Duryea, U.S.A.
[1] preliminary data

astigmatism is regular and correctable. If they are not, the astigmatism is irregular and cannot be corrected easily ([8.4] p. 147). A cylindrical lens can be used to compensate for a defect in corneal curvature that occurs in only one meridian. The vision in an eye that is both astigmatic and either myopic or hyperopic can be corrected with a toric lens which combines both spherical and cylindrical surfaces.

The final configuration of the spectacle lens prescribed for a particular visual deficiency depends upon the power of the lens required, the refractive index of the lens material and the curvatures chosen for the surfaces of the lens. An optimized configuration is chosen to prove the lightest and cosmetically most attractive lens which meets the required prescription. The white crown glass most commonly used in ophthalmic lenses has an n_d equal to 1.523. For extreme myopic and hyperopic eyewear prescriptions which require strong curves, a problem arises with lens centre or edge thickness. In an effort to reduce lens weight and make the spectacles cosmetically more appealing, high refractive index, low density glasses have been introduced (see Table 8.1).

The need for glasses which protect the eye against the effects of harmful ultraviolet radiation and reduce sun glare have fostered a wide variety of coloured ophthalmic lenses. Glasses which block those particular wavelengths of radiation produced during arc welding or those produced by a specific laser have also been developed. These glasses meet all of the physical and optical requirements as previously outlined for clear or "white" ophthalmic glasses. In addition, colour, ultraviolet and near infrared transmission specifications may also be required. The chemistry and applications of such glasses will be discussed in the section dealing with coloured glasses (Sect. 8.8) and eye protection glasses (Sect. 8.9).

8.1.3 Properties of Ophthalmic Glasses

The art and science of ophthalmic lenses has made great strides since its primitive beginnings in the Middle Ages. Raw glass blanks for modern day spectacle lenses are mass produced on automated, computer controlled production lines. The blanks are typically formed directly from melted glass, using high speed automatic presses. Blanks are produced in various sizes and shapes depending upon the application. Blanks for full sized ophthalmic lenses, for multifocal sunglasses, or small segments for use in multifocal prescription eyeglasses can be made in this manner. After pressing, the blanks are annealed in a lehr, given a preliminary inspection and placed in trays. The lens blanks will be more closely inspected for internal defects such as bubbles and striae as well as measured for conformance to dimensional requirements before the glass is released to customers. An alternative method of lens blank manufacture is to first produce the glass in strip form. The glass is lehr annealed, inspected and placed on pallets. The strips are reheated to approximately the softening temperature of the glass and lens blanks or segments are then pressed out in a remolding operation. Producing lenses for the ophthalmic industry requires careful control of pressing dimensions and associated tolerances, surface texture of the lens blank, degree of annealing and internal quality such as bubbles, striae, stones and crystals. The refractive index and dispersion of the glass are typically specified at the green mercury *e*-line (546.07 nm). There are, however, other lines used such as the yellow sodium *D*-line (589.29 nm) in the United States or the yellow helium *d*-line (587.56 nm) in other parts of the world. The refractive indices of ophthalmic glasses are typically specified to three decimal places and controlled to within 0.001 of the nominal value. The tolerance on Abbe number ν_e is typically within 0.5 of the nominal value. Other optical properties including luminous transmittance and transmission at discrete wavelengths may also be specified. Chromaticity is measured for coloured glasses used in applications such as sunglasses and fashion tints.

The softening, annealing and strain point temperatures as well as the coefficient of thermal expansion (20 to 300 °C) are important characteristics of the glass which must be closely controlled to permit the efficient fusion of major and segment glasses for multifocal lenses. Knowledge of these properties is also necessary for the effective chemical or thermal strengthening of plano (no power), single vision, or multifocal lenses.

The generic base composition utilized for most ophthalmic lens applications is a soda lime silicate. The addition of soda or potash to silica lowers the softening point of the glass. Lime, alumina and sometimes magnesia are added to improve chemical durability. Small additions of boric acid improve glass melting. Zinc oxide additions improve chemical durability as well as enhance the chemical strengthening of glasses. The addition of titania can be used to increase refractive index without the large increases in density associated with lead or barium oxide

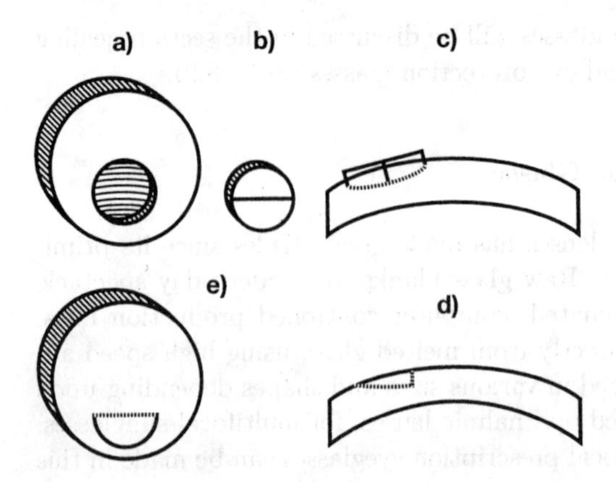

Fig. 8.2. Manufacture of a fused bifocal lens: (a) The inner surface of the major is ground to the prescribed curvature. A countersink is ground and polished in the outer surface; (b) The high refractive index segment glass is edged and then fused to an upper segment of the same glass as the major. The segment button is ground and polished to fit the countersink; (c) The segment button is fused to the major; (d) The outer surface of the major is ground and polished to complete the prescription of the bifocal lens; (e) The finished bifocal lens

additions. Light-weight optical and ophthalmic glasses are frequently formulated with substantially all of the lead or barium oxide replaced by titania (Sect. 2.2.3). These glasses have nearly the same optical properties as their heavier counterparts with a substantial reduction in density. Table 8.1 summarizes nominal ranges in optical and physical property data of typical Schott ophthalmic glasses used in single vision lenses or as the major or carrier blanks in multifocal lenses.

8.1.4 Multifocal Ophthalmic Lenses

Fusing high refractive index segment glasses to the afore-mentioned glass types extends lens design capabilities. Table 8.2 summarizes the properties of various segment glasses used in fused multifocal ophthalmic lenses. Over the temperature range of 20 to 300 °C, soda lime based ophthalmic crown glasses typically have a coefficient of thermal expansion of approximately $8\text{–}9 \times 10^{-6}$/K.

When one is considering the fusibility of a segment/major composite, both the thermal expansion and the viscosity compatibility must be carefully matched. Ideally, the coefficient of expansion of the segment glass should be equal to or slightly lower (approximately 0.5×10^{-6}/K) than the major. This provides a stress free or slight compressive stress on the segment. Viscosity of the two components is also an important consideration since the fusion process typically occurs at or near the softening point of the segment glass. If the softening points of the segment and major are too similar, undesired deformation of the major can occur. Typically, one would like to have a higher viscosity in the major. Annealing points of the segment

Table 8.2. Properties of typical Schott segment glasses

Glass type	BS-558	BaF 52	BaF 51	BS-565	BaF 50	BS-570	BS-575	BS-668	BS-673	BS-676	BS-681	S-1022	BS-775	BS-781	BS-786
n_e	1.5906	1.6099	1.6539	1.6570	1.6840	1.7060	1.7545	1.6855	1.7391	1.7718	1.8097	1.8096	1.7610	1.8171	1.8560
n_d	1.5880	1.6068	1.6505	1.6533	1.6804	1.7013	1.7501	1.6821	1.7344	1.7665	1.8048	1.8070	1.7560	1.8113	1.8500
v_e	51.7	46.1	44.7	42.0	44.3	35.1	39.7	46.9	37.4	34.3	39.0	33.9	34.5	34.1	32.1
v_d	52.0	46.3	44.9	42.3	44.5	35.4	40.0	47.3	37.7	34.6	39.2	34.0	34.7	34.3	32.4
Softening point °C	686	697	693	687	682	685	669	744	745	752	735	669	720	693	691
Annealing point °C	556	544	557	558	549	542	557	613	622	627	642	557	610	584	583
Transformation temperature °C	547	546	556	539	546	520	559	607	615	613	641	546	605	570	582
Density g/cm^3	3.05	3.32	3.42	3.54	3.80	3.88	4.38	3.82	3.88	4.05	4.39	4.51	3.75	4.54	4.40
Coefficient of therm. expansion α_{20-300} 10^{-6}/K	9.4	9.4	9.6	9.4	9.6	9.2	9.4	8.1	8.1	7.9	8.2	9.5	9.3	9.3	8.8
For fusion with	UV-W76							HC white 0389				High Lite BaSF 64			

and major glass should be similar, thereby allowing sufficient stress removal of the composite after fusion. Another concern which must be addressed when considering fusibility between the segment and the major is glass interface compatibility. During fusion, both segment and major glass constituents interdiffuse at the glass interface line. If the composition of this diffusion zone is unstable, undesirable crystallization or phase separation can occur. For multifocal components which are only marginally compatible, one can, in certain cases, circumvent the problem by fusing at as low a temperature and for as short a time as is practically possible while still producing an acceptable fusion. Meticulous cleaning of both the segment and major components prior to fusion is necessary. Any residue left on the surfaces prior to the fusion step can be a source of interfacial defects. Figure 8.2 illustrates the steps used to produce one possible configuration of a fused bifocal ophthalmic lens containing a "D" shaped segment for near vision correction. The rough glass pressing or "blank" which is to become the major is first ground to prescription on the inner surface of the lens. The upper segment, which is the same glass as the major, and lower segment which is a glass of higher refractive index than the major are edge ground and fused together to form a segment button. The segment button is then ground and polished on the surface to be fused to the major. A round countersink for the segment button is ground and polished into the outer surface of the major. The assembled lens is placed on a ceramic support to reduce deformation during thermal treatment. The segment button is fused to the major by passing the assembled components via conveyor belt through a lehr furnace with an appropriate temperature profile to fuse and then anneal the bifocal lens. An alternative method is to place the assembled components in a box furnace programmed with an appropriate time and temperature cycle for fusion and annealing of the lens. The outer surface of the composite bifocal lens is then ground and polished to complete the necessary prescription.

8.1.5 Strengthening of Ophthalmic Lenses

Although pristine glass is an inherently strong material, microcracks which form during the lens processing steps can sharply decrease its strength. The breakage resistance of a glass lens can be dramatically improved by using an appropriate strengthening technique. Two techniques, thermal and chemical strengthening, are commonly used within the ophthalmic industry (see Sect. 4.5). Both techniques create a compressed surface layer on the strengthened article. Since glass breaks under tension rather than compression, the compressive stress induced in the surface must be overcome before the glass will break.

To thermally strengthen a lens, the material must first be heated to slightly less than the transformation point of the glass. As the temperature of the glass increases, the volume increases; therefore, the density decreases. After several minutes at this temperature, the surface of the lens is then rapidly chilled by blasting the lens with a stream of dry air. This freezes the surface of the glass into the less dense, high temperature structure, while allowing structural configurational changes (densification) to take place in the bulk of the lens as it cools more slowly.

This places the interior of the lens under tension and the surface in compression. One can see a distinctive Maltese cross or a similar pattern in a thermally strengthened lens when examined between crossed polarizers.

Chemical strengthening occurs by the diffusion controlled exchange of smaller alkali ions present in the glass with larger alkali ions in a melted salt bath. Typically, sodium ions in the glass are exchanged for the larger potassium ions in a melted salt bath in most white crown and fixed tint ophthalmic glasses. This "stuffing" of potassium ions into the glass surface places the surface of the lens under compression, thereby improving the strength of the material. The depth of the compressed layer and magnitude of the induced surface stress in terms of birefringence is measured via a polarizing microscope. Typical depths of the compressed layer range from 50 to 80 microns. The maximum surface compression achieved for typical white crown lenses is 5500 to 9000 nm/cm. The higher index ophthalmic glasses do not strengthen as well. For example, 1.701 index High-Lite® achieves an average of 3400 nm/cm compression at a depth of 55 microns, while the 1.805 high index light weight flint can only reach a compression of 1800 nm/cm at a depth of 36 microns at optimum strengthening temperatures. The efficiency of this process depends upon the chemical composition of the glass, particulary the sodium content, the viscosity of the glass at the temperature of the salt bath and the presence of ion exchange inhibitors such as barium or lead in the base composition. The bath composition most commonly used for ophthalmic lens strengthening consists of 99.5% potassium nitrate and 0.5% silicic acid. The silicic acid scavenges impurities from the salt bath and causes them to sink to the bottom. A reasonable starting point in determining the optimum bath temperature is 50 to 100 °C below the annealing temperature of the glass. The optimum temperature for ion exchange for a particular glass type cannot be reliably predicted from theory. It is usually determined empirically, for a given strengthening cycle, by varying the ion exchange bath temperature in 10 to 20 °C increments and then measuring the depth of the compressed layer and magnitude of the induced surface stress. The measured magnitudes of the induced surface compression usually pass through a maximum with increasing temperature. This is shown in Figure 8.3(a) for a white crown glass, S-1, and a fixed tint, S-2082 (S-1 and S-2082 are crown glasses produced by Schott Glass Technologies Inc., Duryea), of nominally the same composition. This behaviour can be attributed to the increase in ease of viscous flow and associated stress relaxation at higher temperatures. The depth of the ion exchanged layer can also be obtained from scanning electron microscope profiles of sample cross-sections. It should be noted that the depth of the compressed layer is not necessarily the same as the depth of the ion exchanged layer. This is illustrated in Figure 8.3(b). Note the open triangles designating the depth of the ion exchanged layer at various temperatures. *Stroud* [8.5] found that at higher exchange temperatures, the depth of the ion exchanged layer frequently exceeds that of the compressed layer. This was attributed to the ease of viscous flow and subsequent relaxation of the induced stress at higher temperatures. The depth of ion exchange can also be smaller than that of the compressed layer, especially at lower temperatures.

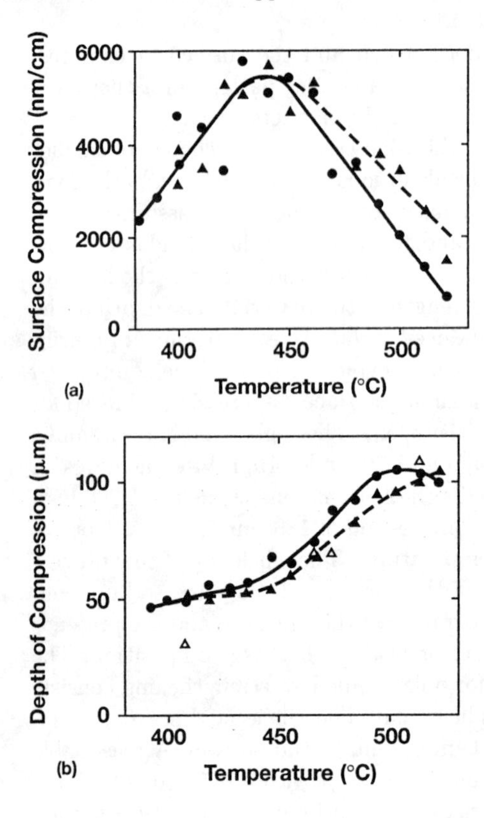

Fig. 8.3. (a) and (b): Surface compression (a) and its depth (b) after ion exchange in strengthening bath for crown glasses: dark triangle = S-1, circle = S-2082, triangle = depth of ion exchange, S-1 [8.5]

Thermal rather than chemical strengthening is more widely used in the processing of white crown and fixed tint ophthalmic lenses. Chemical strengthening is recommended for photochromic lenses in order to maintain their optimum darkening characteristics. The most commonly used chemical strengthening cycle is sixteen hours. Depending upon the size of the salt bath, a few hundred lenses can be strengthened at the same time. Thermal strengthening of lenses is more common in large ophthalmic operations in which thousands of lenses need to be processed daily. Each lens takes only a few minutes to thermally strengthen. The advantages of choosing a particular strengthening technique are as follows: The depth of the compressed surface layer is thicker for a thermally strengthened lens; however, the depth varies with the thickness of the lens. A uniformly thick compressed layer is formed on a chemically strengthened lens. A higher maximum compressive stress is achievable with chemical strengthening rather than with thermal strengthening [8.6]. Chemical strengthening is considered to produce more consistent strengthening results.

The U.S. Food and Drug Administration (FDA) requires that all prescription and nonprescription fashion and sunglass eyewear sold in the United States must pass the test for impact resistance described in the American National Standards Institute (ANSI) standards Z80.1-1987 and Z80.3-1986 [8.7–9]. Certain types of specialty lenses are exempt from this test. This test consists of dropping a steel ball 16 mm in diameter, weighing at least 15.9 g, in free fall from a height of not less than 127 cm onto the horizontal outer surface of a finished and edged glass lens within a 16 mm diameter circle whose centre is approximately the geometric centre of the lens. A lens must not fracture in order to pass the test.

8.2 Photochromic Glasses

Hans-Jürgen Hoffmann

8.2.1 Basic Principles of Photochromic Effects

The interaction of electromagnetic waves with matter is generally characterized by terms like dispersion, absorption, scattering, refraction etc. In the present section we deal with special absorptive effects which are induced in matter by the electromagnetic waves themselves.

Electromagnetic waves can cause electronic transitions from one state into another, create new particles or quasi-particles or excite vibrations of ions and ensembles of charged particles. These effects are well-known absorption mechanisms. Obviously, the absorption spectrum depends on the electronic and ionic configuration. Consequently, the transition from one electronic or ionic configuration into a different configuration necessarily changes the absorption spectrum of the system under consideration. The control of the absorbance spectrum of any material by photons is called photochromism, irrespective of the spectral range in which the absorbance is changed, and irrespective of the energy of the photons which induce the photochromism. Following this definition, a rich variety of possible effects exists, which can be exploited for photochromy (e.g. [8.10–13]): the famous transition of F-centres into α- and F'-centres [8.14] or, more generally, the transformation from one into another colour centre by illumination [8.15], the photon-induced orientation of colour centres (photodichroism) [8.15], the spectral hole burning [8.16] as well as the photon-induced polymerization [8.10, 12] or the decompositon of compounds [8.17], the photon-induced bleaching of absorption bands for Q-switching a laser [8.18], just to name a few.

The change of the absorption spectrum, however, is hard to be seen, if it occurs in a spectral region with a large absorption constant. Therefore, one has to focus on materials and spectral regions, where both the absolute and relative change of the absorbance is large, if one wants to exploit the corresponding effect of photochromism for technical applications. Challenging mass applications of photochromic effects are optical data storage and "smart" windows. However, several qualitative and quantitative requirements have to be fulfilled: the requirement for special spectral ranges optimized both for the inducing photons and for the induced absorbance is highly important in many applications. It is equally important whether the effect is reversible spontaneously or under irradiation, whether it shows fatigue, and how efficient the absorption spectrum changes per incident photon. As a consequence, the photon-induced control of the absorbance, which is obviously very intriguing from a theoretical point of view, is reduced to only a few applications in practice. Until now, the most prominent widespread application of a photochromic effect seems to have occurred in photochromic eye-glasses.

Although photochromic effects have long been known to occur in glasses (e.g. [8.19]), the basis of a material which is useful for photochromic sunglasses was laid by *W. H. Armistead* and *S.D. Stookey* [8.20, 21] in the fifties and sixties of this century. These scientists introduced silver halides into suitable glass melts and succeeded in forming small silver halide precipitations or particles in the glass. The basic effect, namely the photolytical decomposition of silver halides, is very well-known in chemistry. In a test tube, the silver halide is decomposed into Ag clusters, which are responsible for the absorbance in the visible spectral region, and volatile halogens. The photolytical decomposition is not reversible in a test tube, however, if the halogens escape. In a glassy matrix, on the contrary, the photolytic decomposition products stay close together to recombine into silver halides. In the following, we will elaborate on the chemical and physical details of the photochromism of glasses doped with silver halides.

Beside silver halides, many other different compounds can be decomposed by light changing their absorption spectra under irradiation. There is an abundance of references in the literature on photochromic systems investigated until now. To list a small arbitrary selection of inorganic systems, we mention the Pb-, Cu-, and Cd-halides [8.22–26], the transition metal oxides [8.27–30] and some other compounds [8.10, 11, 31–35]. However, most systems are interesting from a scientific point of view only and have no technical potential; or other practical disadvantages reduce their applicability. For ophthalmic applications, for instance, a series of minimum requirements has to be fulfilled in order that a material be useful: The induced absorptivity must match the sensitivity curve of the human eye, the darkening must be sufficient under exposure to sunlight, the "speed" of the induced change of the transmittance should be as fast as possible, i.e. the absorptivity induced by the solar irradiance should change as fast as possible upon variation of the irradiance, the darkening should be insensitive to variations of the ambient temperature, etc. Considering all technical aspects, oxidic glasses doped with silver halides have turned out to best meet the sum of all requirements. Consequently, many investigations have been performed on such systems, of which we will give a survey in the following.

8.2.2 Photochromic Oxidic Glasses Doped with Silver Halides

An arbitrary selection of photochromic glass systems is listed in Table 8.3. One may choose the base glass — or the matrix — from a large variety of different glass compositions. Therefore, the glass system can be selected to fulfill additional requirements, such as stability of the production process, chemical resistivity, or good grinding and polishing characteristics. Mostly alkali-alumo-borosilicate glasses are used. A special reason for that will be explained later. The presence of silver and halogen ions (mostly a mixture of Cl^- and Br^-) in the order of several tenths of a percent and copper ions in the order of several hundredths

Table 8.3. Compositions of different photochromic glasses (quantities are given in parts by the mass, which is for all examples approximately mass%, since the sum of components of each glass is close to 100)

No:	1	2	3	4	5	6	7	8	9	10	11	12	13	14	15
Ref.:	[8.21]	[8.36]	[8.36]	[8.37]	[8.38]	[8.39]	[8.39]	[8.39]	[8.40]	[8.40]	[8.41]	[8.42]	[8.42]	[8.43]	[8.43]
Year:	1962	1979	1979	1982	1985	1970	1970	1970	1966	1966	1969	1977	1977	1972	1972
SiO_2	60.1	56.46	54.0	55.6	55.2	10.5						3.8	9.3	8.7	
B_2O_3	20.0	18.15	16.5	16.4	20.6	30.3	54.0	55.0	81.4	14.9	70.0	21.3	33.0		16.7
Al_2O_3	9.5	6.19	8.9	8.9	7.4	14.9	22.8	8.0		1.98	15.0	1.0		28.8	24.2
P_2O_5						0.8	0.6							34.6	34.2
PbO			0.6	5.0	0.03	23.9		29.6		69.4	1.25				
La_2O_3											2.5	49.7	23.0		
Li_2O		1.81	2.37	2.65	4.3			0.3		0.26					
Na_2O	10.0	4.08	1.88	1.85	0.8		5.1			0.10				6.6	6.38
K_2O		5.72		0.01	6.1	3.5	1.3		6.11	1.02				7.9	8.14
MgO			2.42				14.9	2.0						3.2	
CaO				0.2									0.9	7.0	4.42
SrO															
BaO			9.7	6.7		1.0		2.0			10.0			6.6	
ZnO									10.2	9.90	1.15		44.0		
ZrO_2		4.99	1.9	2.2	4.0	3.2									2.94
TiO_2		2.07			1.4	8.1								0.47	2.94
WO_3												0.8			
CdO												0.7			
Ta_2O_5												23.6			
Ag^+	0.40	0.252	0.14	0.16	0.25	0.5	0.3	0.55	0.3	0.19	0.5	0.75	0.6	0.11	0.079
F^-	0.19		0.19	0.19			0.3	1.4		0.37	0.3	0.5	0.3	0.4	0.47
Cl^-	0.10	0.195	0.59	0.24	0.35	1.2	0.2	0.4				0.30	0.8	0.5	0.47
Br^-	0.17	0.155	0.18	0.145	0.095	2.1	0.5	0.7	1.03	1.0	0.5			0.2	0.35
I^-									1.17	1.14					
CuO	0.017	0.006	0.015	0.035	0.009	0.008	0.01	0.03		0.005	0.1	0.032	0.024	0.2	0.039

of a percent by the mass in the matrix, however, is prerequisite to achieve photochromism. The requirement that the copper ions have to be incorporated into the glass matrix as Cu^+ (cuprous ions) is very important, as we will see later. Therefore, the redox potential of the melt has to be controlled very carefully to render the copper effective for photochromy. The melting temperature of the composition is governed mainly by the composition of the matrix, since the amount of the ingredients responsible for photochromism is small. Depending on the exact compositions, the melting temperature of alkali-alumo-boro-silicates may range from 1200 °C to 1450 °C. Since the melting temperatures are rather high, one has to take into account that silver and halogens are volatile and can, among other components, escape from the melt. Thus, special provisions have to be made in order to avoid such a disadvantageous effect.

The refractive index of the glass for the yellow He line $d = 587.56$ nm is usually adjusted to $n_d = 1.5230$, in order to match the standard tools of the manufacturers of ophthalmic lenses. In recent years, however, photochromic glasses with larger refractive indices have been developed. The reason for this was the aim to reduce the difference in the thickness between the centre and the circumferential parts of large diopter correction lenses. A large difference of the thickness causes large differences in the transmissivity. This is inconvenient to the bearer of large diopter photochromic correction lenses. The refractive index could be increased to about $n_d = 1.6$ without reducing the Abbe number (increasing the dispersion) too much. This has been achieved by the introduction of ions with large optical polarizability into the glass matrix, such as La^{3+}, Ti^{4+}, Zr^{4+}, Pb^{2+}/Pb^{4+}, or Ta^{5+}.

In order to obtain special colouring effects in the darkened and transparent state, traces of palladium and gold in the order of ppm [8.44] or rare earth and transition metal ions are introduced into the batch. The melt may also contain refining agents such as As_2O_3/As_2O_5 or Sb_2O_3/Sb_2O_5 in the order of one tenth of a percent by the mass or less.

Lens blanks are pressed from the melt and cooled rapidly to room temperature. In this status, the blanks do not yet show remarkable photochromism. This is due to the fact that the silver, copper and halide ions are basically dispersed or nearly isolated from each other in the glass matrix. In order to become photochromic, the lens blanks have to be annealed at temperatures around or above the glass transformation temperature, T_g. For alkali-alumo-boro-silicates this heat treatment takes place at temperatures between 500 °C and 730 °C for time intervals in the order of one hour. The exact parameters of this heat treatment depend on the glass composition; along with other details, they are considered as proprietary information of the manufacturer. During the heat treatment the silver and halogen ions precipitate into $AgCl_xBr_{1-x}$ clusters containing Cu^+ ions and additional Cl^- and Br^- ions for charge compensation. Since these clusters are responsible for photochromy in glasses, they are called photochromic centres.

The formation of silver halide clusters has been studied by electron microscopy [8.45–50]. Figure 8.4 shows a typical electron micrograph of a photochromic alkali-alumo-boro-silicate glass. The black dots represent the silver halide precipitations or photochromic centres. For an optimum performance of a self-adjusting sunglass

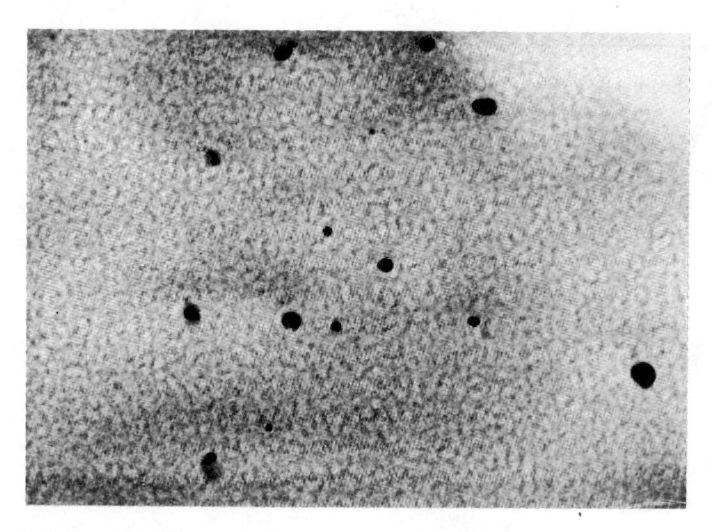

Fig. 8.4. Transmission electron micrograph of a photochromic alkali-alumo-boro-silicate glass after heat treatment: 700 °C for 1 hour (1 cm in the micrograph corresponds to 50 nm, from [8.49–50])

the diameter of the photochromic centres should be between about 10 and 20 nm [8.39, 46, 47]. If the diameter is smaller, the darkening of the photochromic glass is too low, and if the diameter is larger, the glasses scatter the sunlight too much and regenerate too slowly. Assuming that the amount of 0.4 mass % is distributed to photochromic centres of a uniform diameter of 20 nm, one estimates that there are 5×10^{15} of these silver halide clusters per cm^3 of the glass with an average distance of about 125 nm. Each of these photochromic centres contains silver ions in the order of 10^5.

The formation of photochromic centres in alkali-alumo-boro-silicate glasses is driven by phase separation of the material in a twofold sense. The heat treatment takes place at temperatures for which these glasses show phase separation into one phase which is rich in silica and another one which is rich in borates. The latter phase is enriched by all ingredients of the photochromic centres. The borate-rich phase supersaturates considerably with these ingredients since it is much smaller in volume than the phase rich in silica. Thus, a second phase separation — the formation of silver halide particles containing cuprous ions — takes place. *R. Pascova* and *I. Gutzow* have investigated the supersaturation of silver chloride in alkali-borate glasses experimentally [8.51] and showed that the solubility of silver chloride decreased considerably with decreasing temperature.

Because of the formation of silver halide particles, the UV-absorption edge is shifted (see Fig. 8.5). The difference $K_i(\Lambda)$ between the absorption constants before and after the heat treatment can be attributed to the indirect optical transitions between the valence and the conduction band of silver halide particles [8.52, 53]. Taking into account that the silver halides are diluted by the glass matrix, one can

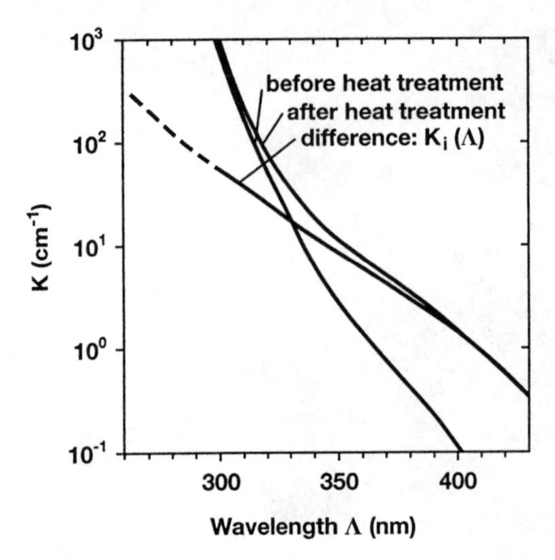

Fig. 8.5. Absorption coefficient K of a photochromic alkali-alumo-boro-silicate glass as a function of wavelength Λ before and after heat treatment (700 °C for 1 hour) and the difference $K_{\mathrm{i}}(\Lambda)$ (from [8.55])

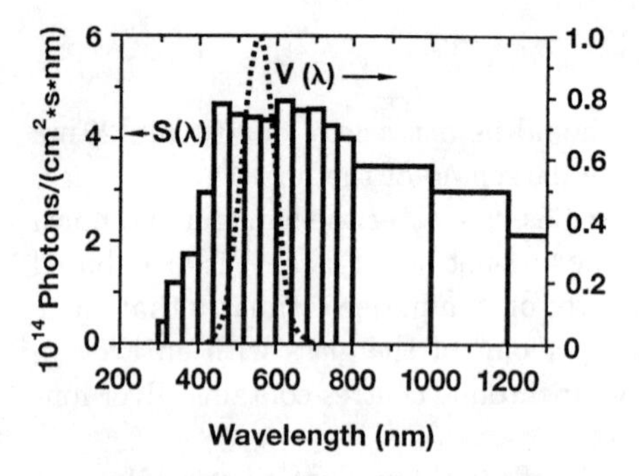

Fig. 8.6. Average photon flux density spectrum $S(\lambda)$ of the sun for air mass 1 under normal incidence on earth as a function of wavelength λ (according to [8.54]) and spectral luminous efficiency $V(\lambda)$ for photopic vision of the human eye

estimate from $K_{\mathrm{i}}(\Lambda)$ that nearly all silver ions added to the melt are precipitated into silver halide particles.

Figure 8.6 shows the solar photon flux density spectrum. Since a photon must be absorbed by a photochromic centre, the photons, essentially those of the UV and blue spectral range, cause the photolysis of the silver halide particles in the glass.

The performance of today's photochromic glasses is characterized by Figs. 8.7 and 8.8 showing the spectrum of the radiation-induced transmittance change of a 2 mm thick commercial photochromic glass and its dependence on time at a wavelength of 555 nm. This wavelength is close to the maximum sensitivity of the human eye for photopic vision (see Fig. 8.6). The transmittance decreases at 20 °C

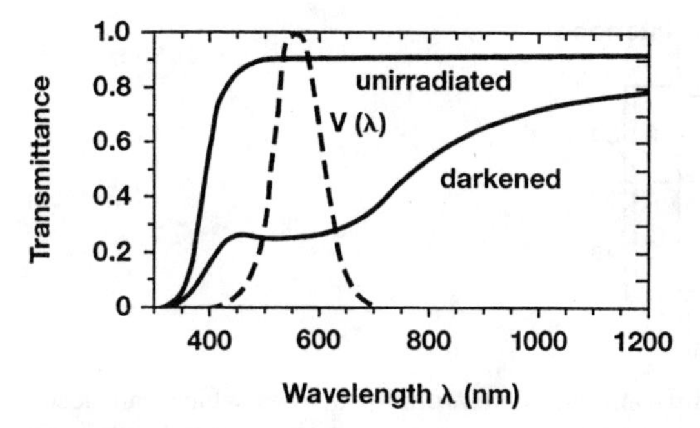

Fig. 8.7. Transmittance of a commercially available photochromic glass as a function of wavelength before and under solar irradiation after about 15 minutes at normal incidence. (Thickness of the sample: 2 mm; temperature: 20 °C). The dashed curve indicates the spectral power sensitivity $V(\lambda)$ for photopic vision of the human eye (from [8.55])

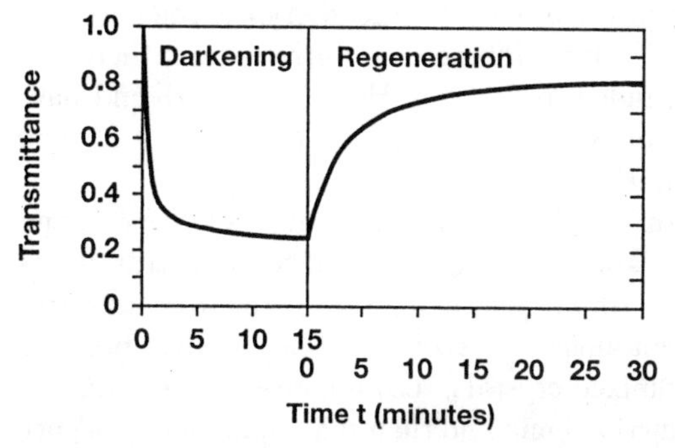

Fig. 8.8. Kinetics of the transmittance of the photochromic glass sample from Fig. 8.7 for the wavelength 555 nm after switching on and off the solar irradiation at 20 °C (from [8.55])

within some minutes from about 90% to less than 25% and increases again from the darkened state within thirty minutes to about 80% if the solar radiation is blocked.

8.2.3 Qualitative Description of the Darkening and the Regeneration Mechanisms of the Photochromic Glasses

The photoelectronic and photochemical processes which occur in a photochromic centre are to be discussed using the simple band diagram shown in Fig. 8.9 [8.55]. The silver halide particles are assumed to be slightly n-type in the dark. Therefore, the position of the Fermi-level E_F is in the upper half of the forbidden gap between E_V and E_C which are the upper edge of the valence band and the lower edge of the conduction band, respectively. A possible band bending in the silver halide particle near the interface photochromic centre/glass matrix has been neglected. This band bending may originate from a negative surface charge of the silver

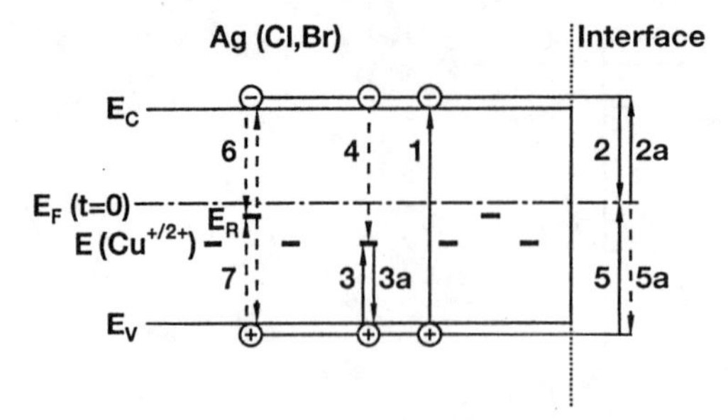

Fig. 8.9. Energy band diagram of different photoelectronic processes which can occur in the silver halide particles and at the interface between silver halide particle/vitreous matrix. For explanation of the transitions see text. The figure characterizes the situation at the onset of UV-irradiation in the transparent state

halide particles, and a compensating positive charge in the interior of the particles. Figure 8.9 represents the situation at the beginning of UV-irradiation; therefore, quasi-Fermi-levels have not been included in Fig. 8.9. However, one should have in mind that the quasi-Fermi-levels for electrons and holes move towards E_C and E_V, respectively, with increasing duration of the irradiance.

Photoelectrons in the conduction band and photoholes in the valence band are created by the absorption of UV-photons causing band-to-band transitions (process 1 in Fig. 8.6), driving the system away from its lowest state of Gibbs' free energy, where photoelectrons and photoholes are recombined. Since silver bromide and silver chloride (and presumably mixed crystals, too) are indirect semiconductors with their minimum of the conduction band and their maximum of the valence band at different wavevectors, the recombination must occur essentially via recombination centres. The level of such recombination centres has been assumed at E_R in Fig. 8.9. The recombination centres have to capture photoholes and photoelectrons in turn in order to become effective (processes 7 and 6 in Fig. 8.9). Once a hole has been captured, the recombination centre must capture an electron in order to complete the recombination. The photohole, on the other hand, can also be captured by Cu^+ ions, which are present in the photochromic centres, and form Cu^{2+} ions. Cu^+ ions are known to be efficient hole traps in silver halides (the capture cross-section is large for holes, whereas it is small for electrons). In this way, the recombination traffic via E_R is blocked and the lifetime of the photoelectrons is increased until they are trapped — preferentially at the interface silver halide particle/glass matrix, which has a large trap density. The trap depth corresponding approximately to the difference between E_C and E_F at that interface (since the Fermi-level is pinned usually to an energy with large density of states) can be of the order of 1 eV. An electron trapped at the interface can neutralize a nearby Ag^+ ion, or an Ag^+ diffuses to the interface. If this process is repeated several times, a silver speck will grow. At the same time, Cu^+ ions are transformed into Cu^{2+} ions. If nearly all Cu^+ have been transformed, further photolytic decomposition of the photochromic centre is less probable, since the lifetime of photoholes is increased in this case. Then, recombination via recombination centres becomes more effective, or the photoholes diffuse to the surface and recombine there with trapped electrons before the silver speck grows further (process 5 in Fig. 8.9). Other mechanisms that can limit the photolytic decomposition are also possible and have already been discussed in [8.55].

To understand the reverse reactions, we consider the situation after switching off the irradiation after the steady state has been attained. Two possible regeneration paths are indicated in Fig. 8.9: an electron is reemitted from the silver speck into the conduction band (process 2a), from which it recombines with a Cu^{2+} ion (process 4), or a hole is reemitted from a Cu^{2+} ion into the valence band (process 3a) and diffuses to the silver speck at the interface, where it is trapped (process 5). In both cases, a neutral silver atom of the silver speck is transformed into an Ag^+, which diffuses back into the interior of the photochromic centre, and is again incorporated into the silver halide lattice.

The reverse reaction can be enhanced by optical excitation [8.45], which increases the emission rate of holes from Cu^{2+} centres into the valence band (optically enhanced process 3a) or the re-emission rate of electrons from the silver speck into the conduction band (optically enhanced process 2a). Thus, the rate to generate Ag^+ ions in the silver specks is increased. From this, the effect of optical bleaching can be understood without difficulty.

As to the photo-induced absorption coefficient, there are at least two species of reaction products which absorb in the visible part of the solar spectrum: the silver specks and the Cu^{2+} ions formed in the silver halide particles.

For spherical silver clusters of large diameter (about 10 to 20 nm) in a vitreous matrix, a rather sharp resonance (Mie resonance) has been observed at wavelengths slightly above 400 nm [8.56–58].

The silver specks in photochromic glasses, however, are much smaller and their shapes are expected to deviate considerably from spheres. The photolytically deposited silver can form only a thin layer, one or few atomic diameters thick at the interface photochromic centre/glass matrix. Therefore, the absorption spectrum due to the silver specks is not sharp but rather extends over quite a broad range of wavelength [8.57–60]. This absorption is superposed by the absorption due to the Cu^{2+} ions in the silver halide particles [8.61, 62]. From the work of *D. C. Burnham* and *F. Moser* [8.62] it is known that the absorption coefficient of the Cu^{2+} ions in silver halides extends over a broad range of wavelength, mainly in the blue spectral region.

In addition, a small contribution to the induced absorption coefficient is due to the Cu^+ ions photo-oxidized into Cu^{2+} in the glass matrix. In boro-silicate glasses this absorption can be observed in the red/infrared spectral region [8.63, 64]. The strength and the position of this absorption band depend on the kind and on the concentration of alkali ions present in the glass matrix [8.64].

8.2.4 Photochemical Reaction Kinetics in Photochromic Glasses

The light intensity at wavelength λ transmitted through a photochromic glass of thickness w is given for normal incidence by [8.65]

$$\begin{aligned} I_w(\lambda,t) &= I_0(\lambda,t)(1-R)^2 \exp\left[-k_0(\lambda)w - \int_0^w k_\mathrm{i}(\lambda,x,t)dx\right] \\ &= I_{w0}(\lambda,t)\exp\left[-\int_0^w k_\mathrm{i}(\lambda,x,t)dx\right] \end{aligned} \qquad (8.2)$$

wherein

$$I_{w0}(\lambda,t) = I_0(\lambda,t)(1-R)^2 \exp\left(-k_0(\lambda)w\right) \qquad (8.3)$$

is the transmitted light intensity for the case where photochromic reactions do not occur; $I_0(\lambda,t)$ is the incident light intensity at wavelength λ and time t, R is Fresnel's reflection loss, $k_0(\lambda)$ is the intrinsic absorption constant of the photochromic glass before the onset of darkening, $k_\mathrm{i}(\lambda,x,t)$ is the induced absorption constant and x is the distance from the surface of the photochromic glass facing the irradiation source.

Multiple reflections as well as possible antireflection coatings have been neglected. These corrections can be included in (8.2) if necessary.

The induced absorption constant, $k_\mathrm{i}(\lambda,x,t)$ comprises all positive and negative changes depending on the distribution of the full incident spectrum as a function of the wavelength. It includes as the dominating positive contributions the absorption constant due to the silver specks deposited photolytically and, additionally, that due to the Cu^{2+} ions both in the photochromic centres and in the glass matrix. It includes also the negative contribution due to the bleaching as well as the decrease of the absorption constant of the photochromic centres, if these centres have been partially photolyzed. Furthermore, changes of the absorption constant in the UV due to the transformation of Cu^+ into Cu^{2+} and changes of the vibrational spectra of the ions in the infrared spectral region have to be taken into account. These changes, however, are small compared to the large intrinsic absorption coefficients of the photochromic glasses in the UV and IR that are due to electronic transitions or excitation of ionic vibrations, respectively. Therefore, it is justified to completely neglect changes of the electronic absorption spectrum in the UV and of the vibrational spectrum of the ions in the IR and to focus on the visible part of the absorption spectrum.

The reaction kinetics of the induced absorption constant $k_\mathrm{i}(\lambda,t)$ has recently been investigated as a function of several parameters, in order to develop a quantitative recombination model [8.66, 67]. Using monochromatic UV-irradiation at wavelength Λ alone, the absorption constant $k_\mathrm{i}(\lambda,t)$ induced by the UV can be calculated from the rate equation

$$\frac{dk_\mathrm{i}(\lambda,t)}{dt} = \alpha(\lambda)\beta_0\left(1-\frac{k_\mathrm{i}(\lambda,t)}{\kappa_\mathrm{i}(\lambda)}\right)K_\mathrm{i}(\Lambda)J(\Lambda) - \frac{r}{\alpha(\lambda)}k_\mathrm{i}^2(\lambda,t) \qquad (8.4)$$

($\alpha(\lambda)$ absorption cross-section of the photolysis products per elementary process, $\beta_0\left(1-\frac{k_\mathrm{i}(\lambda,t)}{\kappa_\mathrm{i}(\lambda)}\right)$ quantum yield, $\kappa_\mathrm{i}(\lambda)$ maximum induced absorption coefficient at wavelength λ, $K_\mathrm{i}(\Lambda)$ absorption constant of the photochromic centres in the UV including the dilution factor due to the vitreous matrix, $J(\Lambda)$ flux density of UV-photons at wavelength Λ, r effective recombination coefficient.)

The absorption constant $K_\mathrm{i}(\Lambda)$ of the photochromic centres can be determined as a function of the wavelength from the difference between the absorption coefficients before and after the heat treatment of the photochromic glass, as can be seen from Fig. 8.5. This difference corresponds to the absorption due to indirect band-to-band transitions in $AgCl_xBr_{1-x}$ mixed crystals [8.52, 53]. For the beginning of the UV-irradiation one can use the approximation $k_i(\lambda, t) = 0$ on the right-hand side of (8.4), which yields

$$\frac{dk_i(\lambda, t=0)}{dt} = \alpha(\lambda)\beta_0 K_i(\Lambda) J(\Lambda) \,. \tag{8.5}$$

According to (8.5) one expects that the initial slope of the induced absorption constant is proportional to the flux density of UV-photons absorbed by the photochromic centres, if there was no induced $k_i(\lambda, 0)$ at the beginning. This expectation is indeed confirmed by experimental results as can be seen from Fig. 8.10.

Equation (8.4) can be solved, if the time dependence of $J(\Lambda)$ and the initial value are known. In the special case $k_i(\lambda, t=0) = 0$, and switching on a constant UV-irradiation, one obtains from (8.4)

$$k_i(\lambda, t) = \frac{\alpha(\lambda)}{r\theta} \tanh\left(\frac{t}{\theta} + C_1\right) - \frac{\alpha^2(\lambda) G_0}{2r\kappa_i(\lambda)} \,, \tag{8.6}$$

with the reciprocal time constant

$$\frac{1}{\theta} = \sqrt{rG_0 + \left(\frac{\alpha(\lambda) G_0}{2\kappa_i(\lambda)}\right)^2} \,, \tag{8.7}$$

and the abbreviations

$$G_0 = \beta_0 K_i(\Lambda) J(\Lambda) \text{ and } C_1 = \operatorname{arcoth}\sqrt{1 + \frac{4r\kappa_i^2(\lambda)}{\alpha^2(\lambda) G_0}} \,.$$

For the regeneration after switching off the irradiation ($J(\Lambda) = 0$) we obtain with $k_i(\lambda, 0)$ as the initial value

$$k_i(\lambda, t) = \frac{k_i(\lambda, 0)}{1 + \frac{r}{\alpha(\lambda)} k_i(\lambda, 0) t} \,. \tag{8.8}$$

Inserting $dk_i(\lambda, t)/dt = 0$ in (8.4) one can calculate the induced absorption constant of the steady state

$$\begin{aligned} k_{i,st}(\lambda) &= \sqrt{\frac{\alpha^2(\lambda) G_0}{r} + \frac{\alpha^4(\lambda) {G_0}^2}{4r^2\kappa_i^2(\lambda)}} - \frac{\alpha^2(\lambda) G_0}{2r\kappa_i(\lambda)} \\ &= \left[\begin{array}{ll} \sqrt{\frac{\alpha^2(\lambda)\beta_0 K_i(\Lambda) J(\Lambda)}{r}} & \text{if } \frac{G_0}{r} \gg \frac{\alpha^2(\lambda) {G_0}^2}{4r^2\kappa_i^2(\lambda)} \\ \kappa_i(\lambda) & \text{if } \frac{G_0}{r} \ll \frac{\alpha^2(\lambda) {G_0}^2}{4r^2\kappa_i^2(\lambda)} \end{array} \right] . \end{aligned} \tag{8.9}$$

Thus, one expects that $k_{i,st}$ is proportional to the square root of $K_i(\Lambda) J(\Lambda)$ for small $K_i J$, whereas it is constant κ_i for large values. This is confirmed by the experimental results shown in Fig. 8.11.

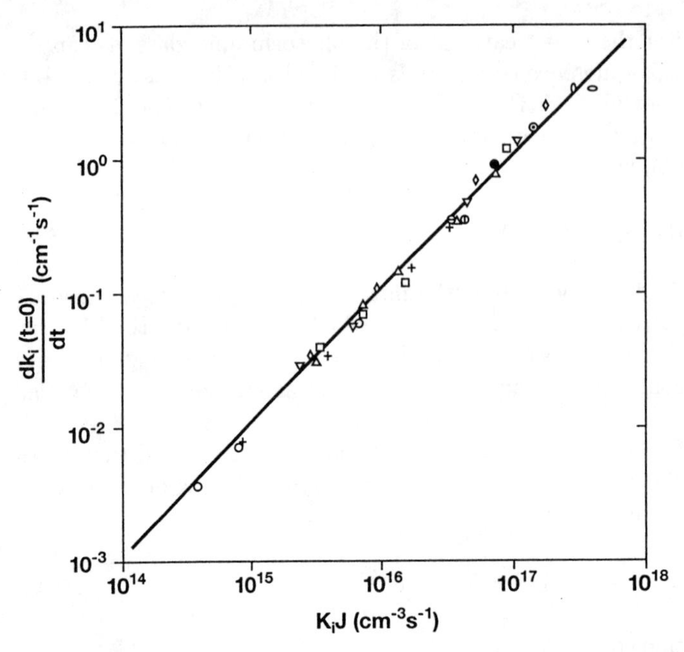

Fig. 8.10. Initial slope of the induced absorption constant, $dk_i(t=0)/dt$, at λ = 540 nm as a function of the density of photons absorbed by the photochromic centres, $K_i(\Lambda)J(\Lambda)$. Different symbols mark the results obtained for different wavelengths Λ in the range between 300 and 410 nm (from [8.55])

Figures 8.10 and 8.11 show experimental results for different wavelengths Λ of the UV-photons. As can be seen from the figures, the results do not depend explicitly on the wavelength Λ of the UV-photons used for these experiments. Thus, the essential characteristics of the induced absorption constant $k_i(\lambda, t)$ are independent of the wavelength or energy of the UV-photons absorbed by the photochromic centres. Furthermore, the spectral shape of k_i does not depend on time. This justifies the assumption that

$$k_i(\lambda, t) = \alpha(\lambda)Z(t) , \tag{8.10}$$

wherein $Z(t)$ represents the concentration of photolytic reaction products as a function of time, t, and $\alpha(\lambda)$ is a function of the wavelength λ. Using (8.10), one can transform (8.4), (8.5), (8.6), (8.8) and (8.9) into equations of the concentrations of photolytic reaction products. This is an important transformation, since it is not necessary to investigate the kinetics of the induced absorption constant $k_i(\lambda, t)$ as a function of both the "writing" wavelength, Λ, and the "reading" wavelength, λ. Instead, it is sufficient to choose just two representative wavelengths Λ and λ to study the characteristics of the kinetics.

From Fig. 8.10 one can determine the product $\alpha\beta_0$ for the wavelength λ = 540 nm used for this set of data. The product $\alpha\beta_0$ corresponds to the absorption cross-section of all photolytic reaction products created by an absorbed UV-photon times the quantum yield at the beginning of the UV-irradiation. Its value

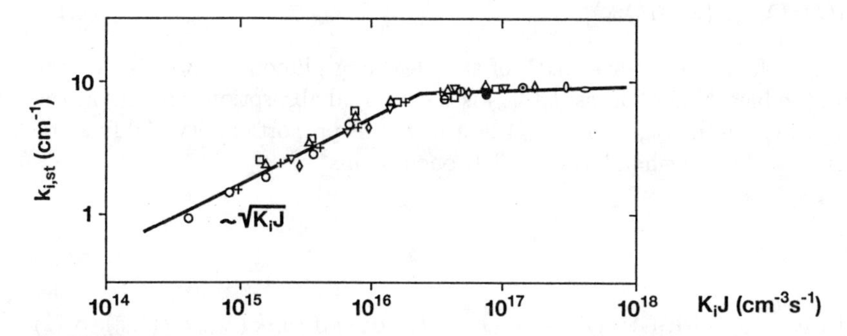

Fig. 8.11. Induced absorption constant in the steady state, $k_{\mathrm{i,st}}$, at $\lambda = 540\,\mathrm{nm}$ as a function of the density of photons absorbed by the photochromic centres, $K_{\mathrm{i}}(\Lambda)J(\Lambda)$. Different symbols mark the results obtained for different wavelengths Λ in the range between 300 and 410 nm (from [8.55])

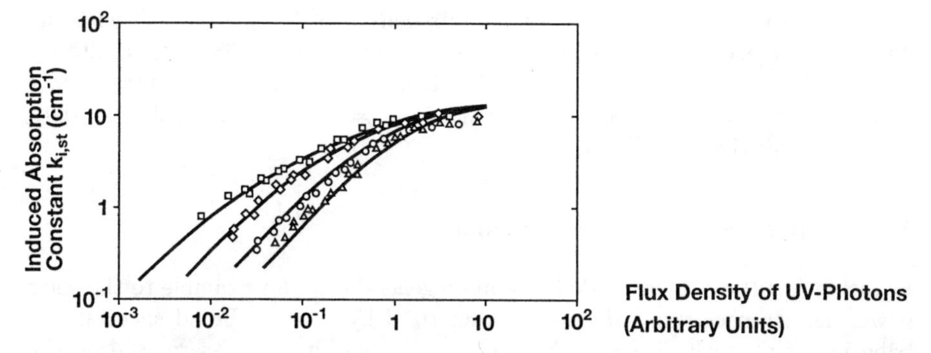

Fig. 8.12. The induced absorption constant in the steady state, $k_{\mathrm{i,st}}$, at $\lambda = 632.8\,\mathrm{nm}$ as a function of the photon flux density in the UV for various intensities of the bleaching light, $I(\lambda_{0\mathrm{b}})$, with $\lambda_{0\mathrm{b}} = 632.8$ nm. Relative photon flux densities of the bleaching light: triangles: $I(\lambda_{0\mathrm{b}}) = I_0$, circles: $I(\lambda_{0\mathrm{b}}) = 0.48\ I_0$, diamonds: $I(\lambda_{0\mathrm{b}}) = 0.15\ I_0$, squares: $I(\lambda_{0\mathrm{b}}) = 0.04\ I_0$. Data from [8.68]. ($I_0$ corresponds to the intensity of an unfocused beam of a 5 mW HeNe laser.)

is about $10^{-17}\,\mathrm{cm}^2$ for $\lambda = 540\,\mathrm{nm}$. Since $k_{\mathrm{i}}(\lambda, t)$ can be determined as a function of wavelength at a given time, such as in the steady state, $\alpha\beta_0$ can be immediately determined as a function of λ. According to the results in [8.66] $\alpha\beta_0$ is nearly independent of the wavelength Λ of the UV-photons. Thus, all UV-photons absorbed by photochromic centres create the same photolysis products, irrespective of their wavelength or energy.

The thermal regeneration rate $-\frac{r}{\alpha}k_{\mathrm{i}}^2(\lambda, t)$ in (8.4) can be enhanced by irradiating the photochromic glass in the visible spectral region, as has been pointed out before. In a recent investigation [8.68] this enhancement has been shown to be given by

$$- \alpha(\lambda)\gamma(\lambda_{0b})k_i(\lambda_{0b})I(\lambda_{0b}) , \tag{8.11}$$

wherein λ_{0b} refers to the wavelength of the bleaching photons, $I(\lambda_{0b})$ is the flux density of the bleaching photons, $k_i(\lambda_{0b})$ is the induced absorption constant at the bleaching wavelength λ_{0b}, and $\gamma(\lambda_{0b})$ is a factor of proportionality. Adding the term (8.11) to the right-hand side of (8.4) one obtains

$$\frac{dk_i(\lambda,t)}{dt} =$$

$$\alpha(\lambda)\beta_0\left(1-\frac{k_i(\lambda,t)}{\kappa_i(\lambda)}\right)K_i(\Lambda)J(\Lambda)-\frac{r}{\alpha(\lambda)}k_i^2(\lambda,t)-\alpha(\lambda)\gamma(\lambda_{0b})k_i(\lambda_{0b},t)I(\lambda_{0b}) . \tag{8.12}$$

This rate equation describes the kinetics of darkening and regeneration if excitation of the photochromic glass occurs both in the UV at Λ and in the visible at λ_{0b}.

In order to test (8.12), the induced absorption constant of the steady state $k_{i,st}$ has been measured as a function of the flux density of UV-photons, J, with the flux density $I(\lambda_{0b})$ of the bleaching photons as a parameter [8.68]. The results are shown in Fig. 8.12 by the data points. The family of curves represents the results of a fitting routine using (8.12) with $dk_i(\lambda,t)/dt = 0$. One can see that all data can be described very well by (8.12).

8.2.5 Generalisations and Consequences

Equation (8.12) can be extended to a more general case, for example to the solar power density spectrum, which is characterized by a rather broad spectral distribution of photons shown in Fig. 8.6. In this case, it is necessary to drop the discrimination between "writing", "reading" and "bleaching" photons, although we still have to consider the corresponding rates.

With the spectral distribution function $S(\lambda,t)d\lambda$ of a broad spectrum one has to replace in (8.12) the product $K_i(\Lambda)J(\Lambda)$ by the integral

$$\int K_i(\lambda)S(\lambda,t)d\lambda , \tag{8.13}$$

and the product $\gamma(\lambda_{0b})k_i(\lambda_{0b},t)I(\lambda_{0b})$ by the integral

$$\int \gamma(\lambda)k_i(\lambda,t)S(\lambda,t)d\lambda . \tag{8.14}$$

Both integrals extend over the whole range of wavelength of the spectrum under consideration. The spectral distribution of the radiation $S(\lambda,t)d\lambda$ has been assumed to depend on time. Inserting (8.13) and (8.14) into (8.12) allows one to calculate the induced absorption constant if $S(\lambda,t)d\lambda$ has before been determined independently. The application of (8.12), however, requires still several refinements. In (8.12), the absorption constant $K_i(\Lambda)$ of the photochromic centres has been assumed to be independent of time. In contrast to this, $K_i(\Lambda)$ can vary slightly, since the photochromic centres are decomposed photolytically which reduces $K_i(\Lambda)$. Therefore, during the darkening, the actual $K_i(\Lambda)$ will be smaller than the data shown in Fig. 8.5, which have been determined for the situation without photolytical decomposition. It has been estimated that $K_i(\Lambda)$ can decrease by about 10% [8.66], which has to be taken into account for a more accurate calculation of $k_i(\lambda, t)$ according to (8.12).

If $S(\lambda, t)d\lambda$ and the initial value $k_i(\lambda, 0)$ are known, one can calculate with (8.12) the induced absorption constant as a function of time only for a thin surface layer of the photochromic glass. For thick samples, one has to take into account for the calculation that the initial spectral distribution $S(\lambda, t)d\lambda$ changes considerably with the penetration depth x into the sample. This is obviously due to the absorption constant of the bulk glass k_0. However, there are further changes due to the induced absorption constant $k_i(\lambda, t)$, which extends over both the visible and the UV spectral regions and effects both the darkening and the bleaching rates. The corresponding modifications of (8.12) have to be included, since for practical applications of photochromic glasses one is not interested in the local induced absorption constant alone, but rather in the transmittance for a given thickness of the glass and a given spectral distribution of the incident radiation. Further complications may arise from the fact that the radiation may be incident under different angles. Such calculations have to be done numerically and have not been dealt with in the present article, which only covers the basic phenomenological model. In this context, one should mention the experimental work on the transmittance of photochromic glasses under natural illumination published by *E. Sutter* and *W. Möller* [8.69].

8.2.6 Conclusions

Glasses can be rendered photochromic by the addition of silver halides to the melt in amounts of some tenths of a percent. In order to achieve photochromy, the silver halide must precipitate in the glass matrix, induced by a heating process, and form small silver halide particles. To increase the sensitivity, one adds to the melt Cu^+ ions which are built in the particles during the heat treatment. The precipitation of small silver halide particles can be observed directly with an electron microscope and indirectly by a shift of the absorption edge in the UV and blue spectral regions.

The darkening of the photochromic glasses is caused by UV photons which decompose or photolyze the silver halide particles. The main reaction products of the photolysis are silver specks near the interface silver halide/glassy matrix and Cu^{2+} ions. Both kinds of reaction products absorb in the visible part of the electromagnetic spectrum. Since the reaction products of the photolysis stay close together in a silver halide particle, regeneration into the initial state is possible. In this case, the silver is redistributed in the silver halide particle as Ag^+ ions, and Cu^{2+} ions are reduced to Cu^+ ions.

The elementary steps of the darkening and the regeneration are shown qualitatively in a simple energy band diagram. The rates of the limiting elementary steps of the electronic and ionic processes are combined to a simple rate equation which is suitable to describe both the darkening and the regeneration kinetics, as well as the optical bleaching quantitatively. In order to test the theoretical results, the kinetics have been investigated spectroscopically using narrow irradiation and bleaching bands. The results can be extended mutatis mutandis to describe also the kinetics for broad band irradiation (i.e. sunlight) which is necessary in order to evaluate the changes of the transmission of sunglasses. The necessary modifications are discussed. For this case, which is of importance for practical applications, one has to consider that the distribution of the irradiance spectrum is a local function of the penetration depth in the sample because of the absorption spectrum of the base glass. In addition, this absorption spectrum changes with that depth under irradiation due to the photolytic decomposition of the silver halide particles and due to the photolytic reaction products. Once these quantities have been determined experimentally, one can calculate for a given irradiance spectrum the transmissivity as a function of time for an arbitrary thickness of the respective type of photochromic glass used in practical applications.

8.3 Ultraviolet-Transmitting Glasses

Monika J. Liepmann, Norbert Neuroth

The fundamentals of transmission and reflection have already been discussed in Sect. 2.3 and will be further explained for different glass types in this and the following section. Mechanisms which lead to optical absorption in the ultraviolet and infrared regions of the electromagnetic spectrum will be described. Some extensive reviews on the subject of optical absorption of glasses were provided in the late 1970s by *Sigel* [8.70] and *Wong* and *Angell* [8.71].

8.3.1 Intrinsic Absorption

Principal causes of optical losses in glass are: intrinsic absorption, absorption resulting from the presence of impurities and light scattering within the bulk glass. In the ultraviolet spectral region transmission is limited by electronic transitions, whereas in the infrared region atomic and molecular vibrations are responsible for absorption. Glasses in useful thicknesses become opaque for wavelengths shorter than visible or near ultraviolet radiation. Because of strong absorption, micrometer thin samples are required for transmission measurements in the UV-region. However, in regions of strong absorption, the reflectance is also high; therefore, reflectance measurements are often utilized at the short wavelength region.

A very good UV-transmitting material is pure silica (SiO_2). The basic building blocks in the structure of this material are corner-sharing tetrahedral $[SiO_4]^{4-}$ units. Differences between vitreous and crystalline silica arise in the glass from a broader distribution of the Si-O-Si bond angles and a more random arrangement of the tetrahedra, resulting in an irregular, more open structure. In Fig. 8.13

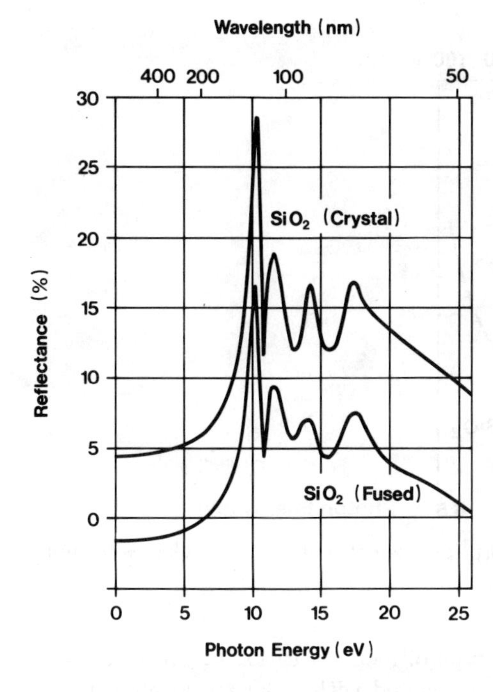

Fig. 8.13. UV-reflectance spectra of crystalline and fused quartz. For clarity, the values for fused quartz have been lowered by 5 percent [8.72]

the UV-reflectance spectra of crystalline and fused quartz are represented [8.72]. The similarity of the spectra has repeatedly been confirmed in subsequent studies [8.73–75]. The results suggest that the UV-absorption in SiO_2 is dominated by the short range order, and it is nearly independent of the long range order of a glassy network. The absorption is caused by the excitation of an electron from the valence band to unoccupied higher energy states. The first absorption peak in SiO_2 does not appear until 10.2 eV (122 nm). This sharp peak has been attributed to a *Wannier* exciton [8.76]. The next peaks at higher energy are due to interband transitions (Oxygen $2p \rightarrow$ Silicon $3d$) [8.76]. In SiO_2 the absorption corresponds to the transition of electrons belonging to tightly bound bridging oxygens. It is well-known that in oxide glasses the intrinsic absorption edge is determined by the interaction of UV-radiation with electrons belonging to the oxygen ions. Glasses containing weakly-bound oxygens are less transparent in the UV-region than vitreous SiO_2. For instance, the features in the reflectance spectra of multicomponent silicate glasses are similar to those of vitreous silica at high energy above 9.5 eV (below 130 nm), but the bands are considerably broader and additional absorption occurs at lower energy [8.76–78]. Several examples are shown in Figs. 2.27 and 2.29 in Sect. 2.3; the strong (exciton) peak is broadened and shifted to somewhat lower energy. The introduction of various metal oxides (network modifier oxides) into the SiO_2 network produces so-called non-bridging oxygens (NBO). These are oxygen ions which are bound to one silicon ion only and do not form a bridge between two Si^{4+} ions. The glass network is disrupted and its average chemical bond strength is weakened. Ultraviolet absorption bands are shifted to lower energies (longer wavelengths) since NBO bind excitable electrons less tightly than do

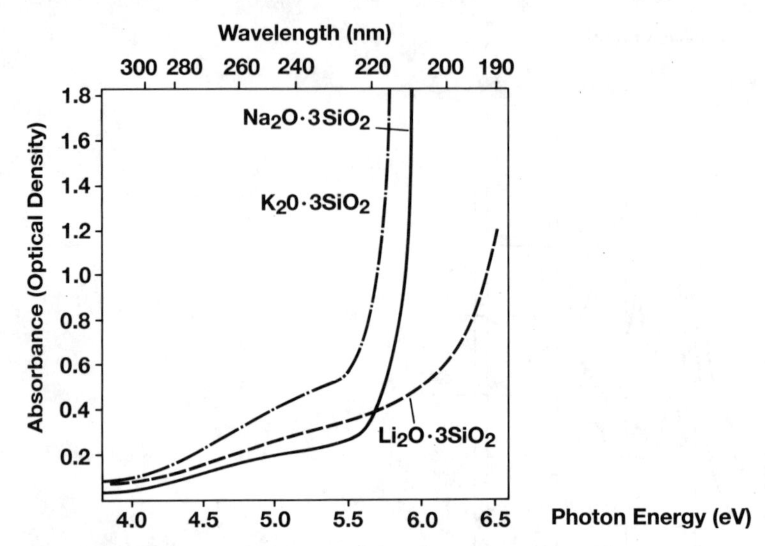

Fig. 8.14. Effect of alkali ions on the absorption edge of binary silicate glasses (sample thickness: 2 mm) [8.79]

bridging oxygens. Depending upon the type of modifier cation, additional peaks or shoulders in the reflectance spectra associated with NBO are located in the range of 8.8 eV (140 nm) down to 4.7 eV (up to 264 nm) for silicate glasses [8.75]. With increasing glass modifier content, the number of NBO is increased and the absorption edge shifts to longer wavelengths. In general, the spectral position of the absorption due to NBO is influenced by several factors such as bond strength, coordination/symmetry, ionic radii and concentration of the particular modifier ions.

For cations having low valence states such as alkali or alkali earth ions, the UV-absorption edge shifts to longer wavelengths with increasing ionic radii. Figure 8.14 illustrates this effect for the series of alkali ions (Li, Na, K) in simple binary silicate glasses [8.79]. The number of NBO is the same in all three glasses. The glass containing the smallest modifier ion (lithium) has the best UV-transparency.

Sigel [8.73] has shown that UV-transmission of SiO_2 glass containing small amounts of alkali can be improved by adding aluminum oxide (Al_2O_3). UV-transmission spectra of SiO_2 glass doped with Na and K and co-doped with equal amounts of Al are shown in Fig. 8.15. Aluminum is able to substitute for silicon in four-fold coordination with oxygen; thereby eliminating NBO as shown schematically in Fig. 8.16 [8.73]. The local charge is compensated through a nearby alkali ion.

The addition of boron oxide to alkali silicate glasses likewise shifts the UV-edge to shorter wavelengths. B_2O_3, which is a glass former, reduces the concentration of NBO. It has been suggested that the first introduction of B_2O_3 will eliminate NBO by formation of four-fold coordinated boron up to comparable amounts of alkali and boron in the silicate glass. If the amount of B_2O_3 is further increased, it participates in the network with three-fold coordination without forming NBO

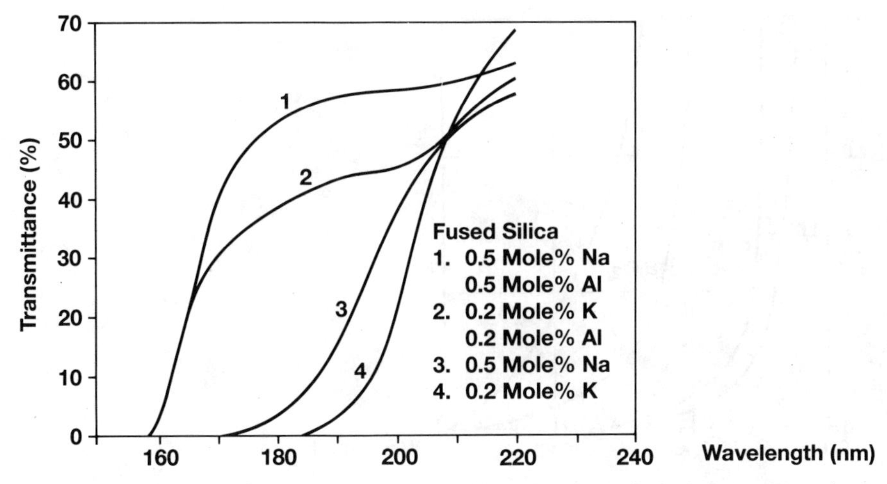

Fig. 8.15. Effect of aluminum on the UV-absorption of alkali-doped silica (sample thickness: 1 mm) [8.73]

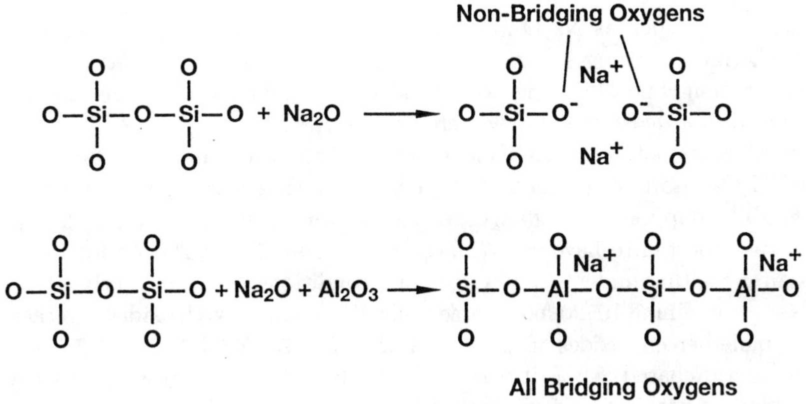

Non-Bridging Oxygens

O–Si–O–Si–O + Na_2O ⟶ O–Si–O^- Na^+ Na^+ O^-–Si–O

O–Si–O–Si–O + Na_2O + Al_2O_3 ⟶ Si–O–Al–O–Si–O–Al–O (Na^+, Na^+)

All Bridging Oxygens

Fig. 8.16. Schematic representation in two dimensions of (1) the creation of non-bridging oxygens by the introduction of Na into the SiO_2 network and (2) the ability of aluminum to restore the bridging in spite of the presence of alkali [8.73]

[8.79]. The transmission of these borosilicate glasses extends to shorter wavelengths than that of the the alkali silicate glass.

Pure B_2O_3 glass also possesses good UV-transparency, but it is, by itself, unsuitable for commercial applications due to its hygroscopic nature. It has been shown that the UV-transparency of vitreous B_2O_3 decreases with increasing water content [8.80]. The structure of boron oxide is based on planar $[BO_3]^{3-}$ units. The boron atoms are three-fold coordinated with oxygen. In vitreous B_2O_3 three $[BO_3]^{3-}$ triangles share corners to form a so-called boroxol ring [8.81]. These $[B_3O_6]^{3-}$ boroxol rings are present together with $[BO_3]^{3-}$ triangles in a random network. From neutron diffraction studies it has been suggested that in pure vitreous B_2O_3, 60 percent of the boron atoms are incorporated in boroxol rings [8.82].

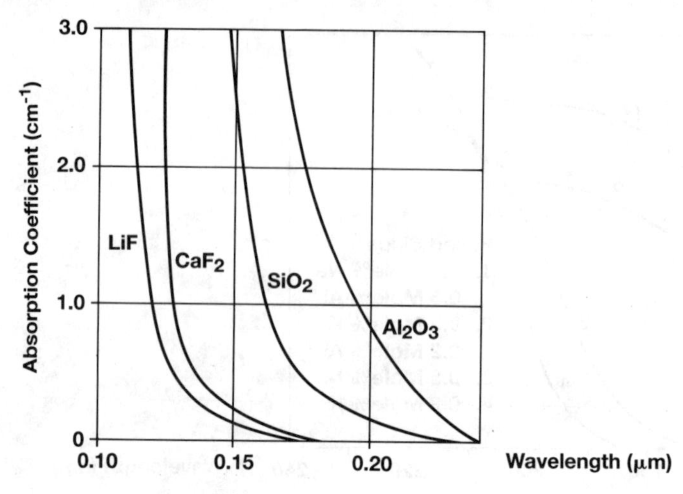

Fig. 8.17. Ultraviolet absorption edge of different crystals [8.97]

When alkali oxides such as Na_2O are added to B_2O_3, the coordination number of some of the borons changes from 3 to 4, thereby forming $[BO_4]^{5-}$ tetrahedron [8.83]. The creation of non-bridging oxygens is avoided and good UV-transparency is preserved up to at least 15–20 mole percent Na_2O [8.84].

Borate glasses containing 15 mole percent alkali oxide and an equivalent amount of Al_2O_3 also have the potential for high UV-transparency since the formation of NBO is suppressed. A glass of the molar composition 1.0 K_2O, 4.5 B_2O_3, 1.0 Al_2O_3 was reported to have over 75 percent transmission at 200 nm for 1 mm thickness after melting under strongly reducing conditions [8.85]. Recently, other borate glasses containing 10–20 mole percent alkali and alkali earth oxides together with 10–20 mole percent oxides of the type X_2O_3 (X = Al, Y, Ga) and 60–70 percent B_2O_3 were prepared [8.86]. It was confirmed by Raman spectroscopy that a boron crown glass within this compositional range contained the same structural elements as pure B_2O_3 glass. The boron was almost exclusively present in the form of boroxol groups. The UV-absorption edge nearly corresponded to that of pure vitreous B_2O_3 but was at longer wavelengths than that of fused quartz. Studies of the effect of different cations on the UV-absorption edge in borate glasses were described by several authors in the late 1970s and early 1980s [8.70,87–90].

An oxide glass former with a potential for excellent UV- transparency is phosphorous oxide. However, the pure oxide is extremely hygroscopic and it becomes difficult to measure the UV-absorption edge of vitreous anhydrous P_2O_5. The absorption edge shifts to longer wavelengths with increasing water content [8.91]. Multicomponent phosphate glasses are less common than borate and silicate glasses, and only a few articles can be found describing the UV-transparency of phosphate glasses [8.91–96].

A different route to improve the UV-transparency of glasses significantly is to utilize fluorides instead of oxides. The electronic transitions of fluorides cause absorption at higher energy (shorter wavelenghts) than those of oxides. It is well-known that fluoride crystals transmit to shorter wavelengths than oxide crystals; some examples are shown in Fig. 8.17 [8.97].

In ionic crystals the energy for electronic excitation can be approximated by [8.97]:

$$h\nu = \frac{Ae^2}{r} + E + I \,, \tag{8.15}$$

where ν is the frequency of the UV absorption and h is Planck's constant. The first term on the right side denotes the energy necessary to overcome the Coulomb forces of the individual lattice charges, where A is Madelung's constant, e the electron charge and r the interionic distance. E is the electron affinity of negative ions and I is the ionization potential of the positive ions. Crystals having small ions and high bond strength will absorb at high frequencies (high energy). This relation is approximately applicable for those glasses, e.g. fluoride glasses, where the short range order of the ions is similar to that of crystals. For instance, fluorine-containing glasses have inherently better UV-transparency than oxide glasses without fluorine [8.75]. From the known vitreous materials, glasses based on beryllium fluoride transmit to the shortest wavelength. Vitreous beryllium fluoride has superior UV-transmission to vitreous silica; its UV cut-off wavelength is below 160 nm. Even beryllium fluoride glasses with up to 30 mole % network modifiers transmit to lower wavelengths than vitreous silica [8.98]. However, beryllium fluoride is very hygroscopic and highly toxic. Special precautions must be taken during melting, forming and handling of these glasses.

Recently, a UV-transmitting alkali earth fluoroaluminate glass was prepared [8.99]. In a thin (0.76 mm) sample the glass with the composition 20 MgF_2 . 50 CaF_2 . 30 AlF_3 transmitted light down to wavelengths of 160 nm. This UV-transparency is nearly as far as that of some specially processed fluorophosphate glasses [8.100–103]. Practically, these multicomponent fluoride glasses transmit nearly as far into the ultraviolet region as vitreous silica. This is consistent with UV-reflectance spectra of fluoroaluminate [8.104] and fluorophosphate [8.75] glasses. In these fluorine-containing glasses absorption bands are located at higher energy than in multicomponent oxide glasses. These fluoride glasses do not show any reflection peaks below 10 eV (above 124 nm). On the other hand, fluoride glasses containing heavy metals, such as glasses based on zirconium fluoride, show reflection peaks at 9.3 eV (133 nm) and even as low as 7.3 eV (170 nm) [8.104]. Their UV-transmission is inferior compared to vitreous silica.

8.3.2 Effects of Impurities

The ultraviolet absorption discussed above pertains to electronic transitions which originate from the glass structure and are intrinsic to the particular glass composition. In the following, the significant effect of impurities on the ultraviolet absorption edge of glasses will be explained. It is well-known that ions of 3d transition metals such as V, Cr, Mn, Fe, Co, Ni, Cu as well as most of the rare earth elements give rise to colour in glasses [8.105, 106]. Figure 8.18 shows ultravio-

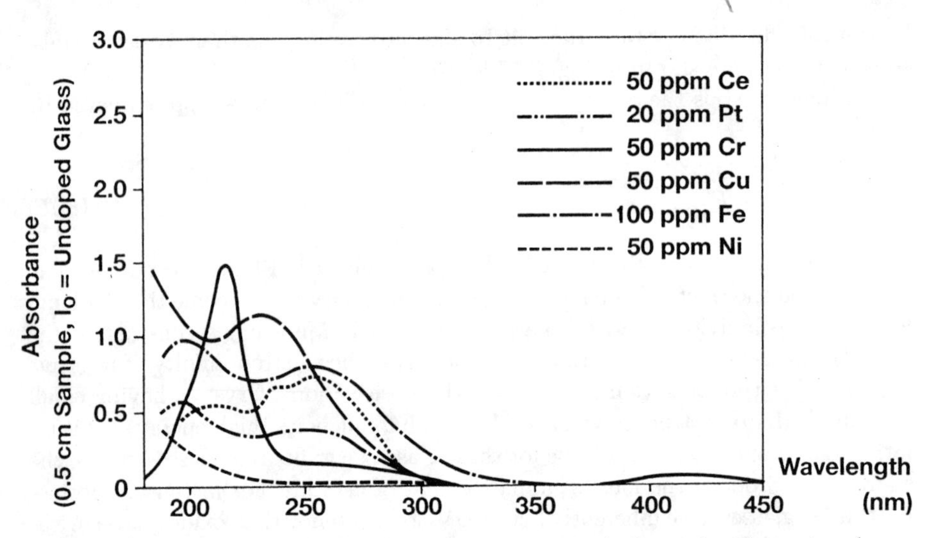

Fig. 8.18. Ultraviolet absorbance of metal ion impurities in fluorophosphate glass (sample thickness 5 mm) [8.107]

let transmission spectra of some impurities in fluorophosphate glasses melted in vitreous carbon crucibles [8.107].

The spectra illustrate that small amounts in the ppm range of these transition elements and platinum already have a significant effect in lowering the ultraviolet transparency. The most common impurities in optical glasses are platinum and iron. Platinum impurities originate from the melting unit, and trace iron impurities are commonly present in all raw materials. The impurities can have different absorption characteristics depending on the melting conditions (oxidizing or reducing) and the glass composition. For example, platinum can be present in the glass in ionic form or as metallic particles. UV-absorption spectra for ferric and ferrous iron in sodium silicate glass are shown in Fig. 8.19 [8.108]. Fe^{3+} has a strong absorption band centred at 230 nm. Fe^{2+} shows only little absorption in this region, but it causes absorption in the infrared. Other ions which have strong absorption in the ultraviolet are from the metals cerium and titanium. The UV-absorption spectra of cerium in silicate glass is shown in Fig. 8.20 [8.109]. Ce^{3+} ions cause a single absorption band at 314 nm. A second absorption band due to Ce^{4+} appears at 240 nm if the glass is melted under oxidizing conditions. In order to improve the ultraviolet transparency of glasses with trace impurities, it is mostly advantageous to melt under reducing conditions since the prevalent impurity is generally iron.

8.3.3 Commercial Glasses

Ultraviolet transmitting optical materials are required in spectroscopy as window materials for lamps and cuvets, optical filters and lenses. They are utilized in

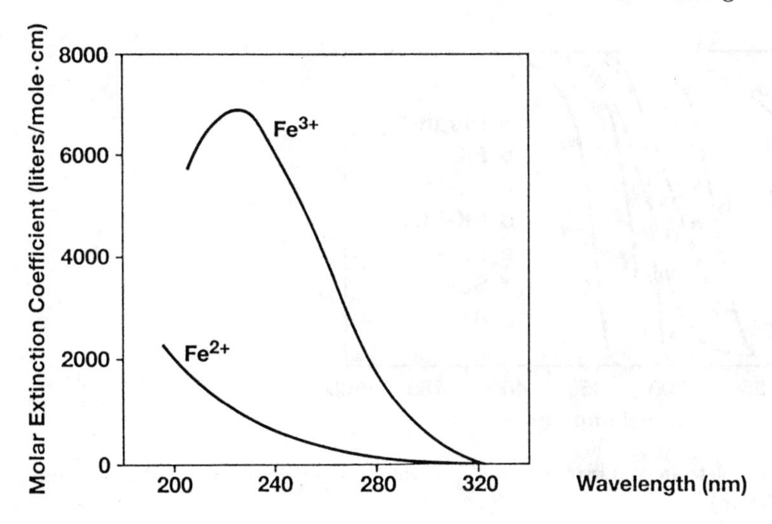

Fig. 8.19. Ultraviolet absorption spectra of Fe^{2+} and Fe^{3+} in soda-silicate glass [8.108]

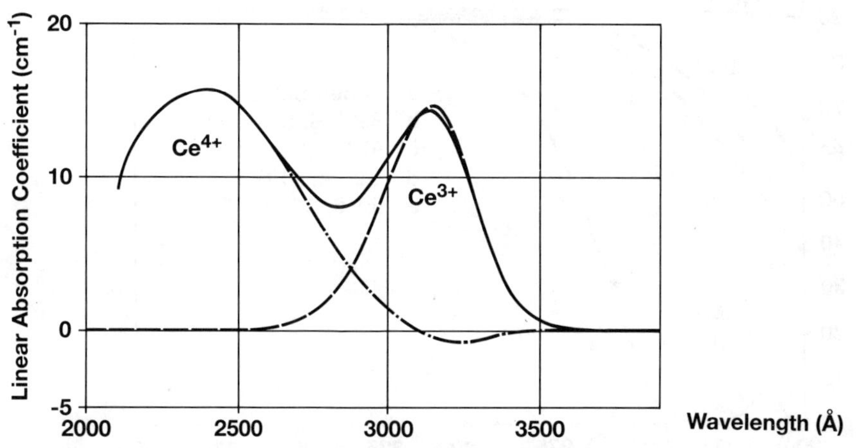

Fig. 8.20. Absorption spectrum of Ce^{4+} in silicate glass with 5.33×10^{18} cerium ions/cm^3. — absorption of cerium when part is present as Ce^{4+}; – – – absorption of 5.33×10^{18} Ce^{3+}/cm^3; – · – · – difference between the other two curves [8.109]

the semiconductor industry, in microlithography and as lens materials in excimer laser system optics. Beside serving in high energy laser applications, ultraviolet materials are also required for fibre optics and space applications.

The best UV-transmitting materials are fluoride crystals such as LiF, CaF_2, MgF_2, etc. Among the vitreous materials, an outstanding oxide glass is "fused quartz", in particular high purity synthetic SiO_2 (vitreous silica).

In commercial optical glasses the UV absorption edge is shifted to longer wavelengths. Optical glasses are multicomponent glasses and the addition of heavier elements as glass modifiers has an adverse effect on the UV-transparency of the

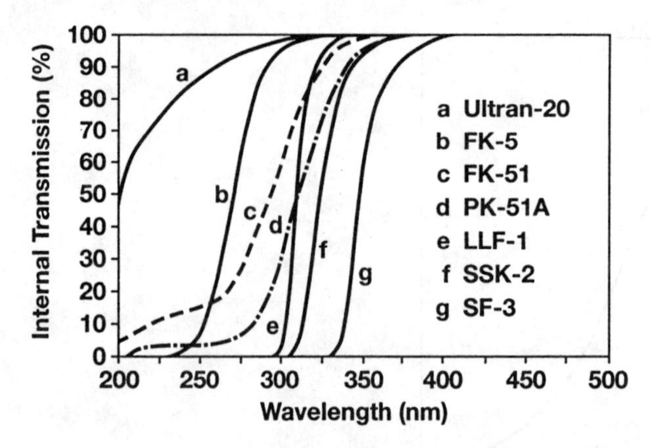

Fig. 8.21. Ultraviolet internal transmission spectra of some Schott optical glasses (sample thickness: 5 mm)

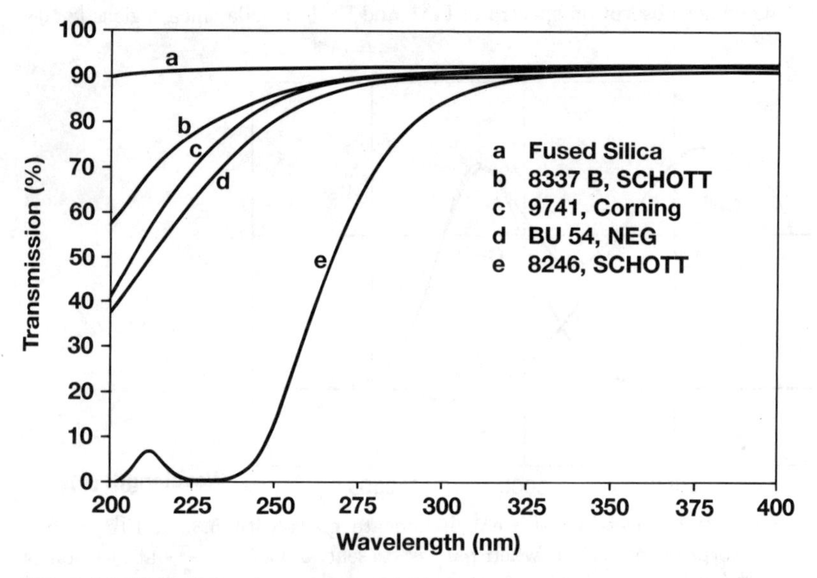

Fig. 8.22. Ultraviolet transmission spectra of some UV-transparent technical glasses for photodiodes, photodetectors and lamps (sample thickness: 1 mm)

glasses. Crown glasses generally transmit further into the ultraviolet than do flint glasses because they contain mostly lighter elements.

Moreover, most optical glasses contain trace impurities which sometimes can obscure the intrinsic UV-absorption edge. Impurities are contained in the raw material batches, or they can enter the melt during the production process by reactions of the glass with the melting unit. Figure 8.21 shows transmission spectra of several optical glasses with good UV-transmission in 5 mm thickness. Here the letters K and F indicate crown and flint glasses, respectively. The 50 percent transmission value for the multicomponent oxide glasses is in the wavelength range of

270 to 350 nm and greater. Superior UV-transmission is shown by recently developed Ultran® glasses, which belong to the fluorophosphate glass family (Ultran® is a registered trademark of Schott Glaswerke for certain UV-transmitting optical glass types). The 50 percent transmission value for 5 mm thickness is located near 200 nm. Through this development in recent years, crown glasses with Abbe values of 68 (vitreous silica) to 84 (Ultran® 20) are available for UV-imaging optics below 300 nm. Fluorophosphate glasses also have an anomalous partial dispersion (see Fig. 2.37 in Sect. 2.3).

In Fig. 8.22, transmission spectra of several ultraviolet transmitting technical glasses in 1 mm thickness for photodetector technique and lamp applications are shown. For comparison the spectrum of synthetic vitreous silica is included in the figure.

8.4 Infrared-Transmitting Glasses

Monika J. Liepmann, Norbert Neuroth

8.4.1 Sources of Absorption

In the infrared region, transparency is restricted by the vibrations of molecules, atoms and ions in the glass network. The long wavelength limit is therefore often referred to as vibrational or multiphonon edge. The absorption results from the interaction of the incident radiation with the vibrational modes of the glass structure which can cause changes of the dipole moment. For simple diatomic molecules the fundamental absorption frequency can be approximated by two point masses, m_1 and m_2, (ions of opposite charge) oscillating in opposition to one another. The vibrational frequency, V, for the simple harmonic motion is given by:

$$V = \frac{1}{2\pi}\sqrt{\frac{f}{M}} \tag{8.16}$$

where f is the force constant and M is the reduced mass:

$$\frac{1}{M} = \frac{1}{m_1} + \frac{1}{m_2} \, . \tag{8.17}$$

With increasing ionic mass the frequency of vibration becomes smaller; thus heavier ions and weaker bonds (small force constant) are found in materials with good infrared transparency (transmit to long wavelengths). In the case of real materials, mostly anharmonic oscillators are present which give rise to a series of overtone vibrational bands at higher frequencies than the fundamental frequency. These multiphonon absorption processes are intrinsic to the material.

Extrinsic absorption can result from different impurities dissolved in the glass network. Traces of light elements and molecules; e.g., combinations of hydrogen, carbon and oxygen such as hydroxyl, water and carbon oxides give rise to absorption bands at specific wavelengths in the infrared. The fundamental OH-stretching vibration occurs in the range of 2.7 to 3.0 μm in oxide and fluoride glasses. Additional OH absorption bands can be present near 3.6 and 4.2 μm depending on the glass composition as shown in Fig. 2.41 of Sect. 2.3 for silicate glasses [8.110, 111]. In glasses which transmit to longer wavelengths a band resulting from OH-bending vibrations is observed at 6 μm; for example, in As_2S_3 glass [8.112]. It is therefore important to reduce the water content in infrared-transmitting glasses by special production methods. In fluorozirconate glass a band at 4.25 μm is assigned to the assymetric stretching vibration of carbon dioxide dissolved in the glass [8.113]. Impurities from transition metal ions should also be avoided because some of these ions have electronic transitions which give rise to absorption bands in the near infrared. The most common of such impurities is iron. Fe^{2+} has a broad absorption band at 1.1 μm and a second band in phosphate glasses at 2.1 μm [8.114].

8.4.2 Oxide Glasses

The molecular vibrational modes of the network former determine the intrinsic limit of infrared transparency in traditional oxide glasses. Molecular vibrations of network modifiers occur at longer wavelengths. For the common oxide glass formers B_2O_3, SiO_2, P_2O_5 and GeO_2, the fundamental infrared absorptions are in the range of 6 μm to 25 μm [8.115] as shown in Fig. 8.23. Several absorption maxima are observed; they originate from different molecular vibrations such as stretching and bending modes. For infrared transparency, the strong fundamental absorption at the short wavelength side is of importance. Its maximum is approximately located at 7.5 μm for B_2O_3 and P_2O_5; at 9.0 μm for SiO_2 and at 12 μm for GeO_2. For heavier, less common glass formers such as As_2O_3, Sb_2O_3, and TeO_2, the corresponding values are in the region of 12.5 μm to 14.5 μm. At the maxima the absorption coefficient reaches values of 10^4 to $10^5\,cm^{-1}$. Such strong fundamental vibrations give rise to significant overtone vibrational bands at approximately double the frequency or half the wavelength as illustrated for fused silica in Sect. 2.3, Fig. 2.31. Thus, for commercial use where thicknesses of the order of millimeters are required, the first overtone determines the limit of infrared transparency.

Considering the oxide glass systems, borate and borosilicate glasses have the least infrared transparency as shown in Fig. 8.24 [8.116]. Silicate and fluorophosphate glasses transmit to longer wavelengths followed by calcium-aluminate, germanate and tellurite glasses. Oxide glasses containing metals with heavy mass and low field strength have the longest wavelength cutoff. Of the currently known systems, glasses based on heavy metals such as lead or cadmium and bismuth together with gallium oxide transmit to the longest wavelengths [8.117–119]. These so-called heavy metal oxide (HMO) glasses are transparent up to 7 or 8 μm in 1 mm thickness (see also Fig. 8.24).

The glass-forming ability of some of these systems is relatively good; however, they have a much higher tendency to crystallize upon forming than silicate glasses. A rough measure to characterize the devitrification tendency of an unstable glass is the difference in degree Celsius between the onset of crystallization, T_x (determined by differential scanning calorimetry [DSC] or differential thermal analysis

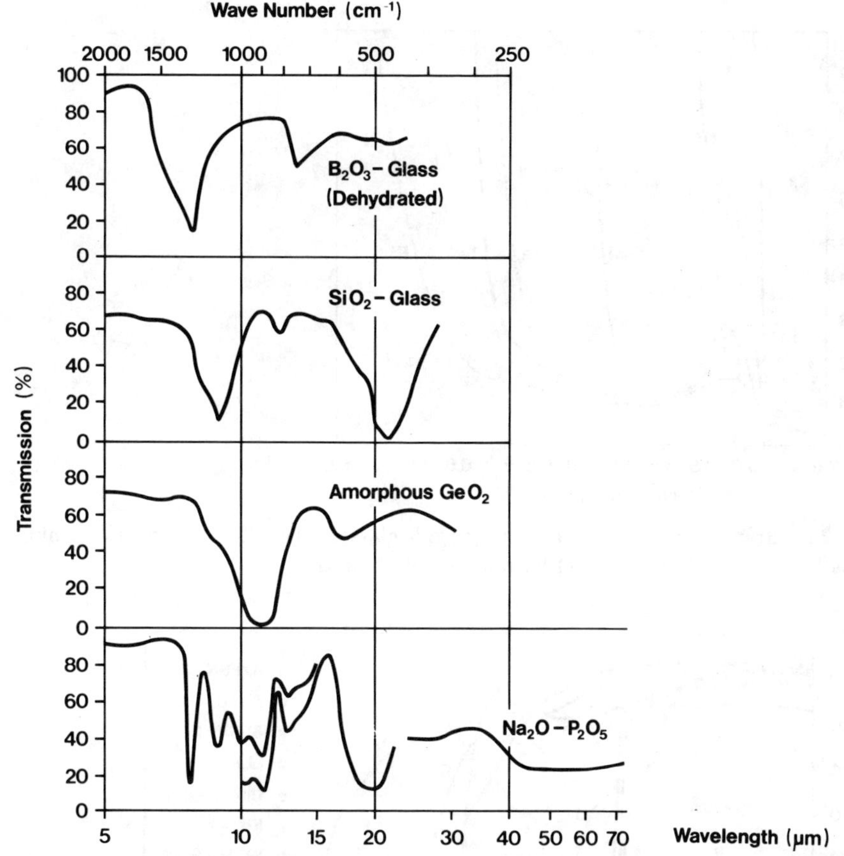

Fig. 8.23. Infrared transmission spectra of different glass formers (powder immersion in KBr or foils one or several microns thick) [8.115]

[DTA]), and the transformation temperature, T_g. Good glass forming ability is expected for materials with large values for $(T_x - T_g)$. However, this method is a very crude approximation and does not always provide a reliable scale for the crystallization tendency. Bulk glass production in thicknesses of millimeters may be possible if the temperature difference is approximately 100 °C. For some HMO glasses $T_x - T_g$ is more than 150 °C, and bulk glasses of 10 mm thickness were prepared [8.117]. Conversely, most silicate glasses are very stable and do not exhibit any crystallization temperature measurable by DSC or DTA. The visible transmission cut-off is near 400–500 nm for HMO glasses, resulting in an orange-to-red appearance. They have expansion coefficients of approximately 10×10^{-6}/K and refractive indices (n_d) in the vicinity of 2.4. HMO glasses are in competition with infrared-transmitting non-oxide glasses because they are, unlike chalcogenide glasses, transparent into the visible region, and they could offer some advantages in terms of better thermal and mechanical properties. Compared to halide glasses

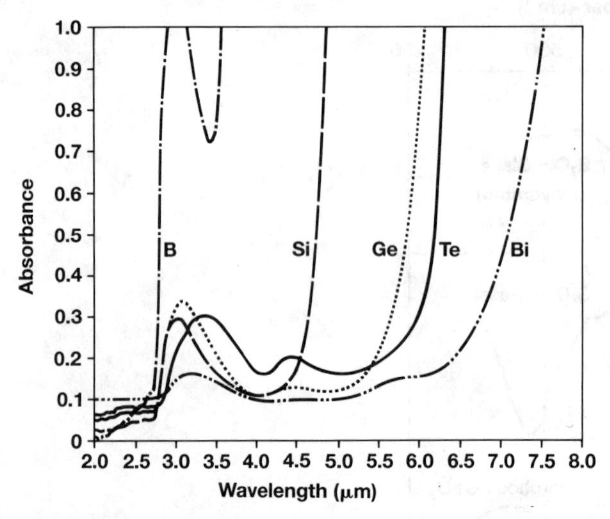

Fig. 8.24. Infrared absorbance of analogous oxide glasses with different network-forming cations [8.116]; sample thickness 1.85 mm, except Bi: 2.0 mm

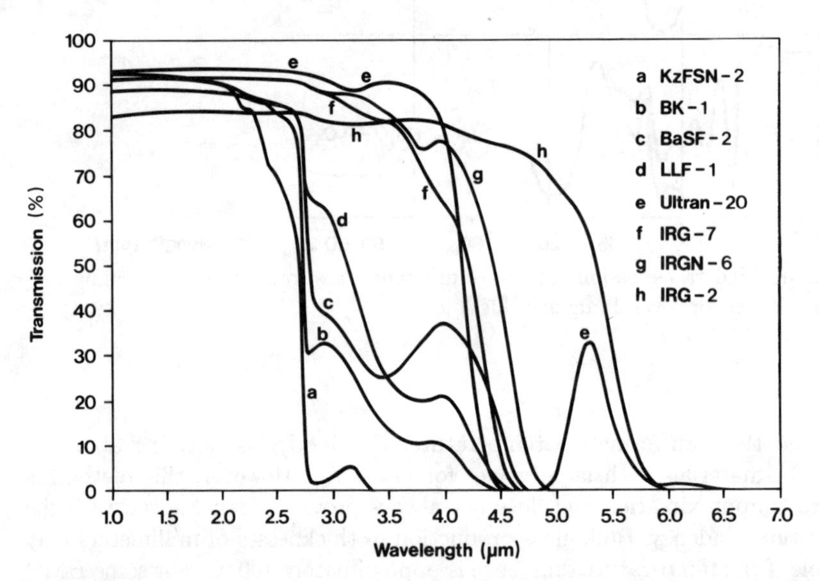

Fig. 8.25. Infrared transmission spectra of some Schott optical glasses and special IR-transmitting glass types (sample thickness 2.5 mm). The OH absorption of the glasses may vary depending upon the melting process

their chemical durability is favorable. HMO glasses have been known for only a few years, and they may, in the future, partially replace chalcogenide glasses.

Figure 8.25 shows infrared transmission spectra of some optical oxide glasses. Borosilicate glasses such as KzFSN 2 and BK 1 transmit to only approximately

2.75 μm (50% internal transmission for 2.5 mm thickness) because of their high OH content (absorption at 2.9 μm) and B-O-overtone vibration (absorption at 3.7 μm). BaSF 2, a barium and lead oxide-containing silicate glass, has weaker OH absorption (at 2.9 and 3.4 μm) and better IR-transmission. Depending on the water content, lead silicate glasses can have reasonable IR-transparency up to 4.2 μm as illustrated by the two lead silicate glasses LLF 1 an IRG 7. Exceptions to pure oxide glasses are fluorine-containing fluorophosphate glasses such as Ultran 20® and IRG 9 which also transmit to 4.2 μm since their water content is low. Even better infrared transparency is exhibited by a Ca-Al-silicate glass (IRG N6) and germanate glasses; for example, IRG 2 (transparent to 5.3 μm).

8.4.3 Halide Glasses

Halide glasses belong to the non-oxide glass category. They contain group VII elements of the periodic table; namely, F, Cl, Br and I (halogens) as anions instead of oxygen. Their transparency extends from the visible into the infrared region. For these glasses the previously mentioned rule also applies: the fundamental molecular vibrational modes are in the range of 15 to 45 μm; therefore, transparency below the first overtone up to approximately 7 μm is expected for fluoride glasses. The multiphonon absorption edge is moved to longer wavelengths the heavier the halogen becomes. Thus, chloride, bromide and iodide glasses have transparencies up to approximately 13 μm, 20 μm and 30 μm, respectively [8.120].

The oldest known halide-glass formers are BeF_2, $ZnCl_2$ and AlF_3. BeF_2 and $ZnCl_2$ are both very hygroscopic materials. BeF_2 is an excellent glass-former, but its toxicity is a severe drawback. Glasses based on AlF_3 have a high devitrification tendency. Several fluoroaluminate glasses were investigated recently [8.121–123]; however, only thin samples could be prepared.

Examples of other metal halides which have been reported to act as glass formers are $ThCl_4$, $BiCl_3$, $CdCl_2$, CdI_2 and $ZnBr_2$. Compared with silicate glasses their glass-forming ability and chemical durability are very poor. Most of these glasses have low transformation temperatures, sometimes close to room temperature, which severely limit their practical application.

In 1974, glasses based on ZrF_4 were discovered by *Poulain* and *Lucas* [8.124]. Subsequently, a series of so-called heavy metal fluoride (HMF) glasses was developed. In these glasses fluorides such as ZrF_4, HfF_4 and ThF_4 act as glass formers in the presence of suitable modifiers such as BaF_2. The stability of the glasses against devitrification is increased if one or more additional metal fluorides e.g. of the alkalis, alkali earths or rare earths are included. An example of a fluorozirconate glass which is relatively resistant to devitrification is ZBLAN. This glass consists of five fluoride compounds and has the molar cation composition $Zr_{55}Ba_{18}La_6Al_4Na_{17}$ [8.120].

Intensive research has been initiated since HMF glasses were expected to be more suitable for optical fibres than silica glass. Based on theoretical models, HMF glasses were predicted to have intrinsic *Rayleigh* scattering lower than that attainable in silica glass fibres. The minimum loss in HMF glass is calculated to

Table 8.4. Property ranges of different infrared-transmitting glass types

Glass type	Trans-parency range[a] μm	Refractive index n_d	Abbe value ν_d	Coeff. of thermal expansion 10^{-6}/K	Micro-hardness Vickers	Trans-formation temperature °C	Density g/cm^3
Fused Silica	0.17–3.7	1.46	67	0.51	800	1075	2.2
Oxide Glasses	0.25–6.0	1.45–2.4	20–100	3–15	330–700	300–700	2–8
Fluorophosphate Gl.	0.2–4.1	1.44–1.54	75–90	13–17		400–500	3–4
Fluoride Glasses	0.3–8.0	1.44–1.60	30–105	7–21	225–360	200–490	2.5–6.5
Chalcogenide Glasses	0.7–25	2.3–3.1[b]	105–185[c]	8–30	100–270	115–370	3.0–5.5

[a] 50% internal transmission for 5 mm pathlength
[b] Infrared refractive index at 10 μm
[c] Infrared Abbe value $v_{8-12} = n_{10} - 1/n_8 - n_{12}$

be 10^{-2}–10^{-3} dB/km in the vicinity of 2–3 μm [8.125], while for silica glass the attenuation is 0.16 dB/km at 1.55 μm. In silica glass fibres the predicted transparency has been achieved in practice. For HMF glass, however, many extrinsic sources of scattering losses such as crystallites, bubbles, inclusions and losses due to impurities are still present.

Compared to conventional silicate and borosilicate glass, HMF glasses have low resistance to moisture corrosion, poor mechanical strength, a low transformation temperature (usually near 300 °C for fluorozirconate glasses [8.126]) and a tendency to form ionic crystals. The devitrification tendency characterized by $T_x - T_g$ is for fluorozirconate glasses in the range of 70 °C to 125 °C. The more stable glass composition ZBLAN has a $T_x - T_g$ value of 123 °C [8.113] and was prepared from 1 l melt volumes into bulk glass specimens with a thickness of 2.5 cm [8.127].

Fluoride glasses have low and medium refractive indices, low dispersions and anomalous partial dispersions. The refractive indices, n_d, for HMF glasses are in the range of 1.45–1.6. The coefficients of thermal expansion are high, typically near 17×10^{-6}/K and the mass densities are mostly above 4.5 g/cm^3. An overview of the properties of fluoride glasses compared to oxide glasses is shown in Table 8.4. Beside the above-mentioned fluorides in the ZBLAN composition, HMF glasses may also contain fluorides of elements such as lithium, lead, indium, zinc, cadmium, yttrium, ytterbium, gadolinium and lutetium. Figure 8.26 is a schematic representation of the spectral transmission of HMF glasses compared to fused silica with a thickness of 5 mm [8.128]. SiO_2 has better transparency in the ultraviolet region than HMF glasses but has limited infrared transmission. Fluorophosphate and fluoroberyllate glasses transmit further into the IR-region. The absorption edge is shifted to longer wavelengths for fluoroaluminate glasses and HMF glasses containing aluminum. Figure 8.26 shows HBLA (57 HfF_4–36 BaF_2–3 LaF_3–4 AlF_3 mole %) which has slightly better IR-transmission than its sister glass, ZBLA (57 ZrF_4–36 BaF_2–3 LaF_3–4 AlF_3 mole %). Extended transparency to more than 10 μm is illustrated by aluminum fluoride-and zirconium fluoride-free HMF glasses

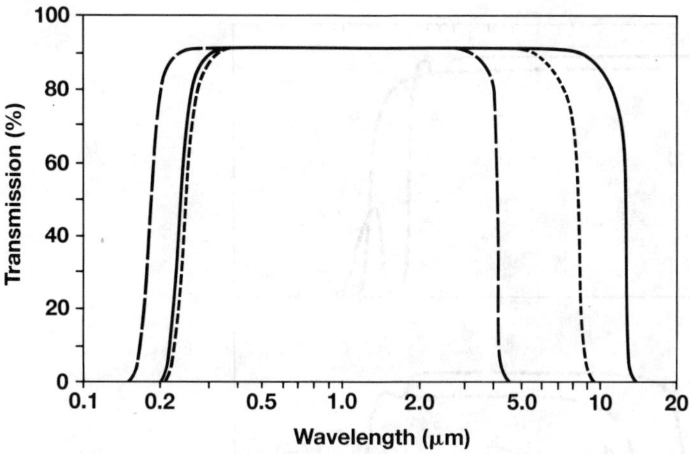

Fig. 8.26. Principal transmission region of heavy metal fluoride glasses and fused SiO_2 (sample thickness approx. 5 mm) [8.128] - - - - HBLA glass (57 HfF_4–36 BaF_2–3 LaF_3–4 AlF_3 mole %); — BZnLuT glass (19 BaF_2–27 ZnF_2–27 LuF_3–27 ThF_4 mole %) and – – SiO_2

such as BZnLuT (19 BaF_2–27 ZnF_2–27 LuF_3–27 ThF_4 mole %) resulting from the high percentage of heavy (high atomic weight) cations.

At present, halide glasses are being researched extensively and a number of recent reviews on these materials can be found in the literature; e.g., [8.128–132].

8.4.4 Chalcogenide Glasses

Another group of non-oxide vitreous materials are the chalcogenide glasses. Instead of oxygen, these glasses contain other elements from group VI of the periodic table; namely sulphur, selenium and tellurium as glass-forming anions. Most chalcogenide glasses are opaque in the visible with the exception of some sulphide glasses which are dark red. They transmit from approximately 1 μm to longer infrared wavelengths than oxide or fluoride glasses. Transmission spectra for chalcogenide glasses with different chemical compositions are shown in Fig. 8.27. Glasses containing heavier anions transmit to longer wavelengths. Sulphide glasses are transparent up to 12 μm, selenide glasses up to 15 μm and telluride glasses transmit to 20 μm.

The classic chalcogenide glass is As_2S_3 which has served since the 50s as infrared transmitting material in commercial applications. Selenide and telluride glasses were developed in the 1960s to 1980s. The three most important ternary systems containing stable glass forming regions are: Ge-As-S, Ge-As-Se and Ge-Sb-Se. The glass forming ability of tellurides is poorer than that of sulphides and selenides. The elements sulphur and selenium can form glass without other elements. Glasses made from Se alone have a T_g of 31 °C; they have been known for a long time [8.133]. These glasses as well as the glasses made from sulphur

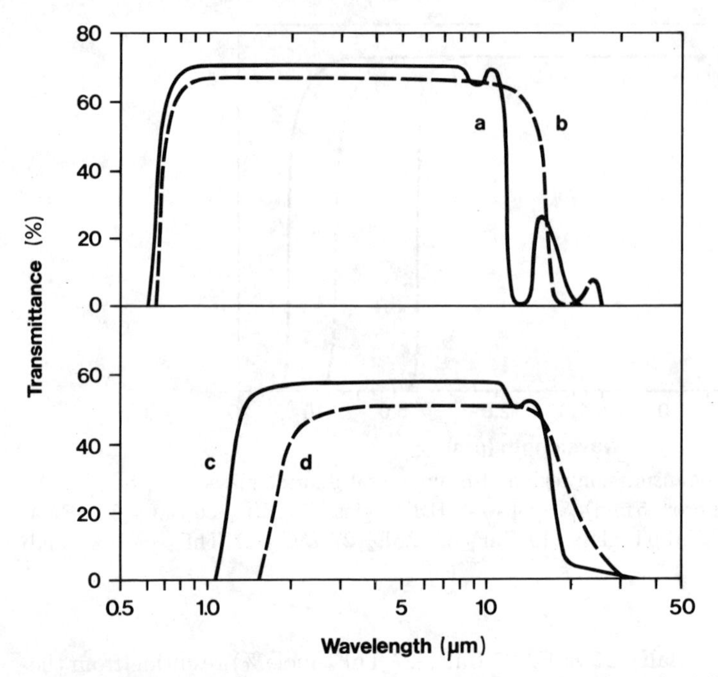

Fig. 8.27. Spectral transmission of chalcogenide glasses with different compositions [8.136]. a: sulphide glass (30 Ge, 20 As, 50 S [atomic %]), sample thickness 1.9 mm, b: selenide glass (34 Ge, 8 As, 58 Se [atomic %]), sample thickness 1.8 mm, c: selenide-telluride glass (30 Ge, 13 As, 27 Se, 30 Te [atomic %]), sample thickness 2.3 mm, d: telluride glass (10 Ge, 50 As, 40 Te [atomic %], sample thickness 1.6 mm

consist of irregularly-arranged chains and 8-membered rings. A chalcogenide glass composition suitable for commercial manufacture is $Ge_{30}As_{15}Se_{55}$ [8.134].

Trace impurities of oxygen are detrimental to the transmission of chalcogenide glasses. Oxides of the elements Ge, As, Se as well as OH and H_2S cause a series of absorption bands in the infrared. For acceptable IR-transparency near 12 μm the oxide impurity levels must be kept below 1 ppm wt. [8.134]. Chalcogenide glasses are typically prepared from the elements in evacuated and sealed silica tubes located inside a rocking furnace. Commercially, these glasses can be produced in volumes up to 2.5 litres; e.g. in the shape of a 30 cm diameter disk with a thickness of 2.5 cm with good optical quality.

For the application of infrared transmitting materials, their refractive index and its dependence on wavelengths in the infrared are of importance. The infrared refractive indices for chalcogenide glasses are in the range of 2 to 3. Reciprocal dispersive powers are typically listed for the mid-IR (3–5 μm) and long-IR (8–12 μm) wavelength region:

$$\nu_{3-5} = \frac{n_4 - 1}{n_3 - n_5}, \quad \nu_{8-12} = \frac{n_{10} - 1}{n_8 - n_{12}}. \tag{8.18}$$

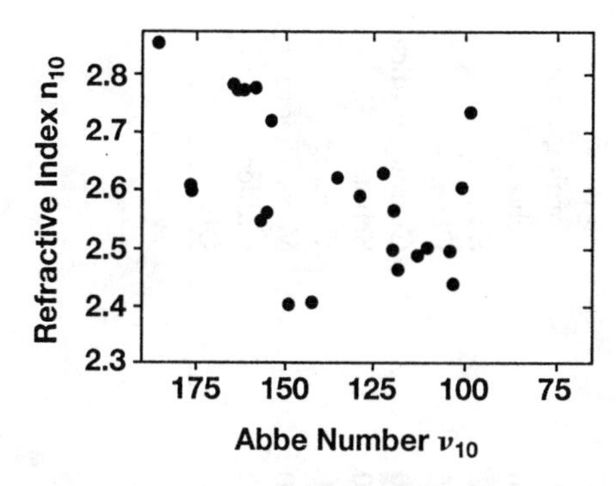

Fig. 8.28. Abbe diagram of chalcogenide glasses [8.135] in the wavelength region of 8–12 µm

For chalcogenide glasses the infrared Abbe values ν_{3-5} are in the range of 140 to 190, and ν_{8-12} are between 105 and 185. These glasses have relative high dispersion (small ν_{8-12} numerical values) in comparison to crystalline germanium (ν_{8-12} = 858). On the other hand, compared to crystalline ZnSe with $\nu_{8-12} = 59$, the dispersion of the chalcogenide glasses is small which makes chalcogenide glasses an interesting material for a combination with Ge or ZnSe for infrared lens systems.

Figure 8.28 shows an Abbe diagram (n_{10} versus ν_{8-12}) for chalcogenide glasses [8.135]. The T_g of these glasses is below 400 °C and with high Se content even below 200 °C. The thermal expansion is between 12 to 20 × 10^{-6}/K. Several surveys have been written on chalcogenide glasses; a more recent review is by *Savage* [8.136]. A group of less common non-oxide glasses is based on a combination of chalcogenides with halogens. Examples of these so-called chalcohalide glasses [8.137] belong to glass systems such as Ge-S-I, As-Se-Br, As-S-I. Typically, these glasses are unstable and cannot be prepared in large bulk pieces. Recently, a family of tellurium halide (TeX) glasses was discovered which have good resistance towards devitrification and water corrosion. Stable ternary glasses were reported with addition of sulphur or selenium. Glasses such as Te_3Cl_2S and Te_2BrSe showed no crystallization measurable by DSC [8.138]. The TeX glasses have been classified according to their IR-transparency into two families: the heavy TeX glasses based on the systems Te-Br-Se and Te-I-Se having an IR-transparency of up to 20 µm and the light TeX glasses containing light elements such as sulphur or chlorine, which limit their transmission to 13 µm. TeX glasses are black, they begin to transmit near 2 µm. These glasses have high coefficients of thermal expansion (in the range of 30–70 × 10^{-6}/K) and low transformation temperatures (near 75 °C) which limits their practical application.

8.4.5 Commercial Glasses

Table 8.4 gives an overview of the physical properties of the different IR-transmitting glass types. Table 8.5 lists some commercially available IR-transmitting glasses together with some of their physical properties. The glasses are of the types dry silicate, fluorophosphate, calcium aluminate and germanate glasses (compare also Fig. 8.25), heavy metal oxide glasses, heavy metal fluoride glasses, and chalcogenide glasses. IR-transmitting glasses are utilized as window ma-

Table 8.5. Commercial Infrared-Transmitting Glasses

Glass type	Name	Transparency range[d] μm	Refractive index	Density g/cm^3	Coeff. of thermal expansion 10^{-6}K	Transformation temperature °C	Manufacturer
Fused Silica	SiO_2	0.17–3.7	1.4585[a]	2.20	0.51	1075	Corning, General Electric Heraeus, Quartz et Silice, Shinetzu et al.
Silicate	IRG 7	0.32–3.8	1.5644[a]	3.06	9.6	413	Schott
	IRG 15	0.28–4.0	1.5343[a]	2.80	9.3	522	Schott
	IRG 3	0.40–4.3	1.8449[a]	4.47	8.1	787	Schott
Fluorophosphate	IRG 9	0.36–4.2	1.4861[a]	3.63	16.1	421	Schott
Ca-Aluminate	9753[e]	0.40–4.3	1.597[c]	2.80	6.0	830	Corning
	IRG N6[e]	0.35–4.4	1.5892[a]	2.81	6.3	713	Schott
	WB37A[e]	0.38–4.7	1.669[a]	2.9	8.3	800	Sasson
	VIR6	0.35–5.0	1.601[a]	3.18	8.5	736	Corning France
	IRG 11	0.38–5.1	1.6809[a]	3.12	8.2	800	Schott
	BS39B	0.40–5.1	1.676[a]	3.1	8.4	–	Sasson
Germanate	9754	0.36–5.0	1.664[a]	3.58	6.2	735	Corning
	VIR 3	0.49–5.0	1.869[a]	5.5	7.7	490	Corning France
	IRG 2	0.38–5.2	1.8918[a]	5.00	8.8	700	Schott
Heavy Metal Oxide	EO	0.5–5.8	2.31[c]	8.2	11.1	320	Corning
Heavy Metal Fluoride	ZBLA	0.30–7.0	1.5195[a]	4.54	16.8	320	Verre Fluoré
	Zirtrex	0.25–7.1	1.50[a]	4.3	17.2	260	Galileo
	HTF-1	0.22–8.1	1.517[c]	3.88	16.1	385	Ohara

Table 8.5. (cont.)

Glass type	Name	Transparency range[d] μm	Refractive index	Density g/cm^3	Coeff. of thermal expansion 10^{-6}K	Transformation temperature °C	Manufacturer
Chalcogenide	AMTIR 1	0.8–12	2.5109[b]	4.4	12	362	Amorph. Mat.
	IG 1.1	0.8–12	2.4086[b]	3.32	24.6	150	Vitron
	AMTIR 3	1.0–13	2.6173[b]	4.7	13.1	278	Amorph. Mat.
	IG 6	0.9–14	2.7907[b]	4.63	20.7	185	Vitron
	IG 2	0.9–15	2.5098[b]	4.41	12.1	368	Vitron
	IG 3	1.4–16	2.7993[b]	4.84	13.4	275	Vitron
	As_2Se_5	1.0–16	2.653[b]	4.53	30	98	Corning France
	IG 5	1.0–16	2.6187[b]	4.66	14.0	285	Vitron
	IG 4	0.9–16	2.6183[b]	4.47	20.4	225	Vitron
	1173	0.9–16	2.616[b]	4.67	15	300	Texas Inst.

[a] Refractive index at 0.587 μm
[b] Refractive index at 5 μm
[c] Refractive index at 0.75 μm
[d] 50% internal transmission for 5 mm pathlength
[e] Glass contains SiO_2

terials for photodetectors, cuvets (IR-spectroscopy), as lenses for IR-objectives (Pyrometer), domes for aircraft and missiles, military IR-optics and prisms (IR-spectroscopy). Further, optical components and fibres are required for lasers such as: CO_2 (10.6 μm), CO (5-6 μm), DF (3.6-4 μm), HF (2.6-3 μm), erbium:YAG (2.94 μm), holmium:YAG (2.1 μm) and erbium:glass (1.54 μm).

8.5 Laser Glasses

Joseph S. Hayden, Norbert Neuroth

The first demonstration of light amplification by a stimulated emission process was achieved by *Maiman* of Hughes Research Laboratories in 1960 [8.139]. His laser was fabricated from a ruby crystal whose ends had been silver coated and the excitation was provided by inserting the crystal into a helical flashlamp. In the following year, *Snitzer* demonstrated that neodymium doped glass also exhibited laser action [8.140]. Ever since that time, solid state laser materials have been the materials of choice for application as the active component in advanced laser systems.

The need for new and improved active crystalline and glass materials for application in solid state laser systems continues even today. Although crystals offer better thermal and mechanical properties, glasses can be manufactured in large volume pieces and a wide variety of shapes with exceptional optical quality. In addition, the ability to adjust glass properties through compositional modification, in comparison to the fixed set of properties offered by crystalline materials, has led to considerable research into the identification of glasses which have specific glass properties optimized for particular applications.

8.5.1 Requirementss for Laser Glass Materials

Typically, laser glass is placed within a pumping chamber containing a number of high intensity flashlamps which deliver energy to excite lasing ions within the glass host. The active material of a solid-state laser has to absorb pump light and convert it into usable laser radiation as efficiently as possible. Much of this flashlamp energy ultimately ends up as heat deposited into the base glass. Since glass surfaces are readily cooled by air or application of a suitable coolant fluid, thermal gradients are established within the laser glass. Even when the excitation energy is supplied by other laser sources tuned to specific transitions of the lasing ions, the influence of the temperature gradient in the laser glass must be taken into account. As a result of these considerations, various combinations of the laser, thermal, physical and optical properties of laser glass materials may become important for a particular application. We will now consider specific important properties of laser glass materials.

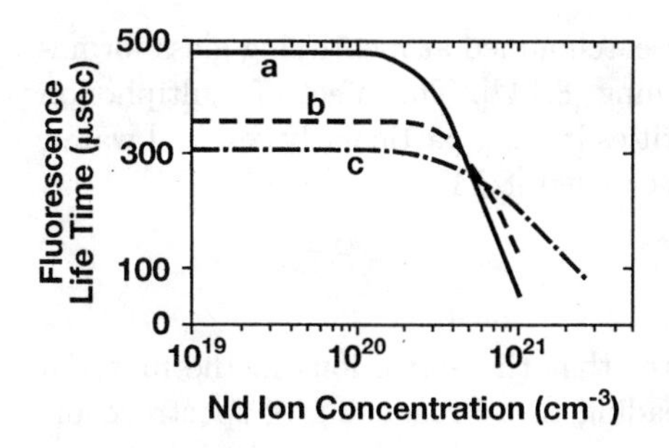

Fig. 8.29. Dependence of the Fluorescence Lifetime on the Nd ion Concentration [8.147] a) silicate glass (LG660), b) phosphate glass (LG750), c) $LiNd_xLa_{x-1}$ phosphate glass

Cross-Section for Stimulated Emission

The extent to which an active material can amplify an incident laser beam is given by the cross-section for stimulated emission, σ. For situations where the propagating laser beam to be amplified is of low fluence and saturation effects are thus not present, if there are N excited state ions per cubic centimetre of glass, the gain is $g = \sigma N$ [8.141–143]. It might then be expected that the highest possible cross-section is desirable. However, this is not always the case; in particular, in laser designs where a high cross-section implies a high rate of amplification of randomly directed light provided by spontaneous emission of excited state ions and, consequently, depletion of energy stored in the upper lasing excited state population. Thus high cross-section materials are preferred when the overall gain is the critical parameter, but low cross-section materials are desired where the most important consideration is maximizing the amount of stored energy, and operation is limited by amplification of spontaneous emission. Clearly, then, the most desirable cross-section value is dependent upon the laser design under consideration.

Fluorescence Lifetime

To ensure maximum utilization of the available pump radiation, long fluorescence lifetimes for the active ion excited state associated with the laser transition are desirable. The longer the flourescence lifetime the better the pumplight is utilized and thus the higher is the efficiency of the pumping. There are a number of factors which reduce or quench the lifetime of an active ion excited state in glass. Three principle quenching processes are:

- self-quenching between two neighboring ions, often referred to as concentration quenching [8.142]
- transfer of excited state energy to vibrational modes in the glass network called nonradiative multiphonon relaxation
- transfer of excited state energy to impurities including transition metal ions and residual (OH^-) groups, present in the glass.

In laser glasses fluorescence lifetime is largely independent of doping levels up to some critical doping level, where concentration quenching becomes significant and excited state lifetime is thus reduced to higher doping concentrations (see

Fig. 8.29). There has been considerable research aimed at identifying glass formulations with reduced concentration quenching [8.144]. The effect of multiphonon relaxation [8.145] and the impact of impurities [8.146], particularly residual water, on fluorescence lifetime have also been researched [8.147].

Emission Bandwidth

The disordered structure of a glass ensures that the active ions in the material assume a distribution of environments leading to a broadening of spectroscopic features in comparison to similarly doped crystals. Although these broad absorption features enable a higher percentage of flashlamp light to be coupled into the active species in the glass, they are correlated with correspondingly broad emission features. In some laser systems the bandwidth of the laser beam from a laser crystal which is to be amplified is smaller than the fluorescence bandwidth of the glass laser amplifier; thus, only a portion of the excited active ions are in the appropriate environment and can contribute to the amplification of the propagating laser beam. On the other hand, laser systems designed for operating with pulse lengths of picoseconds or less can take advantage of the broad fluorescence linewidths found in glasses.

Thermal Shock Resistance

Laser glasses must possess good thermal shock resistance which means a large value of $K_{1c}\kappa(1-\nu)/\alpha E$, where K_{1c} is the fracture toughness, κ the thermal conductivity, ν the Poisson's ratio, α the thermal expansion and E the Young's modulus for the glass. This is important not only from the perspective of fracture under thermal loading during actual laser operation, but also from the point of view of manufacturing, polishing, handling and storage. These latter considerations ultimately influence the cost of an active material for example, due to manufacturing yields.

Temperature Dependence of Optical Pathlength

To minimize degradation of beam quality through distortion of the laser beam by thermal effects, it is necessary that thermally-induced stresses and changes in optical pathlength with temperature must be minimized. One route to this is to employ athermal glasses where the temperature dependence of refractive index, dn/dT, is opposite in sign and of an appropriate magnitude to offset the change with temperature in optical pathlength associated with the bulk thermal expansion of the glass, α. This combined effect of dn/dT and α is described by the following equation for the change in optical pathlength with temperature, $dw/dT = \alpha(n-1) + dn/dT$, where n is the refractive index at the wavelength of interest (see Sect. 2.4.4). Additionally, it is desirable to have laser glasses with a small value of the stress-optic coefficient. This is a measure of the amount of induced birefringence in a glass component as a result of stresses which are thermally induced, as well as stresses which are remaining in the component as a result of the manufacturing process.

Nonlinear Refractive Index

It is well-known that the observed refractive index of an optical material exhibits nonlinear behaviour and is increased during the transmission of a high fluence laser beam (see Sect. 2.6). The extent of this nonlinearity is given by the nonlinear refractive index, n_2, where $n = n_0 + n_2E^2$; n_0 is the linear refractive index in the absence of laser radiation and E is the electric field strength of the applied laser beam. Due to normal spatial variation in the intensity of a propagating laser beam, this nonlinear behaviour leads to a self-focusing of the beam causing higher peak fluence locations and often to a thread-like damage occurring at these locations within the bulk material. Consequently, to minimize beam distortion and bulk damage to laser system components, it is most desirable to have material with a low n_2-value.

Optical Quality

Laser glasses must be available in the highest optical quality possible. Laser glass components must be free of refractive index variations associated with hard, cord-like, striae and with soft striae, also referred to as global inhomogeneity. Additionally, laser glass must be free of bubbles and other inclusions which lead to scattering of pump and laser light and, potentially, to bulk damage sites within the element when it is used with high power laser beams.

Absorption Properties

Obviously, the absorption of a laser glass needs to be small at the lasing wavelength. For the case of neodymium-doped glass which has a principle fluorescence wavelength in the vicinity of 1.06 µm, the impurity elements absorbing at the lasing wavelength which are of principle concern are Fe^{2+} and Cu^{2+}. Additionally, the application of strong pump and laser radiation should not increase the optical loss or, for that matter, induce absorbing centres which compete with the lasing ions for pump radiation. An example of this latter effect is the darkening often referred to as solarization of glass by the UV radiation component of flashlamp radiation. To eliminate this solarization effect, it is common practice to add to glass formulations solarization inhibitors such as Ce, Sb and Ti which function as traps for the electrons and holes created in the glass network by UV exposure during flashlamp irradiation.

Bulk Damage Threshold

Bulk damage due to nonlinear effects has already been described above. An additional damage mechanism is caused by inclusions in glass which absorb at the lasing wavelength and become macroscopic damage sites upon use of a laser glass component in a high fluence application. The most representative example of this is damage sites associated with metallic platinum particles left in laser glass as a direct result of the manufacturing of these materials in platinum melting units. Laser glass for utilization in such high fluence applications must, thus, be free of all absorbing inclusions (Sect. 2.7).

8.5.2 Activating Ions

Over the last 30 years there has been extensive research covering a large number of active ions in every known glass system. Today, the research activity in this area remains at a high extensive level, with emphasis on improved host glasses for active ions with particular emission wavelengths for specific applications, on improved manufacturing techniques and on new glass laser system concepts. Despite this level of research activity, the primary lasing ion of interest for most commercial applications of laser glass today continues to be neodymium. Neodymium fluoresces, in a silicate host glass, at nominally 0.88 μm, 1.06 μm, 1.35 μm and 1.80 μm with the strongest emission at 1.06 μm. Neodymium glasses are utilized in high energy laser systems; the largest is the NOVA laser at Lawrence Livermore National Laboratory in the USA, and in high power laser systems; for example, for the application as a driver of an X-ray source for advanced microlithography. The only other active ion which has emerged in a commercial laser glass product is erbium in military eye-safe laser rangefinders which operate at 1.54 μm.

Recently, research interest has been focused on active ions which will offer laser action both in the visible region and at 2.9 μm, 1.5 μm, 1.3 μm and 0.8 μm. Visible, in particular, blue light laser sources, are of interest since utilization of shorter wavelength light is expected to increase the storage density in optical recording and memory devices and the former three wavelengths are of particular interest as laser sources and amplifiers for telecommunication applications. Much of this activity, then, has dealt with guided wave optical technology; i.e., fibre and planar waveguide structures. Interestingly, for these latter application wavelengths, neodymium still plays a leading role due to its fluorescence band near 1.3 μm. The emission of Er at 2.75 μm is important because this wavelength is strongly absorbed by the biological tissue. Therefore, such a laser is suitable for medical operations. Other ions under investigation include thulium and praseodymium. For visible wavelength sources, ions under consideration include Sm^{3+} with emission near .65 μm and Ho^{3+} with emission near .54 μm.

Another area which has received much attention has been the co-doping of glasses with one or more additional ions which absorb pump radiation and couple this energy into the lasing ions; thus increasing overall pump utilization efficiency. These co-dopants are referred to as sensitizing ions. The most successful commercial example of a sensitized laser glass is Er:Yb:glass, where the ytterbium ions act as a sensitizing ion for the erbium lasing species. Table 8.6 offers a brief listing of some lasing ions with associated emission wavelengths along with some sensitizing ions which have been researched. Wavelengths of operation are dependent on the

Table 8.6. Selected activated ions in glass

Ion	Approximate Emission Wavelengths μm	Sensitizing Ion(s)
Nd^{3+}	0.93 1.06 1.35	Cr^{3+}, Mn^{2+}, Ce^{3+}
Er^{3+}	1.30 1.54 1.72 2.75	Cr^{3+}, Yb^{3+}
Yb^{3+}	1.03	Nd^{3+}
Sm^{3+}	0.65	
Ho^{3+}	0.55 1.38 2.05	
Tm^{3+}	0.80 1.47 1.95 2.25	Er^{3+}, Yb^{3+}
Tb^{3+}	0.54	
Pr^{3+}	0.89 1.04 1.34	

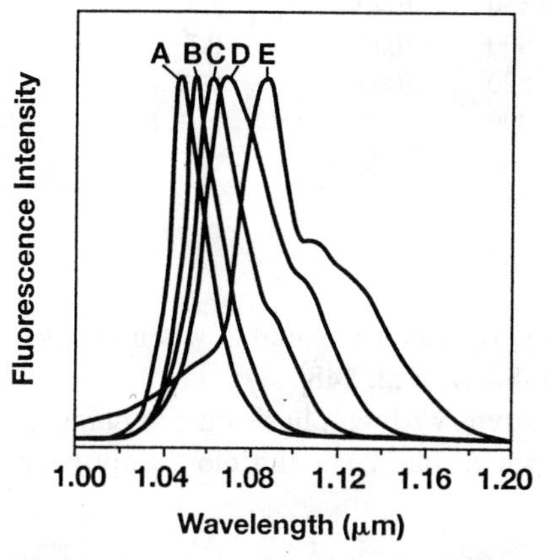

Fig. 8.30. Fluorescence spectra of the ${}^4F_{3/2} \rightarrow {}^4I_{11/2}$ transition of Nd^{3+} in different glass types [8.181]: A) fluoride, B) phosphate, C) silicate, D) aluminate, E) fused silica

base glass. For instance, the strongest emission line for Nd^{+3} has its maximum in silicate glass at 1060 nm and in phosphate glasses at 1054 nm (see Table 8.7 and Fig. 8.30). The list is not intended to be complete. Interested readers can consult reference texts for more detailed compilations [8.141, 142].

8.5.3 Glass Development

Snitzer demonstrated the first functioning glass laser in 1961 using a neodymium-doped barium crown glass [8.140]. In the thirty years that have elapsed since that time, there has been an intense research effort to discover new and better laser glass materials. As a result of this effort, considerable information has been collected on how glass composition effects important laser glass properties as well as how these properties are correlated to one another. In addition to investigating a wide range of dopants and various co-dopants, glass developers have also explored a broad spectrum of glass forming systems. Table 8.7 provides an overview of many

Table 8.7. Overview of neodymium-doped laser glass systems

Glass type	Stimulated emission cross-section, σ 10^{-20} cm^2	Fluorescence Lifetime τ T 10^{-6} s	Wavelength λ_{max} µm	Nonlinear refractive index n_2 10^{-13} esu
Silicate	1.0–3.5	300–1000	1060	1.4–1.6
Borate	1.0–2.0	50–400	1055	0.9
Silicoborate	0.8–3.0	200–800	1058	1.2
Phosphate	1.8–4.8	100–500	1054	0.9–1.2
Tellurite	3.0–5.1	140–240	1060	10
Fluorosilicoborate	1.0–1.7	300	1060	0.8–1.1
Fluorophosphate	1.7–3.8	350–500	1052	0.5–0.7
Fluoroberyllate	1.6–4.0	460–900	1048	0.3
Fluorozirconate	2.9–3.0	430–450	1049	1.2
Fluoroaluminate	2.2–2.9	400–600	1050	0.5
Chloride	6.0–6.3	180–220	1063	> 7
Chlorophosphate	5.2–5.4	290–300	1055	> 1.4
Sulphide	6.9–8.2	60–100	1076	> 7

glass forming systems along with typical property ranges available when utilized as a host for a neodymium-doped laser glass [8.141, 142, 148].

We will now discuss the principle glass systems with emphasis on the perspective of neodymium laser glass development which has been the most important activator ion up to now.

Oxide Glasses

The bulk of laser glass research and most of the presently available commercial laser glasses are oxide based. In this group of glasses the principle glass formers are SiO_2, B_2O_3, GeO_2, P_2O_5 and TeO_2. With silicate laser glasses one finds the largest observed fluorescence lifetimes, up to 1 ms, although these glasses include some of the lowest cross-sections. The largest cross-sections from within the oxide glass families are offered by phosphate glasses (to 4.8×10^{-20} cm^2) [8.142, 149] and tellurite glasses (5.1×10^{-20} cm^2) [8.142, 150]; however, these large cross-section values come at the expense of lower fluorescence lifetime values.

Halogenide Glasses

Glass formers among the halides include BeF_2, CdF_2, AlF_3, ZrF_4, HfF_4, $BiCl_3$, $ThCl_4$ and $BiBr_3$ [8.142, 151, 152]. The fluoroberyllate glasses offer the lowest nonlinear refractive index values (down to 0.3×10^{-13} esu).

The fluorozirconate glasses are under scrutiny as possible hosts for near-IR lasing hosts with ions such as Er^{3+} (emission at 1.54 µm), Tm^{3+} (emission at 1.85 µm), Ho^{3+} (emission at 1.95 µm), and Pr^{3+} and Nd^{3+} (both with emission near 1.30 µm) [8.142, 153, 154].

The fluoroaluminate glasses also offer low nonlinear refractive index values (down to 0.5×10^{-13} esu). Cross-section values for these glasses are considered moderately high (up to $2.9 \times 10^{-20}\,cm^2$). A representative composition in mole percent is: 34.5 CaF_2, 44.5 AlF_3, 20 BaF_2 and 1 NdF_3.

Also among the glasses based on halide glass formers are chloride glasses; e.g. $BiCl_3$-containing glasses which offer the highest observed cross-section values in this group (up to $6.3 \times 10^{-20}\,cm^2$) [8.155]. Unfortunately, these glasses are also characterized by a large nonlinear refractive index, poor chemical durability, crystallization tendency during manufacturing and low transformation temperatures; thus, research with these glasses is largely restricted as available samples are of limited size and optical quality.

Oxihalogenide Glasses

In this group of materials we have glasses involving mixed species of anions such as fluorophosphate and chlorophosphate glasses. The fluorophosphate glass family has the distinction of being the only type of oxihalogen glass group to have yielded commercial laser glass products as discussed further below. Chlorophosphate glasses (for example, in mole percent: 66.6 $NaPO_3$, 33.3 ZnCl) show possibilities of providing high cross-section values ($5.4 \times 10^{-20}\,cm^2$) [8.156].

In the early 1980s a number of neodymium-doped fluorophosphate laser glass compositions were developed and offered on a commercial basis [8.167, 168]. These glasses were characterized by moderate cross-sections, $\sigma = 2.6 \times 10^{-20}\,cm^2$, near athermal behaviour; long fluorescence lifetime of up to 475 µs, and exceptionally low nonlinear refractive index values of the order of 0.5×10^{-13} esu. The motivation for these developments was the anticipated high bulk damage threshold which was forecasted for these compositions by virtue of their low non-linear refractive index. The actual damage thresholds were, because of the content of metallic platinum in these glasses, found to be unacceptably low for high fluence applications; thus eliminating one chief advantage of these glasses. Up to now a production method for a Pt-free version of this glass type has not been developed. Consequently, the commercial demand for these materials has been small, and they have largely vanished from laser glass vendor catalogues.

Chalcogenide Glasses

Another group of materials with glass forming properties is the chalcogenides of which glass forming compounds include As_2S_3, As_2Se_3 and La_2S_3. Glasses based on the sulphides of La, Gd, Ga, etc. have shown extremely high cross-section values in the neighborhood of $8.2 \times 10^{-20}\,cm^2$. The fluorescence lifetimes, however, have not been observed over 100 µs [8.157–159]. These glasses are, however, promising candidates for near-IR laser sources and amplifiers such as those presently being sought for telecommunication applications.

Table 8.8. Comparison of Laser Crystals and Laser Glasses

	Ruby $Cr{:}Al_2O_3$	Nd:YAG $Nd{:}Y_3Al_5O_{12}$	Nd:YLF $Nd{:}YLiF_4$	Nd:YAP $Nd{:}YAlO_3$	Nd:GSGG $Nd{:}Gd_3Sc_2Al_3O_{12}$	Nd: Glass
σ	2.5	40–80	32	37		1.2–4.5
c_1	0.05	0.73	0.8	1	1	0.5–10
c_2	0.158	1.38		1.95		
τ	3000	200	520	180	260	200–500
d_λ	0.53	0.2				20–35
λ_{max}	694	1064	1053	1079	1061	1050–1065
n_2	1.1	3.2				0.5–1.6
dw/dT	13	12.8		18.4	17.4	8 to −1
κ	35	13	6	11	6	0.6–1.3

σ: Cross-section for stimulated emission, $10^{-20}\,cm^2$
c_1: Neodymium concentration expressed as weight percent Nd_2O_3
c_2: Neodymium concentration (10^{20} ions/cm^3)
τ: Fluorescence lifetime (10^{-6} s)
d_λ: Fluorescence linewidth, FWHM (nm)
λ_{max}: Peak fluorescence wavelength (nm)
n_2: Nonlinear refractive index (10^{-13} esu)
dw/dT: Temperature coefficient of optical pathlength (10^{-6}/K)
κ: Thermal conductivity (W/m K)

Comparison with Crystals

With the information collected here, it is possible to make a comparison between crystals and glasses as a host for lasing ions. Table 8.8 presents some key properties of several popular commercial, mostly neodymium-doped, crystals in comparison to those values offered by the more readily available laser glasses.

Crystals can have very high cross-section values and are therefore very attractive as laser materials. Ruby and Nd:YAG are used extensively in lasers for materials working and measurement technique. One can also generate pulses with a high repetition rate as well as CW operation owing to the outstanding thermal and mechanical properties of these materials. The disadvantages of these crystals are that they are limited in size; and larger crystals, when available, can become extremely expensive. Large sizes of active laser material can be necessary if a chain of laser amplifiers has to be built and the cross-section of the beam has to be enlarged successively in order to keep the power density below the damage threshold of the active laser material.

Glass materials, on the other hand, can be produced in large sizes with high optical quality. Glass materials are thus often employed in high power lasers where the power density is limited by the damage limit of the active and passive materials used in the system, and the cross-section of the beam must be enlarged when the energy density of the propagating beam nears the damage threshold of the employed materials. Today, it is possible to fabricate neodymium glass rods; for example, 120 mm diameter by 300 mm long or 64 mm diameter by 900 mm long and plates 770 mm × 430 mm × 45 mm thick.

Glass Ceramics

Neodymium-doped glass ceramics have also been investigated; however, early studied materials have shown only moderate lasing properties and the optical quality was degraded by the differing optical properties of the crystal and glassy phases inherent in these materials (scattering) and their transmisison quality in the near-IR. Alternatively, the thermal mechanical characteristics of these materials are enhanced by the potential of their low thermal expansion; thus, these materials continue to attract attention.

Glass ceramics have been prepared in the system SiO_2-$LiAlO_2$-(Mg,Zn) Al_2O_4 in which the crystallization process was controlled so that the resultant glass ceramic contained crystalline particles less than 25 nm in average size. In such a material, the scattering of light by the crystal phase is reduced, and the overall transmission of the material including in the near-IR is improved. By additional compositional modifications, in particular, by substituting a portion of the SiO_2 for $AlPO_4$ and by including La_2O_3 as an additional glass component, the fluorescence intensity for this neodymium-doped glass ceramic system is increased by a factor of 3 to 4. The resultant materials exhibit a laser effect of the same order of magnitude as that observed in laser glasses; e.g. a cross-section of 2.3×10^{-20} cm^2, a 200 μs lifetime for 2 wt % Nd_2O_3 and a 27 nm wide fluorescence feature centred on 1065 nm [8.160].

8.5.4 Process Development

In addition to developmental efforts at identifying new and improved laser glasses, research efforts have also been directed at developing new manufacturing techniques and processes to improve the quality and performance level of existing laser glasses.

One of the most important critical developments in this area has been the successful development and full implementation of manufacturing techniques which can produce phosphate glass compositions entirely free of absorbing inclusions [8.161]. Prior to 1985, all laser glasses were characterized by a high density of metallic platinum inclusions which became macroscopic damage sites when the glass was used in high fluence applications (Sect. 2.7.3). These inclusions were directly related to the practice of utilizing metallic platinum in the construction of optical glass melters; a practice of considerable importance if the resultant glass castings are to be produced with the high quality; i.e. high optical homogeneity, required for laser applications. Other commonly-employed melter materials are chemically attacked by the glass melt leading to partial dissolution of the melting unit into the melted glass, which with the inevitable incomplete mixing just before casting, leads to hard cord striations in the produced glass. Even with the employment of platinum containers, a residual amount (ppm levels) of platinum does

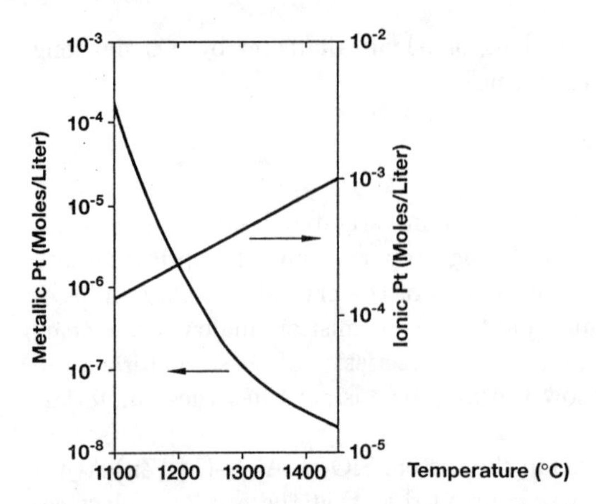

Fig. 8.31. Ionic and metallic platinum in phosphate laser glass as a function of melt process temperature [8.161]

still enter the glass melt, and this has always resulted in glass castings containing small inclusions of platinum. Upon exposure to a high energy laser beam, these particles absorb enough energy to vapourize, and the inclusions then explode in the glass leading to macroscopic damage sites.

Eliminating platinum particles in phosphate laser glasses relies on combining the high solubility limit for platinum in these glasses with adjustment of those process variables which are conducive to driving platinum into the ionic state [8.161, 162]. The key process variables which are readily controlled are the melt temperature and the redox condition. The latter is most directly adjusted by controlling the identity of a cover gas supplied to the melting unit. It has been our experience that selection of high temperatures and oxidizing conditions dramatically reduce the metallic platinum content in a series of half-litre, experimental melts by more than five orders of magnitude (see Figs. 8.31 and 8.32). Corresponding to this decrease in metallic platinum content is an increase in the total content of platinum in the glass (see the plots based on the right ordinate in Figs. 8.31 and 8.32).

Absolute verification of the damaging inclusion content in laser glass is carried out by the deliberate damage testing of each component as an integral part of quality and process control procedures [8.163]. This is accomplished using a commercial 800 mJ YAG laser, Q-switched so that the pulse length is 8 ns. The beam is focused so that within the target to be tested the laser beam has a diameter of 2 mm inside of which the fluence is at least $7 \pm 1\,\mathrm{J/cm^2}$. The target is then scanned so that every point in the sample receives at least ten exposures from the 2 mm diameter laser beam. Under these test conditions, we have found that a 5 μm platinum particle becomes a 0.25 mm bulk damage site. Since trained inspectors can detect inclusions down to 20 μm in size reliably with a focused 100 W lamp, all damage sites of any consequence are then easily located and can be rejected in subsequent processing.

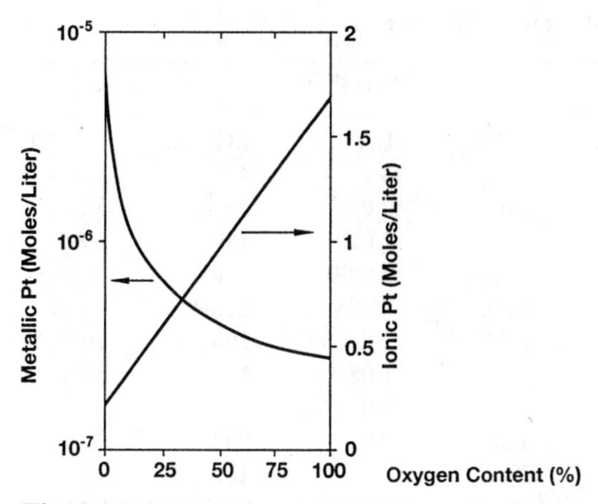

Fig. 8.32. Ionic and metallic platinum in phosphate laser glass as a function of oxygen content in the atmosphere over the melt. (Note, right-hand ordinate is linear) [8.161]

Table 8.9. Absorption weighting factors for phosphate laser glass produced under oxidizing conditions [8.164]

Impurity	Specific Absorption Constant (m^{-1}/ppm of contaminant at 1.05 μm)
Cu	0.270
Fe	0.013
V	0.026
Co	0.022
Ni	0.016

Current levels of metallic platinum in Schott phosphate laser glasses are now low enough that glass volumes of up to seven litres are routinely provided with no platinum particles, and larger volume pieces of up to fifteen litres typically contain less than 0.2 particles per litre, and more than 50% of these pieces are particle free. The average size of the remaining inclusions is less than 5 microns.

The need for producing laser glass components which are also characterized by low absorption at the lasing wavelength should be mentioned here again. In the case of neodymium laser glass for use at 1.05–1.06 μm, the key species which give high absorption are Fe^{2+} and Cu^{2+}. Clearly, the absorption problem is tied closely to the redox condition of the glass during melting. The development of new manufacturing techniques for producing platinum-particle-free glass, which rely heavily on maintaining the melt in as strong as possible an oxidizing environment, has led to a reinvestigation of absorption in these glasses under those conditions appropriate to platinum-particle-free production [8.164]. Table 8.9 summarizes the extent of 1.054 μm absorption constant, expressed in m^{-1}, induced in the phosphate laser glass LG-750 prepared under conditions resulting in platinum particle-free glass for 1 ppm by weight of a number of common contaminants present in glass.

Table 8.10. Property values for Schott commercial laser glasses

Laser glasses	Silicate	Phosphate[b]				
	LG-680	APG-1	APG-2	LG-750	LG-760	LG-770
σ	2.9	3.5	2.5	3.7	4.7	4.2
c_1	2.0	2.0	2.0	2.0	2.0	2.0
c_2	1.82	1.88	1.86	2.02	1.86	1.88
τ	> 350	> 350	> 350	> 360	> 350	> 350
$d\lambda$	28.2	27.8	33.0	25.28	22.64	24.17
λ_{max}	1061.0	1054.5	1055	1053.5	1053.5	1053.5
n_2	1.60	1.13	1.01	1.08	1.02	1.04
dw/dT	8.1	5.2		0.8	−0.4	
κ	1.35	0.83	0.85	0.52	0.60	0.54
α	103	98	70	132	152	142
FOM	1.10	0.90	1.61	0.59	0.54	0.66

[a]: For explanation of symbols, refer to Table 8.8
[b]: Presented data for APG-2 and LG-770 are preliminary
α:Thermal expansion from 20 to 300 °C (10^{-7}/K)
FOM: Figure of merit of strength against thermal shock loading, equal to $\kappa(1-\nu)/\alpha E$, where κ is thermal conductivity; ν is Poisson's ratio, α the thermal expansion, and E the Young's modulus for each glass ($10^{-6}\,m^2/s$)

Another process which has the potential to improve the performance of a laser glass component is strengthening by ion exchange. In this process a compressive layer is created on glass surfaces when smaller mobile ions within the glass surface are exchanged for larger ions initially present in a salt bath in which the component to be strengthened has been immersed [8.165]. The glass article is stronger after the exchange has been completed, since to fracture the glass, sufficient energy has to be available to overcome not just the original strength of the glass article but also the additional compression at the surface (Sect. 4.5). Of course, once a fracture has penetrated beyond the depth of the compression layer, the strength of the glass effectively returns to that present prior to the ion exchange.

8.5.5 Commercial Neodymium Laser Glasses

Despite the extensive research efforts of laser glass manufacturers and the wealth of glass forming systems which have been investigated, the number of commercially available catalogue laser glasses is quite limited. Currently available commercial laser glass compositions are all silicate and phosphate-based glass systems; however, most laser glass suppliers are willing and capable of preparing custom formulations upon customer request. Table 8.10 summarizes typical properties of commercial neodymium-doped laser glasses available from Schott [8.166].

Commercial silicate laser glasses offer cross-section values, σ, between 1.6 and $2.7 \times 10^{-20}\,cm^2$ and nonlinear refractive index values, n_2, of the order of 1.4×10^{-13} esu. As has already been mentioned, the silicate laser glasses are characterized by long fluorescence lifetimes, τ, of up to 500 µs for glasses doped with

2 weight percent Nd_2O_3. These glasses are also characterized by smaller cross section values. The silicate compositions are known for good thermal mechanical attributes indicated here by thermal conductivity values, κ, of greater than 1 W/m K. It should be noted that the silicate glasses are also characterized by a large stress optic constant and a dependence of optical pathlength on temperature given by dw/dT.

Commercial phosphate laser glasses offer high cross-section values, σ, between 3.0 and $4.7 \times 10^{-20}\,\mathrm{cm}^2$ and small, nonlinear refractive index values, n_2, on the order of 1.0 to 1.5×10^{-13} esu. In addition, as mentioned earlier, manufacturing techniques have been developed which allow phosphate glass compositions to be prepared in large castings of high optical quality which are entirely free of damaging inclusions. Thus, because of the low nonlinear refractive index and inclusion-free quality, phosphate glasses are characterized by high laser fluence bulk damage threshold values; e.g. $> 30\,\mathrm{J/cm}^2$ for a 1 ns laser pulse length. The phosphate laser glasses are also characterized by fluorescence lifetimes, τ, of over 350 µs for glasses doped with 2 weight percent Nd_2O_3. Some phosphate compositions are known for good thermal mechanical attributes indicated here by thermal conductivity values, κ, of greater than 0.8 W/m K, but this is accomplished only for those glasses also characterized by the lower range of cross-section values. Of special note is the athermal behaviour offered by many phosphate laser glasses made possible by virtue of their large negative value for the dependence of refractive index on temperature which offsets the change in optical pathlength, dw/dT, associated with the bulk thermal expansion of these glasses.

8.5.6 Functional Forms of Laser Glass

The first glass laser can be described as having been in the form of a multimode fibre [8.140]. Since that time, however, the traditional forms for active laser glass components have been as bulk rods, slabs and disks, often of sizable dimensions (see Fig. 8.33). Although the selected base composition controls the laser properties of the glass, the neodymium ion concentration must also be taken into account when selecting a laser glass for a specific application. In particular, the doping level must be selected so that the active element can be adequately excited over its entire volume. For a given thickness of a rod or plate, the neodymium doping level must be of such an amount that the medium penetration depth of the pump light is of the order of the thickness of the component as viewed by the excitation source. As part of this selection process, concentration effects on fluorescence lifetime of neodymium may also become important.

Laser glasses can be readily prepared in the form of clad and unclad fibres [8.169], and many compositions readily lend themselves to the fabrication of planar laser waveguides by techniques such as chemical ion exchange [8.170, 171]. Although the majority of commercial glasses contain neodymium as the only active dopant, laser glass suppliers are experienced at preparing special melts containing a wide variety of dopants and co-dopants, and experience exists to modify exist-

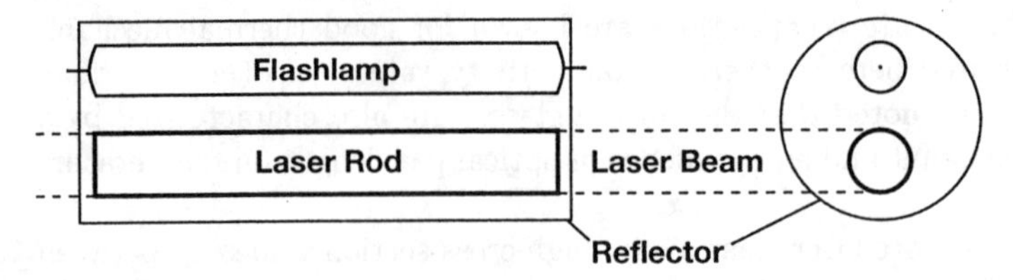

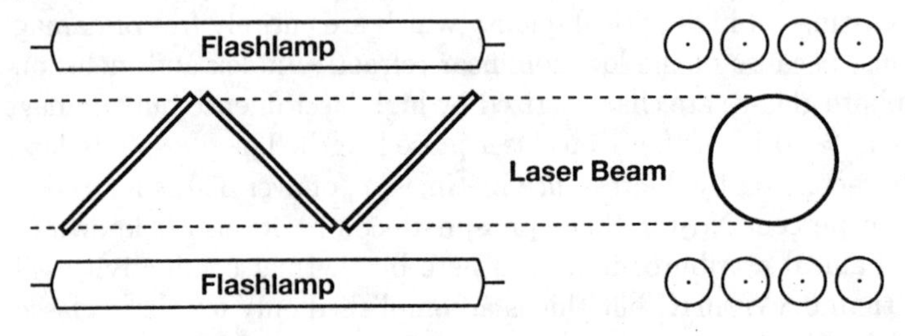

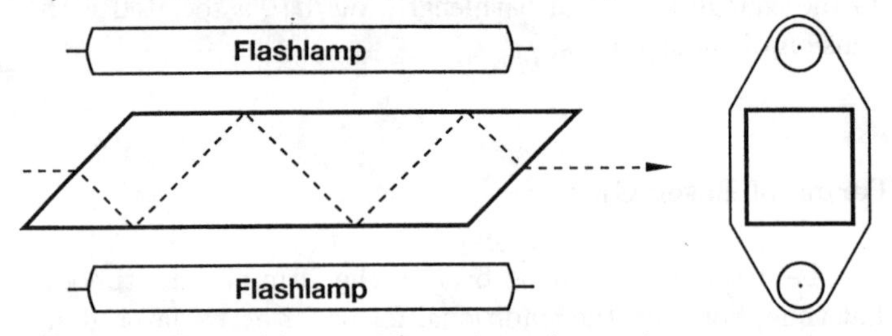

Fig. 8.33. Typical forms of laser glass: a) rod, b) plates mounted at Brewster's angle to the laser beam, and c) plate with zig-zag (total internal reflection) propagation of laser beam

ing compositions to enhance other properties; e.g. ion exchangeability for planar waveguide generation, for particular applications.

8.5.7 Applications of Laser Glass and Future Development Trends

From a commercial perspective, the dominant market for laser glass is in the area of large, high energy (per pulse), solid state laser systems for fusion research. These systems operate at low repetition rates (i.e. a few pulses per day) and, thus, the glasses of choice offer good laser properties at the expense of thermal mechanical characteristics. The laser glass materials of choice for such systems have been platinum particle-free phosphate laser glasses such as LG-750 and LG-760.

The largest such system is the NOVA laser at the Lawrence Livermore National Laboratory in the USA. The NOVA laser contains a series of phosphate laser glass amplifiers, split among ten different beam lines with active apertures up to 46 cm diameter, producing 120 kJ of 1.06 μm laser energy.

Designs for next generation solid state laser systems call for of the order of 12 000 pieces of active laser glass, each with a volume of nominally 7 litres of glass. For such a system a new commercial phosphate-based laser glass has been under development which, combined with improved laser characteristics, offers improvements in those properties which directly impact ease of manufacturing and, ultimately, production yields [8.172–174]. As a result of this development activity a new material, designated LG-770 in Table 8.10, has been created.

Another major area of activity is in high average power laser development for such applications as drivers for X-ray sources for advanced microlithography. These laser systems do not only operate at fluence levels near the laser damage threshold of the employed optical components but also at repetition rates limited by the strength of the active glass element to fracture under thermal loading. Here, thermal mechanical considerations are a major factor in selecting the best active glass. Although silicate-based glasses are well-known for good thermal mechanical properties, they are not available in high optical quality, free of inclusions which can create laser damage sites. In response to this requirement, phosphate laser glass compositions such as APG-1 which can be manufactured in a platinum-particle-free form have been identified. These offer thermal mechanical properties equivalent to silicate glasses with minimal degradation of the good laser characteristics of phosphate-based glasses.

Current next generation high average power solid state laser designs call for even higher repetition rates, and progress in laser system technology has made it possible to adjust designs to compensate for laser material limitations and still achieve required output power specifications [8.175]. Consequently, a new phosphate composition, APG-2 in Table 8.10, with manageable reduction in laser properties, has been developed which has optimized thermal mechanical properties, allowing maximum repetition rate operation without fracture of the active laser glass.

Of extreme current interest is the identification of laser sources and amplifiers for application at 1.3 μm and 1.5 μm, two areas in which silica-based communication fibres have low loss and low dispersion. Thus, much of the activity in this area deals with guided wave systems; i.e. fibre and planar wave- guide structures. The need for an amplifier at 1.55 μm will apparently be satisfied by erbium-doped CVD-formed fibres of the Al_2O_3-GeO_2-SiO_2-P_2O_5 system or its variants [8.176–179]. Materials issues to be addressed in search of a useful 1.3 μm region amplifier are reducing multiphonon quenching of the lasing excited state lifetime and the absorption of the light to be amplified by activated ions in the glass, identifying systems with acceptable levels of concentration quenching, and minimizing the effects of amplified, spontaneous emission and unfavorable emission of the lasing excited state to other energy levels. [8.180]

Acknowledgment

Some of the work described here has been performed under the auspices of the U. S. Department of Energy by the Lawrence Livermore National Laboratory under contract number W-7405-ENG-48.

8.6 Glasses for High Energy Particle Detectors

Susan R. Loehr, Peter Naß, Burkhard Speit

8.6.1 Introduction

Optical glasses are often used as shielding materials, as dosimeters or as detectors for high energy particles and γ-rays because of their homogeneity and availability in large volumes and different shapes. Moreover, by the use of glasses one can cover a large range of relevant material properties, such as density, radiation length and refractive index. Glasses doped with luminescent activator ions like Ce^{3+} and Tb^{3+} are used as scintillating detectors for charged particles and photons. Codoping with isotopes, like ^{6}Li and ^{10}B, makes these glasses sensitive to thermal neutrons as well. The European Nuclear Fusion concept includes glassy Li_4SiO_4-pebbles as breeder material for tritium. Thus, by variation of the composition, the glasses can be tailored to a variety of different applications.

In X-ray sensitive glasses oxides of high-Z elements, such as barium and lanthanum are introduced to increase the X-ray attenuation coefficient. Mostly, Tb^{3+} is used as an activator ion in this case. At present, developments are focussed on X-ray imaging devices for medical and industrial non-destructive testing applications. In addition to other concepts, coherent bundles of Tb-doped multicomponent fibres may find some applications in this field.

In high energy physics, so-called calorimeters are used to measure the total energy of charged particles. These detectors should exhibit a density as large as possible to stop the particles within practical dimensions. Beside crystals like CsI(Tl) and BGO (bismuth germanate), Ce^{3+}-doped glasses with high densities are suitable materials. Undoped glasses which act as Čerenkov counters are also used as calorimeters. It is customary to utilize these glasses as volume detectors.

Detectors consisting of coherent scintillating fibres to obtain additional information on particle tracks were considered as early as 1957 [8.182]; however, only recently have such devices become competitive with other tracking detectors due to progress in material development and optoelectronic technologies. Such fibres may be suitable as tracking devices in central detectors or in fixed target experiments at high energy physics facilities like CERN, DESY and Fermilab. Today, there are three kinds of scintillating fibre detectors under investigation: plastic fibres, glass fibres and glass capillaries filled with liquid scintillators.

Results on recent studies and developments on radiation resistant optical and shielding glasses, dosimeter glasses, Čerenkov glasses, Ce- and Tb-activated glasses and glass fibres and on liquid scintillator filled glass arrays will be given in this section.

8.6.2 X-ray Sensitive Glasses

X-ray imaging techniques are convenient methods for the non-invasive viewing of the interior of both living creatures and inanimate objects. Physicians can use X-ray images to discover a broken bone or distinguish a tumor from healthy tissue. X-rays are also useful in industry for Non-Distructive Testing (NDT) to detect subsurface defects such as cracks in metal parts or disbonds in composite materials.

X-rays cannot be directly detected by the human eye. The X-ray image can be converted into either a fixed (photographic) or real-time visual image through the use of X-ray sensitive materials or scintillators in the form of polycrystalline powders and, more recently, as glasses or transparent ceramics. Useful X-ray scintillator materials luminesce in the ultraviolet, visible, or near infrared wavelengths following the absorption of X-rays. X-ray scintillator materials typically have a high X-ray attenuation coefficient so that the material efficiently absorbs X-rays which, in turn, generate luminescence. Elements such as tantalum, tungsten, lanthanum and barium which have high effective nuclear stopping powers related with high atomic numbers Z are frequently included in the scintillator composition [8.183]. Materials resistant to X-ray radiation damage are also desirable. *Greskovich* et al. [8.184] describe the process as the creation of multiple electron-hole pairs in the host lattice as a result of the absorption of an X-ray photon. The electron-hole defect or exciton migrate through the host material until an activator ion, defect in the lattice, or other impurity which has energy levels within the band gap is encountered. Recombination of the electron-hole pair at the activator site liberates energy which, in turn, may manifest itself through a radiative emission of energy such as luminescence. Nonradiative transfer of energy may also occur between ions. *Blasse* [8.185] notes that for crystalline $NaGdF_4$:Ce,Tb the excitation of the sensitizer Ce^{3+} results in the transfer of energy to the Gd^{3+} sublattice. The activator ion Tb^{3+} is fed through migration of energy in this sublattice with subsequent emission of radiation.

Photographic film used to record X-ray images is only slightly sensitive to X-ray exposure; therefore, a high X-ray dose would be necessary in order to obtain an informative picture. Intensifying screens containing a layer of scintillating material are used in medical X-ray imaging in order to reduce the dose of X-rays to which the patient need be exposed in order to obtain a useful image. The photographic film which has photosensitive emulsion on both sides is sandwiched between two intensifying screens in a cassette. Different photographic films have different spectral sensitivities; therefore, the combination of film and intensifying screen is chosen so that the light output of the screen is high in the spectral region to which the film is particularly sensitive. The original combination used in medical X-ray applications was a silver halide film with a calcium-tungstate-coated intensifying screen which emits in the blue region of the spectrum. More recently, rare earth oxysulphide, phosphors such as terbium-activated gadolinium oxysulphide which luminesces in the green have been introduced. Compatible films have been developed to take advantage of these new scintillators. The patient is placed between the X-ray source and the cassette. The X-rays are selectively attenuated and/or scattered within the patient according to the density of the body part being scanned. Denser material such as bone attenuates the X-ray beam more than fat and muscle. The partially attenuated X-ray beam then strikes the film cassette. The intensifying screens luminesce in proportion to the intensity of the X-ray beam. The differences in luminescence across the intensifying screens produce a useful image by exposing the film, duplicating the attenuation of the X-ray beam as closely as possible. In theory, the absorption of a single high energy X-ray photon by the X-ray scintillating material will yield several thousand visible light photons of lower energy. However, in practice, the conversion efficiency of typical crystalline phosphors ranges from five percent for calcium tungstate to perhaps twenty percent for some of the newer rare earth phosphors [8.186]. *Greskovich* et al. [8.184], describe a new sintered, transparent, polycrystalline phosphor $Y_{1.34}Gd_{0.60}Eu_{0.06}O_3$ which is useful in computer tomography (CT). An array of nearly 1000 elements composed of small bars of this ceramic scintillator and silicon photodiodes has been constructed. This technique allows the reconstruction of detailed cross-sectional images of the human body. Glasses which luminesce in the visible region of the spectrum when exposed to X-rays have also found applications in medical and industrial imaging. *Reade* [8.187] describes terbium-activated silicate glasses to replace particulate phosphor image intensifying screens used to expose film during diagnostic X-ray procedures. The use of a homogeneous glass rather than a polycrystalline phosphor yields an image with better spatial resolution due to reduced light scatter within the intensifying screen. High quality monolithic glass plates as well as fused fibre-optic faceplates have been made of terbium-activated silicate glasses for real-time industrial NDT and medical imaging applications. Effective real-time imaging requires that the X-ray scintillator has as short a lifetime as possible to prevent smearing of the resultant visual image. Terbium activated fibre optic faceplates have been developed by Schott for use as X-ray conversion screens. These plates, which are vacuum tight, can be used directly as cathode plates in image intensifier tubes, as well as in large arrays for use with very high energy X-rays in NDT applications (jet engine testing, for example). See W.P. Siegmund, R.R. Strack, H.M. Voyagis, M. Heming and P. Nass, "Glass Capillary Production Possibilities," In: Proceedings of the "Workshop on Application of Scintillating Fiber in Particle Physics," p. 127, Blossin, 1990. Further materials were recently developed by *Buchanan* and *Bueno* [8.188–190] who found that the addition of small amounts of cerium or europium oxides to a luminescent terbium-activated silicate glass resulted in a glass which exhibited strong visible luminescence yet low phosphorescence when exposed to 4 keV to 16 MeV X-rays.

8.6.3 Scintillating Glass Calorimeter

Introduction

Detectors for electromagnetic showers include a powerful electromagnetic calorimeter which combines good energy and spatial resolution with large efficiency for energies between 10 MeV and 20 000 MeV [8.191]. In addition, the transparent detector material should have a high radiation resistance and the mechanical handling should be easy. Scintillating glasses like HED-1 [8.192] are potentially capable of yielding, in the energy region of interest, a resolution comparable to CsJ(Tl) and BGO (Bismuthgermanate) at a much lower price.

Calorimeter

Conceptually, the calorimeter intercepts the primary particle and is of sufficient material thickness to cause it to interact and deposit all its energy inside the detector volume in a subsequent cascade or shower of increasingly lower energy particles [8.193]. Eventually, most of the incident energy is dissipated and appears in the form of heat. Some fraction of the deposited energy produces a practical signal (e.g. scintillating light, Čerenkov light or ionization charge), which is proportional to the initial energy. Besides the material, the instrumental technique for converting the signal of the showers into a measurable signal is also important for a calorimeter [8.194]. Glass calorimeters are modules of polished glass blocks wrapped separately in aluminum foil and covered by a light tight shrinking tube. Each block is seen from one end by a photomultiplier followed by a multichannel analyser.

Main Properties of HED-1

If a high energy photon or a particle impinges on the Ce^{3+}-activated Ba-silicate glass block, an electromagnetic shower is generated, developed predominantly through bremsstrahlung and pair production [8.195]. The scale for the longitudinal distribution of the shower in the glass block is set by the radiation length (X_0), which is primarily determined by the main glass component BaO. The parameters characterizing the electromagnetic shower development in different materials are collected in Table 8.11.

Two components of light are emitted and contribute to the observed signal: a fast pulse due to Čerenkov radiation (<10% of the total signal) and a slower component due to scintillation light. This Ce^{3+}-fluorescence with a lifetime of 85(±5) ns emits with its maximum at 435 nm [8.196, 197]. Three factors determine the energy resolution of the calorimeter, namely, intrinsic shower fluctuations, leakage fluctuations and the amount of light produced in the shower counter. The number of photons n_γ/MeV energy deposited is, therefore, of great interest [8.198]. To derive n_γ (330 ± 42), the emission spectrum, the optical absorption length (89 cm) and the refractive index of the material ($n = 1.6295$; 400 nm) must

Table 8.11. Properties of different counter materials

	HED-1	SF 6-glass	CsJ(Tl)	BGO
ρ g/cm^3	3.44	5.18	4.53	7.13
X_0 cm	4.12	1.70	1.85	1.12
n_d (587.6 nm)	1.609	1.805	1.80	2.15
decay const. ns	87 ± 5	20	1000	350
Emission[1] nm	435	continuous	550	480
n_γ/MeV	330 ± 42	65 ± 5	4.5×10^4	10^3

[1] wavelength

ρ: density, X_0 = radiation length, n_d = refractive index at the wave length 587.6 nm, n_γ/MeV = number of photons per MeV

be known. The ultimate limit for the energy resolution of a homogeneous sensitive calorimeter like HED-1, expressed as a fraction of the total energy, improves with increasing energy as $E^{-0.5}$. The measured normalized energy resolution σ as a function of electron energy E ($E > 1\,\text{GeV}$) and beam resolution $\sigma_{\text{beam}} = (1.4 \pm 0.1)\%$ results in

$$\frac{\sigma(E)}{E}[\%] = \sqrt{(1.0 \pm 0.06)^2 + \sigma^2_{\text{beam}} + \frac{(1.6 \pm 0.08)^2}{E}} \;. \tag{8.19}$$

The energy dependent term of the resolution function is mainly determined by the low energy data, whereas the constant term, giving an approximate description of the leakage fluctuations, is fixed by the data taken at high electron energies. The deterioration of the scintillating glass by radiation is a potential danger and might inhibit its use as a calorimeter material. No change of the light output within the regions of reproducibility has been observed in tests of HED-1 depositing integrated doses of 6.5×10^3 Gy (electrons). The expected doses in storage ring experiments are much lower than the values given above.

Summary

The HED-1 calorimeter material was tested in the energy region $15\,\text{MeV} < E < 6000\,\text{MeV}$. As far as linearity, energy and spatial resolution are concerned, this glass is an almost ideal calorimeter. Compared to other homogeneous shower detectors such as BGO, CsI(Tl), BaF_2 and NaI(Tl), this glass has the disadvantage of larger radiation length (factor 2 or 3) and less light output (factor 50 to 500). However, this can be compensated at comparable energy resolution (BGO, CsI(Tl)) by a much lower price (factor 20 to 50). An additional advantage is the high radiation resistance of the glass and its short decay constant (factor 3 to 10).

8.6.4 Čerenkov Glass Counters

Introduction

When a high energy photon or particle enters a transparent material, electrons and protons are multiplied to form so-called "showers". The intensity of these emitted light flashes of less than 10^{-10} s duration is proportional to the energy of the primary photon or particle and is detected by special electronic devices where the amplified electronic signal is registered in a computer memory. This total absorption Čerenkov counter is a small part of a bigger calorimeter where the so-called blue Čerenkov radiation is generated when the secondary electrons and protons move in the material at a velocity (v) greater than that of light:

$$v \geq c_0/n$$

c_0 = velocity of light in a vacuum, n = refractive index .

This phenomenon is the electromagnetic analogy to the acoustic head wave (Mach wave) of a missile flying at supersonic speed. The Čerenkov light spreads on the envelope of a cone of an aperture angle 2φ which reveals the particle's energy if its mass is known [8.199].

Demands on Material Properties

One of the conditions desirable for the non-activated material used as a radiator is high optical transmittance, especially in the shorter wavelength region because the number of Čerenkov photons N is inversely proportional to the square of light wavelength λ as shown by the formula

$$dN \cong \frac{1}{\lambda^2} d\lambda \, . \tag{8.20}$$

The discolouration of the glass caused by irradiation with energetic photons and particles is connected to the optical transmission of glasses. This so-called "browning-effect" decreases the number of emitted photons and harms the accuracy in counter results, especially the data on energy resolution. To avoid this, special lead silicate glasses with the highest possible resistance to colouration caused by particle radiation are selected to guarantee a long useful life.

Another condition for the material is the short radiation length X_0, the distance over which the energy of an electron is reduced to $1/\mathrm{e}$ due to bremsstrahlung and pair production. So the spread of the shower in the material is defined by X_0.

Because materials of short radiation length include elements of high atomic numbers Z and high mass numbers A, the desirable materials are of high densities (3.5–6.2 g/cm^3) and show high refractive indices. The high density is desirable because the absorption of the high speed particles is directly proportional to radiator density. The higher the density the smaller the glass block that can be chosen. The high refractive index is needed to allow a low threshold energy for the traceable particles.

Glasses for Čerenkov Counters

The use of lead glass blocks for electromagnetic shower measurements has been widespread in high energy physics experiments [8.200, 201]. Table 8.12 summarizes the important data on the optical glasses most frequently used as Čerenkov radiators. These glasses are melted in large volumes while maintaining excellent optical quality.

Table 8.12. Several lead silicate glasses from Schott suitable for Čerenkov counters

Glass type	Refractive index (435.8 nm)	Density g/cm^3	Radiation length X_0 cm	Internal transm. 400 nm; 100 mm thickn.
F 2	1.64202	3.61	3.22	0.93
SF 5	1.69985	4.07	2.55	0.82
SF 3	1.77444	4.64	2.04	0.74
SF 6	1.84705	5.18	1.70	0.35
SF 57	1.89391	5.51	1.55	0.32

The short radiation length and the visibility of the full shower development make lead glass one of the most precise detectors for electron and photon energy measurements. Compared to other detector concepts (scintillation counters, wire chambers, etc.), the relatively low cost of glass is another reason for its universal use.

Glasses of short radiation length, especially SF 6 and SF 57 [8.202], exhibit good energy resolution $\sigma/E[\%]$ up to 100 GeV electron or gamma energies E. This feature is quite attractive in the design of electromagnetic shower calorimeters for electron–positron collision experiments, where good energy resolution is required over a wide energy range. So with a positron beam, whose energy ranged from 2 to 17.5 GeV, σ/E values of $3/\sqrt{E} + 0.78(\%) - 5/\sqrt{E}(\%)$ were found in $15\,X_0$ to $20\,X_0$ long lead glass blocks of SF 6.

Summary

Čerenkov counters are detectors mostly consisting of a glass radiator combined with an electronic receiver device. They serve for both the identification and the determination of the energy and charge of fast particles. Total absorbing lead-glass Čerenkov counter arrays are widely used as an electromagnetic shower calorimeter because of their good energy resolution due to high transparency, short radiation length, high radiation resistance, the absence of afterglow and low price. The disadvantage of poor light emission compared to other technical concepts does not limit the universal use.

8.6.5 Active Glasses for Scintillating Fibres

Like optical fibres scintillating fibres consist of a low refractive index cladding (n_2) and a high refractive index core (n_1) material which also scintillates. During the passage of ionizing radiation through the fibre, scintillating radiation is captured by total internal reflection within the fibre and is transmitted to the end faces. If one assembles a large number of scintillating fibres into a well-ordered, so-called coherent fibre bundle, one obtains a detector which can be used to observe particle tracks (Fig. 8.34). The image of the particle track, or more precisely, the projection of the track in the plane perpendicular to the fibre axis, is formed on the end face of the fibre bundle. From here it can be recorded by means of image intensification plus an image detector such as a CCD. The limiting spatial resolution which is achieved is determined by the diameter of the individual fibres.

The number of observable photons depends upon the energy conversion efficiency η_E of the active core material, the transmission efficiency η_t of the scintillating light, which is mainly determined by absorption losses in the scintillating core material (Stokes-shift) and reflection losses at the core-cladding-interface, the trapping efficiency η_Ω, which for meridional rays is proportional to the difference of the refractive indices of the core and the cladding material, the (packing) frac-

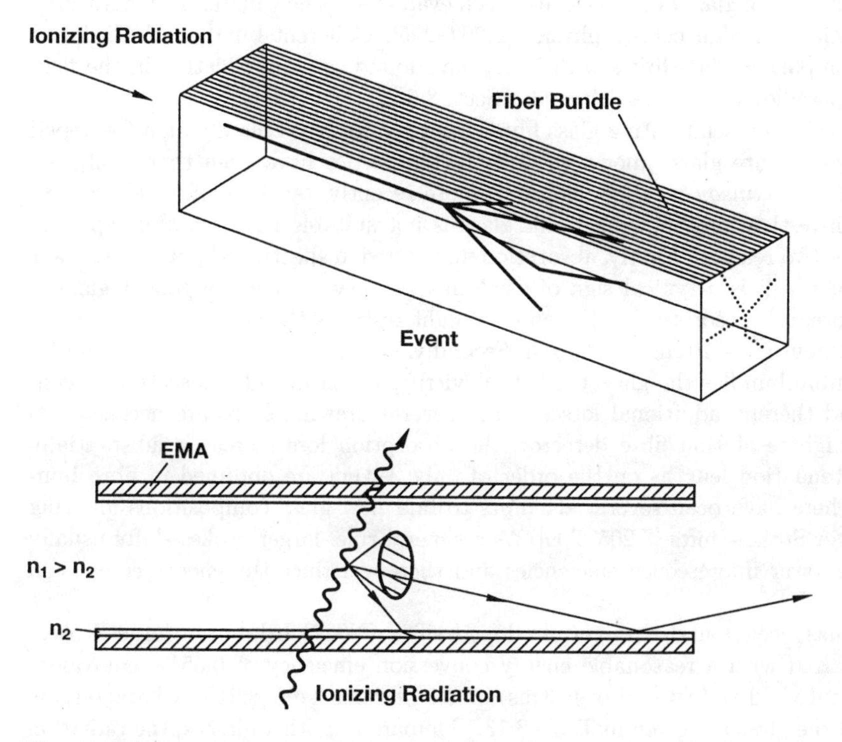

Fig. 8.34. Below: Operation of a scintillating fibre. The extra mural absorber (EMA) suppresses cross talk between neighbouring fibres. The cone indicates the guided light bundle. Above: Schema of a scintillating fibre detector

tion of the active material η_f, the coupling efficiency to the optoelectronic read-out chain η_c and the quantum efficiency of the first photocathode of the read-out, η_Q. The overall efficiency $\eta = \eta_E \eta_t \eta_\Omega \eta_f \eta_c \eta_Q$ of the scintillating fibre detector should be as large as possible.

The transmission efficiency is given by $\eta_t = \exp(-L/L_{att}(L))$ where $L_{att}(L)$ is the attenuation length of the fibre and L the transmitted pathlength. In addition to the sensitivity, the decay rate of the scintillating light, the radiation hardness of the material, the achievable spatial resolution and the noise level (cross talk) are of interest for specific applications.

Today, various scintillating plastic fibres with different dopants are commercially available. Such fibres exhibit fast decay times (ns-region), fairly good attenuation lengths on the order of several tens of cm and overall sensitivities for minimum ionizing particles (m.i.p.) of about two observed photons (hits) per mm at 100 cm detector length (corresponding to $\eta \sim 3 \times 10^{-5}$) [8.203]. So far, no extra mural absorber (EMA) for suppressing cross talk among the fibres can be introduced because no black material has been developed that does not diffuse into the cores during the manufacturing process. Furthermore, the application of these fibres is restricted to environments with rather low radiation levels because they can only sustain doses of a few kGy.

Scintillating glass fibres have also been evaluated as new materials for tracking applications in high energy physics [8.204–206]. Coherent bundles of Tb-doped multicomponent glass fibres with EMA have found some application in the field of X-ray radiography [8.207] (see also Sect. 8.6.2).

Research on scintillating glass fibres has been focussed mainly on a Ce-doped Li-Al-Mg-silicate glass, known as GS1. Its properties have been thoroughly described by *Atkinson* et al. [8.206] and, more recently, by *Angelini* et al. [8.208]. These investigations show that this glass is not suitable for large scale applications for two reasons. Firstly, absorption and emission spectra exhibit a significant overlap, which is a typical sign of Ce^{3+} in most low-Z multicomponent glasses. Therefore, self-absorption of the emitted light restricts the attenuation length to about 20 cm for a 10 cm pathlength. Secondly, during the manufacturing process of the fibre bundles the glass tends to devitrify, which introduces scattering centres and thereby additional losses. Since several drawing steps are necessary to get a high resolution fibre detector, the absorption loss increases substantially and attenuation lengths on the order of only 2–4 cm are obtained in fibre bundles. There have been several attempts to find new glass compositions allowing for larger Stokes-shifts [8.205, 209]. As a general rule, larger Stokes-shifts usually lead to lower fluorescence efficiencies and thereby reduce the energy conversion efficiency.

It has been found that cerium-doped silica (N327) exhibits a large Stokes-shift paired with a reasonable energy conversion efficiency of 0.55 %, equivalent to a total yield (integrated over 3 ms) of 2.1 photons/keV [8.210]. Characteristic data of the glass are given in Table 8.13. Compared to other glasses, the radiation length of about 12 cm is significantly larger, which is of advantage in low event experiments.

Table 8.13. Basic Properties of N327

Ce-doped Silica (0.2 wt-%): N 327	
Absorption maximum	320 nm
Emission maximum	468 nm
Radiation length	11.8 cm
Refractive index	1.46
Energy conversion efficiency (m.i.p.)	0.55 %
Photon yield (m.i.p., within 3 ms)	2.1/keV
Fast decay time	50–100 ns

The decay of the Ce-doped silica shows a fast component of less than about 100 ns; however, as in GS1, an afterglow up to the ms-region has been observed [8.208, 210]. The integrated light emitted up to 10 μs represents about 50 % of the total light. While for neutrino physics this afterglow does not pose a serious problem, Ce-doped silica as well as GS1 seem to be unsuitable for high rate applications.

The Ce^{3+}-absorption peaks at 320 nm as in GS1, but the emission maximum is shifted to longer wavelengths by more than 70 nm. Calculations of the emission spectra for some finite detector pathlengths from the emission point to the readout system prove that even for pathlengths exceeding 1 m a large fraction of the light would be transmitted.

Figure 8.35 shows the hit density as a function of pathlength for N327 silica fibres for various cladding materials. Within the limits of error the attenuation of the F-doped silica cladded fibre is identical with the losses in bulk material. For a 20 cm pathlength one observes an attenuation length of 200 cm. This result is remarkable, since it proves that it is possible to draw the glass into small-sized

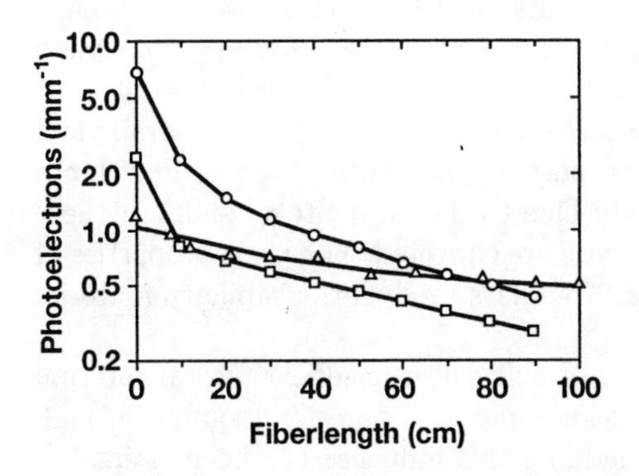

Fig. 8.35. Hit densities for 1 m long incoherent N327 fibre bundles with F^--doped silica cladding (△–△–, 50 μm-fibres), Hostaflon cladding (○–○–, 75 μm-fibres) and a UV-hardenable polymer cladding (□–□–, 150 μm-fibres) as a function of fibre length. The numbers given are based on measured energy conversion efficiencies and refractive indices, a filling factor of 70 %, a coupling efficiency of 75 %, a quantum efficiency of 20 % and a reflectivity of 80 % (Al-coating) on one end of the fibre bundle

fibres without degradation of the transmission properties, and that there are no losses due to insufficient reflectivity at the core/cladding interface. Although the transmission properties of this fibre are unsurpassed until now, the system suffers from a very low trapping efficiency of about 0.5 %. Polymer claddings with low refractive indices lead to higher trapping efficiencies of up to 4.1 %. However, these systems exhibited lower attenuation lengths on the order of 60 to 80 cm. This is attributed to insufficient reflectivity at the core/cladding interface.

In addition to cerium, terbium is a very sensitive activator in inorganic glasses. Moreover, Tb-doped glasses exhibit a very large Stokes-shift, which is a necessary condition to obtain good transmission. Estimates give attenuation lengths of more than 10 m for the bulk material. However, the rather long decay time of 2.5 ms restricts its application to experiments with low repetition rates – like neutrino experiments – and to static X-ray imaging, respectively.

Several Tb-doped glasses applicable as core material for scintillating fibre detectors for neutrino experiments have been developed [8.211]. However, with these glasses it was not possible to preserve the good transmission properties in the fibre bundles due to devitrification of the glasses during the drawing process and insufficient total internal reflection at the core cladding interface caused by contact reactions of the core and cladding material.

8.6.6 Glass Capillaries Filled with Liquid Scintillator for High Resolution Detectors

A different approach to large scale scintillating fibre detectors is the use of capillary arrays filled with a suitable liquid scintillator. Locally emitting solutes such as PMP (1-phenyl-3-mesityl-pyrazoline), BBQ (4,4‴-bis-(2-butyloctyloxy)-p-quaterphenyl) and 3-HF (3-hydroxy-flavone) with high light yield and ns-decay times recently have become available. Solvents with high refractive indices, like PN (phenylnaphthalene, n_1=1.66) and MN (methylnaphthalene, n_1=1.62) yield trapping efficiencies of about 7–8 % in Schott type 8250-capillaries (n_2=1.49) for each direction [8.212, 213]. For an MN/PMP solution, for example, an overall yield of 13 hits per mm for minimum ionizing particles has been observed in a 1 mm single glass capillary of type 8250 [8.214]. Due to the large Stokes-shifts of these dyes attenuation lengths of 100 cm and more are expected. The basic properties of some selected liquid scintillators and of 8250-glass used as the capillary material are summarized in Table 8.14.

Absorption losses in the capillary glass will cause insufficient total internal reflection. This may be critical in rather long and thin fibres as required in high energy physics applications. For 8250-cladding this influence has been estimated theoretically. The calculations prove that even in 100 cm long 20 μm-capillaries the surface loss coefficient $A = 1 - R$ is less than 10^{-5} for wavelengths greater than 380 nm which is sufficient for most applications [8.212].

The manufacturing of capillary arrays is based on a process similar to that used in the production of fibre bundles. The essential difference, however, is the use of a glass tube as the preform in place of the usual rod-in-tube system used

Table 8.14. Basic properties of some selected liquid scintillators and of 8250-glass used as capillary material

	8250	1PN/PMP	1MN/PMP	IBP/3HF
refractive index	1.487	1.664	1.617	1.582
density g/cm^3	2.28	~ 1	1.02	0.98
radiation length cm	12.4	~ 40	45	47
yield n_γ/keV	—	7.2	9.8	7.0
peak emission nm	—	430	430	540
decay time (fast) ns	—	3.45	2.65	3.2

for optical fibres. The basic process consists of a series of individual drawing steps alternating with preform assemblies. This is shown schematically in Fig. 8.36. With each draw the diameter of the glass tube is reduced, and after the second draw the individual tubes, called "monos", are fused together into multis and then multi-multis. This technology allows the production of bundles with lengths up to 200 cm, and by variations of the process parameters it has been possible to attain capillary diameters down to 5 μm and bundle cross-sections up to $10 \times 10\,\text{mm}^2$ (Fig. 8.37). Beside hexagonal cross-sections it is also possible to produce other cross-sections such as square and rectangular ones. The arrays can also be tapered so that the output cross-section is enlarged to reduce any loss of resolution in the coupling of the array to the read-out chain. Capillary arrays can also be "bent" by controlled heating into right angles, S-bends, or other "contours" without loss of resolution or coherency. Thus, for example, the output face can be removed from a region that is "inhospitable" to the read-out chain.

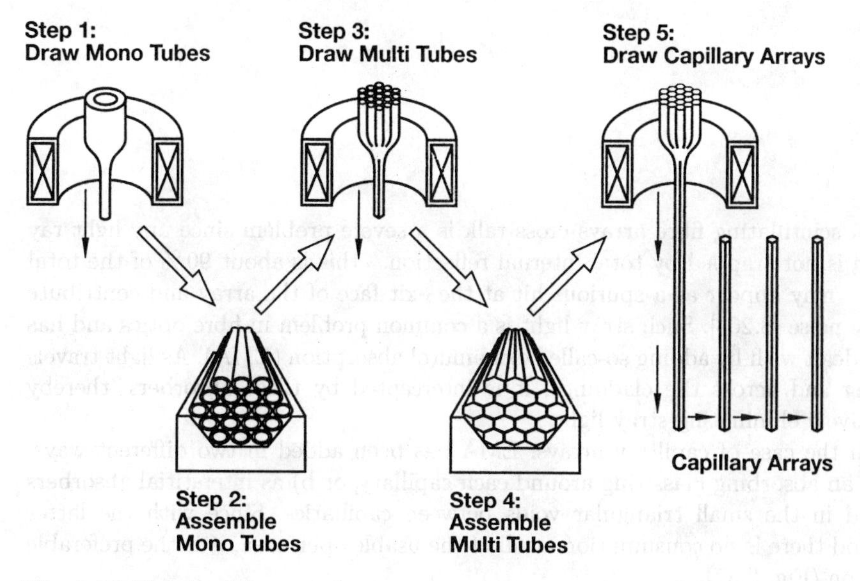

Fig. 8.36. Sketch of the manufacturing process of coherent glass capillary bundles

Fig. 8.37. Multi-capillary bundle (left) and mono-components of a capillary array with interstitial EMA (right)

In scintillating fibre arrays cross talk is a severe problem since any light ray which is not trapped by total internal reflection – this is about 90 % of the total light – may appear as a spurious hit at the exit face of the array and contribute to the noise [8.203]. Such stray light is a common problem in fibre optics and has been dealt with by adding so-called extramural absorption (EMA). As light travels among and across the claddings, it is intercepted by these absorbers, thereby effectively eliminating stray light.

In the case of capillary arrays, EMA has been added in two different ways: a) as an absorbing glass ring around each capillary, or b) as interstitial absorbers placed in the small triangular voids between capillaries. Since with the latter method there is no consumption at all of the usable open area, it is the preferable solution (Fig. 8.37).

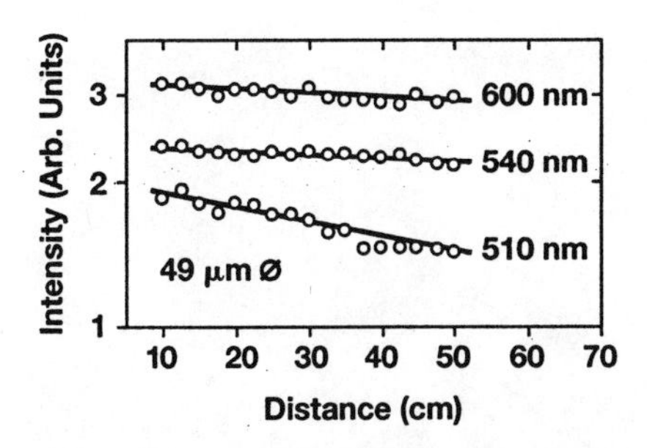

Fig. 8.38. Wavelength dependent attenuation lengths of 49 μm capillaries filled with IBP/3HF. The attenuation lengths are 115, 141 and 215 cm at 510, 540 and 600 nm, respectively [8.212]

Experimental studies with these capillary arrays have been performed at CERN by the CHARM II [8.210, 214, 212, 215] and the WA84 [8.213, 216] collaborations utilizing high energy, i.e., minimum ionizing particles. Investigations with liquid filled capillaries have also been carried out by *Golovkin* et al. at IHEP, Protvino [8.217–219]. Target preparations, i.e. finishing of the capillary endface, cleaning and filling of the arrays, as well as purification of the liquid scintillators have been improved considerably at CERN (details are given in *Adinolfi* et al. [8.213] and *Bähr* et al. [8.215]).

Measurements of light attenuation in single 8250-capillaries of different diameter proved the theoretical prediction that reflection losses at the liquid-core-interface are negligibly small [8.212]. Figure 8.38 shows the wavelength dependent attenuation lengths for 49 μm capillaries with isopropylbiphenyl (IPB) doped with 3-hydroflavon (3HF) as the scintillating liquid. According to these results reflection losses of $A = 1 - R = 1 \times 10^{-5}$ at 600 nm are found. This is a remarkably good result, which proves that thin capillaries can be produced with an excellent glass/liquid scintillator interface. Experiments by *Golovkin* et al. are in agreement with these findings [8.217].

The first investigations using capillary bundles were also performed with IPB/3HF as liquid scintillator [8.215]. Recently, *Adinolfi* et al. studied liquids with MN (methylnaphthalene) and PN (phenylnaphthalene) as solvents and PMP (1-phenyl-3-mesityl-pyrazoline) and 1-(2′-methoxyphenyl)-3-biphenyliyl-5-(4′-methoxyphenyl)-2-pyrazoline (MBMP), as solutes emitting at 500 nm [8.216]. Bundles of 8250-capillaries of 25 μm diameter and 45 cm in length were filled with these "cocktails". With an MN/MBMP mixture a hit density of $5\,\text{mm}^{-1}$ and an attenuation length of 15 cm is observed at short distances (~ 2 cm, Fig. 8.39). At longer distances (≥ 10 cm) the attenuation length increases to 70 cm.

Preliminary studies on the radiation hardness have been performed at IHEP, Protvino, using 110 μm quartz capillaries filled with MN/MBMP [8.219]. Figure 8.40 shows that absorbed doses of up to 640 kGy result in a reduction of light yield of only a factor of 2 at a 1 m detector length. Thus, the radiation hardness of a liquid filled capillary array is significantly better than that of other vertex detectors based on scintillating fibres.

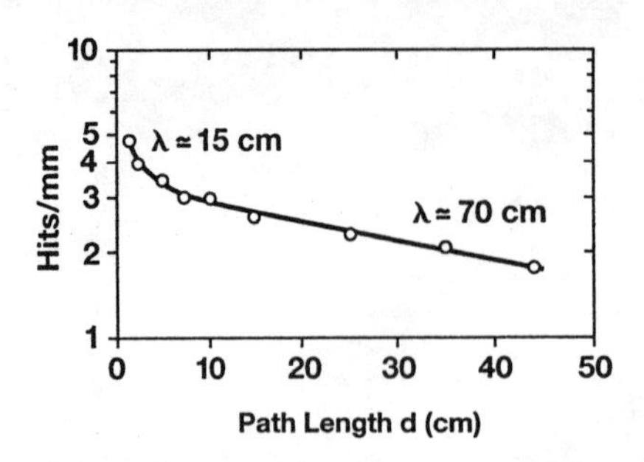

Fig. 8.39. Hit densities for particles crossing a bundle of 25 μm capillaries as a function of pathlength. The capillaries are filled with MN/MBMP [8.216]. The local attenuation lengths are indicated.

The spatial resolution and the influence of EMA on the noise level has been studied in detail by *Adinolfi* et al. [8.213]. In addition to plastic scintillating fibres they also investigated arrays with 20 μm capillaries both with and without an interstitial EMA filled with IBP/PMP. Utilizing the 5 GeV/c hadron beam at the CERN Proton Synchrotron they obtained a spatial resolution of 15 and 17 μm for arrays with and without EMA, respectively. Moreover, the noise contribution is substantially reduced from 39 % to 18 % by introducing the EMA as shown in Fig. 8.41. For the plastic scintillating fibres a spatial resolution of 17 μm at a noise level of 31 % was observed. Due to improvements of the optoelectronic read-out chain, a spatial resolution of 12 μm at a noise level of 15 % was recently obtained with a scintillating capillary array [8.216].

In conclusion, detectors utilizing coherent capillary arrays filled with a liquid scintillator currently offer a better performance than devices based on either massive glass fibres or plastic scintillating fibres. The achievement of high spatial resolution, high yields, low cross talk and high radiation resistance allows for

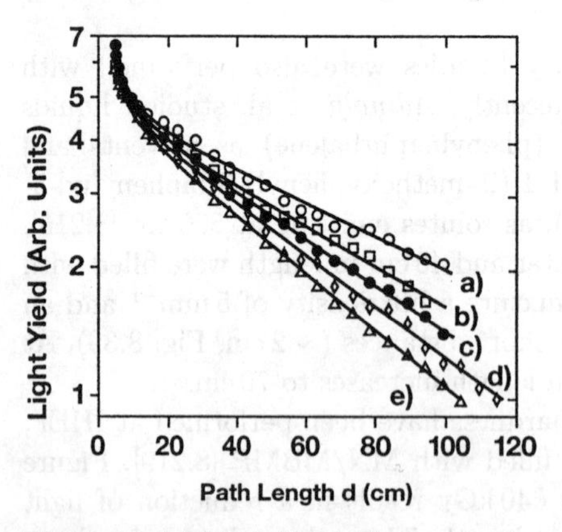

Fig. 8.40. Light yield of 110 μm quartz capillaries filled with MN/MBMP before and after irradiation as a function of pathlength: a) no irradiation, $\lambda_{att} = 118$ cm, b) 125 kGy, 101 cm, c) 160 kGy, 93 cm, d) 320 kGy, 85 cm, e) 640 kGy, 64 cm [8.219]

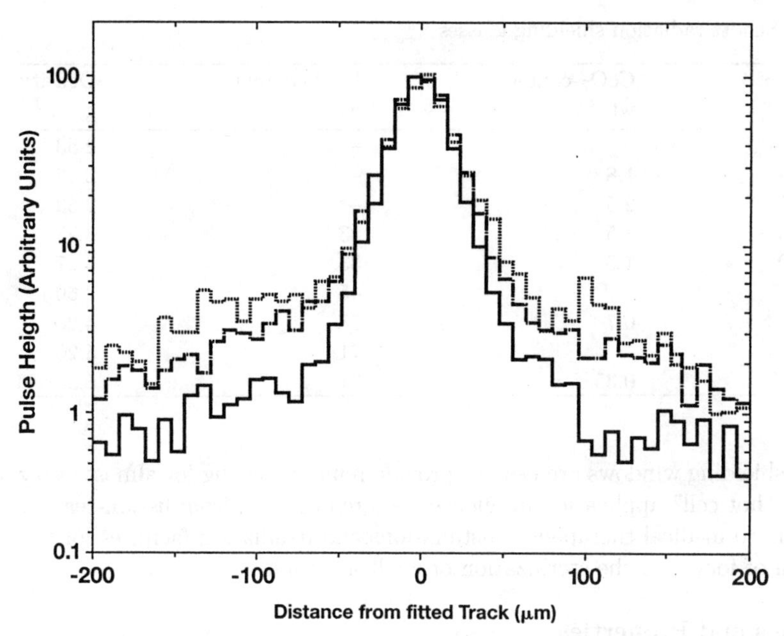

Fig. 8.41. Transverse pulse height distributions for tracks obtained with scintillating plastic fibres (30 μm, PMP-doped polystyrene; dashed line), glass capillaries without (dotted line) and with EMA (solid line) filled with IBP/PMP [8.213]. Noise is substantially reduced by introducing EMA

many applications in existing accelerator facility. They may also be considered for incorporation into central detectors in the planned accelerator at CERN Geneva (LHC) (Large Hadron Collider). The higher energies and the greater frequency of events at these colliders require especially fast, high resolution and radiation-resistant detectors, such as those utilizing glass capillary arrays filled with liquid scintillator.

8.7 Special Glasses for Nuclear Technologies

Burkhard Speit

8.7.1 Radiation Shielding Glasses

Introduction

The two principal radiation-safety-related functions of shielding glass blocks in "hot cell" viewing windows are to shield against the radiation given off by the radioactive materials in the cells and to withstand the radiation concerning discolouration and electrical discharge [8.220, 221].

Table 8.15. Schott radiation shielding glasses

Glass type	CeO_2-content wt. %	PbO content wt. %	Density g/cm^3
RS 253	–	–	2.53
RS 253 G 18	1.8	–	2.53
RS 253 G 25	2.5	–	2.53
RS 323 G 15	1.5	33	3.23
RS 323 G 19	1.9	33	3.27
RS 360	–	45	3.60
RS 420 G 7	0.7	58	4.20
RS 520	–	71	5.20
RS 520 G 3.5	0.35	71	5.20

Today, shielding windows are used to provide remote viewing for almost every conceivable "hot cell" application in nuclear research centres, from fission-reactor fuel handling to medical therapeutic installations and irradiation facilities for the preservation of food and the sterilization of medical supplies.

Glass Types and Properties

Nearly ten different glass types are presented in Table 8.15. These offer a well balanced range of products for clear viewing at a maximum angle of vision and fulfil safety requirements. These radiation shielding glasses are usually employed as a combination of different glass types and glass thicknesses in a shielding window. The different glass blocks are framed individually. The frames are screwed together, and mounted in concrete or lead walls.

The main radiation attenuation is caused by lead glasses containing 30 to 70 weight % of lead oxide (PbO). The additional cerium oxide (CeO_2) doping of these three most common glasses reduces discolouration caused by the applied radiation. Depending on the particular application, these glasses are often positioned in the centres of windows.

Beside this group of glasses, two cerium-doped boron-silicate glasses are suitable for applications involving high doses and high-energy radiation fields. On the "hot side" of the window they are able to moderate the radiation before it reaches the lead glasses. With respect to electrical discharge, the application limits of boron-silicate glasses lie several orders of magnitude higher than those of the most common lead glasses.

A third group of three glasses without cerium doping consists of two different lead glasses and one boron-silicate glass. They are intended as additional attenuators on the "cold side" of a shielding window.

A cerium-doped glass type containing around 30 weight % of lead oxide has been developed for universal use. As a result of its unique electrical discharge stability of more than 10^8 Gy accumulated dose, it is applied even in severe radiation fields. Discharge of the common cerium-doped lead glasses manifests itself beginning at integral doses of 10^4 Gy as a pale blue light flash and the so-called

"Lichtenberg tree" which causes permanent damage to the glass. The discharge can be initiated by external pressure or by impact stress [8.222, 223].

Shielding Properties

When designing radiation shielding windows, parameters like source spectrum, geometry and configuration of the radiation source, specific utilization limits and absorption characteristics of the applied glasses have to be taken into consideration.

This determines a glass combination with optimum performance for the particular application. The interactions of the most common gamma or neutron radiation passing through the windows cause an attenuation of the initial radiation or a deposition of a certain rate of initial energy in the different glasses depending on the radiation energy, E, on the glass composition (atomic number of the elements, Z; density of the glass, ρ) and the individual glass thickness, d. The probability for such interactions to take place is described by the reactive cross-sections of interaction. The linear attenuation coefficient μ (cm^{-1}) for photons like γ-radiation is calculated from atomic interaction cross-sections of the photoelectric effect, Compton effect and pair creation [8.225–227]. In the case of homogeneous materials like glasses and idealized parallel radiation, the general exponential attenuation is given by

$$I(d) = I_0 e^{-\mu(Z,E)d} . \qquad (8.21)$$

But in a real situation the primary absorption is accompanied by emission of secondary radiation of lower photon energy. This makes the remaining radiation level beyond the absorber greater than it would be expected from (8.21). This effect is taken into account by introducing a built up factor B (μ, d, E):

$$I_{eff} = B(\mu d E) I_0 e^{-\mu(Z,E)d} . \qquad (8.22)$$

The exact solution of the integral equations for more complex source geometries with consideration of the secondary radiation is complicated. Thus one normally uses an approximate solution in order to describe the problem with sufficient accuracy.

Radiation Effects

The interaction between high energy photon or particle radiation and glasses will cause electrons to be excited sufficiently to leave their normal positions. These electrons and the holes left by them can cause specific absorption of light. The reason for this macroscopic optical absorption is different microscopic colour centres identified by EPR (Electron Paramagnetic Resonance). Details of this method have been published in [8.228–231] and elsewhere.

The transmission reduction for high radiation doses up to 1×10^4 Gy (1 Gy = 100 rad) is eliminated almost completely even in a typically required glass thickness of some hundreds of millimetres by doping the shielding glasses with a few weight %

of cerium oxide (CeO_2). Adequate transparency of the glasses then still exists even after applied doses of greater than 1×10^8 Gy [8.232, 233].

The addition of CeO_2 to shielding glasses reduces the transmission in the short wavelength region around 380 nm because of the characteristic absorption of the Ce^{4+} ion (typical yellow colouration of radiation shielding lead glasses). The extent of this effect depends on the base glass and the added amount of CeO_2. However, this effect becomes almost insignificant when the glasses, used as "hot cell" windows, are illuminated from the "hot" side with sodium vapour lamps (wavelength = 589 nm).

Besides the internal transmission and radiation resistance with respect to colouration, the application limits of glasses are determined by the discharge stability which has to be considered designing shielding windows. Especially lead glasses with higher gamma radiation attenuation are capable of being charged. By Compton scattering a γ-ray beam produces a secondary electron flux travelling in the same direction. This leaves an irradiated region with an excess negative charge which can accumulate. Even a light impact or other conditions capable of cracking the glass surface, can cause an electrical discharge resulting in the fracture of the glass. A new lead glass overcomes this problem by means of higher ionic and electrical conductivity and, additionally, by higher homogeneity. The mono alkali potassium lead glass is not phase separated, in contrast to nearly all common lead glasses.

Summary

There exists a sufficient number of different glass types for various radiation shielding applications that fulfil all safety requirements and withstand severe radiation over a period of 20 to 30 years. Radiation shielding windows have been designed and constructed for all special applications required in nuclear technology. Many refining processes have been developed to enhance the utility of radiation shielding glasses: reflection losses are prevented by leaching or coating the glass surfaces. To increase the mechanical strength and the resistance to temperature shocks, the glass blocks are thermally pre-stressed. To avoid transmission loss and to achieve viewing angles that are as big as possible, glass blocks are cemented together.

8.7.2 Radiation Resistant Optical Glasses

Introduction

Hostile environments created by short wavelength electromagnetic radiation, e.g. UV, X-ray and γ-radiation, or by particle fluxes, e.g.,α-particles, β-particles, protons and neutrons can produce discolouration within optical glasses. The associated loss in transmission is detrimental to the performance of any optical system and must be eliminated or reduced to a manageable level. For applications within these hostile environments, radiation-stabilized optical glasses have been developed [8.234–236].

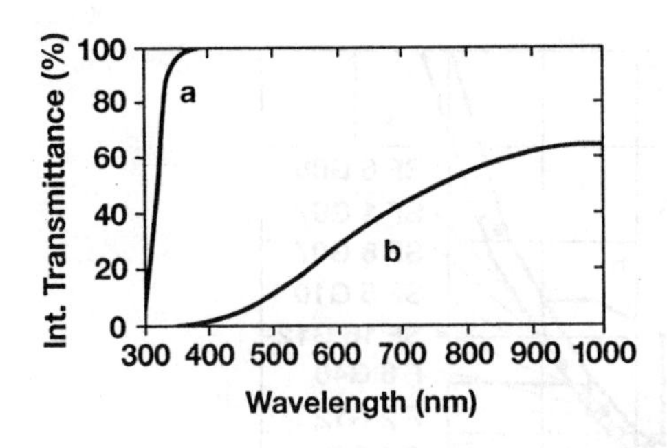

Fig. 8.42. Internal transmittance as a function of wavelength for BK-7: a) unirradiated, b) Co-60 irradiated (10^4 Gy)

Discolouration in Glasses

Ionizing radiation can produce free holes and electrons in glass which can then be trapped, thus forming defect centres. These defect centres cause an increase in the optical absorption in the visible portion of the spectrum, leading to a darkening of the glass. Short wavelength electromagnetic radiation travelling through glass interacts with electrons and atomic cores through their electric fields. The most important possible loss mechanisms are the photoelectric effect, Compton scattering and pair formation. Charged particles can interact with both the electron clouds and the atomic cores, whereas for neutrons only interactions with atomic cores are of importance.

For any particular radiation situation the intensity of discolouration is a function of many parameters, including the type and dose of radiation, the energy of the ionizing radiation and its intensity, and the composition of the exposed glass. For unstabilized optical glasses, total doses of as low as 10 Gy (10^3 rad) of gamma radiation (photon energy of 1.25 MeV) can result in visually detectable colouration. The internal transmittance of normal, unstabilized BK-7 is given as a function of wavelength in Fig. 8.42. After exposure to 10^4 Gy of Co-60, the transmittance is degraded to the point where the glass may no longer perform adequately in an optical system.

Radiation Resistant Glasses

In order to optimize system performance one must select optical glasses, which have been stabilized for applications within the particular radiation environment. If the environment is a mixture of radiation fields, compromises will have to be made.

Studies have shown that cerium inhibits the formation of colour centres [8.237]. Cerium-doped glasses have been developed with increased resistance to radiation fields [8.238, 239]. In Fig. 8.43 these special glasses from all glass families are shown in the $n_d - \nu_d$ glass map, where n_d is the refractive index at 587.56 nm (the He-line) and ν_d is the Abbe value or the dispersion [8.240]. Concerning the Abbe value, see Sect. 2.1.

At Schott, all optical glasses containing cerium oxide have “G” in their designation; e.g. BK-7 G 18. This is the glass type BK-7 with 1.8 % CeO_2 added to the

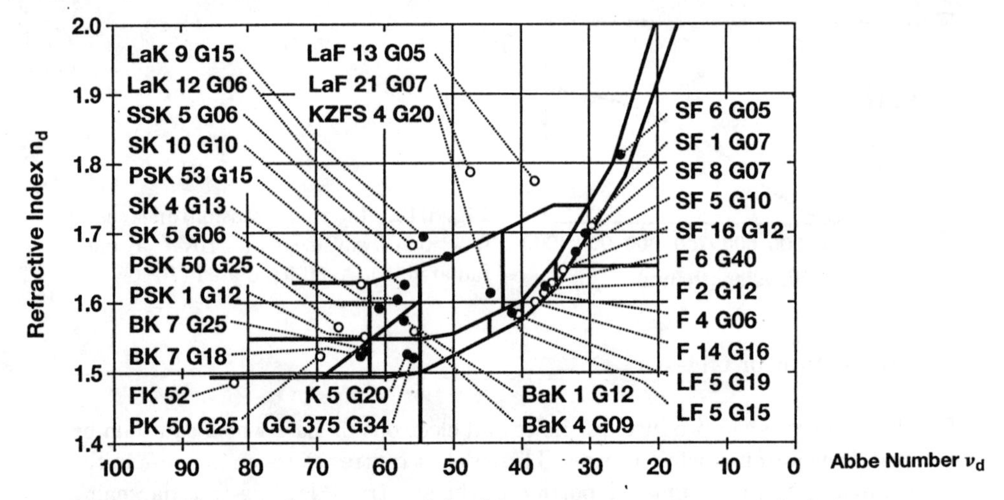

Fig. 8.43. The optical glass map of the Schott radiation resistant optical glasses

normal glass composition. The addition of CeO_2 to the batch changes the cut-on characteristics of the optical glass and is dependent on the amount of cerium and the glass type. So this cut-on shift to longer wavelengths is particularly enhanced for glasses containing constituents of high polarizability, such as Pb, Ba and La. This can be a very critical issue if the glass is intended for use in the near UV.

Discolouration of Radiation Resistant Glasses

In contrast to normal optical glasses, radiation-resistant glasses show only weakly increased absorption when exposed to high doses of high energetic radiation. The addition of cerium is very efficient to prevent the darkening of glass due to gamma radiation. BK7 G 25®, for example, shows no discolouration, indeed, for total doses up to 10^6 Gy (10^8 rad) of exposure to Co-60. However, if one considers particle radiation, one observes that there is a strong dependence of the resultant discolouration on the type of intervening particle. Exposure to protons (total dose 1.4×10^{18} MeV/m^2 with individual energies between 7 and 50 MeV) shows minor discolouration, whereas neutron exposure (10^{20} n/m^2) leads to strong absorption below 550 nm, in this case related to the high cross-section of boron, a major constituent in BK 7®, for neutron interaction.

It depends on the glass composition how efficiently CeO_2 prevents discolouration. The lead containing glass LF-5 G 15 shows good resistance to gamma, proton and neutron irradiation, but does not perform well in an electron flux. In this case, a certain part of the lead exists in this glass in non-bonding sites. During irradiation by an electron beam Pb is also reduced to metallic lead. These small metallic lead particles are responsible for the discolouration.

Colour Centres in Irradiated Resistant Glasses

As pointed out above, the influence of radiation on glass can be studied macroscopically by numerous spectroscopic techniques including optical absorption or emission. Another spectroscopic technique to identify the reason for the absorption microscopically is the electron paramagnetic resonance (EPR). The typical colour centres in irradiated resistant glasses are the so-called oxygen-hole centres (OHC). They can easily be observed by means of EPR in multicomponent glasses and in high-OH silicas when observation is made at sufficiently low temperatures [8.241]. These defects absorb in the visible region and are almost certainly formed in this way:

$$\text{a)}\ \equiv \mathrm{Si-O-Na} \xrightarrow{h\cdot\nu} \equiv \mathrm{Si-O^{\bullet}} + \mathrm{Na^{+}} + \mathrm{e^{-}}$$

$$\text{b)}\ \equiv \mathrm{Si-O-O-Si} \equiv \xrightarrow{h\cdot\nu} \equiv \mathrm{Si-O-O^{\bullet}} + \mathrm{Si} \equiv^{+} + \mathrm{e^{-}} \tag{8.23}$$

where "•" denotes the EPR-active unpaired electron.

The typical OHC-signals on a so-called "nonbonding site" appear in all usual glasses irradiated with photons or particles but they are much less intensive in irradiation resistant glasses. The EPR-signal intensity is a function of colour centre concentration. The colour centre formation is typical of the glass types but not dependent on the species of radiation. Thus, in lead glasses, contamination deduced colour centres are created by a change in redox-equilibrium:

$$\text{a)}\ \mathrm{Fe^{2+}} + \mathrm{h^{+}} \rightarrow \mathrm{Fe^{3+}} \quad (\text{"}h^{+}\text{" denotes an OHC})$$

$$\text{b)}\ \mathrm{Pb^{4+}} + \mathrm{Fe^{2+}} \xrightarrow{h\cdot\nu} \mathrm{Pb^{3+}} + \mathrm{Fe^{3+}} \tag{8.24}$$

Conclusions

Upon review of Fig. 8.43 it becomes apparent that there are CeO_2-stabilized optical glasses from all families found on the glass map. As has been shown in some examples, the glasses of different compositions demonstrate varying resistance to discolouration depending on the type and dose of radiation. The EPR active colour centres, however, do not depend on the type of radiation. The environment to which the glass will be exposed must thus be considered when choosing optical glass. In order to optimize optical system performance, compromises must be made if the environment is a mixture of radiation fields known from applications in space.

8.7.3 Dosimeter Glasses

Introduction

With the extensive use of radiation sources and reactors for civilian and military applications it has been realized that the capabilities of the most commonly used film as a large-scale long-term dosimeter were rather limited. The main reasons were problems of a strong energy dependence, fading (regeneration after exposure to radiation) at higher temperatures as well as by humidity and chemicals, poor reproducibility and high sensitivity to light and pressure.

Radiophotoluminescence (RPL) dosimeters have been developed and crystalline materials like LiF, BeO:Na, $CaSO_4$:Tm and CaF_2 were introduced using the thermoluminescence (TL) to detect radiation doses [8.242].

The main glass based dosimeters have been applied in personal- (or low dose-) and accident- (or high dose-) dosimetry. These dosimeters are sensitive to photon- and particle radiation [8.243].

RPL Dosimeters

The term radiophotoluminescence means that a glass which is originally nonluminescent under visible or ultraviolet light, is made responsive to such excitation by pretreatment with ionizing radiation. The amount of this induced fluorescence is proportional to the absorbed radiation dose. Even after very low dose exposure of 0.5 µGy–10 µGy (1 Gy = 100 rad) which is already significant from the biological point of view, silver-activated phosphate or borate glasses exhibit strong orange luminescence during excitation with the 365 nm mercury line. In addition, a considerably improved photon energy response is observed particularly if the silver content of 4.0 is reduced to 0.2 weight % [8.244]. Whereas the signal in TL-dosimeters is largely destroyed by the readout procedure, the basic advantage of RPL dosimeters is the excellent storage stability, the possibility of integral partial dose accumulation and interval readout and the reusability after thermal annealing.

During irradiation, absorption bands in the near ultraviolet and visible regions are produced. Illumination with this light produces visible polarized luminescence. The RPL is ascribed to neutral silver atoms or Ag^+-ion modified colour centres [8.245]. More complex centres, such as Ag^+-ion + holes + positive ion vacancy and Ag^+-ion pair + electron + negative ion vacancy, are more probable. A mathematical treatment of the kinetics on the basis of a simplified band model has been carried out assuming the following process: ionizing radiation of sufficient energy lifts electrons to the conduction band. Some of them are directly trapped by positive charged silver atoms or aggregates; others are first trapped in centres, which do not contribute to the radiophotoluminescence before being transferred thermally to the effective RPL centres. In Fig. 8.44 the schematic band diagram of the main processes occurring in an Ag-activated dosimeter glass are shown. Fading or ultraviolet bleaching may be explained by a recombination of electrons from the excited state of the RPL centres with electron vacancies in "hole" traps.

High-Dose Glass Dosimeters

The operation of glass-block radiation dosimeters is based on the observation that several normal clear glasses are discoloured at doses exceeding 1 Gy, turn-

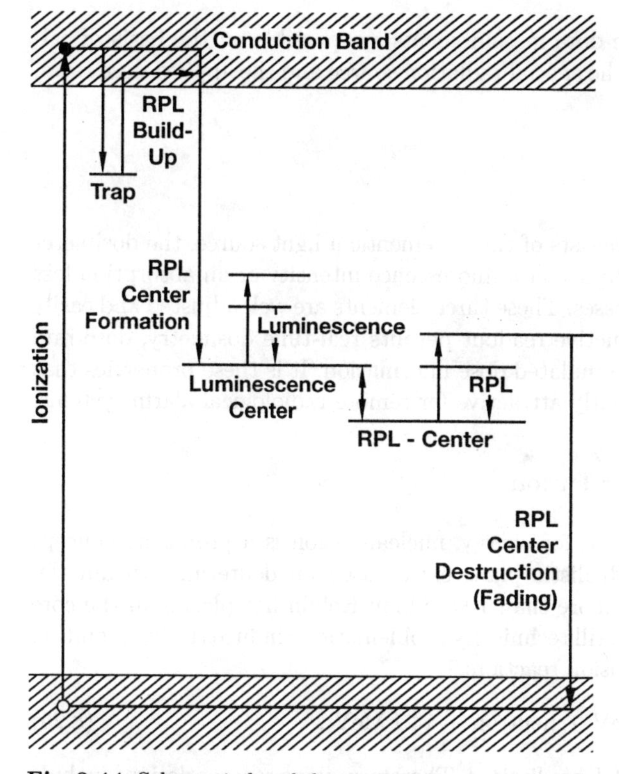

Fig. 8.44. Schematic band diagram of the main processes occurring in an Ag-activated dosimeter glass

ing to yellowish-brown or red upon exposure to ionizing radiation. The colour appearance results from strong optical absorption in the UV and visible. These absorption bands are due to defect centres produced by the ionizing radiation and their intensities are proportional to the applied radiation doses between 1 Gy and 10^7 Gy. Although the exact nature of these defects is often complex, many oxide glasses show absorption bands in the UV and near 550–700 nm associated with holes trapped near non-bridging oxygens [8.246].

The sensitivity of glass can be increased ten thousand times by using multi-mode step index glass fibre wave guides. For example, 10 m long fibres formed by a 50 μm diameter dense flint core surrounded by a thin borosilicate cladding have typical transmission losses of $\approx$ 300 dB/km Gy (X-ray) at 900 nm [8.247]. Further enhancement of sensitivity is achieved by switching over to absorption losses in the UV.

The most common dosimeter glasses are phosphate glasses showing saturation effects at doses higher than 10^7 Gy. Annealing of the dosimeter glasses to 180 °C for 10 minutes overcomes the fading or regeneration of instable colour centres which would otherwise lead to unreproducible results. The intensity of the dominant absorption band at 510 nm is evaluated before the exposure to radiation and after annealing. Precise dose measurements are possible on the basis of calibration curves considering the special radiation source and the applied dose rates.

Dosimeter System

A basic dosimeter system consists of three elements: a light source, the dosimeter glass and a photo-detector to detect a fluorescence intensity or an absorption loss in the exposed dosimeter glasses. These three elements are well adjusted and easily usable as a unit. Non-destructive readout permits real-time dosimetry, dose rate measurements and total accumulated dose information. It is these properties that make these systems particularly attractive for remote radiological alarm systems.

8.7.4 Glasses for Nuclear Fusion

Due to the increasing demand for energy, nuclear fusion is a promising concept for future energy supply. Realistic plans exist based on deuterium-tritium (D-T) reactors working with a more than 100 million Kelvin hot plasma in the core [8.248]. At the moment, it is still technically problematic to induce the high ignition temperatures to start the fusion reaction:

$$\mathrm{D} + \mathrm{T} \rightarrow {}^{4}\mathrm{He} + 3.5\ \mathrm{MeV} + \mathrm{neutron} + 14.1\,\mathrm{MeV}\ . \qquad (8.25)$$

The natural resources of T are limited. Therefore, some reactor designs include helium cooled solid breeder blankets around the plasma based on the use of a mixed bed of beryllium and more or less glassy lithium orthosilicate (Li_4SiO_4 + 2.2 weight % SiO_2) pebbles [8.249, 250]. These small spheres in the diameter range of 0.1–0.5 mm with 60% Li-6 enrichment set free gaseous T when exposed to neutron fluxes coming from the fusion plasma:

$${}^{6}\mathrm{Li} + \mathrm{neutron} \rightarrow \mathrm{T} + 4.78\,\mathrm{MeV} + {}^{4}\mathrm{He}. \qquad (8.26)$$

For reasons of neutron physics, the beryllium is required as a neutron multiplier (Be + neutron$\rightarrow$ 2 neutrons + 2 ^{4}He – 1.5 MeV) to maintain the self-sustaining fusion process.

To make practical use of the fusion technology, the energy transport out of the plasma is achieved by high pressure helium cooling of the blanket. There the neutrons carrying 80% of the fusion energy are thermalized. Helium is also the transport medium for the bred T which has been prepared outside for the fusion process [8.251]. 100 mg/s of a mixture of T and D is sufficient "injected fuel" to maintain the fusion reaction of a 5 GW (GigaWatt) reactor. In each (D-T) reaction 17.6 MeV energy is produced. This is more than a factor 4×10^6 greater than in the well-known reaction $C + O_2 \rightarrow CO_2$.

Production by spraying liquid material has proven one of the best ways to fabricate glassy pebbles. Compared to pebbles prepared by granulating and sintering, glassy products produced from the melt are closer to spherical shape, have a smoother surface and a higher density. These properties are important to satisfy the design requirements.

Pebbles with a surplus of 2.2 weight % silica to the stoichiometric composition that are produced by rapid quenching from the melting temperature have a dendritic structure with grains up to 5 μm diameter and 15 μm length. This structure spreads out over the whole cross-section of the pebble, an indication of a supercooled liquid.

At temperatures above 1024 °C, the melting point of the glassy phase, orthosilicate crystals and a silicon rich-liquid exist simultaneously. That is the reason why these pebbles have a glass cemented grain structure. Quenching from the melting temperature requires special techniques; otherwise, the pebbles may contain cracks and stresses which lower their mechanical stability. By a subsequent thermal treatment above 1050 °C that lasts some minutes, microcracks can be healed, especially in pebbles with a 2.2% surplus of silica.

The key issue for the design of a (D-T) reactor is the T-release behaviour of the breeder material in general and of the Li_4SiO_4 pebbles in particular. Pebbles exposed to neutrons have shown considerably short T-release times compared to other ceramic breeder materials like Li_2O, LiOH, Li_2SiO_3 and $LiAlO_2$. Beds with pebbles of the same diameter or of diameters ranging within a relatively narrow margin do not have a large packing factor (≈ 60%); however, there is a well-defined space between the pebbles to allow for the flow of T in the helium gas.

Li_4SiO_4 + 2.2 weight % SiO_2 is a promising material for nuclear fusion breeder blanket material with a proven shaping technique. Its mechanical stability is improved and its T-release properties and not least the reusability after remelting and ^{6}Li enrichment have been demonstrated. One day, Li_4SiO_4 + 2.2 weight % SiO_2 glassy pebbles might contribute to solving future energy problems.

8.8 Coloured Glasses

Uwe Kolberg

8.8.1 Introduction

Colour in glass arises by the selective attenuation or amplification of the incident light within special regions of wavelength in the visible part of the spectrum, i.e. in the range from about 380 nm to 780 nm. Generally, the shape of the transmission curve against the wavelength λ is described by (8.27)

$$\tau_i(\lambda) = \exp \sum_m (- \epsilon_{\lambda,c,m} c_m d) \tag{8.27}$$

where ϵ and c are the extinction coefficient and the concentration of the colouring species m, respectively, and d is the thickness of the probe. The internal transmission τ_i and the relation between internal transmission and transmission τ has been defined in Sect. 2.3. The extinction coefficient for a given base glass is a function of the wavelength, the colouring species m and, in some cases, the concentration of this species. Keep in mind that one colouring ion in two different oxidation states acts like two different species (e.g. Fe^{2+}, Fe^{3+}).

Unfortunately, there is no way to calculate ϵ from theoretical considerations but it must be taken from experiments for every single colouring species. Some data have been compiled by *Bamford* [8.252], however, not for all elements and combinations. Additionally, all data taken from the literature depend on the special base glass composition. So, for a given new problem, the extinction coefficients must first be established experimentally.

In many cases, (8.27) can be simplified. Very often, glasses are coloured only by one species in one oxidation state and the extinction coefficient is not a function of concentration. This is especially true if the concentration of the species is low. Then (8.27) can be written as (Lambert-Beer law):

$$\tau_{\mathrm{i}}(\lambda) = \exp\left(- \epsilon_{\lambda} c d \right) . \tag{8.28}$$

From this equation, the transmission plots for different concentrations of the colouring species and different probe thicknesses can easily be calculated from one experiment, since ϵ is then a constant for a given wavelength.

As already mentioned, the colour of a glass can be taken from a plot of its internal transmission against wavelength. A translucent body, showing an internal transmission $\tau_{\mathrm{i}} = 1$ at every wavelength in the visible range is called colourless (WG 280; Fig. 8.48). These are, for example, the usual optical glasses. When the internal transmission at a range of wavelengths is attenuated by any of the mechanisms discussed in Sect. 8.8.2, a colour arises. (Amplification of light has already been discussed in Sect. 8.5 "Laser Glasses". Therefore, the following discussion is restricted to the case of attenuation). The kind of colour depends on the part of the spectrum missing:

In Figs. 8.45, 8.46, 8.47, 8.48 examples of glasses representing the main colours are shown. Figure 8.45 shows the transmission plot of two yellow glasses: GG 19 is coloured with CrO_4^{2-} (see the section on "Charge Transfer Effects"). Glasses of this type are more or less replaced by semiconductor (Cd(S,Se)) doped glasses (GG19 is no longer available) like GG 475 with a superior spectral characteristic (see the section on "Semiconductor Doped Glasses").

Figure 8.46 gives the spectrum of a blue glass (BG 39) and a green glass (VG 14). The colour of BG 39 is obtained by doping with Cu^{2+}. This type of colouration is explained in the section on "Ligand Field Effects". The colour of VG 14 is caused by Cu^{2+} and a mixture of Cr^{3+} and Cr^{6+} (see the sections below on "Ligand Field Effects" and "Charge Transfer Effects").

Figure 8.47 shows two examples of red glasses. The glass RG 6 is more or less of historical interest (RG 6 is no longer available). It is coloured by gold metal colloids (gold ruby, see the section on "Metal Colloids"). Red glasses of this type have been displaced by semiconductor (Cd(S,Se)) doped glasses with superior spectral characteristics (RG 610). The colouration mechanism of these glasses is discussed in the section on "Semiconductor Doped Glasses". In addition to an optical, colourless glass (WG 280), Fig. 8.48 shows a transmission plot where all the wavelengths of the visible range are more or less equally attenuated (NG 4).

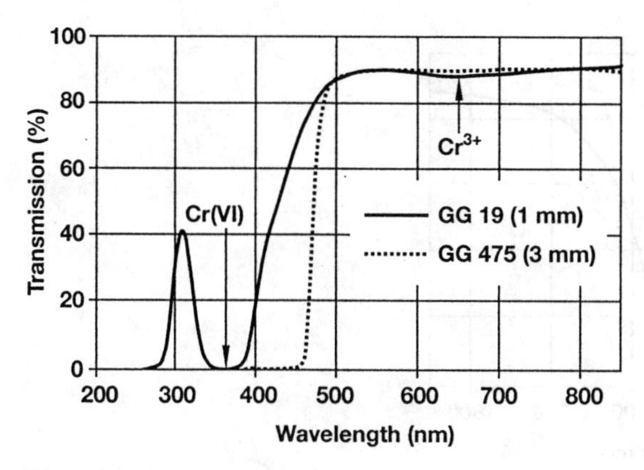

Fig. 8.45. Transmission spectra of yellow glasses

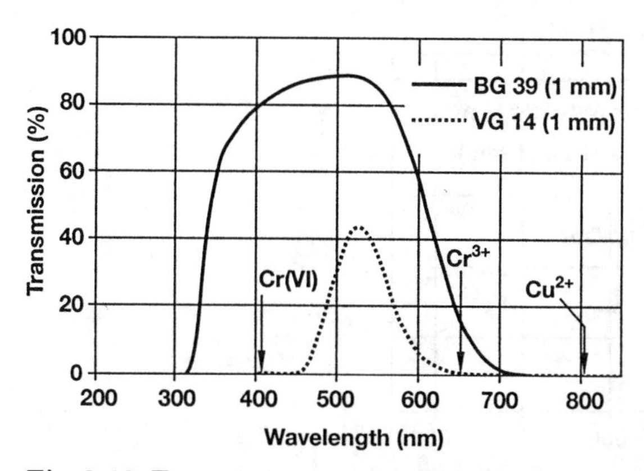

Fig. 8.46. Transmission spectra of blue and green glasses

This results in a grey, or, at low transmission rates, black colour of the glass. The attenuation is achieved by a combination of the absorption of Co^{2+}, Fe^{2+}, Fe^{3+} (see Sect. 8.8.2 under "Ligand Field Effects" and "Charge Transfer Effects").

This kind of representation is chosen for scientific applications and construction of optical instruments because it advantageously shows the spectral distribution of transmitted light. Another way to describe the colour of the glass is the use of colour coordinates. When this method is used, the colour is defined by two coordinates x and y which are obtained by weighting the $\tau(\lambda)$ curve of the glass with the intensity distribution of the standardized light source for the colours red, green and blue. In the ideal case, the colour portions of the light source add to clear white light. The procedure has been extensively described by *Bamford* [8.252].

Figure 8.49 shows the distribution coefficients for the red (P_x), green (P_y), and blue (P_z) part of a special standardized light source (illuminant D65) as has been set up by the International Commission on Lighting (CIE) [8.253]. After multiplication of the transmission curve with the distribution coefficients, and normalizing the sum to unity, one obtains the colour coordinates shown in Fig.

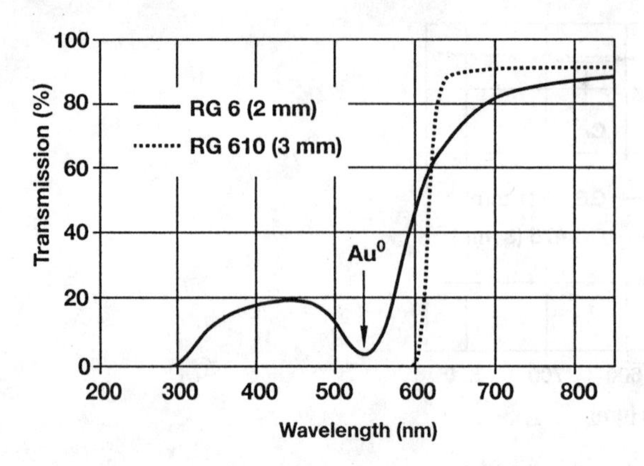

Fig. 8.47. Transmission spectra of red glasses, see next page

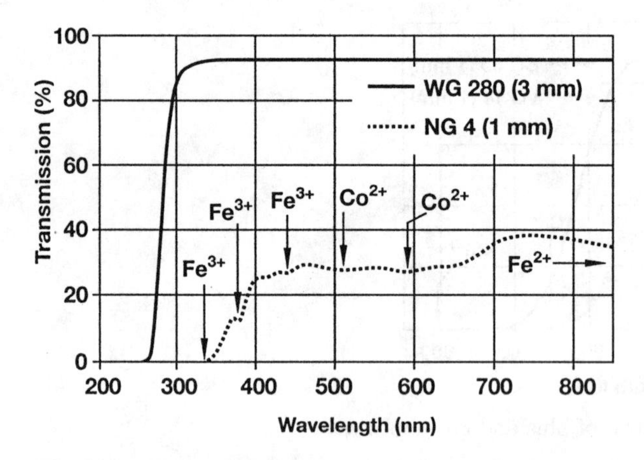

Fig. 8.48. Transmission spectra of grey and white

8.50 for the examples presented in Figs. 8.45, 8.46, 8.47, 8.48. The "z"-coordinate has been omitted because it can easily be calculated from the condition that $x + y + z = 1$.

The advantage of this method is that one obtains immediately the real colour of the glass from a reference chart (a coloured form of Fig. 8.50). Therefore, this method is used when the colour impression is needed; e.g. for traffic signs and glasses for artists.

From the chemical point of view, the base glass composition of coloured glasses ranges within the compositional region of normal optical glasses, as discussed in Sect. 2.2. Historically, the coloured glasses have been developed from the optical glasses as the so-called "colouring elements" were added empirically. There has been a lot of experimental work on this item of which a survey has been given by *Weyl* [8.254] and, to some extent, by *Vogel* [8.255]. The colouring elements additionally present in the base glasses, transition metals and lanthanides, contain

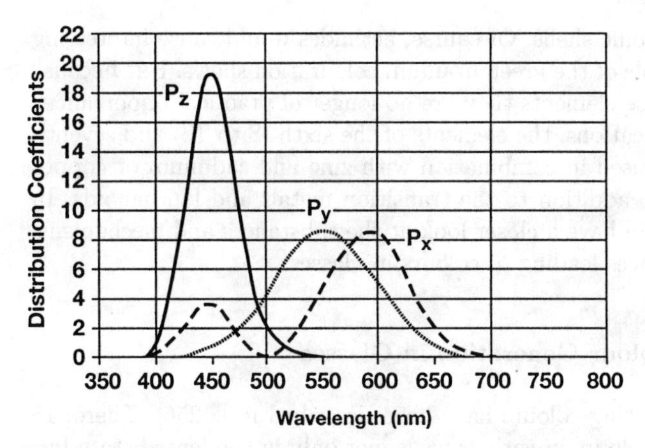

Fig. 8.49. Distribution coefficients for red (P_x), green (P_y) and blue (P_z) part of light of illuminant D65 (standardized light source)

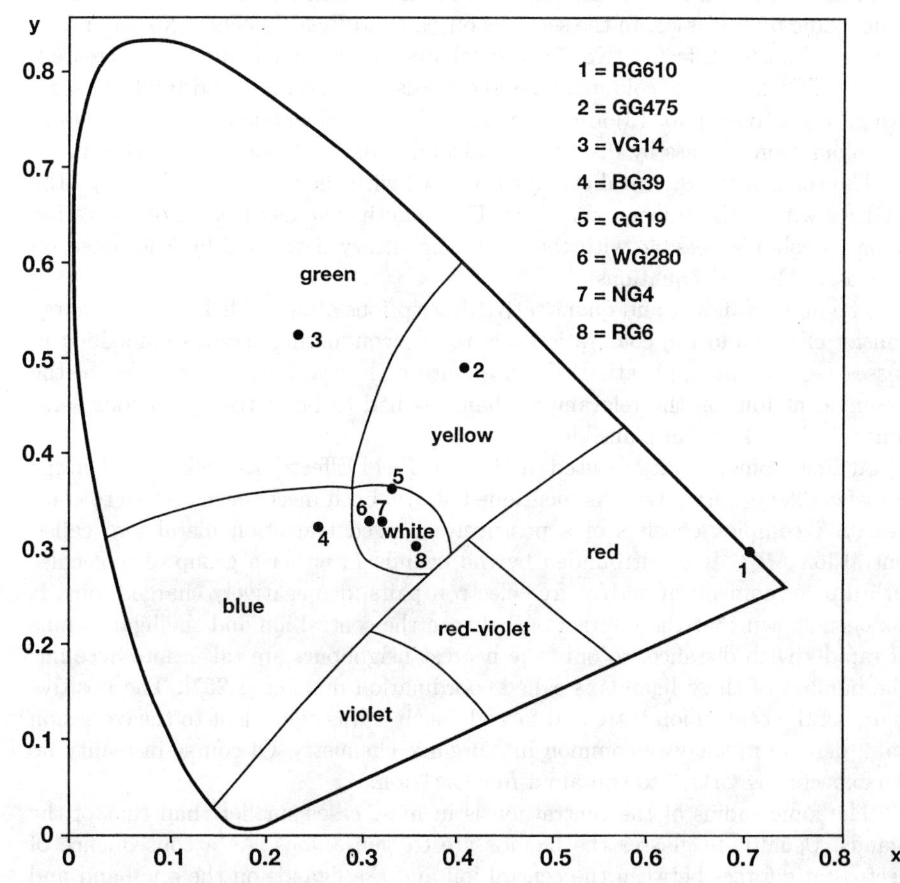

Fig. 8.50. Colour chart (colour coordinates) with the glass types of Figs. 8.45, 8.46, 8.47, 8.48 and illuminant D65 (standardized light source)

partly filled d and f electronic shells. Of course, actinides would cause interesting colours, too, as the example of the green uranium colouration shows. But because of the radioactivity of these elements they are no longer of practical importance.

For some special applications, the elements of the sixth (S to Te) and seventh (Cl to I) main group are used in combination with zinc and cadmium or copper and silver, respectively, in addition to the transition metals and lanthanoids. In the next subsection, we will have a closer look at the substances and mechanisms related with these substances leading to colours in glasses.

8.8.2 The Basics of Colour Generation in Glasses

The mechanisms of generating colours have been described in [8.256]. There, 15 mechanisms that lead to colours in various ways, not only in condensed state but also in gases, are discussed.

From these 15 mechanisms, only four are important for practical use in inorganic oxidic bulk glasses. In the sections on "Ligand Field Effects" (No. 5) and on "Charge Transfer Effects" (No. 7), the colours caused by complexes are treated. In "Metal Colloids" the colours of metal colloids, based on Mie and Rayleigh scattering, respectively (No. 13) are reported, and in "Semiconductor Doped Glasses" the colouration of glass by dispersed semiconductor crystals (No. 9) is shown.

The colouration by metal colloids is caused with the scattering of light by the particles within the transparent glass. The quantitative description of scattering by metal colloids possible with the scattering theory developed by Mie based on the classic Maxwell equations.

The understanding and quantitative descriptions of ligand field effects, charge transfer effects and the colouration due to semiconducting crystals embedded in glasses require the application of the quantum theory. Within the scope of the present contribution the relevant mechanisms had to be restricted to considerations which make them plausible.

At first, some definitions used in "Ligand Field Effects" as well as in "Charge Transfer Effects" are given. As mentioned above, both mechanisms relate to complexes. A complex consists of a positively charged transition metal ion, called central ion M^{n+}. It is surrounded by the ligands L, either a group of molecules with dipole momentum and/or free electron pairs, or negatively charged ions. It has been shown that the interaction between the central ion and the ligands falls off rapidly with distance, so only the nearest neighbours are taken into account. The number of these ligands is called coordination number [8.257]. The positive charge of the central ion is treated formally as if it was equivalent to the oxidation state, derived in the way common in inorganic chemistry. Of course in reality no one expects the CrO_4^{2-} to contain a free Cr^{6+}-ion.

The ionic radius of the central ion is in most cases smaller than that of the ligands. Usually, in glasses the ligands are oxygen anions. As a consequence of electrostatic forces between the central ion and the ligands on the one hand and inter-ligand-interaction on the other hand, the ligands occupy places that are as close as possible to the central ion and as far as possible from each other. This

Table 8.16. Usual colours of some ions in glasses, according to [8.254]

Element	Valency	Colour
Fe	2+	green, sometimes blue
Fe	3+	brown
Cu	2+	blue, turquoise
Cr	3+	green
Ni	2+	violet (tetrahedral)
Ni	2+	yellow (octahedral)
Co	2+	blue
Mn	2+	pale yellow
Mn	3+	violet
Pr	3+	green
Nd	3+	violet
Er	3+	pale red

results in a coordination in the form of a regular polyhedra. The type of the polyhedron is determined by the ratio of the radii of central ion and ligand. Radius ratios of central ions to ligands (r_M/r_L) smaller than 0.23 lead to coordination number four (tetrahedra), ratios between 0.23 and 0.73 to coordination number six (octahedra) and ratios greater than 0.73 give coordination numbers of eight (cubes) [8.258]. Although there are deviations, these rules are sufficient to explain most of the phenomena observed in glass.

Ligand Field Effects

The colouration by ligand field effects is a very important mechanism in the generation of colour in condensed materials. It takes place in hydrous solutions, in crystals as well as in glasses. There is a lot of empirical information about the colours of different ions in glass. The usual colours of some of the more important ions are listed in Table 8.16 [8.254]. Special cases and deviations can also be found in this literature. In the following paragraph, the ligand field theory is introduced in a simplified way. The reader interested in more details should refer to literature [8.257, 259–262].

Let us now consider an ion of the d-series with one electron in the d-shell, e.g. Ti^{3+}. The orbitals involved are the d_{xy}, d_{xz}, d_{yz}, $d_{x^2-y^2}$, d_{z^2} orbitals which are oriented in a special way oriented relative to the axes of an arbitrarily chosen cartesian coordinate system.

In case of the free, undisturbed central ion, the orbitals with different angular momentum are degenerate (Fig. 8.51). Now, this ion is brought into an octahedral surrounding of negatively charged ligands, in glass normally oxygen anions located at $\pm X_L$, $\pm Y_L$, $\pm Z_L$. The $d_{x^2-y^2}$ and the d_{z^2} orbitals point directly to the ligands (Fig. 8.52). The energy of the electrons in these orbitals is strongly increased by the repulsion with the negatively charged ligand. The d_{yz}, d_{xz}, and d_{xy} orbitals point to the gap between the ligands. Their energy is increased, too, but not as much as the energy of the electrons in the two other orbitals [8.257].

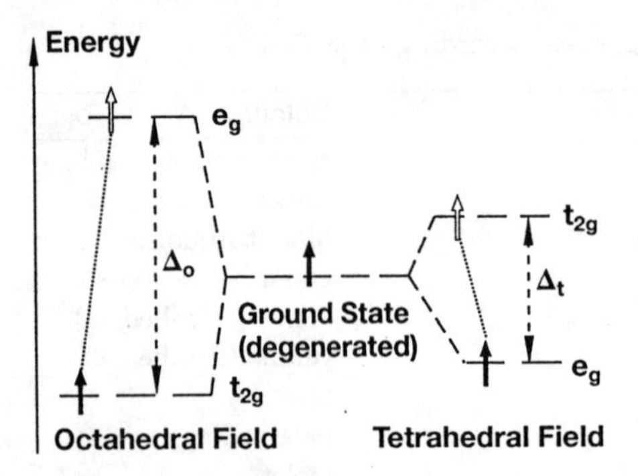

Fig. 8.51. Splitting in the ligand field: Energy levels of an ion (Ti^{3+}) with d^1-configuration without ligands and in the case of an octahedral and tetrahedral ligand field [8.257]

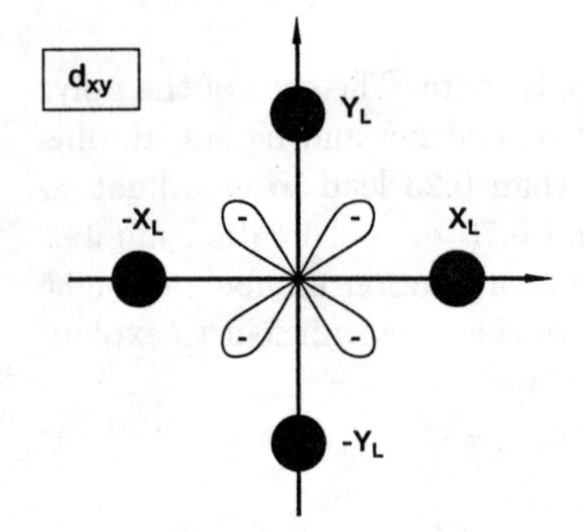

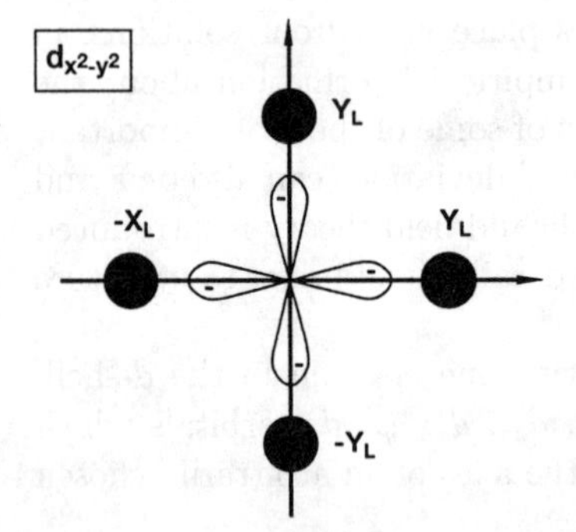

Fig. 8.52. Orientation of the d-orbitals relative to the ligands in an octahedral complex [8.257]. The minus sign indicates the negative charge of an electron occupying the orbital. The repulsive force due to the likewise negatively charged ligands is strong in the $d_{x^2-y^2}$ case and weak in the d_{xy} case

The degeneracy of the orbital energies is broken, resulting in two groups of states, one with elevated (e_g) and one with reduced (t_{2g}) energy (Fig. 8.51). The lower state, t_{2g}, can be occupied by six electrons. The higher state, e_g, can obtain four electrons. The two states are separated by an energy gap named Δ [8.259]. Notice that the mean value of the five orbital energies is chosen as the zero level of energy. Therefore, the distance of t_{2g} from zero level is -0.4Δ and the distance of e_g is $+0.6\Delta$. When there is only one electron as in Ti^{3+}, the ground state is

t_{2g} occupied with this electron. If energy in form of light is added, the electron is raised to the e_g state. No other transition is possible: The absorption spectrum shows only one line. The position (wavelength) of the line is determined by the value of Δ.

In Fig. 8.51 the splitting of the orbitals is also shown for tetrahedral coordination. The main differences are that the energy levels t_{2g} and e_g are inverted and that the energy difference Δ between them is smaller by the factor of 4/9 than in the octahedral case. A cubic surrounding leads to the same energy scheme as in the tetrahedric case but with $\Delta_{\text{cub}} = 8/9\Delta_{\text{oct}}$. All other geometries yield more complicated diagrams, but fortunately they are rare in glasses and it is therefore not necessary to discuss them here.

In the case of ions with more than one electron, the spectra become much more complicated. The reason for this is the interaction between the electrons that leads to a breakdown of the degeneracy of the orbitals already in the free ion. The orbitals are further split when brought into the field of ligands. The complexity of the possible electronic states and, resulting from this, the spectra, are beyond the scope of this article. They have been compiled and interpreted by *Bates* for a number of colouring ions [8.257]. Another compilation and interpretation of spectroscopic data has been given by *Kumar* [8.263].

The degree of the splitting (Δ) for a given central ion in given coordination is determined by the ligand. This can be seen from the so-called spectrochemical series, that has been found empirically [8.264]: $I^- < Br^- < Cl^- < OH^-\ NO_3^- < F^- < C_2O_4^{2-} < H_2O < SCN^- < NH_3 < NO_2^- \ll CN^-$.

The ions on the left side cause a weak splitting (absorption at long wavelengths) and the ions on the right side cause strong splitting (absorption at short wavelengths) of the energy levels. The position of the O^{2-} ligand is between F^- and H_2O. Data on the splitting of some ions in glass and in water have been compiled in [8.265]. The splitting in glass is about 12% lower than in water except for ions with d^4 and d^9 configuration, where the Δ-values are nearly equal to the values in water. This relation is relatively constant and can be used to estimate the splitting in glass from the data on splitting in aqueuos solutions if there are no values available for glass.

Fluoride ions, the only species that can be introduced into oxide glasses in greater amounts instead of oxygen ligands, seem to show a peak-shift to longer wavelengths; i.e., a smaller splitting [8.266].

The influence of the base glass to the spectra can be attributed to two items. The first one is the intensity of the absorption band caused by the transition. This is influenced by the probability of the electron transfer. On a static model, *d*-*d* transitions are forbidden by the so-called Laporte selection rule ($\Delta l = 0$: no intensity). But, by coupling with molecular vibrations, a relaxation of this rule takes place and some transitions are possible. The extinction coefficients are small, resulting in relatively weak bands compared to the allowed charge transfer transitions or absorption by metal colloids or semiconductors [8.257]. Different base glasses show different coupling modes and therefore different extinction coefficients for a given

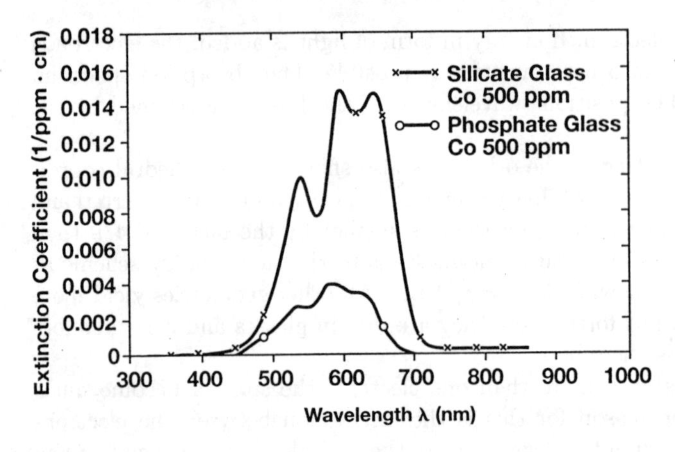

Fig. 8.53. Extinction spectra of Co^{2+} in a phosphate and a silicate glass (CoO content 500 ppm)

colouring ion. Commonly, phosphate glasses show lower extinction and the colour of a given ion is fainter than in silicate glasses.

The second reason is the difference of absorption peak maxima between silicate, phosphate and borate glasses. It is only about 1–2% provided the arrangements of the ligands of the colouring ion is unchanged.

These two points will be illustrated by the following two examples. First we will look at the extinction spectra of the Co^{2+}-ion in a phosphate and a silicate base glass. The maxima of the extinction spectra of these glasses in the visible range (Fig. 8.53) can be taken from Table 8.17. The observed bands belong to Co^{2+} in a tetrahedral environment as follows from the comparison with the spectra of ZnO crystal doped with CoO, where the fourfold coordination has been proved [8.257]. The differences between the band positions are situated within the above mentioned limits of 2%.

The difference between the two base glasses is demonstrated by the peak heights. The extinction coefficients of the silicate glass are greater by a factor 3–4 compared to those of the phosphate glass. Because the peak height relations of the bands are not the same in the two different base glasses and the peaks

Table 8.17. Peak maxima of the extinction spectra of the ions Co^{2+} and Ni^{2+} in a silicate and a phosphate plass

Species	Silicate Glass	Phosphate Glass	Difference
	541 nm	542 nm	0.2%
Co^{2+}	594 nm	584 nm	1.7%
	642 nm	635 nm	1.1%
	467 nm	438 nm	6.6%
Ni^{2+}	557 nm	–	–
	631 nm	–	–
	770 nm	857 nm	10.2%

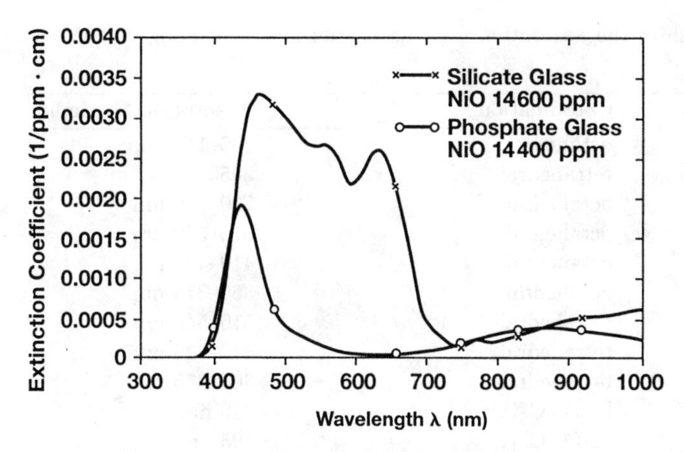

Fig. 8.54. Extinction spectra of Ni^{2+} in a shosphate (14 400 ppm) and a silicate (14 600 ppm) glass

are not exactly in the same spectral positions, there is a difference in the colour appearence even if the integral light transmission has been equaled by variation of the probe thickness or Co-concentration. It can be observed that the Co^{2+} shows a more reddish blue in a phosphate base glass.

The second example shows the influence of the base glass on the coordination and, therefore, on the spectra. The extinction spectra of Ni^{2+} in the same two base glasses are shown in Fig. 8.54, the maxima are listed in Table 8.17.

The differences between the peak maxima positions are much more pronounced. Some transitions show no correspondence at all. The explanation follows from the different coordination. The coordination of Ni^{2+} in the silicate glass is unexpectedly fourfold (tetrahedral), resulting in a purple colour, whereas the dark yellow of Ni^{2+} in the phosphate glass results from the normal octahedral coordination [8.257]. This colour is also observed in some crystalline compounds of Ni. For instance, $NiTiO_3$ shows a similar yellow colour.

In phosphate glasses, the intensities of the transitions are lower than in silicate glasses as has already been described for Co^{2+} ions.

The spectra of Ni^{2+} and Co^{2+} are discussed extensively not only because of academic interest but also because of the great importance for the design of filters. These two ions are the only ones that do not absorb in the UV region while showing strong absorptions in the visible range. Therefore, a combination of Ni and Co is used in various UV-transmitting filter glasses with low transmission in the visible. The regions of absorption and transmission can be extracted from Figs. 8.53 and 8.54 (high extinction coefficient = low transmission and vice versa). In addition to the utilizations known up to now, a rapidly growing need for measuring the intensity of additional UV radiation caused by the vanishing ozone shield is expected. Thus, some work is being done today to achieve a better understanding of the properties of these filters.

Another colouring species of high technological importance is the Cu^{2+} ion. Its broad band centred around 850 nm (see Fig. 8.46, BG 39) is well suited to adjust

Table 8.18. Wavelength of the absorption maxima of iron ions depending on valency and coordination

Valency	Coordination	Absorption Maximum
2+	octahedral	950–1100 nm
2+	tetrahedral	2050–2200 nm
3+	octahedral	700–780 nm
3+	octahedral	510–540 nm
3+	octahedral	410–425 nm
3+	octahedral	365–375 nm
3+	tetrahedral	510–540 nm
3+	tetrahedral	410–425 nm
3+	tetrahedral	365–375 nm
3+	Fe-O (C.T.)	240 nm
3+	Fe-O (C.T.)	195 nm
3+	Fe-S (C.T.)	410 nm*
3+	Fe-S (C.T.)	300 nm*

C.T. = Charge Transfer
*: see the section on "Colouration by Charge-Transfer Effects"

the sensitivity of the Si-CCD (Charge Coupled Device) used in video cameras to that of the human eye. The red tint of the pictures due to the unfiltered CCD is avoided. Compared with the bands of other ions, the absorption band of Cu^{2+} is broad. This results from the fact that there is an elongation of the octahedron of oxygen ligands surrounding the copper ion. This leads to a fine splitting of energy bands and therefore to a great number of transitions with overlapping bands.

The interpretation of the spectra of iron is one of the most complicated tasks. One has not only to consider two oxidation states with different peak positions but also that these two species appear each in tetrahedral as well as in octahedral positions. For further confusion, these bands can occur partially at nearly the same wavelengths. Additionally, charge-transfer bands of Fe^{3+} with oxygen occur. The interactions with sulphur, often present as an impurity in raw materials, is discussed in the next subsection. The assignments that can be made are given in Table 8.18 [8.267]: The colour of the glass depends strongly on the redox conditions during melting. By the application of a low oxygen partial pressure the glass becomes very pale green to blue according to the transition at about 1100 nm. Under oxidizing conditions the glass turns brown because of the manifold bands of trivalent iron present all over the visible and UV range of the spectrum.

A very important technological application is based on the fact that glasses with divalent iron absorb the near infrared heat radiation and transmit the visible without great losses. Therefore, these glasses are used for heat protection, e.g. in slide projectors. To maintain the nearly undisturbed transmission in the visible, Fe^{3+} must be avoided. The atmosphere has to be controlled exactly during melting. Information about the Fe^{2+}/Fe^{3+} ratio depending on the oxygen partial pressure is given in [8.268]. Another application results from the fact that there are many bands when Fe^{2+} and Fe^{3+} are simultaneously present. Fe^{3+} absorbs in the blue region, the offshoot of Fe^{2+} absorbs in the red. If Co^{2+}, absorbing in the yellow-green region, is added, this results in a nearly neutral grey glass. A glass of this type is shown in Fig. 8.48.

Colouration by Charge-Transfer Effects

In the previous section we were concerned with the electron activations that take place within the *d* (or *f*) subshell of a given ion. Now, however, we will look at electron transitions between the shells of two different ions. Usually, these are transitions from the *p*-orbitals of oxygen anions to the *d*-orbitals of the central ion.

The distinction between the two mechanisms can be made by measuring the intensity of the bands. Because charge transfer bands obey the selection rule $\Delta l = \pm 1$, they have a high transition probability and show a very intense band ("allowed transitions"). The intensity of the *d*-*d* bands are lower by a factor of 100–1000, as $\Delta l = 0$ for this type of transition ("forbidden transitions").

The resulting bands are usually located in the UV region, but in some cases they extend far into the visible range of the spectrum. This mechanism is responsible for the colours of ionic chromophors with d^0-configuration. The intense yellow colour of CrO_4^{2-}, the red violet of MnO_4^- and the yellowish tints of TiO_2 and CeO_2 are the best known examples of this type of colouration. Therefore, they are most widely used in the glass-making industry. As an example we will look at the yellow coloured glass type GG 19. The spectrum of this glass was already shown in Fig. 8.45. The band centred at about 370 nm is the lowest energy charge tranfer transition of the Cr-O-system. In spite of the strongly oxidizing melting conditions of the glass, there is always an equilibrium between Cr(VI+) and Cr(III+). The weaker band at about 650 nm thus results from the *d*-*d*-transition of Cr^{3+}. Charge-transfer processes, however, are not restricted to chromophores with empty *d*-shells. They can, in addition to the *d*-*d*-bands, also arise in systems with partially occupied *d*-shell. In this context, the intense colours of the Fe^{3+}-S charge transfer listed in Table 8.18 should be mentioned. In the literature, there is a controversy about the nature of the amber colour of the "carbon yellow" glasses. Many authors have regarded it as a transition of an electron from sulphur to iron in an $[FeO_3S]^{5-}$-complex, because it was only observed in the presence of both Fe^{3+} and S under moderately reducing conditions [8.252,269–273]. According to a recent investigation this is not the chromophore [8.274]. It was pointed out that the band, located at about 400 nm, belongs to an intramolecular excitation of the S_2^--ion that is normally produced by the reaction $3Fe^{3+} + 2S^{2-} = 3Fe^{2+} + S_2^-$. This ion could be observed spectroscopically in an iron free system as an intermediate, when the appropriate reducing oxygen partial pressure was applied to a sulphur containing system. Therefore, the iron ion is only needed to stabilize the chromophore by maintaining the optimal redox conditions.

The S_2^- is one of the rare examples of a nonmetallic chromophore in glasses. Only little attention has been paid to this type of colouration in optical glasses. The use of this chromophore was mainly restricted to technical glasses (container glasses). In the age of environmental protection, however, chromophores like that obtain more and more significance as substitutes for the heavy metal colourants used up to now.

Metal Colloids

The topic of the previous sections was colouring systems distributed homogenously in the base glass. In the next two sections, heterogenously distributed colourants will be treated. This means that, to achieve a colour, one has to segregate a second phase from the base glass. This can be done by an annealing process: after the glass has been cast and cooled down to room temperature, it is reheated to about glass transformation temperature. Then, if the glass is oversaturated with a special component, this component will diffuse to always present nuclei and form colloids or crystals.

In the case of metal colloid glasses, the segregating phase consists of very small metal particles with a usual diameter of some 10 nm. In principle, every metal can form colloids. In practice, however, only metals that can easily be held in the metallic form by a reducing melting process are suitable. Therefore, only those metals can be used for which the reducing potential does not affect the constituents of the base glass. These are mainly the members of the first and second group of the transition metals and the group of the platinum metals, and selenium and tellurium. The majority of these elements give only grey or dirty brown colours [8.255]. Only Se (pink), Cu (red), Ag (yellow) and Au (red) have found wider application. The intense red to purple colour of the Au colloid (gold ruby) has been known for hundreds of years. It is used in the glass type RG 6 shown in Fig. 8.47.

The origin of colouration by metal colloids can be described by Mie's theory of scattering, or, for smaller particles ($r < 0.05\lambda$), by the simpler theory according to Rayleigh. The exact deduction and application of the theory has been described in the literature [8.275–277]. It is beyond the scope of this article, so only some qualitative aspects will be discussed here.

The loss of transmission of metal colloid doped glass can be split into a part due to scattering (α_S) and a part due to absorption (α_A) according to:

$$-\ln \tau = (\alpha_S + \alpha_A)^* d \tag{8.29}$$

with

$$\alpha_S = \text{const}\ n^* r^6/\lambda^{4*}\ \text{Re}\{(m^2-1)^2/(m^2+2)^2\}\ , \tag{8.30}$$

$$\alpha_A = \text{const}\ n^* r^3/\lambda^* \text{Im}\{(m^2-1)^2/(m^2+2)^2\}\ . \tag{8.31}$$

n = refractive index of the glass
r = radius of colloidal particle
λ = wavelength
m = complex ratio of the refractive index of the colloidal particle to the refractive index of the glass
$m, n = f(\lambda)$.

Table 8.19. Colours and absorption maxima of some metal colloids

Element	Peak position	Refr. ind.	Colour	Source
Ag	410 nm	1.5	yellow	[8.278, 279]
Cu	530–560 nm	?	red	[8.280]
Au	550 nm	1.55	red	RG 6, Fig. 8.47
Se	500 nm	?	pink	[8.252]

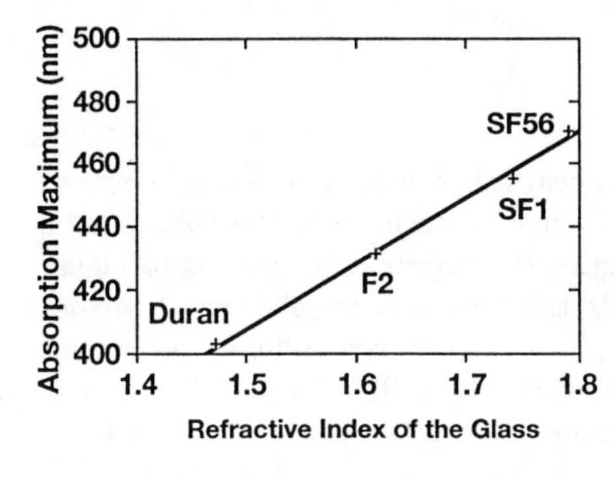

Fig. 8.55. Dependence of the maximum absorption peak position on the refractive index for glasses with Ag metal colloids

Usually there is a monotonic dependence of scattering on the wavelength. According to Rayleigh, scattering falls off sharply towards longer wavelengths. But at some wavelengths (the resonance wavelengths), the square of the complex ratio of the refractive indices m^2 equals -2. The functions α_S and α_A go to infinity. This results in intense, broad bands in the spectra.

In the case of Se, Cu, Ag and Au, the positions of the resonance wavelengths are in the visible. This leads to the known colours. The exact positions of the peaks depend strongly on the refractive index of the glass and moderatly on the particle size of the colloids. Therefore the positions given in Table 8.19 are only approximate values.

The dependence of the peak positions on the refractive index of the embedding glass has been researched for some Schott glass types [8.279]. The results are shown in Fig. 8.55. With respect to experimental errors, the points are situated on a straight line. The experimental error is mainly determined by the unknown size of the colloidal particle. The time for segregation of colloids was short in order to obtain sharp peaks. This allows a good determination of the peak position. For longer annealing times, a parallel shifted line (upward) is expected from numerical calculations [8.281]. An additional interesting property of the colloids appears if elliptical forms of metal particles instead of spheres are chosen. The peak at 400 nm splits then into two, one of them remaining at about 400 nm. The other one shifts to longer wavelengths, finally into IR, with increasing difference between the two half axes of the ellipsoid.

In the literature, there are two ways mentioned how to obtain elliptical particles. The process of *Berg* et al. [8.282] deformates spherical colloids by extensive stretching of the glass in one direction. The ratio of half axis depends on applied stress.

Stookey et al. [8.283] have achieved their aim by incorporating needles of a crystalline phase of NaF into the glass. These needles force the Ag colloids, depositing on them, into an elongated habit. The ratio of the half axes of the ellipsoids can be chosen by varying temperature and time of the silver deposition process. All colours can be obtained by varying the ratio of the half axes. Therefore, the colouration by metal colloids is as variable as the colouration by ions despite the limited number of chemical species. Nevertheless, the wider use of this method is limited by the expensive preparation of these glasses.

Semiconductor-Doped Glass

Glasses coloured by ions or metal colloids always show a more or less pronounced band structure. At certain wavelengths, absorption occurs and the transmission becomes low. At lower or higher wavelengths, the transmission returns to higher values. Often, in addition to the main bands, there are smaller bands which are not always desired. The semiconductor-doped glasses show a very different behaviour, as can be seen from Figs. 8.45 (GG 475) or 8.47 (RG 610).

There is a wavelength λ_c, dividing the spectrum into two parts. In the lower part, all incident light will be absorbed ($\tau_i, \tau \approx 0$). In the upper part above λ_c, all the incident light will be transmitted ($\tau_i = 1$; $\tau = 0.91$). In reality, the changeover is not so sharp but extends to about 60–80 nm. λ_c is now defined as the middle of this region ($\tau_i = 0.5$; $\tau = 0.46$).

These interesting spectral properties are evoked by microcrystals of (Zn, Cd) (S, Se, Te), that are separated from the base glass [8.255, 284–286]. The commata between the elements mean that they can replace each other in an arbitrary manner. To facilitate the process of separation of the crystal phase, the base glass should tend to demix. Frequently, a glass consisting of K_2O-ZnO-SiO_2 is used. K_2O may be replaced by other alkaline or alkaline earth oxides [8.255].

The dimensions of the crystals are so small (1–10 nm) that colour cannot be evoked by scattering. The ions of Cd, Zn, S, Se, Te are colourless, too. This can be seen when the glass has been cast and cooled down rapidly to room temperature. In this case no colour can be observed.

So, why does a colour arise? The separated crystals consist of semiconductor materials. For an explanation of the colour one has to look at the band model. Each semiconductor possesses a valence band (VB), completely filled with electrons, and a completely empty conduction band (CB). These two bands are separated by an energy gap called E_g. This energy has to be spent to lift an electron from the valence band to the conduction band. It can be taken from electromagnetic radiation (light), as it is shown schematically for CdS and CdSe in Fig. 8.56.

Every photon with an energy equal to or greater than E_g can lift an electron and will, therefore, be absorbed ($\tau_i = 0$). Photons with a lower energy pass undisturbed ($\tau_i = 1$). According to

$$\lambda_c = \frac{hc}{E_g} = \frac{1240\ (\text{nm})}{E_g(\text{eV})} \tag{8.32}$$

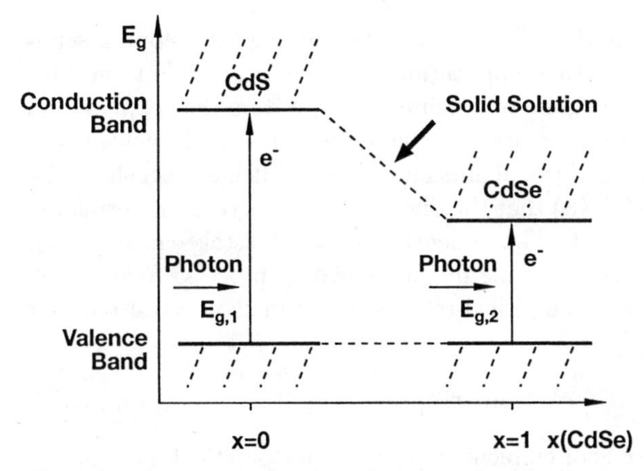

Fig. 8.56. Semiconductor-doped glasses: Band structure and absorption of light for II-VI semiconductors (The electron transition is induced by wavelengths equal to or smaller than $\lambda_c = hc/E_{g,i}$ (see text)

the energy can be expressed by wavelength (h is the Planck's constant, c the velocity of light). The bandgap energy correponds to λ_c, as defined above. The spectra of GG 475 (Fig. 8.45) and RG 610 (Fig. 8.47) can now be explained without problems.

The commercially available filter glasses of this series have absorption edges from about 400 nm to 850 nm. As indicated in Fig. 8.56 the band gap and, therefore, the absorption edge depend on the chemical composition of the cystals. The transition from Zn to Cd and from S over Se to Te leads to an increasing red shift of λ_c.

The exact energies of the band gap can be taken from Table 8.20.

The band gap for mixed crystals is within the values for the pure components [8.288]. The compounds ZnS and CdS [8.289] as well as CdS and CdSe [8.288, 290] form a complete solid solution. This should be true also for ZnS and ZnSe, ZnSe and CdSe, because all compounds have the same hexagonal crystal structure. CdTe (cubic) is differing: At 750 °C 15% CdTe is soluble in CdS and 50% in CdSe [8.290]. Nothing is known about the mixing properties of ZnTe.

By mixing appropriate amounts of the pure compounds, every value between 350 nm and 850 nm can be achieved for λ_c.

Table 8.20. Band gap energies for some semiconductor crystals (macrocrystals)

Compound	E_g eV / λ_c nm [8.287]	E_g eV / λ_c nm [8.286]
ZnS	3.7/335	3.65/340
ZnSe	2.7/459	–/–
ZnTe	2.2/563	–/–
CdS	2.4/517	2.42/512
CdSe	1.7/729	1.73/717
CdTe	1.5/827	–/–

Nevertheless, this cannot be the only possibilty for varying the colour of semiconductor doped glasses. As practical application shows, it is possible to get different absorption edges with one glass of a definite composition. For explanation we should remember that the glasses are colourless after casting. To achieve the final colour, there has to be an additional annealing process. It has been shown by small angle X-ray scattering (SAXS) that the diameter of the crystals is correlated with the position of λ_c [8.291, 292]. The dimensions of the crystals can be varied easily by adjusting temperature and time of the annealing process. *Rehfeld* and *Katzschmann* found the following empirical relation between the crystal radius r and λ_c for CdSe [8.286]:

$$\lambda_c = 1240\,\mathrm{nm}/[1.72\ \mathrm{eV} + 1.70\ \mathrm{eV}\ (\mathrm{nm})^2 r^{-2}]\ . \tag{8.33}$$

The variation of λ_c with r is of eminent importance for practical use, since it makes it possible to use one synthesis for 4 or 5 glass types, covering a range of 60–80 nm for λ_c. This will be discussed further in Sect. 8.8.3.

The extraordinary spectral properties of the semiconductor doped glasses have led to wide fields of application from which the following examples are drawn.

Glasses with an absorption edge near 400 nm absorb the complete UV-radiation without influencing the visible region. This can be used in photography to avoid disturbing exposure of films, in protecting chemical substances and in protecting the human eye from injuries by intense UV radiation, a point of growing importance.

Another broad field of application is signal lights. Yellow and red traffic signs are made of semiconductor doped glasses. Especially for red glasses there is, up to now, no way to get brilliant colours with high transmission by the use of other glasses.

Lastly, the application of glasses with a $\lambda_c \geq 780$ for IR photography and measurements should be mentioned. The nonlinear optical effects of semiconductor doped glasses have been referred to in Sect. 2.6.

8.8.3 Trends in the Development of Coloured Glasses at Schott

It has been shown in the previous sections that there are many possibilities of generating colours in glass. They are sufficient to cover a broad field of applications. Beside the ability to control the principles of colouration, it becomes more and more important to attain a high reproducibility in the manufacture of glass within narrow limits of specification. Moreover, it is important to make available small amounts of glass with slightly modified properties compared to the standard type, when required by customers. This should be done within a short period of time.

To achieve this, it has been necessary to develop some programmes to control fabrication on the one hand, and, on the other hand, to have a basic data set providing the transmitting properties of the elements of the periodic system to accelerate the development of modified filters. Especially for the semiconductor doped glasses, where the position of λ_c is influenced by both synthesis and annealing conditions, exact control of the processes is essential. It has been shown in the section on "Semiconductor Doped Glasses" that the radius of the microcrystals is the quantity influencing the position of λ_c in the annealing process (8.33).

The crystals develop from the primary existing nuclei by growth as a consequence of diffusion. The crystal radius can be determined only with great effort, a method which is therefore not practicable for daily use. One has thus to look for a relationship of λ_c with a quantity which can easily be measured. From experience it is known that diffusion processes are mainly influenced by temperature and time. The higher the temperature and the longer the time of the annealing process, the higher is the transport rate and, as a consequence, the larger are the crystals. Both can be fixed easily in experiments. In industrial practice, the annealing process is performed at a constant temperature, the fixing of λ_c takes place by variation of the annealing time. The relation between crystal radius and annealing time has to be quantified via an appropriate diffusion model. There are three models for the diffusional growth of semiconductor crystals discussed in literature:

$$\begin{array}{l} r = f(\ln t)[8.286] \\ r = f(t^{\frac{1}{3}})[8.293] \\ r = f(t^{\frac{1}{2}})[8.294] \end{array} . \qquad (8.34)$$

These models are combined with (8.33) to obtain a relationship between λ_c and t. These relationships have been tested by a programme that fits a curve based on each model with experimentally determined (λ_c, t) points. It was found that the first model of (8.34) gives the best fit. The exact relationship in this case is given in by

$$\lambda_c = A/[(B + C \ln t)^{-2} + 1] . \qquad (8.35)$$

The constants A, B and C depend, among other things, on composition and temperature. They are not known, but can be determined easily from a few experiments and the above mentioned regression programme.

Provided with this knowledge it is possible to choose the appropriate experimental conditions to obtain exactly the desired cut off position within a narrow target. In Fig. 8.57, examples of the setup of the regression curves for three different temperatures are given along with the final result of the production annealing process. It can be seen from this figure, too, that a number of glass types can be produced from one synthesis, only by a variation of the annealing time.

Another problem that can be mastered with the aid of computers is the calculation of spectra of new types of filters. For this purpose, a collection of data on the extinction coefficients has been established.

For the setup of data, a greater number of melts with the colouring ions in different concentrations has been conducted. These data were used to prove whether there was any concentration dependence of the extinction coefficients. One example of an element that is dependent on the concentration is copper. In most other cases, no dependence could be observed.

According to (8.28), the extinction coefficients should not be influenced by the thickness. Therefore, errors of measurement were minimized by matching the data of samples with different thicknesses.

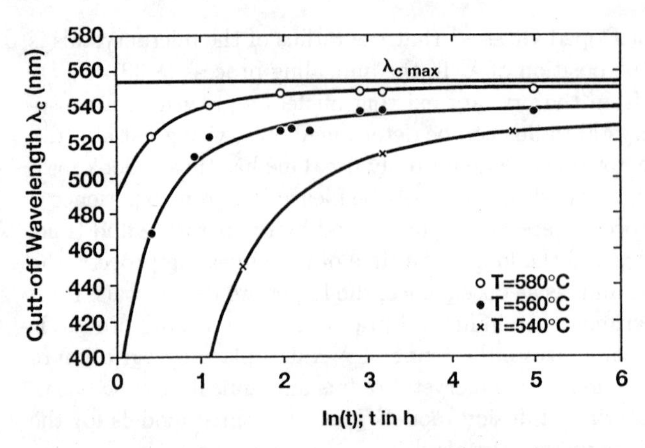

Fig. 8.57. Variation of cut-off position (λ_c) with annealing time (t) (points: experiments; solid lines: regression curves) in a semiconductor doped glass

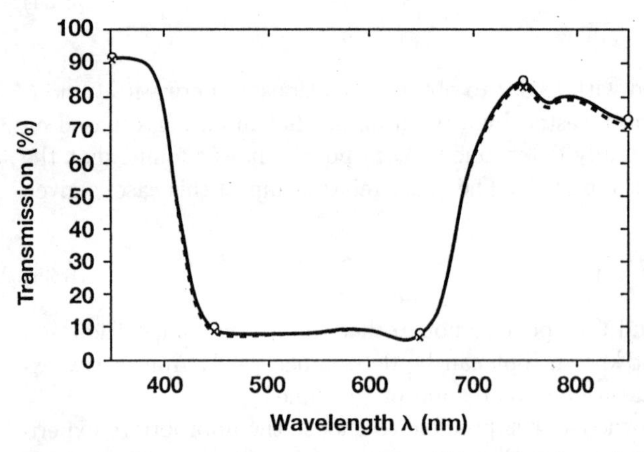

Fig. 8.58. Transmission spectrum of a silicate glass containing Co and Ni. (dashed line: calculated from extinction coefficient data set; solid line: experimental)

It is known from the section on "Charge Transfer Effects" that different base glasses lead to different extinction coefficients for the same ions. For the sake of expense, the investigation has been restricted to a widely used phosphate and silicate base glass. As an example of results, the extinction coefficient spectra of Co^{2+} and Ni^{2+} are shown in Figs. 8.53 and 8.54. They are discussed in "Ligand Field Effects".

The data obtained in this way are used in a programme to calculate the expected transmission spectra for different combinations of ions, concentrations, and sample thicknesses. A comparison between a computed and an experimentally determined spectrum is shown in Fig. 8.58 for a combination of Co^{2+} and Ni^{2+} in a silicate base glass. The two curves show a very good correspondence. The data collection is, therefore, well suited for the construction of new filters. Possible transmission curves can be tried out without time- and cost- extensive experiments to obtain the best combination for a given purpose. It can then be realized with only a low number of experiments.

8.9 Glasses for Eye Protection

Ewald Hillmann, Norbert Neuroth

8.9.1 The Eye

The human eye consists of imaging elements such as cornea, lens and vitreous body, as well as of vasculum-carrying tissue to supply the sensory organ (see also Sect. 8.1). The conversion of the stimuli of light takes place in the multi-layered retina, which terminates the eye to the rear, towards the inner head. Due to effects from outside, the eye can be damaged in multiple ways. These injuries can be caused by both fast-flying particles and high-energy radiation from the visible region of the spectrum and the area adjacent to the same. Mechanical damage will essentially be confined to the cornea and the eyeball, whereas high-energy radiation can, above all, impair the vasculum-carrying shreds of tissue and the retina.

Mankind started early to defend itself against sources of danger: Nero is said to have protected himself against the dazzling sunlight with a cut emerald. Glasses for the mechanical protection of the eyes have been introduced only in recent times. To protect the eyes against dangerous radiation, the industry has developed protection glasses with varying properties.

8.9.2 Protective Glasses

Inspection Glasses for Eye Protection

Every ordinary glass offers the eye protection against mechanical effects, which is, however, insufficient in many cases. Due to a blow from outside the bending strength of the glass is stressed on the opposite side of the plate. As in any brittle material, stress peaks in the surface are not reduced by ductile flow; excessive stresses surpassing the stresses applied occur at the endpoint of the crack. For this reason, glass already breaks under bending stresses far below the theoretical value of 10^6 N/cm^2. This is substantiated in the theory of strength by the existence of flaws called Griffith flaws, which extend from the surface into the glass down to a depth of 5 µm. They reduce the tensile bending strength to the practical value of 10^4 N/cm^2 and lower (Sect. 4.4). Due to Griffith flaws glass breaks only under tensile stresses, as external compressive stresses counteract the effect of Griffith flaws.

Concepts to successfully suppress the effect of the Griffith flaws include the set-up of compressive stresses in the surface. These compressive stresses must first be overcome, before the surface reaches the range of dangerous tensile stresses and breaks. The oldest method of generating compressive stresses in the outer skin of glass articles is thermal strengthening. First of all, the article is heated up to a temperature of about $T_g + 100\,°C$ and then abruptly cooled down. The cooling of products of standard shapes such as ophthalmic glasses is performed by air and that of products with intricate shapes by the use of oil baths. The rapid cooling solidifies first the surface and only then, due to the poor heat conduction, the core of the glass body, thereby putting the surface under compressive stresses. It is easy to understand that the level of compressive stresses is significantly determined by the cooling rate. Therefore, it is of importance for the performance of technical processes, to combine transfer times as short as possible from the heating to the cooling unit with high cooling rates. Thermally toughened glasses attain values of flexural tensile stress of about $5 \times 10^4\,N/cm^2$; thus offering five times the fracture values of untreated glasses. The drawback of thermal strengthening lies in the risk of deformation of the glass article at the high heating-up temperatures. During the technical implementation it is therefore essential to watch out that the maximum temperature is not exceeded. On the other hand, a high heating-up temperature is prerequisite for the generation of high compressive stresses. The optimum temperature of the individual workpieces can thus only be determined experimentally.

A more recent method to introduce compressive stresses into the outer skin of glass articles is chemical strengthening. It is based on the exchange of smaller Na ions in the glass for larger K ions of a salt melt. The ion radii are: Na = 0.096 nm and K = 0.133 nm. The exchange is substantially influenced by chemical potential gradients of the two ions [8.295]. Technically, the ion exchange proceeds as follows: an Na-containing glass is immersed in a KNO_3 melt at temperatures of $(T_g - 100)\,°C$, i.e., about 400 °C, where it is left for 16 hours. Chemically strengthened glasses attain an increase in strength of about a factor of 10.

Glasses are normally considered as isotropic materials. However, they become anisotropic by the introduction of stresses. Stresses in the glass can thus be proved by measuring the phase difference of the ordinary ray and the extraordinary ray, i.e. stress birefringence (Sect. 2.4). Figure 8.59 shows the stress profiles over the thickness of a glassplate, which were generated by thermal and chemical strengthening [8.296]. Thermal strengthening or chemical strengthening hence bring about a considerable increase in the strength of the protective glasses. For the optimum set-up of a stress distribution in glass articles, the development of glasses with special properties is of advantage. Special glasses showing a good chemical strengthening effect have primarily been developed by Schott.

A third method of manufacturing safety-glass panels consists in bonding two glass panels by means of a plastic sheet to form a laminated safety glass.

The strengths of all inspection glasses for eye protection are checked according to the paper of the German Institute of Standards DIN 4646 [8.297], part 3.

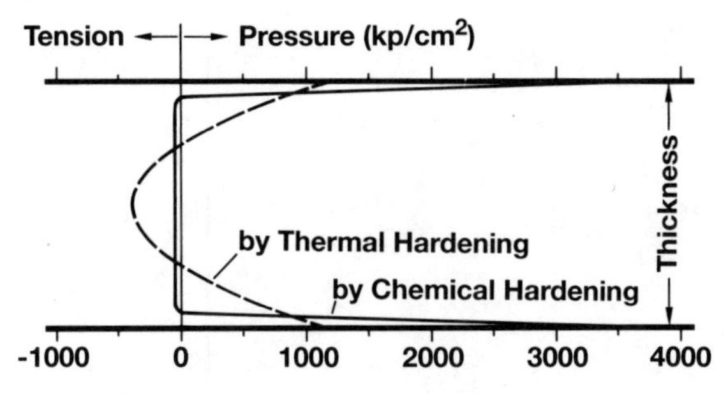

Fig. 8.59. Stress profile of hardened glasses [8.296]

Welding Protection Filters

The human eye is, in general, sufficiently protected to be exposed to all natural radiation sources occurring on earth. Only the appearance of artificial radiators whose radiation energy clearly exceeded the level of natural radiators made it necessary to develop protective glasses. The spectral absorption of the protective glasses should be designed so that, on the surface facing the eyes, the energy level of natural radiators is not surpassed. Protection must above all be ensured in the ultraviolet and infrared regions of the spectrum, because there the eye has no sensors such as the retina in the visible region.

Standard DIN 4647 [8.297] has divided welding protection glasses into graduated classes of protection. The graduation has been made as a function of the luminance of the welding light source in a way that the viewing eye is not dazzled. Mathematically, there is a logarithmic correlation between the luminosity of the light source and the physiological dazzling. Therefore, the graduation of the protective glasses follows a logarithmic law:

$$N = 1 + \frac{7}{3} \log \frac{1}{\tau} \tag{8.36}$$

where N = protection class number; τ= light transmission of protection glass (medium value of the visible spectrum). With the glare shield offered by the filter glass, a protection of at least identical quality in the adjoining regions of the spectrum, i. e., the ultraviolet and the infrared one, must be combined. For safety reasons, however, distinctly higher protection factors are adopted for these ranges of wavelengths (Table 8.21).

The protective effect of the filter glasses is based essentially on the incorporation of large quantities of FeO/Fe_2O_3 into the glass. The equilibrium between Fe (II) and (III) must be established via the conduct of the melting operation. The absorption of the wave ranges responsible for glare and IR protection is provided for by Fe (II), whereas Fe (III) ensures UV protection. The graduation of the protection effect is made in large steps by varying the Fe_3O_4 content in the melt, whereas the fine graduation is set up by the thickness of the glass. A typical composition of a welding protection glass is:

Table 8.21. Transmission values in the different spectral regions of protective filter glasses Athermal A® of different classes of protection

Protection class	Ultraviolet region Maximum transmission % at 313 nm [a]	at 365 nm [b]	Visible region Region for transmission % from	to	Infrared region Maximum of the medium spectral transmission from 1400 to 780 nm %
1.2	0.0003	50	100	74.4	69
1.4	0.0003	35	74.4	58.1	52
1.7	0.0003	22	58.1	43.2	40
2.0	0.0003	14	43.2	29.1	28
2.5	0.0003	6.4	29.1	17.8	23
3.0	0.0003	2.8	17.8	8.5	15
4.0	0.0003	0.95	8.5	3.2	9.4
5.0	0.0003	0.30	3.2	1.2	5.6
6.0	0.0003	0.10	1.2	0.44	3.4
7.0	0.0003	0.037	0.44	0.16	1.8
7.0 w[c]	0.0003	0.037	0.44	0.16	0.23
8.0	0.0003	0.013	0.16	0.061	1.0
8.0 w	0.0003	0.013	0.16	0.061	0.14
9.0	0.0003	0.0045	0.061	0.023	0.36
10.0	0.0003	0.0016	0.023	0.0085	0.14
11.0	0.0003	0.00060	0.0085	0.0032	0.050
12.0	0.002	0.00020	0.0032	0.0012	0.025
13.0	0.000076	0.000076	0.0012	0.00044	0.017
14.0	0.000027	0.000027	0.00044	0.00016	0.011
15.0	0.0000094	0.0000094	0.00016	0.000061	0.0085
16.0	0.0000034	0.0000034	0.000061	0.000023	0.0070

[a] Hg-line in the UVB [b] Hg-line in the UVA [c] protection filter for warm welding

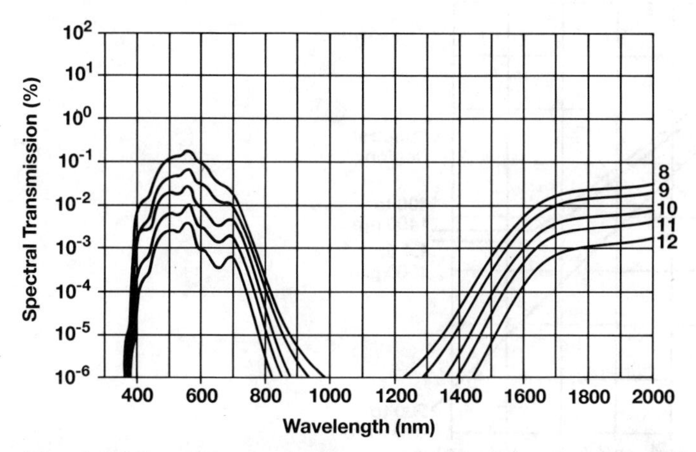

Fig. 8.60. Spectral transmission of the heat protecting glass Athermal A® of the protection classes 8 to 12 according to DIN 4647, part 7, from DESAG, Grünenplan

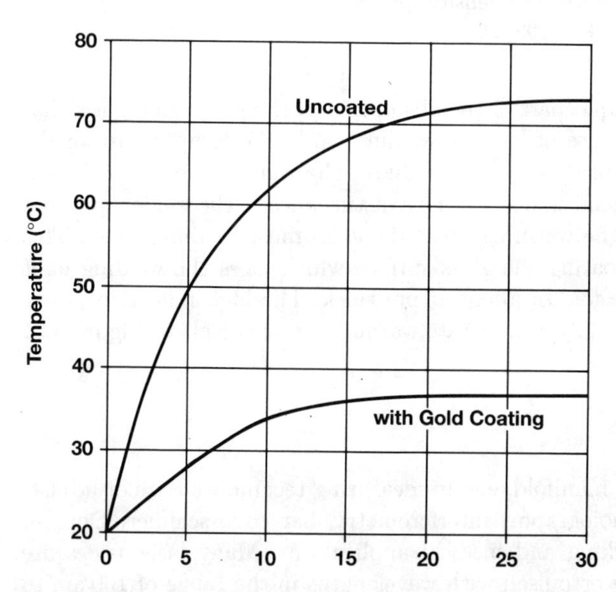

Fig. 8.61. Temperature of the filter glass Athermal A® of protection class 11; uncoated and with gold coating as a function of the welding time

SiO_2:	70%	MgO:	2%
Al_2O_3 :	3%	CaO:	3%
Na_2O :	14%	Fe_3O_4:	8%

Figure 8.60 shows the spectral transmittance of athermal glasses belonging to the protection classes 8 to 12.

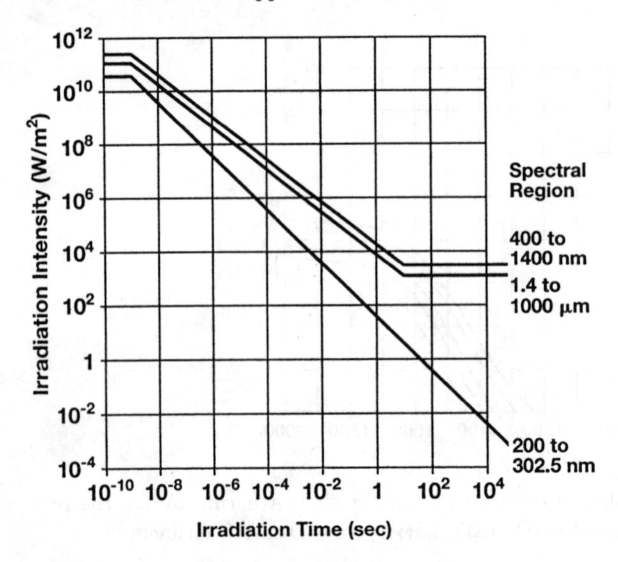

Fig. 8.62. Limits of laser irradiation intensity for the human skin as function of the irradiation time and wavelength [8.298, 299]

It is especially the infrared part of the absorption of the protecting glass that entails its warming up. In case of lengthy welding work, the temperature of the protecting glass reaches considerably high values. The radiation of the inherent heat of the glass onto the facial skin is a serious disturbance to the welder. In order to reduce this annoyance, the warming up of the glass must be diminished. This can be best obtained by coating the glass surface which faces the welding light source with a thin layer of gold of about 10 nm thick. The high reflection power of the gold in the IR allows only a moderate warming up of the glass. Figure 8.61 shows this effect.

Laser Protection Filters

The application of lasers is manifold, e.g. in measuring techniques (size or adjustment of machine parts), holography, interferometry, bar code scanner, Doppler anemometry, material working and medical applications. Many laser types are used, continuously working or pulsed with wavelengths in the range of 100 nm to 100 μm and a huge range of intensities (continuous radiation with power up to 10^3W, pulses with power up to 10^{12}W). This means that protection against laser radiation has to meet the most varied demands. A laser beam can be focused very well and thus represents a greater danger than light from other sources.

The biological effects of laser beams on the human skin and on the eye are different and depend on the type of laser beam. In Fig. 8.62 the maximum permissible irradiation intensity of laser radiation for the human skin is given as function of the irradiation time and the spectral region. Irradiation in the visible and near infrared region (400–1400 nm) has the highest values. The longer wavelength infrared region has smaller values by a half (because of different penetration depth

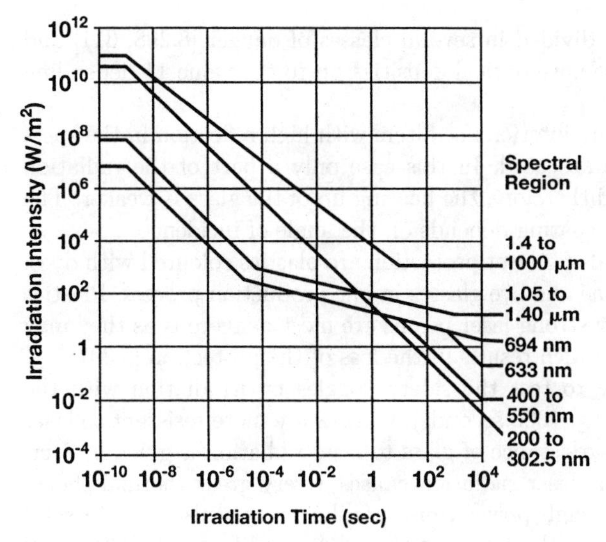

Fig. 8.63. Limits of laser irradiation intensity for the human eye as function of the irradiation time and the wavelength; valid for point like sources [8.298, 299]

of the radiation) [8.298]. The most critical spectral region is the UV region below 300 nm (the cause of sunburn).

We have to distinguish two spectral regions: a) the region whose irradiation penetrates the eye (near UV, visible and near infrared) and b) the part of the spectrum whose radiation does not penetrate. The radiation of a) is focused by the lens onto the retina (see Fig. 8.1). For this region the maximum permissible irradiation intensity is lower than for radiation of b).The values of Figs. 8.62 and 8.63 have been taken from the Association of German Electrical Engineers (Verein Deutscher Elektrotechniker, VDE) and German Industry Standards (DIN) [8.299–301]. Further literature on laser safety is [8.302–312].

Laser safety filters are used either as cabin windows in safety cabins of the laser apparatus or in goggles. The safety filter has to suppress the laser radiation and to transmit the visible light as much as possible so that a good visual orientation is achieved. For lasers emitting in the UV one has to use a filter glass which suppresses the UV and is transparent in the visible; e.g., Schott filter glasses GG 10, VG 10, or GG 400 are suitable. For lasers emitting in the different regions in the visible, filter glasses with absorption bands in the region of the laser line are required. If the laser line is in the blue region, the filter glass types FG 13 or OG 570 may be used. For laser lines in the yellow or red, the filter glass type BG 3 may be suitable. In order to suppress the red and near infrared region, one can use the filter glass types BG 18, BG 39 or KG 1, KG 3 or KG 5. The filter glass BG 36 has several strong absorption bands (at about 360, 525, 580, 740, 800 nm). For suppression of the infrared region from 2.7 µm to longer wavelenghts strongly OH-containing glasses are suitable: PK 50, KzFS N5.

In the infrared region with wavelengths greater than 5 µm all usual oxide glasses (containing SiO_2, B_2O_3 or P_2O_5) are opaque.

The laser apparatus are divided in several classes of danger [8.298, 311] and the protecting material must have optical densities up to the value 11 depending on the class in question.

To increase the protection, interference filters with high reflection in the spectral region of the laser line are used. In this case only a part of the radiation penetrates into the glass and therefore, the heating up of the glass is weaker. The reflection of the interference coating depends on the angle of incidence.

Other important materials for laser protection are plastics coloured with dyes. It is possible to produce frame and eye glasses in one production process. Plastics are lighter than glasses. Very strong laser beams are most dangerous as they may melt holes in the eyeglasses which results in the loss of the protection [8.307].

It is therefore necessary to test the safety goggles by irradiation with the laser beam for a long time (e.g. 1000 seconds). Glasses are more resistant to laser radiation. Details about the resistance of glass to laser radiation is given in Sect. 2.7. Glasses may break if the laser radiation causes a very great thermal shock. Short laser pulses with very high power cause non-linear optical effects in solid material (Sect. 2.6). Especially the long pass filter glasses like the Schott types GG 400–495, OG 515–590, RG 610–850 doped with semiconducting micro-crystals change their absorption if they are irradiated with high power laser pulses [8.313]. The diminution of their absorption is particularly marked in the region near the absorption edge.

One must be cautious: Filter glasses from glass producers are not automatically safety filters! Safety filters have to be tested by an official institute like the National Institute of Standards and Technology (NIST), National Physical Laboratory, or Physikalisch Technische Bundesanstalt and must have a declaration for which wavelength region and which radiation intensity they are to be used. There are several producers of laser safety goggles and cabin windows which comply with the specifications of the national and international standards. A list of producers is given in [8.306–311].

Ultraviolet and Infrared Protection Filters

Corresponding to the different biological influence the ultraviolet spectral region is devided in three areas: A: 380-315, B: 315-280, C: 280-110 nm. The portions of energy of the daylight in these areas are 3.9/0.4/0%, respectively. The radiation of areas B and C is the most critical one. Strong irradiation from these wavelengths may cause inflammation of the cornea and conjunctiva. Strong irradiation with ultraviolet A may make the lens turbid. The sensibility of the eye to UV-radiation can be increased if one takes certain medicines. In this case persons working in the open air should wear protecting goggles. The producers of eye glasses have special glass types which cut off the greatest part of the UV radiation. Figure 8.64 shows the spectral transmission of some types [8.314]. If strong UV-lamps are used at a working area one has to use protecting glasses with strong UV-absoption (protection classes 3–5) [8.306]. The UV portion of daylight causes decoloration of textiles, paper and other goods presented in shop windows. This bleaching

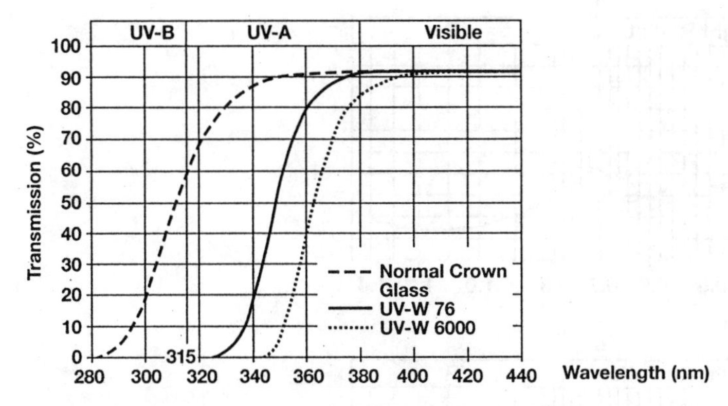

Fig. 8.64. Spectral transmission of UV-absorbing eye glasses in comparison with a standard eye glass [8.314]

effect can be suppressed by the use of special window glasses with reduced UV-transmission [8.315].

Sunglasses must generally provide UV protection. Beside this they have diminished transmission in the visible. In Fig. 8.65 the spectral transmission of three sunglasses with different colour tints is given which diminish the dazzling of the eyes. The colour-character is only a matter of taste.

The cornea of the human eye is transparent at a wavelength shorter than 3 μm. The whole eye - including the lens and glassbody – is infrared transparent until 1.4 μm. The infrared radiation is divided in three regions: A: 780–1400 nm, B: 1.4–3.0 μm, C: 3.0–1000 μm. Strong infrared radiation of type A can injure the retina and the lens can be made turbid (cataract). Also infrared radiation of type B can cause a cataract. Persons working in glass or metal foundries have to protect their eyes if they look into furnaces or at hot material.

There are several coloured glasses with strong infrared absorption (Schott types KG5, KG3, KG1, or DESAG Uro H9® and others). During the irradiation the filter glass is heated up, but this can be diminished effectively by a gold coating [8.316]. The degree of protection of an infrared absorbing glass depends on the temperature of the heat source [DIN 4646, 8.9.3 part 3].

Strong infrared radiation of type C is only generated by lasers; e.g. the CO_2 laser (emission at 10.6 μm). Oxide glasses have very strong absorption in this region and can be used as protecting material.

Acknowledgement

We thank Dipl.-Phys. Klaus-Dieter Loosen (Schott Glaswerke, Mainz) and Dr. H. W. Paysan (C. Zeiss, Aalen) for a critical reading of the manuscript.

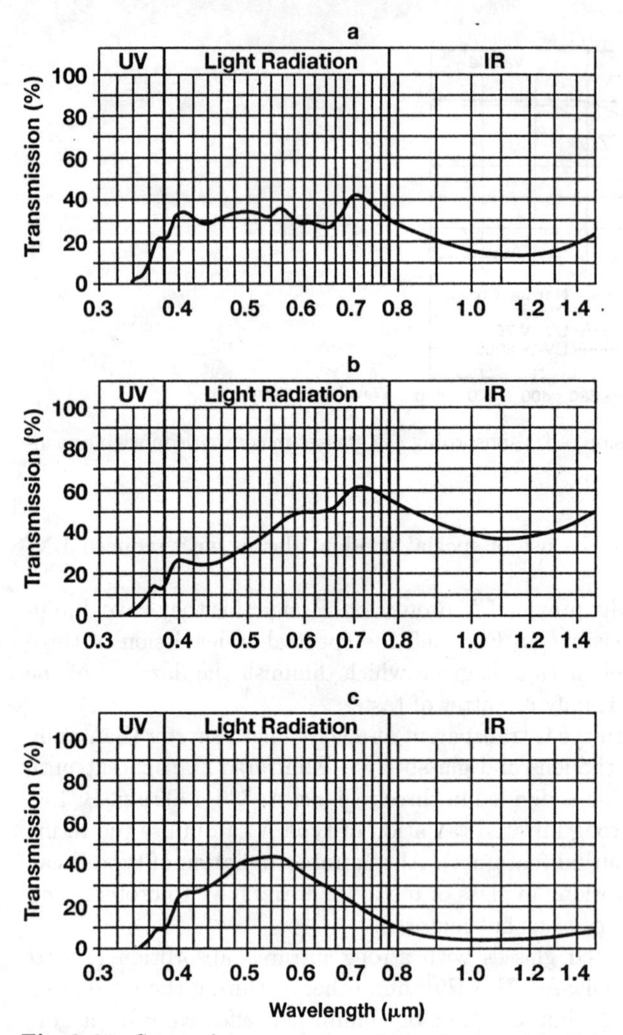

Fig. 8.65. Spectral transmission of sunglasses with different colour tints: a) grey, b) brown, c) green-grey

References

8.1 M.W. Morgan: *Optics of Ophthalmic Optics* (Professional Press, Inc., Chicago, IL 1978) pp. 1–60

8.2 R. Drew: *Professional Ophthalmic Dispensing* (Professional Press, Inc., Chicago, IL 1970) pp. 3–14

8.3 A.C. Guyton: *Textbook of Medical Physiology,* Fifth Edition (W.B. Saunders Co, Philadelphia, PA 1976) pp. 784–792

8.4 E. Hecht and A. Zajac: *Optics* (Addison-Wesley Pub. Co. Reading, MA 1976) pp. 99–147

8.5 J.S. Stroud: "The strengthening of some commercial ophthalmic and filter glasses by ion exchange", Glass Techn. **29**, pp. 108–114 (1988)

8.6 R.P. Krause: *Chemtempering Today* (Corning Glass Works, Corning, NY 1974) pp. 2–4

8.7 ANSI Z80.1-1987: "Prescription Ophthalmic Lenses – Recommendations", American National Standards Institute, Inc. (1987)

8.8 ANSI Z80.3-1986: "Nonprescription Sunglasses and Fashion Eyewear – Requirements", American National Standards Institute, Inc. (1986)

8.9 European Standard EN 166: *Persönlicher Augenschutz: Anforderungen* (Beuth Verlag, Berlin 1994)

8.10 H. Dürr and H. Bouas-Laurent (eds.): *Photochromism – Molecules and Systems,* Studies in Organic Chemistry, Vol. **40** (Elsevier Science Publishers B.V., Amsterdam, The Netherlands 1990)

8.11 C.M. Lampert and C.G. Granqvist (eds.): *Large-Area Chromogenics: Materials and Devices for Transmittance Control,* SPIE Institute Series Vol. **IS4** (1990)

8.12 C.B. McArdle (ed.): *Applied Photochromic Polymer Systems* (Blackie, Glasgow, London 1992)

8.13 G.H. Brown (ed.): *Photochromism,* Techniques of Chemistry, Vol. **III**, Series Editor: A. Weissberger (Wiley, New York 1971)

8.14 H. Pick: "Über den Einfluß der Temperatur auf die Erregung von Farbzentren", Ann. Physik **31**, 365–376 (1938)

8.15 H. Pick: "Struktur von Störstellen in Alkalihalogenidkristallen", in *Springer Tracts Mod. Phys.*, Vol. 38, ed. by G. Höhler (Springer, Berlin, Heidelberg 1965) pp. 1–83

8.16 J. Friedrich and D. Haarer: "Photochemisches Lochbrennen und optische Relaxationsspektroskopie in Polymeren und Gläsern", Angew. Chem. **96**, 96–123 (1984)

8.17 S.K. Deb and L.J. Forrestal: "Photochromism in Inorganic Systems", Ref. [8.13], pp. 633–665 (1971)

8.18 G. Bret and F. Gires: "Giant-Pulse Laser and Light Amplifier Using Variable Transmission Coefficient Glasses as Light Switches", Appl. Phys. Lett. **4**, 175–177 (1964)

8.19 W.A. Weyl: *Coloured Glasses* (Society of Glass Technology, Sheffield 1951)

8.20 W.H. Armistead and S.D. Stookey: "Photochromic silicate glasses sensitized by silver halides", Science **144**, 150–158 (1964)

8.21 W.H. Armistead and S.D. Stookey: "Phototropic material and article made therefrom", US patent 3,208,860 filed July 31, 1962

8.22 A.B. Buckman and N.H. Hong: "On the origin of the large refractive index change in photolyzed PbI_2 films", J. Opt. Soc. Am. **67**, 1123–1125 (1977)

8.23 H. Marquez, J.Ma. Rincon, and L.E. Celaya: "Photochromic $CdCl_2$-CuCl coatings obtained by vacuum deposition", Thin Solid Films **189**, 139–147 (1990)

8.24 H. Hecht and G. Müller: "Über die Phototropie des Kupfer(I)chlorids und des Kupfer(I)bromids", Z. physik. Chemie **202**, 403–423 (1953)

8.25 D.L. Morse: "Copper Halide Containing Photochromic Glasses", Inorganic Chemistry **20**, 777–780 (1981)

8.26 D.M. Trotter, J.W. Schreurs, and P.A. Tick: "Copper cadmium halide photochromic glasses: Evidence for a colloidal darkening mechanism", J. Appl. Phys. **53**, 4657–4672 (1982)

8.27 H. Kawazoe, R. Suzuki, S. Inoue, and M. Yamane: "Mechanism of photochromism in oxide glasses containing a large amount of CdO or ZnO", J. Non-Crystalline Solids **111**, 16–28 (1989)

8.28 G.S. Meiling: "Photochromism in cadmium borosilicate glasses", Phys. Chem. Glasses **14**, 118–121 (1973)

8.29 W.O. Williamson: "The Reversible Darkening in Daylight of Some Glazes containing Titanium", Trans. Brit. Ceram. Soc. **39**, 345–362 (1940)

8.30 S.K. Deb: "Optical and Photoelectric Properties and Colour Centres in Thin Films of Tungsten Oxide", Phil. Mag. **27**, 801–822 (1973)

8.31 Takushi Hirono and Tomoaki Yamada: "Photochromic properties of silver complex oxides", J. Appl. Phys. **55**, 781–785 (1984)

8.32 Yasutaka Suemune: "Photochromic Coloration Enhancement for Rare Earth Orthoniobates", Jap. J. Appl. Phys. **23**, 253–258 (1984)

8.33 Sumio Sakka: "Phototropy of Alkaline Earth Tungstates Doped with Bismuth", J. Am. Ceram Soc. **52**, 69–73 (1969)

8.34 H. Hosono: "Photochromism of reduced calcium aluminate glasses", Mater. Res. Bull. **23**, 171–176 (1988)

8.35 L. Kovács, E. Moya, K. Polgár, F.J. López, and C. Zaldo: "Photochromic Behaviour of Doped $Bi_4Ge_3O_{12}$ Single Crystal Scintillators", Appl. Phys. **A 52**, 307–312 (1991)

8.36 G.B. Hares, D.L. Morse, T.P. Seward III, and D.W. Smith: "Photochromic glass", US patent 4,190,451 filed Feb.28, 1979

8.37 D.J. Kerko, D.W. Morgan, and D.L. Morse: "Very fast fading photochromic glass", US patent 4,407,966 filed Sept. 16, 1983

8.38 D.J. Kerko, D.W. Morgan, and D.L. Morse: "Photochromic glass compositions for lightweight lenses", US patent 4,608,349 filed Nov. 28, 1985

8.39 G. Gliemeroth and K.-H. Mader: "Phototropic Glass", Angew. Chem. internat. Edit **9**, 434–445 (1970)

8.40 G. Gliemeroth: "Phototropes Glas", German patent 1 596 847 filed Dec. 24, 1966

8.41 Yoshio Murakami and Makoto Kume: "Schnell reagierendes phototropes Glas hoher Stabilität auf Borat- oder Borosilikatbasis sowie Verfahren zu seiner Herstellung", German patent 1 924 493 filed May 13, 1969

8.42 R.J. Araujo, N.F. Borelli, J.B. Chodak, G.B. Hares, G.S. Meiling, and T.P. Seward III: "Photochrome Gläser mit Dichroismus, Doppelbrechung und Farbanpassung", German patent 2 747 919 filed Oct. 26, 1977

8.43 S. Lythgoe: "Alumino-phosphate photochromic glasses", US patent 3,876,436 filed July 12, 1972

8.44 G.B. Hares: "Photochromic glasses which darken to a brown hue", US patent 4,251,278 filed Jan. 21, 1980

8.45 G.P. Smith: "Photochromic Glasses: Properties and Applications", J. Mater Sci. **2**, 139–152 (1967)

8.46 A. Krauth and H. Oel: "Elektronenmikroskopische Beobachtungen an Ultramikrotomschnitten von Fotochrom- und Kupferrubingläsern", Glastechn. Ber. **42**, 139–141 (1969)

8.47 H. Bach and G. Gliemeroth: "Über Entmischungen und Phototropie in silberhalogenidhaltigen Gläsern", Glastechn. Ber. **44**, 305–314 (1971)

8.48 Yin Baozhang: "The size of silver halide crystallites precipitated in photochromic glasses", J. Non-Crystalline Solids **52**, 567–572 (1982)

8.49 F. Zörgiebel, H.P. Zeindl, and G. Haase: "New experiences with ultramicrotomy in electron microscopic investigation of photochromic glasses", Ultramicroscopy **16**, 115–118 (1985)

8.50 H.P. Zeindl: "Elektronenmikroskopische Untersuchungen an photochromen Gläsern", Diploma thesis, Institut für Wissenschaftliche Photographie der Technischen Universität München (1984)

8.51 R. Pascova and I. Gutzow: "Modelluntersuchung zum Bildungsmechanismus fototroper Silberhalogenidphasen in Gläsern", Glastechn. Ber. **56**, 324–330 (1983)

8.52 Y. Okamoto: "Optische Absorption von Silberchlorid- und Silberbromid-Kristallen", Nachr. Akad. Wiss. Göttingen IIa **14**, 275–283 (1956)

8.53 V.I. Saunders: "On the absorption and vacuum photolysis of silver halides", J. Opt. Soc. Am. **67**, 830–832 (1977)

8.54 R. Schulze: "Strahlenklima der Erde", Meteorologische Rundschau **23** (2), 56 (1970) cited in: Recommendations for the integrated irradiance and the spectral distribution of simulated solar radiation for testing purposes, CIE Publication No. 20 (TC-2.2) (Bureau Central de la CIE, 4 Av. du Recteur Poincaré, 75-Paris 16e, France 1972)

8.55 H.J. Hoffmann: "The Use of Silver Salts in Photochromic Glasses", Ref. [8.10], pp. 822–854 (1990)

8.56 G. Mie: "Beiträge zur Optik trüber Medien, speziell kolloidaler Metallösungen", Ann. Phys. **25**, 377–445 (1908)

8.57 M.A. Smithard: "Size effect on the optical and paramagnetic absorption of silver particles in a glass matrix", Sol. State Commun. **13**, 153–156 (1973)

8.58 U. Kreibig: "Small Silver Particles in Photosensitive Glass: Their Nucleation and Growth", Appl. Phys. **10**, 255–264 (1976)

8.59 C.L. Marquardt and G. Gliemeroth: "Formation of metallic silver on silver-halide surfaces: Flash-photolysis experiments on silver-halide photochromic glasses", J. Appl. Phys. **50**, 4584–4590 (1979)

8.60 T.P. Seward III: "Coloration and optical anisotropy in silver-containing glasses", J. Non-Crystalline Solids, **40,** 499–513 (1980)

8.61 C.L. Marquardt, J.F. Giuliani, and R.T. Williams: "Darkening mechanisms in silver-halide photochromic glasses: Flash photolysis and ESR studies", J. Appl. Phys. **47**, 4915–4925 (1976)

8.62 D.C. Burnham and F. Moser: "Electron spin resonance and optical studies on copper-doped AgCl.", Phys. Rev. **136**, A744–A750 (1964)

8.63 Dingkun Shen, Kaitai Wang, Xihuai Huang, Youxin Chen, and Jinhua Bai: "Optical absorption and ESR of Cu^{2+} in sodium borosilicate glasses", J. Non-Crystalline Solids **52**, 151–158 (1982)

8.64 Jeong-Hoon Lee and R. Brückner: "Optische und magnetische Eigenschaften Cu^{1+}- und Cu^{2+}-haltiger Alkaliborat-, -germanat- und -silicatgläser mit Bezug auf die Borsäure- und Germanatanomalie", Glastechn. Ber. **57**, 30–43 (1984)

8.65 H.J. Hoffmann: "Photochromic glass", Ref. [8.11], pp. 86–101 (1990)

8.66 H.J. Hoffmann and G. Krämer: "Zur Physik phototroper, silberhalogenid-haltiger Gläser", in DGPh-Intern Sonderausgabe zum 2. Münchener Symposium über Wissenschaftliche Photographie vom 27.–28. Oktober 1983, pp. 78–92

8.67 H.J. Hoffmann and G. Krämer: "Absorption Kinetics of Photochromic Glasses Doped with Silver Halides", XIV Intl. Congr. on Glass (New Delhi), Collected Papers Vol. **II**, 110–117 (1986)

8.68 T. Flohr, R. Helbig, and H.J. Hoffmann: "Rate equation for photochromic glasses considering both thermal and optical regeneration", J. Mater. Sci. **22,** 2058–2062 (1987)

8.69 W. Möller and E. Sutter: "Reaktionskinetik von phototropen Gläsern", Optik **75**, 37–46 (1987)

8.70 G.H. Sigel, Jr.: "Optical Absorption of Glasses" in *Treatise on Materials Science and Technology*, ed. by M. Tomozawa, R.H. Doremus, Vol. 12, (Academic Press, Inc., Orlando, London 1977) pp. 5–89

8.71 J. Wong, C.A. Angell: *Glass Structure by Spectroscopy* (Marcel Dekker, New York, Basel 1976)

8.72 H.R. Philipp: "Optical Transitions in crystalline and fused quartz", Solid State Commun. **4**, 73–75 (1966)

8.73 G.H. Sigel, Jr.: "Vacuum ultraviolet absorption in alkali-doped fused silica and silicate glasses", Phys. Chem. Solids **32**, 2373–2383 (1971)

8.74 P.L. Lamy: "Optical constants of crystalline and fused quartz in the far ultraviolet", Appl. Opt. **16**, 2212–2214 (1977)

8.75 S. Hirota, T. Izumitani, R. Onaka: "Reflection spectra of various kinds of oxide glasses and fluoride glasses in the vacuum ultraviolet region", J. Non-Crystalline Solids **72**, 39–50 (1985)

8.76 E. Ellis, D.W. Johnson, A. Breeze, P.M. Magee, P.G. Perkins: "The electronic structure and optical properties of oxide glasses. I. SiO_2, Na_2O: SiO_2 and Na_2O:CaO:SiO_2", Philosophical Magazine B **40**, 105–124 (1979)

8.77 E. Ellis, D.W. Johnson, A. Breeze, P.M. Magee, P.G. Perkins: "The electronic band structure and optical properties of oxide glasses II. Lead silicates", Philosophical Magazine B **40**, 125–137 (1979)

8.78 B.G. Bagley, E.M. Vogel, W.G. French, G.A. Pasteur, J.N. Gan, J. Tauc: "The optical properties of a soda-lime-silica glass in the region from 0.006 to 22 eV", J. Non-Crystalline Solids **22**, 423–436 (1976)

8.79 J.R. Hensler, E. Lell: "Ultraviolet Absorption in Silicate Glasses" in *Frontiers in Glass Science and Technology*, ed. by S. Bateson and A.G. Sadler (Proc. Ann. Meeting International Commission on Glass, Toronto 1969) pp. 51–57

8.80 W. Poch: "Vollständige Entwässerung einer B_2O_3-Schmelze und einige Eigenschaftswerte des daraus erhaltenen Glasses", Glastechn. Ber. **37**, 533–535 (1964)

8.81 J. Krogh-Moe: "The structure of vitreous and liquid boron oxide", J. Non-Cryst. Solids **1**, 269–284 (1969)

8.82 P.A.V. Johnson, A.C. Wright, R.N. Sinclair: "A neutron diffraction investigation of the structure of vitreous boron trioxide", J. Non-Cryst. Solids **50**, 281–311 (1982)

8.83 P.J. Bray: "Nuclear magnetic resonance studies of glass structure", J. Non-Cryst. Solids **73**, 19–45 (1985)

8.84 B.D. McSwain, N.F. Borrelli, G.-J. Su: "The effect of composition and temperature on the ultraviolet absorption of glass", Physics and Chemistry of Glasses **4**, 1–10 (1963)

8.85 A.M. Bishay: "Ultraviolet transmitting glass", J. Opt. Soc. Amer. **51**, 702 (1961)

8.86 K. Gerth, Th. Kloss, H.-J. Pohl: "Optical and physical properties of a boron crown glass transmitting in the ultraviolet region B", J. Non-Cryst. Solids **129**, 12–18 (1991)

8.87 S. Hirota, T. Izumitani: "Effect of cations on the inherent absorption wavelength and the oscillator strength of ultraviolet absorptions in borate glasses", J. Non-Cryst. Solids **29**, 109–117 (1978)

8.88 J. Nicula, L. Cociu, J. Milea, A. Nicula: "UV transmission of borate glasses", Studia Univ. Babes-Bolyai, Physica **23**, 32–34 (1978)

8.89 H. Bach, J.A. Duffy: "Effect of lead (II) ions on the ultraviolet transparency of sodium borate glass", Physics and Chemistry of Glasses **22**, 86–89 (1981)

8.90 J.-H. Lee, R. Brueckner: "The electrochemical series of the 3d transition metal ions in alkali borate glasses", Glastech. Ber. **59**, 233–251 (1986)

8.91 E. Kordes, R. Nieder: "Die ultraviolette Durchlässigkeit binärer Phosphatgläser", Glastechn. Ber. **41**, 41–47 (1968)

8.92 E. Kordes: "Lichtdispersion, ultraviolette Eigenschwingung und Lichtdurchlässigkeit im Ultraviolett von oxydischen Gläsern I", Glastechn. Ber. **38**, 242–249 (1965)

8.93 J.A. Duffy: "Ultraviolet transparency of sodium phosphate glass containing metal ion impurities", Physics and Chemistry of Glasses **13**, 65–68 (1972)

8.94 J.M. Buzhinskii, N.J. Emelyanova, O.P. Pechneva: Phosphate glasses for manufacturing a glass fiber that is transparent in the ultraviolet region", Sov. J. Opt. Technol. **44**, 609–610 (1977)

8.95 A. Klonkowski: "Bond characteristics of phosphate glasses of the M(II)O-P_2O_5 type. Part 1. The extreme position of strontium phosphate glasses", Physics and Chemistry of Glasses **26**, 11–16 (1985)

8.96 A. Klonkowski: "Bond characteristics phosphate glasses of the M(II)O-P_2O_5 type. Part 2. Binary phosphate glasses with ZnO, CdO or PbO", Physics and Chemistry of Glasses **26**, 31–34 (1985)

8.97 A. Smakula: "Synthetic crystals and polarizing materials", Optica Acta **9**, 205–222 (1962)

8.98 W.H. Dumbaugh, D.W. Morgan: "Preliminary ultraviolet transmission data for beryllium fluoride glasses", J. Non-Crystalline Solids **38 & 39**, 211–216 (1980)

8.99 N. Kitamura, J. Hayakawa, H. Yamashita: "Optical properties of fluoroaluminate glasses in the UV region", J. Non-Crystalline Solids **126**, 155–160 (1990)

8.100 L.M. Cook, M.J. Liepmann, A.J. Marker III: "Large Scale Melting of Fluorophosphate Optical Glasses", Materials Science Forum **19–20**, 305–314 (1987)

8.101 M.J. Liepmann, A.J. Marker III, J.M. Melvin: "Optical and physical properties of UV-transmitting fluorocrown glasses", Proc. SPIE **1128**, 213–224 (1989)

8.102 D. Ehrt, W. Seeber: "Glass for high performance optics and laser technology", J. Non-Cryst. Solids **129**, 19–30 (1991)

8.103 D. Ehrt: "Structure and properties of fluoride phosphate glasses", Proc. SPIE **1761**, 213–222 (1992)

8.104 T. Izumitani, S. Hirota, K. Tanaka, H. Onuki: "Dispersion and reflection spectra of fluoride and oxide glasses in the vacuum and extreme ultraviolet region", J. Non-Crystalline Solids **86**, 361–368 (1986)

8.105 W.A. Weyl: *Coloured Glasses* (Society of Glass Technology, Sheffield 1951), reprinted 1978

8.106 C.R. Bamford: *Colour Generation and Control in Glass* (Elsevier, Amsterdam, Oxford 1977)

8.107 L. Cook, K.-H. Mader: "Ultraviolet transmission characteristics of a fluorophosphate laser glass", J. Amer. Ceramic Soc. **65**, 597–601 (1982)

8.108 G.H. Sigel, Jr., R.J. Ginther: "The effect of iron on the ultraviolet absorption of high purity soda-silica glass", Glass Technol. **9**, 66–69 (1968)

8.109 J. Stroud: "Photoionization of Ce^{3+} in glass", J. Chem. Phys. **35**, 844–850 (1961)

8.110 H. Scholze: "Der Einbau des Wassers in Glas", Glastech.Ber. **32**, 81–88, 142–152, 278–281, 314–320, 381–386 (1959)

8.111 W. Mueller-Warmuth, G.W. Schulz, N. Neuroth, F. Meyer, E. Deeg: "Protonen in Gläsern. Kernmagnetische und infrarot-spektroskopische Untersuchungen zur Glasstruktur und zum Wassergehalt", Z. Naturforschung **20a**, 902–917 (1965)

8.112 H. Schroeder, N. Neuroth: "Optische Materialien für den ultravioletten und infraroten Spektralbereich", Optik **26**, 381–401 (1967)

8.113 M.G. Drexhage: "Optical properties of fluoride glasses", in *Halide Glasses for Infrared Fiberoptics*, ed. by R.M. Almeida (Martinus Nijhoff Publishers, Dordrecht, Boston 1987) pp. 219–234

8.114 K. Zirkelbach, R. Brueckner: "Spectroscopic investigations of barium aluminophosphate glasses containing vanadium, iron and manganese oxides", Glastech. Ber. **60**, 312–323 (1987)

8.115 N. Neuroth: "Zusammenstellung der Infrarotspektren von Glasbildnern und Gläsern", Glastech Ber. **41**, 243–253 (1968)

8.116 W.H. Dumbaugh: "Infrared-transmitting oxide glasses", Proc. SPIE **618**, 160–164 (1986)

8.117 W.H. Dumbaugh: "Heavy metal oxide glasses containing Bi_2O_3", Phys. and Chemistry of Glasses **27**, 119–123 (1986)

8.118 J.E. Shelby: "Lead Galliate Glasses", J. Am. Ceram. Soc. **71**, C-254–C-256 (1988)

8.119 J.C. Lapp, W.H. Dumbaugh, M.L. Powley: "Recent advances in heavy metal oxide glass research", Proc. SPIE, Vol. **1327**, 162–170 (1990)

8.120 J. Lucas, J.-J. Adam: "Halide glasses and their optical properties", Glastech. Ber. **62**, 422–440 (1989)

8.121 N. Kitamura, J. Hayakama, H. Yamashita: "Optical properties of fluoroaluminate glasses in the UV region", J. Non-Cryst. Solids **126**, 155–160 (1990)

8.122 T. Izumitani, Y. Yamashita, M. Tokida, K. Miura, H. Tajima: "Physical, chemical properties and crystallization tendency of the new fluoroaluminate glasses", in *Halide Glasses for Infrared Fiberoptics*, ed. by R. M. Almeida (Martinus Nijhoff Publishers, Dordrecht, Boston 1987) pp. 187–197

8.123 D. Ehrt, C. Erdmann, W. Vogel: "Fluoroaluminat Gläser: System CaF_2-SrF_2-AlF_3", Z. Chem. **23**, 37–38 (1983)

8.124 M. Poulain, M. Poulain, J. Lucas, et al.: "Nouveaux verres fluorés dopés au neodyme", Mater. Res. Bull. **10**, 242–247 (1975)

8.125 J. Lucas: "The first ten years", in *Halide Glasses for Infrared Fiberoptics*, ed. R. M. Almeida (Martinus Nijhoff Publisher, Dordrecht, Boston 1987) pp. 1–8

8.126 M. Poulain: "Halide Glasses", J. Non-Cryst. Solids **56**, 1–14 (1983)

8.127 M.G. Drexhage, L.M. Cook, T. Margraf, R. Chaudhuri: "Large scale fluoride glass synthesis", Proc. SPIE, Vol. **970**, 120–127 (1988)

8.128 M.G. Drexhage: "Heavy-Metal Fluoride Glasses" in *Treatise on Materials Science and Technology*, ed. M. Tomozawa, R.H. Doremus, Glass IV, Vol. **26**, (Academic Press, Orlando, San Diego 1985)

8.129 J.M. Parker, P.W. France: "Optical properties of halide glasses", in *Glasses and Glass-ceramics*, ed. by M.H. Lewis (Chapman and Hall, London, New York 1989) pp. 156–202

8.130 P. Klocek, G.H. Sigel: "Halide Glasses", in *Infrared Fiber Optics*, SPIE Press Vol. **TT2**, 33–60 (1989)

8.131 A.E. Comyns: "Fluoride Glasses", in *Critical Reports on Applied Chemistry*, Vol. 27, publ. on behalf of the Society of Chemical Industry (John Wiley & Sons, Chichester, New York 1989) pp. 87–122

8.132 J. Lucas, J.-J. Adam: "Optical properties of halide glasses", in *Optical Properties of Glass*, ed. by D. R. Uhlmann, N. J. Kreidl (The American Ceramic Society, Inc., Westerville 1991) pp. 37–85

8.133 H. Scholze: *Glass* (Springer, New York, Berlin 1990) pp. 151–152

8.134 J.A. Savage, P.J. Webber, A.N. Pitt: "Infrared optical glasses for applications in 8–12 μm thermal imaging systems", Appl. Optics **16**, 2938–2941 (1977)

8.135 A. Feltz, W. Burckhardt, B. Voigt, D. Linke: "Optical glasses for IR transmittance", J. Non-Cryst. Solids **129**, 31–39 (1991)

8.136 J.A. Savage: *Infrared Optical Materials and Their Antireflection Coatings* (Adam Hilger Ltd., Bristol, Boston 1985)

8.137 J.S. Sanghera, J. Heo, J.D. Mackenzie: "Chalcogenide glasses", J. Non-Cryst. Solids **103**, 155–178 (1988)

8.138 J. Lucas, X.H. Zhang: "The tellurium halide glasses", J. Non-Cryst. Solids **125**, 1–16 (1990)

8.139 T.H. Maiman: "Stimulated Optical Radiation in Ruby", Nature **187**, 493–494 (1960)

8.140 E. Snitzer: "Optical Laser Action of Nd^{+3} in a Barium Crown Glass", Phys. Rev. Lett. **7**, 444–446 (1961)

8.141 S.E. Stokowski: "Glass Lasers", in *Handbook of Laser Science and Technology, Vol. 1, Lasers and Masers*, ed. by M. J. Weber, (CRC Press 1982) pp. 215–264

8.142 S.E. Stokowski, R.A. Savoyan and M.J. Weber: "Nd-doped laser glass spectroscopy and physical properties", Lawrence Livermore National Laboratory Report **M-95**, 2nd rev. (1981)

8.143 W. Koecher: "Solid-State Laser Engineering", in *Springer Series in Optical Science Vol. 1* (Springer, New York, Heidelberg, Berlin 1976)

8.144 L.M. Cook, A.J. Marker, and S.E. Stokowski: "Compositional effects on Nd^{+3} concentration in the system $R_2O.Al_2O_3.Ln_2O_3.P_2O_5$", Proc. SPIE **505**, 102–111 (1984)

8.145 C.B. Layne, W.H. Lowdermilk, M.J. Weber: "Multiphonon Relaxation of Rare-Earth Ions in Oxide Glasses", Phys, Rev. B **16**, 10–16 (1977)

8.146 S.E. Stokowski and D. Krashkevich: "Transition-Metal Ions in Nd-doped Glasses: Spectra and Effects on Nd Fluorescence", Mat. Res. Soc. Symp. Proc. **61**, 273–282 (1986)

8.147 N. Neuroth: "Laser Glass: Status and Prospects", Opt. Eng. **26**, 96–101 (1987)

8.148 M.J. Weber, "Recent development in laser glasses", in *Proc. Int. Conf. on Lasers 1982* (New Orleans 1983) pp. 55–63

8.149 R. Jacobs and M.J. Weber: "Dependence of the $^4F_{3/2}$-$^4I_{11/2}$ induced-emission cross section for Nd^{3+} on glass composition", IEEE J. Quantum Electron. **QE-12**(2), 102–111 (1976)

8.150 M.J. Weber, J.D. Myers, and D.H. Blackburn: "Optical properties of Nd^{3+} in tellurite and phosphotellurite glasses", J. Appl. Phys. **52(4)**, 2944–2949 (1981)

8.151 C.F. Cline and M.J. Weber: "Beryllium fluoride optical glasses: preparation and optical properties", Wiss. Z. Friedrich-Schiller-Univ. Jena, Math.-Nat. H **28**, 351 (1979)

8.152 J.J. Videau, J. Fava, C. Fouassier and P. Hagenmueller: "Elaboration et propriétés optiques de nouveaux verres fluorés actives au néodyme", Mater. Res. Bull. **14(4)**, 499–506 (1979)

8.153 M. Poulain, M. Poulain, J. Lucas and P. Brun: "Verres fluorés au tetrafluorure de zirconium propriétés optiques d'un verre dopé au Nd^{3+}", Mater. Res. Bull. **10**, 243–246 (1975)

8.154 J. Lucas, M. Chanthanasinh, M. Poulain, and M. Poulain: "Preparation and optical properties of neodymium fluorozirconate glasses", J. Non-Cryst. Solids **27**, 273–283 (1978)

8.155 M.J. Weber, D.C. Ziegler, and C.A. Angell: "Tailoring stimulated emission cross sections of Nd^{3+} laser glass: observation of large cross sections for $BiCl_3$ glasses", J. Appl. Phys. **53**, 4344–4350 (1982)

8.156 M.J. Weber and R.M. Almeida: "Large stimulated emission cross section of Nd^{3+} in chlorophosphate glass", J. Non-Cryst. Solids **43**, 99–104 (1981)

8.157 G. Lucazeau, S. Barnier and A.M. Loireau-Lozac'h: "Spectres vibrationnels, transitions électroniques et structures à courtes distances dans les verres: sulfures de terres rares–sulfure de gallium", Mater. Res. Bull. **12**, 437–448 (1977)

8.158 F. Auzel, J.C. Michel, J. Flahaut, A.M. Loireau-Lozac'h and M. Guittard: "Comparison des forces d'oscillateurs et sections efficaces laser de verres oxysulfure, sulfure et oxydes dopés au neodyme trivalent", C. R. Acad. Sci. Ser. **C:291**, 21–24 (1980)

8.159 A. Bornstein and R. Reisfeld: "Laser emission cross section and threshold power for laser operation at 1077 nm and 1370 nm; chalcogenide mini-lasers doped by Nd^{3+}". J. Non-Cryst. Solids **50**, 23–27 (1982)

8.160 G. Müller and N. Neuroth: "Glass ceramic – a new laser host material", J. Appl. Phys. **44**, 2315–2318 (1973)

8.161 J.S. Hayden, D.L. Sapak and A.J. Marker: "Elimination of metallic platinum in phosphate laser glasses", Proc. SPIE **895**, 176–181 (1988)

8.162 J.H. Campbell, E.P. Wallerstein, J.S. Hayden, D.L. Sapak, D.E. Warrington, A.J. Marker III, H. Toratani, H. Meissner, S. Nakajima, and T. Izumitani: "Elimination of Platinum Inclusions in Phosphate Laser Glasses", Lawrence Livermore National Laboratory Rept. **UCRL-53932** (1989)

8.163 C.L. Weinzapfel, G.J. Greiner, C.D. Walmer, J.F. Kimmons, E.P. Wallerstein, F.T. Marchi, J.H. Campbell, J.S. Hayden, K. Komiya and T. Kitayama: "Large Scale Damage Testing in a Production Environment", in *Proc. of the Boulder Damage Symposium, Oct. 26–28, 1987, Boulder, CO*, National Institute of Standards and Technology Special Pub. **756**, Boulder, CO, 112 (1987) pp. 112–122

8.164 D.L. Sapak, J.M. Ward, and J.E. Marion: "Impurity absorption coefficient measurements in phosphate glass melted under oxidizing conditions", Proc. SPIE **970**, 107–112 (1988)

8.165 S.D. Stookey: "Strengthening Glass and Glass Ceramics by Built in Surface", in *High Strength Material*, ed. by V.F. Zackay (J. Wiley and Sons Inc., New York, London, Sydney 1965) pp. 669–681

8.166 See, for example, Schott Laser Glass Brochure **2303/90**, Schott Glass Technologies, Duryea, PA (1990)

8.167 R.K. Sandwick, R.J. Scheller, and K.H. Mader: "Production of high homogeneous fluorophosphate laser glass", Proc. SPIE **171**, 161–164 (1979)

8.168 O. Deutschbein, M. Faulstich, W. Jahn, G. Krolla, N. Neuroth: "Glasses with a large laser effect: Nd-phosphate and Nd-fluorophosphate", Appl. Opt. **17**, 2228–2232 (1978)

8.169 P.A. Nass and H.J. Hoffmann: "Nd-Glass multimode laser systems", Proc. SPIE **1128**, 308–317 (1989)

8.170 E.K. Mwarznia, L. Reekio, J. Wang and J.S. Wilkinson: "Low-threshold monomode ion-exchanged waveguide lasers in neodymium-doped BK 7 glass", Electron. Lett. 2, **26**, 1317–1318 (1990)

8.171 S.J. Najafi, W. Wang, J.F. Currie, R. Leonelli, and J.L. Brebner: "Ion exchanged rare earth doped wave-guides", Proc. SPIE **1128**, 142–145 (1989)

8.172 S.A. Payne, M.L. Elder, J.H. Campbell, G.D. Wilke, M.J. Weber, and Y.T. Hayden: "Spectroscopic Properties of Nd^{+3} Dopant Ions in Phosphate Laser Glass, Lawrence Livermore National laboratory Rept. **UCRL-JC-105473** (1991)

8.173 M.L. Elder, Y.T. Hayden, J.H. Campbell, S.A. Payne, and G.D. Wilke: "Thermal-Mechanical and Physical-Chemical Properties of Phosphate Laser Glasses", Lawrence Livermore National Laboratory Rept. **UCRL-JC-105474** (1991)

8.174 Y.T. Hayden, J.H. Campbell, S.A. Payne, and G.D. Wilke: "Platinum Solubility in Phosphate Laser Glass", Lawrence Livermore National Laboratory Rept. **UCRL-JC-105475** (1991)

8.175 J.S. Hayden, D.L. Sapak, and H.J. Hoffmann: "Advances in glasses for high average power laser systems", Proc. SPIE **1021**, 36–41 (1988)

8.176 S.B. Poole, D.N. Payne, R.J. Mears, M.E. Fermann, and R.I. Laming: "Fabrication and characterization of low loss fibers containing rare-earth ions", J. Lightwave Techn. **LT-4**, 870–876 (1986)

8.177 H. Namikawa, K. Arai, K. Kumata, Y. Ishi and H. Tanaka: "Preparation of Nd-doped SiO_2-glass by plasma torch CVD", Jap. J. Appl. Phys. **21**, L360–L362 (1982)

8.178 *Rare-Earth-Doped Fiber Laser Sources and Amplifiers*, ed. by M.J.F. Diggonnet, SPIE Milestones Series **M537** (1991)

8.179 K. Arai, H. Namikawa, K. Kumata, T. Honda, Y. Ishii, and T. Handa: "Aluminum or phosphorus co-doping effects on the fluorescence and structural properties of neodymium-doped silica glass", J. Appl. Phys. **59(10)**, 3430–3436 (1986)

8.180 P.W. France (ed.): *Optical Fibre Lasers and Amplifiers* (Blackie, Glasgow and London 1991)

8.181 M.J. Weber: "Science and Technology of Laser Glass", J. Non Cryst. Sol. **123**, 208–222 (1990)

8.182 G.T. Reynolds and P.E. Condon: "Filament Scintillating Counter", Rev. Sci. Instr. **28**, 1098–1099 (1957)

8.183 C. Bueno, R.A. Buchanan and H. Berger: "Luminescent Glass Design for High Energy Real-Time Radiography" in *Properties and Characteristics of Optical Glass II*, ed. by A.J. Marker III, Proc. SPIE **1327**, 79–91 (1990)

8.184 C.D. Greskovich, D. Cusano, D. Hoffmann and R.J. Riedner: "Ceramic Scintillators for Advanced, Medical X-ray Detectors", Bull. Am. Cer. Soc. 71 No. **7**, 1120–1130 (1992)

8.185 G. Blasse: "Nonradiative Processes in Luminescent Materials: A Material Scientist's View", in *Advances in Nonradiative Processes in Solids*, ed. by B. DiBartolo (Plenum Press, New York 1991) pp. 287–330

8.186 T.S. Curry III, J.E. Dowdey and R.C. Murry, Jr.: *Christensen's Physics of Diagnostic Radiology*, Fourth ed. (Lea & Febiger, Philadelphia, PA 1990) Chapter 9

8.187 R.F. Reade: *Terbium activated radio luminescent silicate glasses*, U.S. Patent 3,654,172, April 4, 1972

8.188 R.A. Buchanan and C. Bueno: *Terbium activated borate luminescent glasses coactivated with gadolinium oxide*, U.S. Patent 5,108,959, April 28, 1992

8.189 R.A. Buchanan and C. Bueno: *X-ray image intensifier tube and x-ray conversion screen containing terbium activated silicate luminescent glasses*, U.S. Patent 5,120,970, June 9, 1992

8.190 R.A. Buchanan and C. Bueno: *Terbium activated silicate luminescent glasses for use in converting x-ray radiation into visible radiation*, U.S. Patent 5,122,671, June 16, 1992

8.191 C.W. Fabjan et al: "Calorimetry in High Energy Physics", Am. Rev. Nuc. Part **32**, 201–214 (1986)

8.192 Schott Glaswerke: "Szintillationsglas HED-1", 3161 d IX (1988)

8.193 A.D. Bross: "Properties of New Scintillator Glasses and Scintillating Fibres", Nucl. Instr. Methods A **247**, 319–326 (1986)

8.194 D.E. Wagoner et al.: "A Measurement of the Energy Resolution and Related Properties of an SCG-1-C Scintillating Glass Shower Counter Array for 1–25 GeV Positrons", Nucl. Instr. Methods A **238**, 315–320 (1985)

8.195 A.R. Spowart: "Energy Transfer in Cerium Activated Silicate Glasses", J. Phys. C **12**, 3369–3374 (1979)

8.196 U. Buchner et al.: "Performance of a Scintillating Glass Calorimeter for Electromagnetic Showers", Nucl. Instr. Methods A **272**, 695–706 (1988)

8.197 P. Hartmann: "Untersuchungen an Glas-Szintillatoren", Ph. D. Thesis, MPI Mainz (FRG) (1983)

8.198 M. Chiba et al.: "Performance of a Scintillator Glass Calorimeter for 10.1–84.3 MeV Photons", Nucl. Instr. Methods A **234**, 267–270 (1985)

8.199 B. Rossi: *High Energy Particles* (Englewood Cliffs, N. J. Prentice Hall 1952)

8.200 S. Orito et al.: "Dense Lead Glass Shower Counters At High Energies", Nucl. Instr. Methods **215**, 93–101 (1983)

8.201 K. Ogawa et al.: "A Test Of Lead Glass Shower Counters", Jap. J. of Appl. Physics **23**, 897–903 (1984)

8.202 Schott Glaswerke: "Optical Glasses for Čerenkov Counters", Produktinformation **3158** e VI/87

8.203 C. Angelini, W. Beusch, A. Cardini, D.J. Crennell, M. De Vincenzi, G. Di Vita, A. Duane, J.-P. Fabre, V. Flaminio, A. Frenkel, T. Gys, K. Harrison, E. Lamanna, H. Leutz, G. Martellotti, J.G. McEwen, D.R.O. Morrison, G. Penso, S. Petrera, C. Roda, A. Sciubba, E. Vicari, D.M. Websdale and G. Wilquet: "Sources of noise in high-resolution tracking with scintillating fibres", Nucl. Instr. and Methods **A289**, 356–364 (1990)

8.204 R. Ruchti, B. Baumbaugh, N. Biswas, N. Cason, R. Erichsen, V. Kenney, A. Kreymer, R.J. Mountain, W. Shephard and A. Rogers: "Scintillating glass, fiber-optic plate detectors for active target and tracking applications in high energy physics experiments", IEEE Trans. Nucl. Sci. **NS-32**, 69–73 (1984)

8.205 A.D. Bross: "Properties of new scintillator glasses and scintillating fibers", Nucl. Instr. Methods Phys Res. **A247**, 319–326 (1986)

8.206 M.N. Atkinson, J. Fent, C. Fisher, P. Freund, P. Hughes, J. Kirkby, A. Osthoff and K. Pretzel: "Initial test of a high resolution scintillating fibre (SCIFI) tracker", Nucl. Instr. Methods Phys. Res. **A254**, 500–514 (1987)

8.207 W.P. Siegmund: "Fiber optic tapers in electronic imaging", Prepared for Electronic Imaging, West Pasadena 1989, 16 pages, available from Schott Fiber Optics, Southbridge, MA, U.S.A.

8.208 C. Angelini, W. Beusch, D.R. Crennell, M. De Vincenzi, A. Duane, J.-P. Fabre, V. Flaminio, A. Frenkel, T. Gys, K. Harrison, E. Lamanna, H. Leutz, G. Martellotti, J.G. McEwen, D.R.O. Morrison, G. Penso, S. Petrera, M. Primout, C. Roda, A. Sciubba, E. Vicari: "Decay time of light emission from cerium-doped scintillating glass", Nucl. Instr. and Methods **A281**, 50–54 (1989)

8.209 D. Puseljic, B. Baumbaugh, J. Bishop, J. Busenitz, N. Cason, J. Cunningham, R. Gardner, C. Kennedy, E. Mannel, R.J. Mountain, R. Ruchti, W. Shephard, M. Zanabria, A. Baumbaugh, K. Knickerbocker, A. Rogers, B. Kinchen and C.G.A. Hill: "A new scintillating glass for high energy physics applications", IEEE Trans. Nucl. Sci. **35**, 475–476 (1988)

8.210 M. Heming, P. Nass, B. Eckhart, V. Zacek und R. Nahnhauer: "Towards high resolution fiber detectors utilizing glass as active or passive component", in *Proceedings of the ECFA Study Week on Instrumentation Technology for High Luminosity Hadron Colliders*, ed. by E. Fernandez and G. Jarlskog (Barcelona 1989) pp. 231–237

8.211 M. Heming, P.A. Nass, W.P. Siegmund, R.R. Strack and H.M. Voyagis: "Scintillating Glass Fibers", in *Proceedings of the Workshop on Application of Scintillating Fibers in Particle Physics*, ed. by R. Nahnhauer (Blossin 1990) pp. 121–125

8.212 A. Artamonov, J. Bähr, E. Birckner, B. Eckart, W. Flegel, M. Heming, K. Hiller, R. Nahnhauer, P. Nass, F. Russo, S. Schlenstedt, G. Wilquet, K. Winter and V. Zacek: "Investigations on capillaries filled with liquid scintillator for high resolution particle tracking", Nucl. Instr. and Meth. **A300**, 53–62 (1991)

8.213 M. Adinolfi, C. Angelini, J. Bähr, A. Cardini, C. Cianfarani, C. Da Vià, D. De Pedis, M. De Vincenzi, A. Duane, J.-P. Fabre, V. Flaminio, W. Flegel, A. Frenkel, M. Gruwé, K. Harrison, P. Lendermann, D. Lucchesi, G. Martellotti, C. Mommaert, D.R.O. Morrison, R. Nahnhauer, G. Penso, E. Pesen, C. Roda, A. Sciubba, D.M. Websdale, G. Wilquet and V. Zacek: "A high-resolution tracking detector based on capillaries filled with liquid scintillator", Nucl. Instr. and Methods **A311**, 91–97 (1992)

8.214 CHARM II-collaboration and IHEP Berlin-Zeuthen: "Thin scintillating optical fibers for high resolution active target detectors", presented by V. Zacek on the Workshop on Physics at UNK, Serpukhov 1989

8.215 J. Bähr, E. Birckner, A. Capone, D. DePedis, W. Flegel, B. Friend, M. Gruwe, K. Hiller, P. Lendermann, C. Mommaert, R. Nahnhauer, E. Pesen, G. Stefanini, G. Wilquet, K. Winter and V. Zacek: "Liquid scintillator filled capillary arrays for particle tracking", Nucl. Instr. and Methods **A306**, 169–176 (1991)

8.216 M. Adinolfi, C. Angelini, A. Cardini, C. Cianfarani, C. Da Vià, A. Duane, J.-P. Fabre, V. Flaminio, A. Frenkel, S.V. Golovkin, A.M. Gorin, M. Gruwé, K. Harrison, M. Kanerva, E.N. Kozarenko, A.E. Kushnirenko, G. Martellotti, J.G. McEwen, D.R.O. Morrison, G. Penso, A.I. Peresypkin, C. Roda, D.M. Websdale, G. Wilquet and A.A. Zaichenko: "Progress on high-resolution tracking with scintillating fibers: a new detector based on capillaries filled with liquid scintillator", Nucl. Instr. and Methods **A315**, 177–181 (1992)

8.217 S.V. Golovkin, A.M. Gorin, A.V. Kuhchenko, A.F. Kushnirenko, A.I. Peresypkin, A.I. Pyshchev, V.I. Rykalin and A.A. Zaichenko: "Development of tracking detectors based on capillaries with liquid scintillator", Nucl. Instr. and Methods **A305**, 385–390 (1991)

8.218 A.G. Denisov, S.V. Golovkin, A.M. Gorin, F.N. Kozarenko, A.E. Kushnirenko, A.M. Medvedkov, A.I. Peresypkin, S.V. Petrenko, Yu.P. Perukhov, V.I. Rykalin, V.E. Tyukov, V.G. Vasil'chenko and A.A. Zaichenko : "High resolution tracking detector based on capillaries with a liquid scintillator", Nucl. Instr. and Methods **A310**, 479–484 (1991)

8.219 Private communication by R.Yu. Elokhin

8.220 W. Jahn: "Die Einwirkung von radioaktiver Strahlung auf Glas", Glastech. Berichte **31**, 41–69 (1958)

8.221 F.C. Hardtke et al.: "The Fracture by Electrical Discharge of Gamma-Irradiated Shielded Window Glass", 11th Hot Lab Proceedings, ANS, New York, 369–381 (1963)

8.222 Schott Glaswerke: "Radiation Shielding Windows from Schott Quality for Highest Demands", Produktinformation **3150e** XI/86(1986)

8.223 F.M. Ernsberger et al.: "Gamma-Radiation-Induced Conductivity in Nuclear Shielding Glasses as Determined by Space Charge Decay", 11th Hot Lab Proceedings, ANS, New York, 383–393 (1963)

8.224 T.W. Eckels et al.: "Further Data on Gamma-Induced Electrical Charge and Coloration of Shielding Glasses", Proceedings of the 18th Conference on Remote Systems Technology, 143–147 (1970)

8.225 R.G. Jaeger: *Engineering Compendium on Radiation Shielding*, Vol. 1, Chap. 6 (Springer, Berlin, Heidelberg, New York 1968)

8.226 T. Rockwell: *Reactor Shielding Design Manual* , TID-7004 (Mc Graw Hill & Van Nostrand 1956)

8.227 W.W. Engle, jr: "Users Manual for ANISN", K-1693 (1967), available from RSIC as CCC-82/ANISN

8.228 A. Bishay: "Radiation Induced Color Centers in Multicomponent Glasses", J. of Noncryst. Sol. **3**, 54–114 (1970)

8.229 G.S. Bogdanova et al.: " Mechanism of the Coloration of Cerium-Containing Glasses", Sov. J. of Glass Phys. and Chem. **6**, 776–780 (1979)

8.230 B. Mc Grath et al.: "Effects of Nuclear Radiation on the Optical Properties of Cerium-Doped Glasses", Nuc. Instr. Methods **135**, 93–97 (1976)

8.231 F.L. Galeener: "Defects in Glasses", Materials Research Society Symposia Proceedings **61**, 161–213, 319–359 (1986)

8.232 B. Speit et al.: "Radiation Resistant Optical Glasses", Nuc. Instr. Methods **B 65**, 384–386 (1992)

8.233 B. Speit et al.: "Irradiation Energy Dependence of Discolouration in Radiation Shielding Glasses", Proc. SPIE, Vol **1327**, 92–99 (1990)

8.234 W. Jahn: "Strahlenresistente optische Gläser", Glastechn. Berichte **35**, 479–483 (1962)

8.235 B. Mc Grath: "Effects of Nuclear Radiation on the Optical Properties of Cerium Doped Glass", Nuc. Instr. Methods **135**, 93–97 (1976)

8.236 B. Speit et al.: "Radiation Resistant Optical Glasses", Proceedings of the Sixth International Conference on Radiation Effects in Isolators, Nuc. Instr. Methods, B **65**, 384–387 (1992)

8.237 A. Paul et al.: "Cerous-Ceric Equilibrium in Binary Alkali Borate and Alkali Silicate Glasses", Phys. and Chem. of Glasses **6**, 212–215 (1965)

8.238 E.J. Friebele et al.: "Radiation Effects in Glass" in *Treatise on Mat. Sc. Techn.* **17**, *Glass* II, (ed. by M. Tomozawa et al, Academic Press, New York 1979) pp. 257–351

8.239 A.A. Stepanov et al.: "Effect on Gamma Radiation on the Optical Losses of Multicomponent Ultrapure Cerium Glasses", Sov. J. Glass Phys. Chem. **16**, 502–506 (1990)

8.240 Schott Glaswerke: "Radiation Resistant Glasses", Produktinformation **10017** e (1989)

8.241 F.L. Galeener et al.: "Defects in Glasses", Materials Research Society Symposia Proceedings **61**, 161–223, Pittsburgh, Pennsylvania (1986)

8.242 K. Becker: *Solid State Dosimetry* (CRC Press, 18901 Cranwood Parkway, Cleveland, Ohio 44128, 1973) pp. 1–26, 141–173

8.243 Schott Glaswerke: "Gläser zur Dosimetrie von Röntgen- und Kernstrahlung", Produktinformation **3103** (1987)

8.244 R. Yokota and H. Imagawa: "Radiophotoluminescent centers in silver-activated phosphate glasses", J. Phys. Soc. Jap. **23**, 1083–1090 (1967)

8.245 E. Piesch: "Development in RPL Dosimetry", in *Topics in Radiation Dosimetry*, Vol 1, ed. by F.H. Attix (Academic Press, New York 1972) pp. 1–88

8.246 E. Lell, N.J. Kreidl, J.R. Hensler: "Radiation Effects in Quartz, Silica and Glasses", in *Progress in Ceramic Science*, Vol. 4, (Pergamon Press, New York 1966) pp. 1–47

8.247 B.D. Evens, G.H. Sigel: "Radiation Resistant Fiber Optic Materials and Wave guides", IEEE Trans. on Nucl. Sci. NS **22**, 2462–2468 (1975)

8.248 R. Klingelhöfer: "Fortschritte in der Kernfusion", Chemiker Zeitung **111**, Nr. **3**, 85–93 (1987)

8.249 M. Dalle Donne: "Heat Transfer in Pebble Beds for Fusion Blankets", Fusion Technology Vol. **17**, 597–635 (1990)

8.250 G. Schumacher, M. Dalle Donne, S. Dorner: "Properties of Orthosilicate Spheres", J. Nuc. Mater. **155–157**, 451–454 (1987)

8.251 M. Dalle Donne: "Conceptual Design Of A Helium Cooled Solid Breeder Blanket Based On The Use Of A Mixed Bed Of Beryllium And Li_4SiO_4 Pebbles", 17. Soft Conference Rome (Sept., 4–18, 1992)

8.252 C.R. Bamford: *Colour generation and control in glass* (Elsevier Scient. Publ. Co. Amsterdam 1977)

8.253 Comm. Int. de l' Eclairage, Compte Rendu, 1931

8.254 W.A. Weyl: *Coloured Glasses* (Society of Glass Technology, Sheffield 1951)

8.255 W. Vogel: *Glaschemie* (Springer, Berlin 1992) pp. 51–313

8.256 K. Nassau: "The varied causes of colour in glass", Mat. Res. Soc. Symp. Proc. **61**, 427–439 (1986)

8.257 T. Bates: "Ligand field theory and absorption spectra of transition-metal ions in glasses", in: *Modern aspects of the vitreous state*, Vol 2, ed. by J.D. Mackenzie (Butterworth, London 1962) pp. 195–254

8.258 H. D. Hardt: *Die periodischen Eigenschaften der chemischen Elemente* (Thieme, Stuttgart 1974) pp. 153–157

8.259 L.E. Orgel: *An introduction to transition metal chemistry, Ligand field theory* (Methuen and Co Ltd, London 1960)

8.260 H.L. Schläfer, G. Gliemann: *Einführung in die Ligandenfeldtheorie* (Akademische Verlagsgesellschaft, Frankfurt/Main 1967)

8.261 C.J. Ballhausen: *Introduction to ligand field theory* (Mc Graw Hill, New York 1962)

8.262 C.K. Jorgensen: *Absorption spectra and chemical bonding in complexes* (Pergamon Press, Oxford 1962)

8.263 S. Kumar: "Optical absorption of glasses containing ions with partially filled 3d-orbitals, part I–III", Glass and Ceramic Bulletin **6**, 99–126 (1959)

8.264 R. Tsuchida: "Absorption spectra of co-ordination compounds I.", J. Chem. Soc. Japan **13**, 388–400 (1938)

8.265 D. Krause: "Aussagen der Absorptionsspektroskopie im Sichtbaren und UV zur Struktur von Glas", Fachausschußbericht Nr. **70**, Verlag der Deutschen Glastechnischen Gesellschaft, Frankfurt, 188–218 (1974)

8.266 M.V. Ramana, P.S. Lakshmi, G.S. Sastry: "Optical absortion spectra of copper in oxy-fluoro borate glasses", J. Mat. Sci. Letters **11**, 541–542 (1992)

8.267 A. Bishay, A. Kinawi: "Absorption spectra of iron in phosphate glasses and ligand field theory", Phys. of Noncryst. Sol. **2**, 589–605 (1965)

8.268 W.D. Johnston: "Oxidation-reduction equilibria in iron-containing glasses", J. Am. Ceram. Soc. **47**, 198–201 (1964)

8.269 J. Hlavac: *The technology of glass and ceramics* (Elsevier Scient. Publ. Co., Amsterdam 1983)

8.270 A. Paul: Chemistry of glasses (Chapman and Hall, London 1982)

8.271 R.W. Douglas, M.S. Zaman: "The chromophore in iron-sulfur amber glasses", Phys. Chem. Glasses **10**, 125–132 (1969)

8.272 F.L. Harding: "Effect of base glass composition on amber colour", Glass Technol. **13**, 43–49 (1972)

8.273 F.L. Harding, R.J. Ryder: "Amber colour in commercial silicate glasses", J. Can. Ceram. Soc . **39**, 59–63 (1970)

8.274 H.D. Schreiber, S.J. Kozak, C.W. Schreiber, D.G. Wetmore: "Sulfure chemistry in a borosilicate melt. Part 3. Iron-sulfur interactions and the amber chromophore", Glastech. Ber. **63**, 49–60 (1990)

8.275 C.F. Bohren, D.R. Huffman: *Absorption and scattering of light by small particles* (Wiley, New York 1983)

8.276 M. Kerker: *The scattering of light and other electromagnetic radiation* (Academic Press, New York 1969)

8.277 M. Born, E. Wolf: *Principles of optics* (Pergamon Press, Oxford 1986)

8.278 R.N. Dwivedi, P. Nath: "Mechanism of formation and growth of silver colloids in photosensitive glasses", Trans. Ind. Ceram. Soc. **38**, 96–99 (1979)

8.279 H.G. Krolla, U. Kolberg: unpublished results

8.280 D.G. Galimov, A.M. Gubaidullina, A.I. Neich: "Optical properties of colloidal particles of copper in glasses", Fizika i Khimiya Stekla **13**, 50–54 (1987)

8.281 M. Heming, F.T. Lentes, P. Naß, O. Becker: unpublished results

8.282 K.J. Berg, M. Mennig: private communications 1991 (experimental work at the university of Halle)

8.283 S.D. Stookey, G.B. Beall, J.E. Pierson: "Full Colour photosensitive glass", J. Appl. Phys. **49**, 5114–5123 (1978)

8.284 H.P. Rooksby: "The colour of selenium ruby glasses", Journal of the Society of Glass Technology **16**, 171–181 (1932)

8.285 G. Schmidt: "Optische Untersuchungen an Selenrubingläsern", Silikattechnik **14**, 12–18 (1963)

8.286 A. Rehfeld, R. Katzschmann: "Farbbildung und Kinetik von Steilkanten-Anlaufgläsern", Silikattechnik **29**, 298–302 (1978)

8.287 N. Riehl (ed.): *Einführung in die Lumineszenz* (Karl Thiemig, München 1971)

8.288 F.L. Pedrotti, D.C. Reynolds: "Spin-orbit splitting in CdS:Se single crystals", Physical Review **127**, 1584–1586 (1962)

8.289 Levin, Robbins, McMurdie: "Phase diagrams for the ceramists", Suppl. American Ceramic Society (1969)

8.290 M. Clement: "Untersuchungen zur Thermodynamik und Diffusion im Halbleitersystem CdS-CdSe-CdTe"; Ph. D. Thesis, Universität Münster (1987)

8.291 R. Kranold, G. Becherer: "Röntgen-Kleinwinkeluntersuchungen an Anlauf-Farbgläsern"; Universität Rostock (1977)

8.292 G. Walter, R. Kranold, U. Lemke: "Small angle X-ray scattering characterization of inorganic glasses", Makromol. Chem. Makromol. Symp. **15**, 361–372 (1988)

8.293 T. Yanagawa, Y. Sasaki. H. Nakano: "Quantum size effects and observation of microcrystallites in coloured filter glasses", Appl. Phys. Lett. **54**, 1495–1497 (1989)

8.294 A. Uhrig: "Zur optischen Spektroskopie an $CdS_{1-x}Se_x$"; Diplomarbeit, Universität Kaiserslautern (1989)

8.295 H. Scholze: *Glas*, 2nd edition (Springer, Berlin, Heidelberg, New York 1977)

8.296 W.J. Behr, G. Gliemeroth, O. Gott: "Brillenglas mit höherer Festigkeit", Glas-Email-Keramo-Technik **24**, 149–158 (1973)

8.297 German Industry Standard DIN 4646: *Sichtscheiben für Augenschutzgeräte*, Teil 1–8 (Beuth Verlag, Berlin) pp. 1976–1990, German Industry Standard DIN 4647: *Sichtscheiben für Augenschutzgeräte*, Teil 4–8 (Beuth Verlag, Berlin) pp. 1977–1990

8.298 E. Sutter, P. Schreiber, G. Ott: *Handbuch Laserstrahlenschutz* (Springer, Berlin, Heidelberg 1989)

8.299 DIN VDE paper 0837: *Strahlungssicherheit von Lasereinrichtungen* (Beuth Verlag, Berlin 1986)

8.300 DIN 58215 *Laserschutzfilter und Laserschutzbrillen – Sicherheitstechnische Anforderungen und Prüfung* (Beuth Verlag, Berlin 1986), DIN 58219 *Laser adjusting goggles, safety requirements and testing* (Beuth Verlag, Berlin 1982)

8.301 EN 60825 "Radiation Safety of Laser Product, Equipment Classification Requirements and User's Guide, 1992" – amended IEC 825 (1984)

8.302 D.H. Sliney, M.L. Wolbarsht: *Safety with Lasers and other Optical Sources*, 4th edition (Plenum Press, New York 1985)

8.303 D.H. Sliney: "Laser Safety", in *Handbook of Laser Science and Technology*, ed. by M.J. Weber (CRC Press, Boca Raton, Fl, USA 1982)

8.304 D.C. Winburn: *Practical Laser Safety* (Marcel Dekker, New York 1985)

8.305 Z 136.1: "Safe Use of Lasers" (American National Standards Institute ANSI, 1430 Broadway, New York, NY 10018, USA 1986) Z 136.4: "Safe Use of Lasers" (American National Standards Institute ANSI, 1430 Broadway, New York, NY 10018, USA 1986)

8.306 E. Sutter: *Alles über Augenschutz* (Wirtschaftsverlag NW, Verlag für neue Wissenschaft GmbH, Bremerhaven 1990)

8.307 M. McLear: "Modern Laser-protective eyewear is safe and stylish", Laser Focus World, 111–116 (Aug. 1992)

8.308 D.C. Winburn: "Laser protective eyewear: how safe it is", Laser Focus/Electro-Optics, 136–140 (April 1987)

8.309 W. Grimm, H. Schürle: "Brillen für den Schutz am Arbeitsplatz", Optometrie, Heft **5**, 3–10 (1987)

8.310 H.W. Paysan: "Hoffnung auf Leistung, Sorge ums Leben", Sicherheitsingenieur **11/87**, 32–38 (1987)

8.311 R. Yeo: "New laser safety rules tailor guards to hazards", Opto & Laser Europe, 35–38 (Sept. 1992)

8.312 J. Tron: "Augenschaden trotz Laserschutzbrille – wegen falscher Auswahl", Laser und Optoelektronik **24**, 32–35 (1992)

8.313 T.L. Lyon, W.J. Marshall: "Nonlinear properties of optical filters-implications for laser safety", Health Physics **51**, 95–96 (1986)

8.314 Types UV-W 76 and UV-W 6000 from Deutsche Spezialglas AG, Grünenplan, Germany

8.315 Types Uvilex® 390 and Oralan® from Deutsche Spezialglas AG, Grünenplan, Germany

8.316 Neotherm® Ofenschauglas from Deutsche Spezialglas AG Grünenplan, Germany

List of Contributing Authors

Marc K. Th. Clement
Schott Glaswerke, Mainz[1]

Knut Holger Fiedler
Carl Zeiss, 73446 Oberkochen, Germany

Ulrich Fotheringham
Schott Glaswerke, Mainz[1]

Yuiko T. Hayden
YTH Translation, 107 Fox Run Circle, Clarks Summit, PA 18411, USA

Joseph S. Hayden
Schott Glass Technologies, Duryea[2]

Wilfried Heimerl
Schott Glaswerke, Mainz[1]

Ewald Hillmann
Deutsche Spezialglas AG, Postfach 2032, 31073 Grünenplan, Germany

Hans-Jürgen Hoffmann
Fachhochschule Hildesheim-Holzminden, Fachbereich Physik-, Meß- und Feinwerktechnik in Göttingen, Carl-Zeiss-Straße 6, 37081 Göttingen, Germany

Uwe Kolberg
Schott Glaswerke, Mainz[1]

David Krashkevich
Schott Glass Technologies, Duryea[2]

Frank-Thomas Lentes
Schott Glaswerke, Mainz[1]

Monika J. Liepmann
Schott Glass Technologies, Duryea[2]

Susan R. Loehr
Gentex Corporation, P.O. Box 315, Carbondale, PA 18407, USA (formerly at Schott Glass Technologies, Duryea[2])

Alexander J. Marker III
Schott Glass Technologies, Duryea[2]

Peter Naß
Schott Glaswerke, Mainz[1]

Kurt Nattermann
Schott Glaswerke, Mainz[1]

Norbert Neuroth
Schott Glaswerke, Mainz[1]

Arnd Peters
Schott Glaswerke, Mainz[1]

Robert J. Scheller
Schott Glass Technologies, Duryea[2]

Burkhard Speit
Schott Glaswerke, Mainz[1]

[1] Hattenbergstraße 10, 55122 Mainz, Germany
[2] 400 York Avenue, Duryea, PA 18642-2036, USA

Sources of Figures

We are indebted to the following editors and authors, respectively, for the kind permission to reproduce copyrighted materials.

Material	Source	Original Publisher
Fig. 2.3	[2.6]	McGraw-Hill Inc., New York, U.S.A
Fig. 2.5	[2.6]	McGraw-Hill Inc., New York, U.S.A
Fig. 2.11	[2.34]	Chapman & Hall, Andover, U.K.
Fig. 2.17	[2.57]	Deutsche Glastechnische Gesellschaft, Frankfurt, Germany
Fig. 2.18	[2.55]	Deutsche Glastechnische Gesellschaft, Frankfurt, Germany
Fig. 2.19	[2.55]	Deutsche Glastechnische Gesellschaft, Frankfurt, Germany
Fig. 2.20	[2.59]	Dr. V. N. Polukhin, State Optical Institute, Leningrad, Russia
Table 2.5	[2.79–85]	Plenum Publishing Corp., New York, U.S.A
Table 2.6	[2.86]	Gosudarst. Izdatel. Oboronnoi., Prom. Moscow, Russia
Tables 2.7 and 2.8	[2.87]	The American Ceramic Society, Westerville, U.S.A
Tables 2.7 and 2.8	[2.88]	Optical Society of America, Washington, U.S.A
Table 2.9	[2.89]	Society of Glass Technology, Sheffield, U.K.
Fig. 2.27	[2.106]	Elsevier Science Publ. B.V., Amsterdam, Netherlands
Fig. 2.28	[2.107]	Elsevier Science Ltd., The Boulevard, Langford Lane, Kidlington OX5 1GB, U.K.
Fig. 2.29	[2.106]	Elsevier Science Publ. B.V., Amsterdam, Netherlands
Fig. 2.30a	[2.109]	Dr. J. D. Mackenzie, University of California, Los Angeles, U.S.A
Fig. 2.30b	[2.110]	Society of Glass Technology, Sheffield, U.K.
Fig. 2.31	[2.111]	Institute of Physics Publishing Ltd., Bristol, U.K.
Fig. 2.36	[2.121]	Society of Photo-Optical Instrumentation Engineers (SPIE), Bellingham, U.S.A.
Fig. 2.37	[2.121]	Society of Photo-Optical Instrumentation Engineers (SPIE), Bellingham, U.S.A.
Fig. 2.41	[2.122]	Deutsche Glastechnische Gesellschaft, Frankfurt, Germany

Material	Source	Original Publisher
Fig. 8.39	[8.216]	Elsevier Science Publ. B.V., Amsterdam, Netherlands
Fig. 8.40	[8.219]	Dr. R. Yu Elokhin, Institute for High Energy Physics, Protvino, Russia
Fig. 8.41	[8.213]	Elsevier Science Publ. B.V., Amsterdam, Netherlands
Fig. 8.51	[8.257]	Dr. J. D. Mackenzie, University of California, Los Angeles, U.S.A
Fig. 8.52	[8.257]	Dr. J. D. Mackenzie, University of California, Los Angeles, U.S.A
Fig. 8.62	[8.298, 299]	Springer-Verlag, Heidelberg, Berlin, Germany
Fig. 8.63	[8.298, 299]	Springer-Verlag, Heidelberg, Berlin, Germany

Index

Springer-Verlag and the Environment

We at Springer-Verlag firmly believe that an international science publisher has a special obligation to the environment, and our corporate policies consistently reflect this conviction.

We also expect our business partners – paper mills, printers, packaging manufacturers, etc. – to commit themselves to using environmentally friendly materials and production processes.

The paper in this book is made from low- or no-chlorine pulp and is acid free, in conformance with international standards for paper permanency.